Quantum Probability and Randomness

Quantum Probability and Randomness

Special Issue Editors

Andrei Khrennikov
Karl Svozil

MDPI • Basel • Beijing • Wuhan • Barcelona • Belgrade

MDPI

Special Issue Editors

Andrei Khrennikov
Linnaeus University
Sweden

Karl Svozil
Institute for Theoretical Physics of the
Vienna Technical University
Austria

Editorial Office
MDPI
St. Alban-Anlage 66
4052 Basel, Switzerland

This is a reprint of articles from the Special Issue published online in the open access journal *Entropy* (ISSN 1099-4300) from 2018 to 2019 (available at: https://www.mdpi.com/journal/entropy/special_issues/Probability_Randomness)

For citation purposes, cite each article independently as indicated on the article page online and as indicated below:

LastName, A.A.; LastName, B.B.; LastName, C.C. Article Title. *Journal Name* **Year**, *Article Number*, Page Range.

ISBN 978-3-03897-714-8 (Pbk)
ISBN 978-3-03897-715-5 (PDF)

Cover image courtesy of R.C.-Z. Quehenberger.

Contents

About the Special Issue Editors

Andrei Khrennikov was born in 1958 in Volgorad and spent his childhood in the town of Bratsk, in Siberia, north from the lake Baikal. In the period between 1975–1980, he studied at Moscow State University, department of Mechanics and Mathematics, and in 1983, he received his PhD in mathematical physics (quantum field theory) at the same department. In 1990, he became full professor at Moscow University for Electronic Engineering. Since 1997, he has been a professor of applied mathematics at Linnaueus University, South-East Sweden, and since 2002, the director of the multidisciplinary research center at this university, as well as the International Center for Mathematical Modeling in Physics, Engineering, Economics, and Cognitive Science. His research interests are multidisciplinary, e.g., foundations of quantum physics, information, and probability, cognitive modeling, ultrametric (non-Archimedean) mathematics, dynamical systems, infinite-dimensional analysis, quantum-like models in psychology, and economics and finances. He is the author of approximately 500 papers and 20 monographs in mathematics, physics, biology, psychology, cognitive science, economics, and finances.

Karl Svozil is a professor of theoretical physics at Vienna's University of Technology. He earned a Dr. Phil. while studying philosophy and sciences in the old, "Humboldtian" tradition in Heidelberg and Vienna, emphasizing the unity of knowledge. After attending the Lawrence Berkeley Laboratory and UC Berkeley, he worked as a physicist in Vienna, with many shorter stays abroad—among them, the Lomonosov Moscow State University, Lebedev Physical Institute and ICPT Trieste. He recently held an honorary position at the University of Auckland and served as president of the International Quantum Structures Association. Svozil's main interests include quantum logic, issues related to (in)determininsism in physics, and "relativizing" relativity theory in the spirit of Alexandrov's theorem of incidence geometry.

.

entropy

MDPI

Editorial

Quantum Probability and Randomness

Andrei Khrennikov [1],* and Karl Svozil [2]

[1]   International Center for Mathematical Modeling in Physics, Engineering, Economics, and Cognitive Science, Linnaeus University, 351 95 Växjö, Sweden
[2]   Institute for Theoretical Physics, Vienna University of Technology, Wiedner Hauptstrasse 8-10/136, 1040 Vienna, Austria; svozil@tuwien.ac.at
*   Correspondence: Andrei.Khrennikov@lnu.se

Received: 2 January 2019; Accepted: 3 January 2019; Published: 7 January 2019

Keywords: quantum foundations; probability; irreducible randomness; random number generators; quantum technology; entanglement; quantum-like models for social stochasticity; contextuality

The recent quantum information revolution has stimulated interest in the quantum foundations by perceiving and re-evaluating the theory from a novel information-theoretical viewpoint [1–5]. Quantum probability and randomness play the crucial role in foundations of quantum mechanics.

It might not be totally unreasonable to claim that, already starting from some of the earliest (in hindsight) indications of quanta in the 1902 Rutherford–Soddy exponential decay law and the small aberrations predicted by Schweidler [6], the tide of indeterminism [7,8] was rolling against chartered territories of fin de siécle mechanistic determinism. Riding the waves were researchers like Exner, who already in his 1908 inaugural lecture as *rector magnificus* [9] postulated that irreducible randomness is, and probability theory therefore needs to be, at the heart of all sciences; natural as well as social. Exner [10] was forgotten but cited in Schrödinger's alike "Zürcher Antrittsvorlesung" of 1922 [11]. Not much later Born expressed his inclinations to give up determinism in the world of the atoms [12], thereby denying the existence of some inner properties of the quanta which condition a definite outcome for, say, the scattering after collisions.

Von Neumann [13] was among the first who emphasized this new feature which was very different from the "in principle knowable unknowns" grounded in epistemology alone. Quantum randomness was treated as *individual randomness*; that is, as if single electrons or photons are sometimes capable of behaving acausally and irreducibly randomly. Such randomness cannot be reduced to a variability of properties of systems in some ensemble. Therefore, quantum randomness is often considered as *irreducible randomness*.

Von Neumann understood well that it is difficult, if not outright impossible in general, to check empirically the randomness for individual systems, say for electrons or photons. In particular, he proceeded with the *statistical interpretation* of probability based on the mathematical model of von Mises [14,15] based upon *relative frequencies* after *admissible place selections*.

At the same time, it is just and fair to note that the aforementioned tendencies to ground physics, and by reductionism, all of science, in ontological indeterminism, have been strongly contested and fiercely denied by eminent physicists; most prominently by Einstein. Planck [16] (p. 539) (see also Earman [17] (p. 1372)) believed that causality could be neither generally proved nor generally disproved. He suggested to postulate causality as a working hypothesis, a heuristic principle, a sign-post (and for Planck the most valuable sign-post we possess) *"to guide us in the motley confusion of events"*.

This is a good place to remark that random features of an individual system can be discussed in the framework of *subjective probability theory*. The individual (irreducible) interpretation of quantum randomness due to von Neumann matches well with the subjective probability interpretation of quantum mechanics (QBism, see, e.g., [18,19]).

The main reason for keeping the statistical interpretation was that the aforementioned individual randomness of quantum systems was considered by von Neumann as one of the basic features of nature (and not of the human mind!). Von Neumann was sure that such a natural phenomenon must be treated statistically (by the same reason Bohr also treated quantum randomness statistically, see [20] for details).

In particular, von Neumann remarked [13] (pp. 301–302), that, for measurement of some quantity R for an ensemble of systems (of any origin),

> *It is not surprising that R does not have a sharp value ..., and that a positive dispersion exists. However, two different reasons for this behavior a priori conceivable:*
>
> 1. *The individual systems S1,..., SN of our ensemble can be in different states, so that the ensemble [S1,..., SN] is defined by their relative frequencies. The fact that we do not obtain sharp values for the physical quantities in this case is caused by our lack of information: we do not know in which state we are measuring, and therefore we cannot predict the results.*
> 2. *All individual systems S1,..., SN are in the same state, but the laws of nature are not causal. Then, the cause of the dispersion is not our lack of information, but nature itself, which has disregarded the principle of sufficient cause.*

These are characterizations of epistemic and ontic indeterminism, respectively. Von Neumann favored the second, ontic, case which he considered "important and new" (and which he believed to be able to corroborate [21]). Therefore, for von Neumann, quantum randomness is essentially a statistical exhibition of violation of causality, a violation of the principle of sufficient cause.

We compare this kind of randomness with classical interpretations of randomness, see, e.g., Chapter 2 [22]:

1. unpredictability (von Mises),
2. complexity-incompressibility (Kolmogorov, Solomonof, Chaitin),
3. typicality (Martin-Löf).

It seems that the interpretation of randomness as unpredictability (von Mises) is very close to the interpretation of quantum randomness as an exhibition of acausality.

The article by Pavicic and Megill [23], *Vector Generation of Quantum Contextual Sets in Even Dimensional Hilbert Spaces*, is a novel contribution to quantum contextuality theory. As is well known, the most elaborated contextual sets, which offer blueprints for contextual experiments and computational gates, are the Kochen–Specker sets. In this paper, a method of vector generation that supersedes previous methods is presented. It is implemented by means of algorithms and programs that generate hypergraphs embodying the Kochen-Specker property and that are designed to be carried out on supercomputers.

Recent years were characterized by the tremendous development of quantum technology. Quantum random generators are among the most important outputs of this development. As is pointed out in the review by Martínez et al. [24], *Advanced Statistical Testing of Quantum Random Number Generators*, the natural laws of the microscopic realm provide a fairly simple method to generate non-deterministic sequences of random numbers, based on measurements of quantum states. In practice, however, the experimental devices on which quantum random number generators are based are often unable to pass some tests of randomness. In this review, two such tests are briefly discussed, the challenges that have to be encountered in experimental implementations are pointed out. Finally, the authors present a fairly simple method that successfully generates non-deterministic maximally random sequences.

The connection between quantum logic and quantum probability is highlighted by Dalla Chiara et al. [25] in the paper entitled *Probabilities and Epistemic Operations in the Logics of Quantum Computation*. The authors stress that quantum computation theory has inspired new forms of quantum logic, called quantum computational logics. In this article, they investigate the epistemic operation (which is informally used in a number of interesting quantum situations): the operation "being probabilistically informed".

In the paper entitled *Enhancing Extractable Quantum Entropy in Vacuum-Based Quantum Random Number Generator*, Guo et al. [26] commit to enhancing quantum entropy content in the vacuum noise based quantum RNG. They have taken into account main factors in this proposal to establish the theoretical model of quantum entropy content, including the effects of classical noise, the optimum dynamical analog-digital convertor (ADC) range, the local gain and the electronic gain of the homodyne system.

The work by Enríquez et al. [27], *Entanglement of Three-Qubit Random Pure States*, is devoted to studying entanglement properties of generic three-qubit pure states. There are obtained the distributions of both the coefficients and the only phase in the five-term decomposition of Acín et al. for an ensemble of random pure states generated by the Haar measure on $U(8)$. Furthermore, the authors analyze the probability distributions of two sets of polynomial invariants. One of these sets allows us to classify three-qubit pure states into four classes. Entanglement in each class is characterized using the minimal Renyi–Ingarden–Urbanik entropy. The numerical findings suggest some conjectures relating some of those invariants with entanglement properties to be ground in future analytical work.

In the article *New Entropic Inequalities and Hidden Correlations in Quantum Suprematism Picture of Qubit States*, Margarita A. Man'ko and Vladimir I. Man'ko [28] considered an analog of Bayes' formula and the nonnegativity property of mutual information for systems with one random variable. For single-qubit states, they presented new entropic inequalities in the form of the subadditivity and condition corresponding to hidden correlations in quantum systems. Qubit states are represented in the quantum suprematism picture, where these states are identified with three probability distributions, describing the states of three classical coins, and illustrating the states by Triada of Malevich's squares with areas satisfying the quantum constraints.

In the article by Plotnitsky [29], *"The Heisenberg Method": Geometry, Algebra, and Probability in Quantum Theory*, quantum theory is reconsidered in terms of the following principle, which can be symbolically represented as QUANTUMNESS→PROBABILITY→ALGEBRA. The principle states that the quantumness of physical phenomena, that is, the specific character of physical phenomena known as quantum, implies that our predictions concerning them are irreducibly probabilistic, even in dealing with quantum phenomena resulting from the elementary individual quantum behavior (such as that of elementary particles), which in turn implies that our theories concerning these phenomena are fundamentally algebraic, in contrast to more geometrical classical or relativistic theories, although these theories, too, have an algebraic component to them.

The work by Delgado [30], *SU(2) Decomposition for the Quantum Information Dynamics in 2d-Partite Two-Level Quantum Systems*, presents a formalism to decompose the quantum information dynamics in $SU(2^{2d})$ for 2d-partite two-level systems into 2^{d-1} $SU(2)$ quantum subsystems. It generates an easier and more direct physical implementation of quantum processing developments for qubits.

The paper by Marius Nagy and Naya Nagy [31], *An Information-Theoretic Perspective on the Quantum Bit Commitment Impossibility Theorem*, proposes a different approach to pinpoint the causes for which an unconditionally secure quantum bit commitment protocol cannot be realized, beyond the technical details on which the proof of Mayers' no-go theorem is constructed.

In the Copenhagen approach to quantum mechanics as characterized by Heisenberg, probabilities relate to the statistics of measurement outcomes on ensembles of systems and to individual measurement events via the actualization of quantum potentiality. In the review by Jaeger [32], *Developments in Quantum Probability and the Copenhagen Approach*, brief summaries are given of a series of key results of different sorts that have been obtained since the final elements of the Copenhagen interpretation were offered and it was explicitly named so by Heisenberg—in particular, results from the investigation of the behavior of quantum probability since that time, the mid-1950s. This review shows that these developments have increased the value to physics of notions characterizing the approach which were previously either less precise or mainly symbolic in character, including complementarity, indeterminism, and unsharpness.

A new way of orthogonalizing ensembles of vectors by "lifting" them to higher dimensions is introduced by Havlicek and Svozil [33] entitled *Dimensional Lifting through the Generalized Gram-Schmidt Process*. This method can potentially be utilized for solving quantum decision and computing problems.

Recently the mathematical formalism and methodology of quantum theory started to be widely applied outside of physics, especially in psychology, decision making, social and political science (see, e.g., [34]). This special issue contains one paper belonging to this area of research, the article of Khrennikov et al. [35], *On Interpretational Questions for Quantum-Like Modeling of Social Lasing*. The formalisms of quantum field theory and theory of open quantum systems are applied to modeling socio-political processes on the basis of the social laser model describing *stimulated amplification of social actions*. The main aim of this paper is establishing the socio-psychological interpretations of the quantum notions playing the basic role in lasing modeling.

The article by Paul Ballonoff [36], *Paths of Cultural Systems*, is also devoted to applications outside physics, namely to anthropology. A theory of cultural structures predicts the objects observed by anthropologists. A viable history (defined using pdqs) states how an individual in a population following such history may perform culturally allowed associations, which allows a viable history to continue to survive. The vector states on sets of viable histories identify demographic observables on descent sequences.

We hope that the reader will enjoy the present issue, which will be useful to experts working in all domains of quantum physics and quantum information theory, ranging from experimenters, to theoreticians and philosophers.

The cover of this electronic book was created by Renate Quehenberg and the editors would like to thank her for the graphical contribution to this special issue.

Acknowledgments: We express our thanks to the authors of the above contributions, and to the journal *Entropy* and MDPI for their support during this work.

Conflicts of Interest: The authors declare no conflict of interest.

References

1. Khrennikov, A.; Weihs, G. Preface of the special issue Quantum foundations: Theory and experiment. *Found. Phys.* **2012**, *42*, 721–724. [CrossRef]
2. Bengtsson, I.; Khrennikov, A. Preface. *Found. Phys.* **2011**, *41*, 281–281. [CrossRef]
3. D'Ariano, G.M.; Jaeger, G.; Khrennikov, A.; Plotnitsky, A. Preface of the special issue Quantum theory: Advances and problems. *Phys. Scr.* **2014**, *T163*, 010301.
4. Khrennikov, A.; de Raedt, H.; Plotnitsky, A.; Polyakov, S. Preface of the special issue Probing the limits of quantum mechanics: Theory and experiment, Volume 1. *Found. Phys.* **2015**, *45*, 707–710; doi:10.1007/s10701-015-9911-8. [CrossRef]
5. Khrennikov, A.; de Raedt, H.; Plotnitsky, A.; Polyakov, S. Preface of the special issue Probing the limits of quantum mechanics: Theory and experiment, Volume 2. *Found. Phys.* **2015**, doi:10.1007/s10701-015-9950-1. [CrossRef]
6. Von Schweidler, E. *Über Schwankungen der Radioaktiven Umwandlung*; H. Dunod & E. Pinat: Paris, France, 1906; pp. 1–3. (In German)
7. Hiebert, E.N. Common frontiers of the exact sciences and the humanities. *Phys. Perspect.* **2000**, *2*, 6–29. [CrossRef]
8. Stöltzner, M., Vienna indeterminism: Mach, Boltzmann, Exner. *Synthese* **1999**, *119*, 85–111. [CrossRef]
9. Exner, F.S. *Über Gesetze in Naturwissenschaft und Humanistik: Inaugurationsrede Gehalten am 15. Oktober 1908*; A. Hölder: Wien, Austria, 2016.
10. Exner, F.S. *Vorlesungen über die Physikalischen Grundlagen der Naturwissenschaften*; F. Deuticke: Leipzig und Wien, Germany, 1922.
11. Schrödinger, E. Was ist ein Naturgesetz? *Naturwissenschaften* **1929**, *17*, 1. [CrossRef]

12. Born, M. Zur Quantenmechanik der Stoßvorgänge. *Z. Phys.* **1926**, *37*, 863–867. [CrossRef]
13. Von Neuman, J. *Mathematical Foundations of Quantum Mechanics*; Princeton University Press: Princeton, NJ, USA, 1955.
14. Von Mises, R. Grundlagen der Wahrscheinlichkeitsrechnung. *Math. Z.* **1919**, *5*, 52–99. [CrossRef]
15. Von Mises, R. *The Mathematical Theory of Probability and Statistics*; Academic Press: London, UK, 1964.
16. Planck, M. The concept of causality. *Proc. Phys. Soc.* **1932**, *44*, 529–539. [CrossRef]
17. Earman, J. Aspects of determinism in modern physics. In *Part B: Philosophy of Physics, Handbook of the Philosophy of Science*; Butterfield, J., Earman, J., Eds.; North-Holland: Amsterdam, The Netherlands, 2007; pp. 1369–1434.
18. Fuchs, C.A. Quantum mechanics as quantum information (and only a little more). In *Quantum Theory: Reconsideration of Foundations*; Växjö University Press: Växjö, Schweden, 2002; pp. 463–543.
19. Fuchs, C.A.; Schack, R. QBism and the Greeks: why a quantum state does not represent an element of physical reality. *Phys. Scr.* **2014**, *90*, 015104. [CrossRef]
20. Plotnitsky, A.; Khrennikov, A. Reality without realism: On the ontological and epistemological architecture of quantum mechanics. *Found. Phys.* **2015**, *45*, 1269–1300. [CrossRef]
21. Dieks, D. Von Neumann's impossibility proof: Mathematics in the service of rhetorics. *Stud. Hist. Philos. Mod. Phys.* **2017**, *60*, 136–148. [CrossRef]
22. Khrennikov, A. *Probability and Randomness: Quantum Versus Classical*; Imperial College Press: London, UK, 2016.
23. Pavicic, M.; Megill, N.D. Vector Generation of Quantum Contextual Sets in Even Dimensional Hilbert Spaces. *Entropy* **2018**, *20*, 928. [CrossRef]
24. Martínez, A.C.; Solis, A.; Rojas, R.D.H.; U' Ren, A.B.; Hirsch, J.G.; Castillo, I.P. Advanced Statistical Testing of Quantum Random Number Generators. *Entropy* **2018**, *20*, 886. [CrossRef]
25. Dalla Chiara, M.L.; Freytes, H.; Giuntini, R.; Leporini, R.; Sergioli, G. Probabilities and Epistemic Operations in the Logics of Quantum Computation. *Entropy* **2018**, *20*, 837. [CrossRef]
26. Guo, X.; Liu, R.; Li, P.; Cheng, C.; Wu, M.; Guo, Y. Enhancing Extractable Quantum Entropy in Vacuum-Based Quantum Random Number Generator. *Entropy* **2018**, *20*, 819. [CrossRef]
27. Enríquez, M.; Delgado, F.; Zyczkowski, K. Entanglement of Three-Qubit Random Pure States. *Entropy* **2018**, *20*, 745. [CrossRef]
28. Man'ko, M.A.; Man'ko, V.I. New Entropic Inequalities and Hidden Correlations in Quantum Suprematism Picture of Qubit States. *Entropy* **2018**, *20*, 692. [CrossRef]
29. Plotnitsky, A. The Heisenberg Method: Geometry, Algebra, and Probability in Quantum Theory. *Entropy* **2018**, *20*, 656. [CrossRef]
30. Delgado, F. $SU(2)$ Decomposition for the Quantum Information Dynamics in 2d-Partite Two-Level Quantum Systems. *Entropy* **2018**, *20*, 610. [CrossRef]
31. Nagy, M.; Nagy, N. An Information-Theoretic Perspective on the Quantum Bit Commitment Impossibility Theorem. *Entropy* **2018**, *20*, 193. [CrossRef]
32. Jaeger, G. Developments in Quantum Probability and the Copenhagen Approach. *Entropy* **2018**, *20*, 420. [CrossRef]
33. Havlicek, H.; Karl Svozil, K. Dimensional Lifting through the Generalized Gram-Schmidt Process. *Entropy* **2018**, *20*, 284. [CrossRef]
34. Khrennikov, A. *Ubiquitous Quantum Structure: From Psychology to Finances*; Springer: Berlin/Heidelberg, Germany; New York, NY, USA, 2010.
35. Khrennikov, A.; Alodjants, A.; Trofimova, A.; Tsarev, D. On Interpretational Questions for Quantum-Like Modeling of Social Lasing. *Entropy* **2018**, *20*, 921. [CrossRef]
36. Ballonoff, P. Paths of Cultural Systems. *Entropy* **2018**, *20*, 8. [CrossRef]

Article

Vector Generation of Quantum Contextual Sets in Even Dimensional Hilbert Spaces

Mladen Pavičić [1,2,*,†] and **Norman D. Megill** [3,†]

1   Nano Optics, Department of Physics, Humboldt University, 12489 Berlin, Germany
2   Center of Excellence for Advanced Materials and Sensors, Research Unit Photonics and Quantum Optics,
     Institute Ruđer Bošković, 10000 Zagreb, Croatia
3   Boston Information Group, Lexington, MA 02420, USA; nm@alum.mit.edu
*   Correspondence: mpavicic@irb.hr
†   These authors contributed equally to this work.

Received: 29 October 2018; Accepted: 24 November 2018; Published: 5 December 2018

Abstract: Recently, quantum contextuality has been proved to be the source of quantum computation's power. That, together with multiple recent contextual experiments, prompts improving the methods of generation of contextual sets and finding their features. The most elaborated contextual sets, which offer blueprints for contextual experiments and computational gates, are the Kochen–Specker (KS) sets. In this paper, we show a method of vector generation that supersedes previous methods. It is implemented by means of algorithms and programs that generate hypergraphs embodying the Kochen–Specker property and that are designed to be carried out on supercomputers. We show that vector component generation of KS hypergraphs exhausts all possible vectors that can be constructed from chosen vector components, in contrast to previous studies that used incomplete lists of vectors and therefore missed a majority of hypergraphs. Consequently, this unified method is far more efficient for generations of KS sets and their implementation in quantum computation and quantum communication. Several new KS classes and their features have been found and are elaborated on in the paper. Greechie diagrams are discussed.

Keywords: quantum contextuality; Kochen–Specker sets; MMP hypergraphs; Greechie diagrams

1. Introduction

Recently, it has been discovered that quantum contextuality might have a significant place in a development quantum communication [1,2], quantum computation [3,4], and lattice theory [5,6]. This has prompted experimental implementation of 4-, 6-, and 8-dimensional contextual experiments with photons [7–13], neutrons [14–16], trapped ions [17], solid state molecular nuclear spins [18], and paths [19,20].

Experimental contextual tests involve subtle issues, such as the possibility of noncontextual hidden variable models that can reproduce quantum mechanical predictions up to arbitrary precision [21]. These models are important because they show how assignments of predetermined values to dense sets of projection operators are precluded by any quantum model. Thus, Spekkens [22] introduces generalised noncontextuality in an attempt to make precise the distinction between classical and quantum theories, distinguishing the notions of preparation, transformation, and measurement of noncontextuality and by doing so demonstrates that even the 2D Hilbert space is not inherently noncontextual. Kunjwal and Spekkens [23] derive an inequality that does not assume that the value assignments are deterministic, showing that noncontextuality cannot be salvaged by abandoning determinism. Kunjwal [24] shows how to compute a noncontextuality inequality from an invariant derived from a contextual set/configuration representing an experimental Kochen-Specker (KS) setup. This opens up the possibility of finding contextual sets that provide the best noise robustness in demonstrating contextuality. The large number of such sets that we show in the present work can provide a rich source for such an effort.

Quantum contextual configurations that have been elaborated on the most in the literature are the KS sets, and, in this paper, we consider just them. In order to obtain KS sets, so far, various methods of exploiting correlations, symmetries, geometry, qubit states, Pauli states, Lie algebras, etc., have been found and used for generating master sets i.e., big sets which contain all smaller contextual sets [25–37].

All of these methods boil down either to finding a list of vectors and their n-tuples of orthogonalities from which a master set can be read off or finding a structure, e.g., a polytope, from which again a list of vectors and orthogonalities can be read off as well as a master set they build. In the present paper, we take the simplest possible vector components within an n-dimensional Hilbert space, e.g., $\{0, \pm 1\}$, and via our algorithms and programs exhaustively build all possible vectors and their orthogonal n-tuples and then filter out KS sets from the sets in which the vectors are organized. For a particular choice of components, the chances of getting KS sets are very high. We generate KS sets for even-dimensional spaces, up to 32, that properly contain all previously obtained and known KS sets, present their features and distributions, give examples of previously unknown sets, and present a blueprint for implementation of a simple set with a complex coordinatization.

2. Results

The main results presented in this paper concern generation of contextual sets from several basic vector components. Previous contextual sets from the literature made use of often complicated sets of vectors that the authors arrived at, following particular symmetries, or geometries, or polytope correlations, or Pauli operators, or qubit states, etc. In contrast, our approach considers McKay–Megill–Pavičić (MMP) hypergraphs (defined in Section 2.1) from n-dimensional (nD) Hilbert space ($\mathcal{H}^n$, $n \geq 3$) originally consisting of n-tuples (in our approach represented by MMP hypergraph edges) of orthogonal vectors (MMP hypergraph vertices) which exhaust themselves in forming configurations/sets of vectors (MMP hypergraphs). Already in [38], we realised that hypergraphs massively generated by their non-isomorphic upward construction might satisfy the Kochen–Specker theorem even when there are no vectors by means of which they might be represented (see Theorem 1), and finding coordinatizations for those hypergraphs which might have them, via standard methods of solving systems of non-linear equations, is an exponentially complex task solvable only for the smallest hypergraphs [38]. It was, therefore, rather surprising to us to discover that the hypergraphs formed by very simple vector components often satisfied the Kochen–Specker theorem. In this paper, we present a method of generation of KS MMP hypergraphs, also called KS hypergraphs, via such simple sets of vector components.

Theorem 1 (MMP hypergraph reformulation of the Kochen–Specker theorem).
There are nD MMP hypergraphs, i.e., those whose each edge contains n vertices, called KS MMP hypergraphs, to which it is impossible to assign 1s and 0s in such a way that

(α)   *No two vertices within any of its edges are both assigned the value 1;*
(β)   *In any of its edges, not all of the vertices are assigned the value 0.*

In Figure 1, we show the smallest possible 4D KS MMP hypergraph with six vertices and three edges. We can easily verify that it is impossible to assign 1 and 0 to its vertices so as to satisfy the conditions (α) and (β) from Theorem 1. For instance, if we assign 1 to the top green-blue vertex, then, according to the condition (α), all of the other vertices contained in the blue and green edges must be assigned value 0, but, herewith, all four vertices in the red edge are assigned 0s in violation of the condition (β), or, if we assign 1 to the top red-blue vertex, then, according to the condition (α), all the other vertices contained in the blue and red edges must be assigned value 0, but, herewith, all four vertices in the green edge are assigned 0s in violation of the condition (β). Analogous verifications go through for the remaining four vertices. We verified that there is neither a real nor complex vector solution of a corresponding system of nonlinear equations [38]. We have not tried quaternions as of yet.

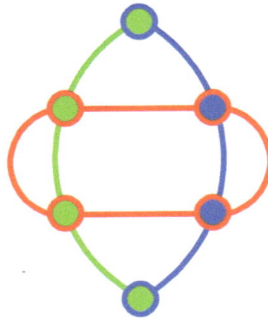

Figure 1. The smallest 4D KS MMP hypergraph without a coordinatization.

When a coordinatization of a KS MMP hypergraph exists, its vertices denote n-dimensional vectors in $\mathcal{H}^n$, $n \geq 3$, and edges designate orthogonal n-tuples of vectors containing the corresponding vertices. In our present approach, a coordinatization is automatically assigned to each hypergraph by the very procedure of its generation from the basic vector components. A KS MMP hypergraph with a given coordinatization of whatever origin we often simply call a KS *set*.

2.1. Formalism

MMP hypergraphs are those whose edges (of size n) intersect each other in at most $n - 2$ vertices [26,37]. They are encoded by means of printable ASCII characters. Vertices are denoted by one of the following characters: 1 2 ...9 A B ...Z a b ...z !   " # $ % & ' ( ) * - / : ; < = > ? @ [\ ] ^ _ ' { | } ~ [26]. When all of them are exhausted, one reuses them prefixed by '+', then again by '++', and so forth. An n-dimensional KS set with k vectors and m n-tuples is represented by an MMP hypergraph with k vertices and m edges which we denote as a k-m set. In its graphical representation, vertices are depicted as dots and edges as straight or curved lines connecting m orthogonal vertices. We handle MMP hypergraphs by means of algorithms in the programs SHORTD, MMPSTRIP, MMPSUBGRAPH, VECFIND, STATES01, and others [5,30,38–41]. In its numerical representation (used for computer processing), each MMP hypergraph is encoded in a single line in which all m edges are successively given, separated by commas, and followed by assignments of coordinatization to k vertices (see 18-9 in Section 2.2).

2.2. KS Vector Lists vs. Vector Component MMP Hypergraphs

In Table 1, we give an overview of most of the k-m KS sets (KS hypergraphs with m vertices and k edges) as defined via lists and tables of vectors used to build the KS master sets that one can find in the literature. These master sets serve us to obtain billions of non-isomorphic smaller KS sets (KS subsets, subhypergraphs) which define k-m classes. In doing so (via the aforementioned algorithms and programs), we keep to minimal, *critical*, KS subhypergraphs in the sense that a removal of any of their edges turns them into non-KS sets. Critical KS hypergraphs are all we need for an experimental implementation: additional orthogonalities that bigger KS sets (containing critical ones) might possess do not add any new property to the ones that the minimal critical core already has. The smallest hypergraphs we give in the table are therefore the smallest criticals. Many more of them, as well as their distributions, the reader can find in the cited references. Some coordinatizations are over-complicated in the original literature. For example (as shown in [37]), for the 4D 148-265 master, components

$\{0, \pm i, \pm 1, \pm \omega, \pm \omega^2\}$, where $\omega = e^{2\pi i/3}$, suffice for building the coordinatization, and for the 6D 21-7 components $\{0, 1, \omega\}$ suffice. In addition, $\{0, \pm 1\}$ suffice for building the 6D 236-1216.

Table 1. Vector lists from the literature; we call their masters *list-masters*. We shall make use of their vector components from the last column to generate master hypergraphs in Section 2.3 which we call *component-masters*. ω is a cubic root of unity: $\omega = e^{2\pi i/3}$.

dim	Master Size	Vector List	List Origin	Smallest Hypergraph	Vector Components
4D	24-24	[25,42,43]	symmetry, geometry	18–9	$\{0, \pm 1\}$
4D	60-105	[28,37]	Pauli operators	18–9	$\{0, \pm 1, \pm i\}$
4D	60-75	[27,30,37,41]	regular polytope 600-cell	26–13	$\{0, \pm(\sqrt{5}-1)/2, \pm 1, \pm(\sqrt{5}+1)/2, 2\}$
4D	148-265	[36,37]	Witting polytope	40–23	$\{0, \pm i, \pm 1, \pm \omega, \pm \omega^2, \pm i\omega^{1/\sqrt{3}}, \pm i\omega^{2/\sqrt{3}}\}$
6D	21-7	[19]	symmetry	21-7	$\{0, 1, \omega, \omega^2\}$
6D	236-1216	Aravind & Waegell 2016, [37]	hypercube →hexaract Schäfli $\{4, 3^4\}$	34–16	$\{0, \pm 1/2, \pm 1/\sqrt{3}, \pm 1/\sqrt{2}, 1\}$
8D	36-9	[37]	symmetry	36–9	$\{0, \pm 1\}$
8D	120-2025	[35,37]	Lie algebra E8	34–9	as given in [35]
16D	80-265	[37,44,45]	Qubit states	72–17	$\{0, \pm 1\}$
32D	160-661	[37,46]	Qubit states	144–11	$\{0, \pm 1\}$

Some of the smallest KS hypergraphs in the table have ASCII characters assigned and some do not. This is to stress that we can assign them in an arbitrary and random way to any hypergraph and then the program VECFIND will provide them with a coordinatization in a fraction of a second. For instance,

18-9: 1234,4567,789A,ABCD,DEFG,GHI1,I29B,35CE,68FH.
{1={0,0,0,1},2={0,0,1,0},3={1,1,0,0},4={1,-1,0,0},5={0,0,1,1},6={1,1,1,-1},
7={1,1,-1,1}, 8={1,-1,1,1},9={1,0,0,-1},A={0,1,1,0},B={1,0,0,1},C={1,-1,1,-1},
D={1,1,-1,-1},E={1,-1,-1,1},F={0,1,0,1},G={1,0,1,0},H={1,0,-1,0},I={0,1,0,0}}.

(To simplify parsing, this notation delineates vectors with braces instead of traditional parentheses in order to reserve parentheses for component expressions.)

However, a real finding is that we can go the other way round and determine the KS sets from nothing but vector components $\{0, \pm 1\}$.

2.3. Vector-Component-Generated Hypergraph Masters

We put simplest possible vector components, which might build vectors and therefore provide a coordinatization to MMP hypergraphs, into our program VECFIND. Via its option -master, the program builds an internal list of all possible non-zero vectors containing these components. From this list, it finds all possible edges of the hypergraph, which it then generates. MMPSTRIP via its option -U separates unconnected MMP subgraphs. We pipe the obtained hypergraphs through the program STATES01 to keep those that possess the KS property. We can use other programs of ours, MMPSTRIP, MMPSHUFFLE, SHORTD, STATES01, LOOP, etc., to obtain smaller KS subsets and analyze their features.

The likelihood that chosen components will give us a KS master hypergraph and the speed with which it does so depends on particular features they possess. Here, we will elaborate on some of them and give a few examples. Features are based on statistics obtained in the process of generating hypergraphs:

(*i*)   the input set of components for generating two-qubit KS hypergraphs (4D) should contain number pairs of opposite signs, e.g., ± 1, and zero (0); we conjecture that the same holds for 3, 4, . . . qubits; with 6D it does not hold literally; e.g., $\{0, 1, \omega\}$ generate a KS master; however, the following combination of ω's gives the opposite sign to 1: $\omega + \omega^2 = -1$;

(*ii*) mixing real and complex components gives a denser distribution of smaller KS hypergraphs;

(*iii*) reducing the number of components shortens the time needed to generate smaller hypergraphs and apparently does not affect their distribution.

Feature (*i*) means that, no matter how many different numbers we use as our input components, we will not get a KS master if at least to one of the numbers, the same number with the opposite sign is not added. Thus, e.g., $\{0, 1, -i, 2, -3, 4, 5\}$ or a similar string does not give any, while $\{0, \pm 1\}$, or $\{0, \pm i\}$, or $\{0, \pm(\sqrt{5}-1)/2\}$ do. Of course, the latter strings all give mutually isomorphic KS masters, i.e., one and the same KS master, if used alone. More specifically, they yield a 40-32 master with 40 vertices and 32 edges as shown in Table 2. When any of them are used together with other components, although they would generate different component-masters, all the latter masters of a particular dimension would have a common smallest hypergraph as also shown in Table 2.

Table 2. Component-masters we obtained. List-masters are given in Table 1. In the last two rows of all but the last column, we refer to the result [33] that there are 16D and 32D criticals with just nine edges. According to the conjectured feature (i) above, the masters generated by {0, ±1} should contain those criticals; they did not come out in [37], so, we do not know how many vertices they have. The smallest ones we obtained are given in Table 1. The number of criticals given in the 4th column refer to the number of them we successfully generated although there are many more of them except in the 40-32 class.

dim	Vector Components	Component-Master Size	Nº of KS Criticals in Master	Smallest Hypergraph	Contains List-Masters
4D	{0, ±1} or {0, ±i} or {0, ±$(\sqrt{5}-1)/2$} or ...	40-32	6	18–9	24-24
4D	{0, ±1, ±i}	156-249	7.7×10^6	18–9	24-24, 60-105
4D	{0, ±$(\sqrt{5}-1)/2$, ±1, ±$(\sqrt{5}+1)/2$, 2}	2316-3052	1.5×10^9	18–9	24-24, 60-75
4D	{0, ±1, ±i, ±ω, ±ω^2}	400-1012	8×10^6	18–9	24-24, 60-105 148-265
6D	{0, ±1, ω, ω^2}	11808-314446	3×10^7	21–7	21-7, 236-1216
8D	{0, ±1}	3280-1361376	7×10^6	34–9	36-9, 120-2025
16D	{0, ±1}	computationally too demanding	4×10^6	?–9 [33].	80-265
32D	{0, ±1}	computationally too demanding	2.5×10^5	?–9 [33].	160-661

We obtained the following particular results which show the extent to which component-masters give a more populated distribution of KS criticals than list-masters. We also closed several open questions:

- As for the features (ii) and (iii) above, components {0, ±1, ω} generate the master 180-203 which has the following smallest criticals 18-9, 20...22-11, 22...26-13, 24...30-15, 30...31-16, 28...35-17, 33...37-18, etc. This distribution is much denser than that of, e.g., the list-master 24-24 with real vectors which in the same span of edges consists only of 18-9, 20-11, 22-13, and 24-15 criticals or of the list-master 60-75 which starts with the 26-13 critical. In Appendix A, we give a detailed description of a 21-11 critical with a complex coordinatization and give a blueprint for its experimental implementation;
- In [19], the reader is challenged to find a master set which would contain the "seven context star" 21-7 KS critical (shown in Tables 1 and 2). We find that {0, 1, ω} generate the 216-153 6D master which contains just three criticals 21-7, 27-9, and 33-11. {0, 1, ω, ω^2} generate 834-1609 master from which we obtained 2.5×10^7 criticals, and {0, ±1, ω, ω^2} generate 11808-314446 master from which we obtained 3×10^7 criticals, all of them containing the seven context star. Some of the obtained criticals are given in Appendix B;

- The 60-75 list-master contains criticals with up to 41 edges and 60 vertices, while the 2316-3052 component-master generated from the same vector components contains criticals with up to close to 200 edges and 300 vertices;
- The 60-105 list-master contains criticals with up to 40 edges and 60 vertices, while the 156-249 component-master generated from the same vector components contains criticals with up to at least 58 edges and 88 vertices;
- Components $\{0, \pm1\}$ generate 332-1408 6D master which contains the 236-1216 list-master while originally components $\{0, \pm1/2, \pm1/\sqrt{3}, \pm1/\sqrt{2}, 1\}$ were used;
- In [37], we generated 6D criticals with up to 177 vertices and 87 edges from the list-master 236-1216, while, now, from the component-master 11808-314446, we obtain criticals with up to 201 vertices and 107 edges;
- We did not generate 16D and 32D masters because that would take too many CPU days and we already generated a huge number of criticals from submasters which are also defined by means of the same vector components in [37]. See also Section 3.

3. Methods

Our methods for obtaining quantum contextual sets boil down to algorithms and programs within the MMP language we developed to generate and handle KS MMP hypergraphs as the most elaborated and implemented kind of these sets. The programs we make use of, VECFIND, STATES01, MMPSTRIP, MMPSHUFFLE, SUBGRAPH, LOOP, SHORTD, etc., are freely available from our repository http://goo.gl/xbx8U2. They are developed in [5,29,30,38–40,47,48] and extended for the present elaboration. Each MMP hypergraph can be represented as a figure for a visualisation but more importantly as a string of ASCII characters with one line per hypergraph, enabling us to process millions of them simultaneously by inputting them into supercomputers and clusters. For the latter elaboration, we developed other dynamical programs specifically for a supercomputer or cluster, which enable piping of our files through our programs in order to parallelize jobs. The programs have the flexibility of handling practically unlimited number of MMP hypergraph vertices and edges as we can see from Table 2. The fact that we did not let our supercomputer run to generate 16D and 36D masters and our remark that it would be "computationally too demanding" do not mean that such runs are not feasible with the current computers, but that they would require too many CPU days on the supercomputer and that we decided not to burden it with such a task at the present stage of our research; see the explanation in Section 2.3.

4. Conclusions

The main result we obtain is that our vector component generation of KS hypergraphs (sets) exhaustively use all possible vectors that can be constructed from chosen vector components. This is in contrast to previous studies, which made use of serendipitously obtained lists of vectors curtailed in number due to various methods applied to obtain them. Hence, we obtain a thorough and maximally dense distribution of KS classes in all dimensions whose critical sets can therefore be much more effectively used for possible implementation in quantum computation and communication. A comparison of Tables 1 and 2 vividly illustrates the difference.

In Appendix A, we present a possible experimental implementation of a KS critical with complex coordinatization generated from $\{0, \pm1, \omega\}$. What we immediately notice about the 21-11 critical from Figure A1 is that the edges are interwoven in more intricate way than in the 18-9 (which has been implemented already in several experiments), exhibiting the so-called δ-feature of the edges forming the biggest loop within a KS hypergraph. The δ-feature refers to two neighbouring edges which share two vertices, i.e., intersect each other at two vertices [37]. It stems directly from the representation of KS configuration with MMP hypergraphs. Notice that the δ-feature precludes interpretation of practically any KS hypergraph in an even dimensional Hilbert space by means of so-called Greechie diagrams, which by definition require that two blocks (similar to hypergraph edges) do not share more than one atom (similar to a vertex) [6], on the one hand, and that the loops made by the blocks must be of order five or higher (which is hardly ever realised in even dimensional KS hypergraphs—see examples in [37]), on the other.

Our future engagement would be to tackle odd dimensional KS hypergraphs. Notice that, in a 3D Hilbert space, it is possible to explore similarities between Greechie diagrams and MMP hypergraphs because then neither of them can have edges/blocks which share more than one vertex/atom (via their respective definitions) and loops in both of them are of the order five or higher [26,39].

Author Contributions: Conceptualization, M.P.; Data Curation, M.P.; Formal Analysis, M.P. and N.D.M.; Funding Acquisition, M.P.; Investigation, M.P. and N.D.M.; Methodology, M.P. and N.D.M.; Project Administration, M.P.; Resources, M.P.; Software, M.P. and N.D.M.; Supervision, M.P.; Validation, M.P. and N.D.M.; Visualization, M.P.; Writing—Original Draft, M.P.; Writing—Review and Editing, M.P. and N.D.M.

Funding: Supported by the Croatian Science Foundation through project IP-2014-09-7515, the Ministry of Science and Education (MSE) of Croatia through the Center of Excellence for Advanced Materials and Sensing Devices (CEMS) funding, and by grants Nos. KK.01.1.1.01.0001 and 533-19-15-0022. This project was also supported by the Alexander or Humboldt Foundation. Computational support was provided by the cluster Isabella of the Zagreb University Computing Centre, by the Croatian National Grid Infrastructure (CRO-NGI), and by the Center for Advanced Computing and Modelling (CNRM) for providing computing resources of the supercomputer Bura at the University of Rijeka in Rijeka, Croatia. The supercomputer Bura and other information and communication technology (ICT) research infrastructure were acquired through the project *Development of research infrastructure for laboratories of the University of Rijeka Campus*, which is co-funded by the European regional development fund.

Acknowledgments: Technical supports of Emir Imamagić and Daniel Vrčić from Isabella and CRO-NGI and of Miroslav Puškarić from CNRM are gratefully acknowledged.

Conflicts of Interest: The authors declare no conflict of interest.

Abbreviations

The following abbreviations are used in this manuscript:

KS   Kochen–Specker; defined in Section 1
MMP   McKay-Megill-Pavičić; defined in Section 2.1

Appendix A. 21-11 KS Critical with Complex States from $\mathcal{H}^2 \otimes \mathcal{H}^2$

Below, we present a possible implementation of a KS critical 21-11 with complex coordinatization shown in Figure A1.

The vector components of the first qubit on a photon correspond to a linear (horizontal, H, vertical, V, diagonal, D, antidiagonal A) and circular (right, R, left L) polarization, and those of the second qubit to an angular momentum of the photon $(+2, -2)$ and (h, v). One-to-one correspondence between them is given below.

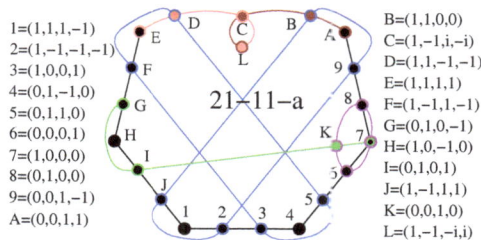

$$
\begin{array}{ll}
1=(1,1,1,-1) & B=(1,1,0,0) \\
2=(1,-1,-1,-1) & C=(1,-1,i,-i) \\
3=(1,0,0,1) & D=(1,1,-1,-1) \\
4=(0,1,-1,0) & E=(1,1,1,1) \\
5=(0,1,1,0) & F=(1,-1,1,-1) \\
6=(0,0,0,1) & G=(0,1,0,-1) \\
7=(1,0,0,0) & H=(1,0,-1,0) \\
8=(0,1,0,0) & I=(0,1,0,1) \\
9=(0,0,1,-1) & J=(1,-1,1,1) \\
A=(0,0,1,1) & K=(0,0,1,0) \\
& L=(1,-1,-i,i)
\end{array}
$$

Figure A1. 21-11 KS set with complex coordinatization.

An example of a tensor product of two vectors/states from $\mathcal{H}^2 \otimes \mathcal{H}^2$ is:

$$|01\rangle = |0,1\rangle = |0\rangle_1 \otimes |1\rangle_2 = \begin{pmatrix} 1 \\ 0 \end{pmatrix}_1 \otimes \begin{pmatrix} 0 \\ 1 \end{pmatrix}_2 = \begin{pmatrix} 1\begin{pmatrix} 0 \\ 1 \end{pmatrix} \\ 0\begin{pmatrix} 0 \\ 1 \end{pmatrix} \end{pmatrix} = \begin{pmatrix} 0 \\ 1 \\ 0 \\ 0 \end{pmatrix}.$$

This is our vector 8 from Figure A1. Since we are interested in the qubit states, we are going to proceed in reverse—from 4-vectors to tensor products of polarization and angular momentum states. Let us first define them:

$$|H\rangle = \begin{pmatrix} 1 \\ 0 \end{pmatrix}_1 \; ; \quad |V\rangle = \begin{pmatrix} 0 \\ 1 \end{pmatrix}_1 \; ; \quad |D\rangle = \frac{1}{\sqrt{2}}\begin{pmatrix} 1 \\ 1 \end{pmatrix}_1 \; ; \quad |A\rangle = \frac{1}{\sqrt{2}}\begin{pmatrix} -1 \\ 1 \end{pmatrix}_1 \; ; \quad |R\rangle = \frac{1}{\sqrt{2}}\begin{pmatrix} 1 \\ i \end{pmatrix}_1 \; ;$$

$$|L\rangle = \frac{1}{\sqrt{2}}\begin{pmatrix} 1 \\ -i \end{pmatrix}_1 \; ; \quad |+2\rangle = \begin{pmatrix} 1 \\ 0 \end{pmatrix}_2 \; ; \quad |-2\rangle = \begin{pmatrix} 0 \\ 1 \end{pmatrix}_2 \; ; \quad |h\rangle = \frac{1}{\sqrt{2}}\begin{pmatrix} 1 \\ 1 \end{pmatrix}_2 \; ; \quad |v\rangle = \frac{1}{\sqrt{2}}\begin{pmatrix} 1 \\ -1 \end{pmatrix}_2 .$$

Now, one can read off our vertex states as follows:

$$1 = \begin{pmatrix} 1 \\ 1 \\ 1 \\ -1 \end{pmatrix} \to \frac{1}{2}\begin{pmatrix} 1 \\ 1 \\ 1 \\ -1 \end{pmatrix} = \frac{1}{2}\left(\begin{pmatrix} 1 \\ 1 \\ 0 \\ 0 \end{pmatrix} + \begin{pmatrix} 0 \\ 0 \\ 1 \\ -1 \end{pmatrix}\right) = \frac{1}{\sqrt{2}}\left(\frac{1}{\sqrt{2}}\begin{pmatrix} 1\begin{pmatrix} 1 \\ 1 \end{pmatrix} \\ 0\begin{pmatrix} 1 \\ 1 \end{pmatrix} \end{pmatrix} + \frac{1}{\sqrt{2}}\begin{pmatrix} 0\begin{pmatrix} 1 \\ -1 \end{pmatrix} \\ 1\begin{pmatrix} 1 \\ -1 \end{pmatrix} \end{pmatrix}\right)$$

$$= \frac{1}{\sqrt{2}}(\begin{pmatrix} 1 \\ 0 \end{pmatrix}_1 \otimes \frac{1}{\sqrt{2}}\begin{pmatrix} 1 \\ 1 \end{pmatrix}_2 + \begin{pmatrix} 0 \\ 1 \end{pmatrix}_1 \otimes \frac{1}{\sqrt{2}}\begin{pmatrix} 1 \\ -1 \end{pmatrix}_2 = \frac{1}{\sqrt{2}}(|H\rangle|h\rangle + |V\rangle|v\rangle) = \frac{1}{\sqrt{2}}(|D\rangle|+2\rangle - |A\rangle|-2\rangle).$$

We will now skip real states and go directly to those with imaginary components, C and L, to illustrate how they can be implemented via circular polarization:

$$C = \begin{pmatrix} 1 \\ -1 \\ i \\ -i \end{pmatrix} \to \frac{1}{2}\begin{pmatrix} 1\begin{pmatrix} 1 \\ -1 \end{pmatrix} \\ i\begin{pmatrix} 1 \\ -1 \end{pmatrix} \end{pmatrix} = \frac{1}{\sqrt{2}}\begin{pmatrix} 1 \\ i \end{pmatrix}_1 \otimes \frac{1}{\sqrt{2}}\begin{pmatrix} 1 \\ -1 \end{pmatrix}_2 = |R\rangle|v\rangle,$$

$$L = \begin{pmatrix} 1 \\ -1 \\ -i \\ i \end{pmatrix} \to \frac{1}{2}\begin{pmatrix} 1\begin{pmatrix} 1 \\ -1 \end{pmatrix} \\ -i\begin{pmatrix} 1 \\ -1 \end{pmatrix} \end{pmatrix} = \frac{1}{\sqrt{2}}\begin{pmatrix} 1 \\ -i \end{pmatrix}_1 \otimes \frac{1}{\sqrt{2}}\begin{pmatrix} 1 \\ -1 \end{pmatrix}_2 = |L\rangle|v\rangle.$$

Thus, in order to handle a complex coordinatization, we need a fifth degree of freedom (circular polarization), but, as we can see, it is manageable.

Appendix B. 6D Criticals from the Masters Containing the Seven Context Star.

The 216-153 KS master generated from $\{0, 1, \omega\}$ contains 21-7 and 27-9, which can be viewed as 21-7 with a pair of δ-triplets interwoven with 21-7, as shown in Figure A2. The 834-1609 KS master generated from $\{0, 1, \omega, \omega^2\}$, which were used for a construction of 21-7 in [19], contains 39-13 as well. Equally so, the 11808-314446 master generated from $\{0, \pm1, \omega, \omega^2\}$.

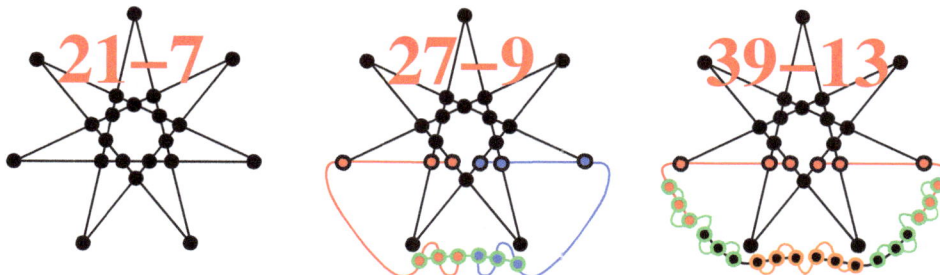

Figure A2. 21-11 KS set from [19] and 27-9 are contained in three different master sets, 39-13 in two (together with 21-11 and 27-9); see the text.

References

1. Cabello, A.; D'Ambrosio, V.; Nagali, E.; Sciarrino, F. Hybrid Ququart-Encoded Quantum Cryptography Protected by Kochen-Specker Contextuality. *Phys. Rev. A* **2011**, *84*, 030302(R). [CrossRef]
2. Nagata, K. Kochen-Specker Theorem as a Precondition for Secure Quantum Key Distribution. *Phys. Rev. A* **2005**, *72*, 012325. [CrossRef]
3. Howard, M.; Wallman, J.; Veitech, V.; Emerson, J. Contextuality Supplies the 'Magic' for Quantum Computation. *Nature* **2014**, *510*, 351–355. [CrossRef] [PubMed]
4. Bartlett, S.D. Powered by Magic. *Nature* **2014**, *510*, 345–346. [CrossRef] [PubMed]
5. Pavičić, M.; McKay, B.D.; Megill, N.D.; Fresl, K. Graph Approach to Quantum Systems. *J. Math. Phys.* **2010**, *51*, 102103. [CrossRef]
6. Megill, N.D.; Pavičić, M. Kochen-Specker Sets and Generalized Orthoarguesian Equations. *Ann. Henri Poincare* **2011**, *12*, 1417–1429. [CrossRef]
7. Simon, C.; Żukowski, M.; Weinfurter, H.; Zeilinger, A. Feasible Kochen-Specker Experiment with Single Particles. *Phys. Rev. Lett.* **2000**, *85*, 1783–1786. [CrossRef] [PubMed]
8. Michler, M.; Weinfurter, H.; Żukowski, M. Experiments towards Falsification of Noncontextual Hidden Variables. *Phys. Rev. Lett.* **2000**, *84*, 5457–5461. [CrossRef]
9. Amselem, E.; Rådmark, M.; Bourennane, M.; Cabello, A. State-Independent Quantum Contextuality with Single Photons. *Phys. Rev. Lett.* **2009**, *103*, 160405. [CrossRef]
10. Liu, B.H.; Huang, Y.F.; Gong, Y.X.; Sun, F.W.; Zhang, Y.S.; Li, C.F.; Guo, G.C. Experimental Demonstration of Quantum Contextuality with Nonentangled Photons. *Phys. Rev. A* **2009**, *80*, 044101. [CrossRef]
11. D'Ambrosio, V.; Herbauts, I.; Amselem, E.; Nagali, E.; Bourennane, M.; Sciarrino, F.; Cabello, A. Experimental Implementation of a Kochen-Specker Set of Quantum Tests. *Phys. Rev. X* **2013**, *3*, 011012. [CrossRef]
12. Huang, Y.F.; Li, C.F.; Zhang, Y.S.; Pan, J.W.; Guo, G.C. Experimental Test of the Kochen-Specker Theorem with Single Photons. *Phys. Rev. Lett.* **2003**, *90*, 250401. [CrossRef] [PubMed]
13. Cañas, G.; Etcheverry, S.; Gómez, E.S.; Saavedra, C.; Xavier, G.B.; Lima, G.; Cabello, A. Experimental Implementation of an Eight-Dimensional Kochen-Specker Set and Observation of Its Connection with the Greenberger-Horne-Zeilinger Theorem. *Phys. Rev. A* **2014**, *90*, 012119. [CrossRef]
14. Hasegawa, Y.; Loidl, R.; Badurek, G.; Baron, M.; Rauch, H. Quantum Contextuality in a Single-Neutron Optical Experiment. *Phys. Rev. Lett.* **2006**, *97*, 230401. [CrossRef] [PubMed]
15. Cabello, A.; Filipp, S.; Rauch, H.; Hasegawa, Y. Proposed Experiment for Testing Quantum Contextuality with Neutrons. *Phys. Rev. Lett.* **2008**, *100*, 130404. [CrossRef] [PubMed]

16. Bartosik, H.; Klepp, J.; Schmitzer, C.; Sponar, S.; Cabello, A.; Rauch, H.; Hasegawa, Y. Experimental Test of Quantum Contextuality in Neutron Interferometry. *Phys. Rev. Lett.* **2009**, *103*, 040403. [CrossRef] [PubMed]
17. Kirchmair, G.; Zähringer, F.; Gerritsma, R.; Kleinmann, M.; Gühne, O.; Cabello, A.; Blatt, R.; Roos, C.F. State-Independent Experimental Test of Quantum Contextuality. *Nature* **2009**, *460*, 494–497. [CrossRef]
18. Moussa, O.; Ryan, C.A.; Cory, D.G.; Laflamme, R. Testing Contextuality on Quantum Ensembles with One Clean Qubit. *Phys. Rev. Lett.* **2010**, *104*, 160501. [CrossRef]
19. Lisoněk, P.; Badziag, P.; Portillo, J.R.; Cabello, A. Kochen-Specker Set with Seven Contexts. *Phys. Rev. A* **2014**, *89*, 042101. [CrossRef]
20. Cañas, G.; Arias, M.; Etcheverry, S.; Gómez, E.S.; Cabello, A.; Saavedra, C.; Xavier, G.B.; Lima, G. Applying the Simplest Kochen-Specker Set for Quantum Information Processing. *Phys. Rev. Lett.* **2014**, *113*, 090404. [CrossRef]
21. Barrett, J.; Kent, A. Noncontextuality, Finite Precision Measurement and the Kochen-Specker. *Stud. Hist. Philos. Mod. Phys.* **2004**, *35*, 151–176. [CrossRef]
22. Spekkens, R.W. Contextuality for Preparations, Transformations, and Unsharp Measurements. *Phys. Rev. A* **2005**, *71*, 052108. [CrossRef]
23. Kunjwal, R.; Spekkens, R.W. From the Kochen-Specker Theorem to Noncontextuality Inequalities without Assuming Determinism. *Phys. Rev. Lett.* **2015**, *115*, 110403. [CrossRef] [PubMed]
24. Kunjwal, R. Hypergraph Framework for Irreducible Noncontextuality Inequalities from Logical Proofs of the Kochen-Specker Theorem. *arXiv* **2018**, arXiv:1805.02083.
25. Cabello, A.; Estebaranz, J.M.; García-Alcaine, G. Bell-Kochen-Specker Theorem: A Proof with 18 Vectors. *Phys. Lett. A* **1996**, *212*, 183–187. [CrossRef]
26. Pavičić, M.; Merlet, J.P.; McKay, B.D.; Megill, N.D. Kochen-Specker Vectors. *arXiv* **2005**, arXiv:quant-ph/0409014.
27. Waegell, M.; Aravind, P.K. Critical Noncolorings of the 600-Cell Proving the Bell-Kochen-Specker Theorem. *J. Phys. A* **2010**, *43*, 105304. [CrossRef]
28. Waegell, M.; Aravind, P.K. Parity Proofs of the Kochen-Specker Theorem Based on 60 Complex Rays in Four Dimensions. *J. Phys. A* **2011**, *44*, 505303. [CrossRef]
29. Megill, N.D.; Fresl, K.; Waegell, M.; Aravind, P.K.; Pavičić, M. Probabilistic Generation of Quantum Contextual Sets. *Phys. Lett. A* **2011**, *375*, 3419–3424. [CrossRef]
30. Pavičić, M.; Megill, N.D.; Aravind, P.K.; Waegell, M. New Class of 4-Dim Kochen-Specker Sets. *J. Math. Phys.* **2011**, *52*, 022104. [CrossRef]
31. Waegell, M.; Aravind, P.K.; Megill, N.D.; Pavičić, M. Parity Proofs of the Bell-Kochen-Specker Theorem Based on the 600-cell. *Found. Phys.* **2011**, *41*, 883–904. [CrossRef]
32. Waegell, M.; Aravind, P.K. Proofs of Kochen-Specker Theorem Based on a System of Three Qubits. *J. Phys. A* **2012**, *45*, 405301. [CrossRef]
33. Waegell, M.; Aravind, P.K. Proofs of the Kochen-Specker Theorem Based on the N-Qubit Pauli Group. *Phys. Rev. A* **2013**, *88*, 012102. [CrossRef]
34. Waegell, M.; Aravind, P.K. Parity Proofs of the Kochen-Specker Theorem Based on 120-Cell. *Found. Phys.* **2014**, *44*, 1085–1095. [CrossRef]
35. Waegell, M.; Aravind, P.K. Parity Proofs of the Kochen-Specker Theorem Based on the Lie Algebra E8. *J. Phys. A* **2015**, *48*, 225301. [CrossRef]
36. Waegell, M.; Aravind, P.K. The Penrose Dodecahedron and the Witting Polytope Are Identical in $\mathbb{CP}^3$. *Phys. Lett. A* **2017**, *381*, 1853–1857. [CrossRef]
37. Pavičić, M. Arbitrarily Exhaustive Hypergraph Generation of 4-, 6-, 8-, 16-, and 32-Dimensional Quantum Contextual Sets. *Phys. Rev. A* **2017**, *95*, 062121. [CrossRef]
38. Pavičić, M.; Merlet, J.P.; McKay, B.D.; Megill, N.D. Kochen-Specker Vectors. *J. Phys. A* **2005**, *38*, 1577–1592; 3709 (corrigendum). [CrossRef]
39. McKay, B.D.; Megill, N.D.; Pavičić, M. Algorithms for Greechie Diagrams. *Int. J. Theor. Phys.* **2000**, *39*, 2381–2406. [CrossRef]
40. Pavičić, M.; Megill, N.D.; Merlet, J.P. New Kochen-Specker Sets in Four Dimensions. *Phys. Lett. A* **2010**, *374*, 2122–2128. [CrossRef]
41. Megill, N.D.; Fresl, K.; Waegell, M.; Aravind, P.K.; Pavičić, M. Probabilistic Generation of Quantum Contextual Sets. *arXiv* **2011**, arXiv:1105.1840.
42. Peres, A. Two Simple Proofs of the Bell-Kochen-Specker Theorem. *J. Phys. A* **1991**, *24*, L175–L178. [CrossRef]

43. Kernaghan, M. Bell-Kochen-Specker Theorem for 20 Vectors. *J. Phys. A* **1994**, *27*, L829–L830. [CrossRef]
44. Harvey, C.; Chryssanthacopoulos, J. *BKS Theorem and Bell's Theorem in 16 Dimensions*; Technical Report PH-PKA-JC08; Worcester Polytechnic Institute: Worcester, MA, USA, 2012. Available online: https://web.wpi.edu/Pubs/E-project/Available/E-project-042108-171725/unrestricted/MQPReport.pdf (accessed on 26 November 2018).
45. Planat, M. On Small Proofs of the Bell-Kochen-Specker Theorem for Two, Three and Four Qubits. *Eur. Phys. J. Plus* **2012**, *127*, 86. [CrossRef]
46. Planat, M.; Saniga, M. Five-Qubit Contextuality, Noise-Like Distribution of Distances Between Maximal Bases and Finite Geometry. *Phys. Lett. A* **2012**, *376*, 3485–3490. [CrossRef]
47. Pavičić, M.; Megill, N.D. Quantum Logic and Quantum Computation. In *Handbook of Quantum Logic and Quantum Structures*; Engesser, K., Gabbay, D., Lehmann, D., Eds.; Elsevier: Amsterdam, The Netherlands, 2007; pp. 751–787.
48. Megill, N.D.; Pavičić, M. New Classes of Kochen-Specker Contextual Sets (Invited Talk). In Proceedings of the 2017 40th International Convention on Information and Communication Technology, Electronics and Microelectronics (MIPRO 2017), Opatija, Croatia, 22–26 May 2017; Biljanović, P., Ed.; Institute of Electrical and Electronics Engineers (IEEE), Curran Associates, Inc.: Red Hook, NY, USA, 2017; pp. 195–200, ISBN 9781509049691.

Article

Advanced Statistical Testing of Quantum Random Number Generators

Aldo C. Martínez [1], Aldo Solis [2], Rafael Díaz Hernández Rojas [3], Alfred B. U'Ren [2], Jorge G. Hirsch [2] and Isaac Pérez Castillo [4,5,*]

[1]   Department of Physics, Center for Research in Photonics, University of Ottawa, 25 Templeton St, Ottawa, ON K1N 6N5, Canada; mbac@ciencias.unam.mx
[2]   Instituto de Ciencias Nucleares, Universidad Nacional Autónoma de México, Apdo. Postal 70-543, Cd. Mx., C.P. 04510 Mexico, Mexico; aldo.solis@correo.nucleares.unam.mx (A.S.); alfred.uren@correo.nucleares.unam.mx (A.B.U.); hirsch@nucleares.unam.mx (J.G.H.)
[3]   Dipartimento di Fisica, Sapienza University of Rome, P.le Aldo Moro 5, I-00185 Rome, Italy; rafael.diaz.hr@gmail.com
[4]   Departamento de Física Cuántica y Fótonica, Instituto de Física, Universidad Nacional Autónoma de México, Apdo. Postal 20-364, Cd. Mx., C.P. 04510 Mexico, Mexico
[5]   London Mathematical Laboratory, 8 Margravine Gardens, London W6 8RH, UK
*   Correspondence: isaacpc@fisica.unam.mx

Received: 20 October 2018; Accepted: 14 November 2018; Published: 17 November 2018

Abstract: Pseudo-random number generators are widely used in many branches of science, mainly in applications related to Monte Carlo methods, although they are deterministic in design and, therefore, unsuitable for tackling fundamental problems in security and cryptography. The natural laws of the microscopic realm provide a fairly simple method to generate non-deterministic sequences of random numbers, based on measurements of quantum states. In practice, however, the experimental devices on which quantum random number generators are based are often unable to pass some tests of randomness. In this review, we briefly discuss two such tests, point out the challenges that we have encountered in experimental implementations and finally present a fairly simple method that successfully generates non-deterministic maximally random sequences.

Keywords: Bell inequalities; algorithmic complexity; Borel normality; Bayesian inference; model selection; random numbers

---

1. Introduction

Monte Carlo methods are one of the essential staples of the basic sciences in the modern age. Although these gained prominence during the early 1940s, thanks to secret research projects carried out in Los Alamos Scientific Laboratory by Ulam and von Neumann [1,2], their origins may be traced back to the famous Buffon's needle problem, posed by Georges-Louis Leclerc, Comte de Buffon, in the 18th century. In the present day, Monte Carlo "experiments" are seen as a broad class of computational algorithms that use repeated random sampling to obtain numerical estimates of a given natural or mathematical process. In order to use these methods efficiently, fully random sequences of numbers are needed. Back in the 1940s, this was a tall order, and various methods to generate random sequences were used (some of them literally using roulettes), until von Neumann pioneered the concept of computer-based random number generators. During the following years, these became the standard tool in Monte Carlo methods and are still generally well-suited for many applications. However, these computer-based methods generate pseudo random numbers [3], which means that the generated sequence can be determined given an algorithmic program and an initial seed, two ingredients which are hardly random. Thus, in order to achieve a truly unpredictable source of random numbers, we must eliminate these two deterministic aspects. The former is easy to overcome using, for example, a pattern of keystrokes typed on a computer keyboard as a random seed. On the other hand, the algorithmic program could be replaced, for instance, by a classical chaotic system [4]. Examples of the latter abound in the area of weather prediction and climate sciences.

In recent years, however, the community has been moving towards using the fundamental laws dictating the behaviour of the quantum realm for the generation of sequences of truly random numbers. This seems, at a first glance, to be at odds with the following rather naïve thought: if the natural laws of the microscopic world are considered to be a computer program under which a system evolves from an initial state (a seed), should not its corresponding generated sequence also be predictable? It turns out that Quantum Mechanics, in its current standard view, related to the Copenhagen interpretation, has a special ingredient that makes the random sequence inherently unpredictable for both the generator and the observer. Such a strange behaviour has been eloquently recast over the years in various forms, famously by the quote "spooky action at a distance" due to Einstein, or mathematically by the celebrated work of Bell [5,6]. The application of quantum randomness in cryptography has given rise to the concept of device independent randomness certification, which, in a nutshell, corresponds to those processes that violate Bell's inequalities [7,8]. However, there seems to be some confusion in the literature regarding two different properties of a given sequence of random numbers. The first one, rather important as we have argued above, is whether the sequence is truly random, meaning that it is unpredictable. In contrast, the second one is related to the issue of assessing whether or not it is biased. It is crucial to keep in mind that these two properties are independent, as evidenced by the random number generator Quantis [9], which is based on a quantum system and is able to pass the standard tests of randomness (NIST (National Institute of Standards and Technology) suite) [10] but has difficulties with other tests [11].

Due to recent advances in quantum technologies, and since the NIST suite has been examined in other works [12], together with a critical view on the use of p-values on which the NIST suite relies [13], it becomes necessary to consider other criteria for measuring the performance of quantum random number generators. Thus, we focus solely on two recently introduced approaches: the first one is based on algorithmic complexity theory evaluating incompressibility and bias at the same time, since an incompressible sequence is necessarily an unbiased one [14], while the second one relies on Bayesian model selection. Both methods are based on solid structures which lead to a definition of randomness that is very intuitive and which arises independently of the development of random number generators. We apply them to analyze sequences of random bits generated in our laboratory using quantum systems. We also address the issue about the origin of the biases observed when utilizing these types of devices.

2. Tests of Randomness

A simple criterion for assessing the predictability of a sequence is the presence of patterns in it. For example, for the sequence 01010101..., we can ask ourselves whether the next number is either 1 or 0. The natural answer is 0 based on the pattern observed in the previous bits. In general, we would like to find any possible regularity that helps us to predict the next bit. Within the framework of algorithmic information theory, it is possible to address this problem by noting that any sequence which exhibits regularity can be compressed using a short algorithm which can produce as output precisely such patterns. Thus, a sequence of this type could be reproduced using fewer bits than the ones contained in its original form. Therefore, whenever a sequence lacks regularity, we refer to it as "algorithmically" random.

We now introduce a remarkable result from algorithmic information theory: the Borel-normality criterion due to Calude [14], which allows us to asymptotically check whether a sequence is not "algorithmically" random. Assuming we are given a string $\ell = \{1001010110110\cdots\}$ of $|\ell| = n$ bits (We will only consider binary sequences, but our results are easily generalizable to other alphabets), the idea of the Borel-normality criterion consists primarily of dividing the original sequence ℓ into consecutive substrings of length i and then computing the frequencies of occurrence of each of them. For brevity and later use, let us define $\Omega^{(i)}$ as the set of 2^i substrings that can be formed with i characters, let ℓ_i be the sequence obtained after dividing it into substrings, and $|\ell|_i \equiv [|\ell|/i]$. Additionally, let $N_i^j(\ell)$ be the number of times the j-th substring of length i appears in ℓ. For example, when considering substrings of length $i = 1$, we are looking at the frequencies of the symbols $\Omega^{(1)} = \{0, 1\}$ that conform the original string ℓ, while for $i = 2$, we have to consider the frequencies of four substrings, namely $\Omega^{(2)} = \{00, 01, 10, 11\}$. According to Calude, a necessary condition for a sequence to be maximally random is that the deviations of these frequencies with respect to the expected values in the ideal random case should be bounded as follows [14,15]:

$$\left| \frac{N_i^j(\ell)}{|\ell|_i} - \frac{1}{2^i} \right| < \sqrt{\frac{\log_2(n)}{n}}, \qquad j = 0, \ldots, 2^i - 1. \tag{1}$$

This condition must be satisfied for all substrings of lengths from $i = 1$ up to $i_{max} = \log_2(\log_2(n))$. Intuitively, this criterion "compresses" the original sequence by reading i bits at a time and tests whether the substrings appear with a frequency that differs from what would be expected in the random case, thus indicating the presence of some regularity. We emphasize that since Borel-normality is not a sufficient criterion for randomness, it can only be used to assess whether a given sequence is not random. In other words, even if a sequence satisfies Equation (1) for all substrings and allowed values of i, the Borel-normality condition cannot guarantee that it is indeed random.

Recently, a Bayesian criterion has been introduced [16,17] by some of the authors of the present article to test, from a purely probabilistic point of view, whether a sequence is maximally random as understood within information theory [18]. The method works by exploiting the Borel-normality compression scheme and then recasting the problem of finding possible biases in the sequence as an inferential one in which Bayesian model selection can be applied. Specifically, for a fixed value of i, we need to consider all the possible probabilistic models, henceforth denoted as $\{\mathcal{M}_\alpha^{(i)}\}_\alpha$, that could have generated the sequence ℓ. Each such model determines a unique probability assignment to the elements of $\Omega^{(i)}$, which depends on a set of prior parameters $\boldsymbol{\theta}$. For these parameters, the Jeffreys' prior, $P_{\text{Jeff}}(\boldsymbol{\theta})$, turns out to be a convenient choice of prior parameter distribution, as it entails the "Occam Razor principle" in which more complex models are penalized, as well as being mathematically convenient for the case at hand; some other advantages are pointed out in [16,17].

Next, the question of finding all the generative models that can produce a sequence ℓ is ultimately solved by noticing that all the possible probabilities assignments are in a one-to-one correspondence with all possible partitions of $\Omega^{(i)}$. Since obtaining the partitions of any set is a straightforward combinatorial task [19], we are able to determine all the relevant models when searching for possible biases in the generation of ℓ. For instance, when $i = 1$, there are two possible models: one in which the two elements of $\Omega^{(1)}$ are equiprobable, corresponding to the partition $\{\{0, 1\}\}$ of $\Omega^{(1)}$ into one subset—i.e., the same set—and another model with probabilities $p_0 = \theta$, $p_1 = 1 - \theta$ corresponding to the partition $\{\{0\}, \{1\}\}$ of $\Omega^{(1)}$ into two subsets. Even though it might seem that the first model is just a particular case of the second one (by letting $\theta = 1/2$), we should keep in mind that the prior distributions are different in both cases, $\delta(\theta - 1/2)$ and $\frac{1}{\pi\sqrt{\theta(1-\theta)}}$, respectively, thus yielding two different models. Analogously, for $i = 2$, there is a single unbiased model, which corresponds to the partition of $\Omega^{(2)}$ into one subset (with probabilities $p_j = 1/4$, for $j = 00, 01, 10, 11$), and 14 additional models associated with the different ways of dividing $\Omega^{(2)}$ into subsets. The latter are related to the number of ways of distinguishing among the elements of $\Omega^{(2)}$ during the assignation of probabilities, and thus any of these models would entail some bias when generating a sequence. Note that, in general, for any value of i, we will face a similar situation in which a single model can produce an unbiased, and hence maximally random, sequence by means of a uniform distribution, while the rest of them will be some type of categorical distribution.

Once all the models have been identified, the remaining part is the computation of the posterior distribution $P\left(\mathcal{M}_\alpha^{(i)}\middle|\ell\right)$, which from an inferential point of view is the most relevant distribution as it gives the probability that the model $\mathcal{M}_\alpha^{(i)}$ has indeed produced the given sequence ℓ. Note that, since a generative approach was adopted, we have direct access to the distribution $P\left(\ell\middle|\theta, \mathcal{M}_\alpha^{(i)}\right)$, which can be combined with the parameters' prior $P_{\text{Jeff}}(\theta)$ to obtaining the distribution $P\left(\ell\middle|\mathcal{M}_\alpha^{(i)}\right) = \int d\theta\, P\left(\ell\middle|\mathcal{M}_\alpha^{(i)},\theta\right) P_{\text{Jeff}}(\theta)$. One of the most important results of [16] is that this marginalization can be accomplished exactly for all the models and any value of i. Therefore, we can obtain the posterior distribution by a simple application of Bayes' rule:

$$P\left(\mathcal{M}_\alpha^{(i)}\middle|\ell\right) = \frac{P\left(\ell\middle|\mathcal{M}_\alpha^{(i)}\right) P\left(\mathcal{M}_\alpha^{(i)}\right)}{\sum_\gamma P\left(\ell\middle|\mathcal{M}_\gamma^{(i)}\right) P\left(\mathcal{M}_\gamma^{(i)}\right)}, \tag{2}$$

with $P\left(\mathcal{M}_\gamma^{(i)}\right)$ being the prior distribution in the space of models that can generate sequences of strings of length i. Therefore, the best model $\alpha^\star$ that describes the dataset ℓ is quite simply given by

$$\alpha^\star = \arg\max_\alpha P\left(\mathcal{M}_\alpha^{(i)}\middle|\ell\right). \tag{3}$$

If the best model $\mathcal{M}_{\alpha^\star}^{(i)}$ turns out to be the unbiased one for all possible lengths i of the substrings, then we can say that the process that generated that dataset was maximally random. However, it remains to discuss how large the length i of substrings can be for a given dataset of n bits. To answer this, we first note that, for any set containing N elements, the possible number of partitions is given by the N-th Bell number B_N [19]. Thus, for a given i, all possible partitions of the set $\Omega^{(i)}$ will result in B_{2^i} models to be tested, and, therefore, it is expected for them to be sampled at least once when observing ℓ. This means that $B_{2^{i_{max}}} \sim n$, which, for sufficiently large n, yields $i_{max} = \log_2(\log_2(n))$, precisely as in the Borel-normality criterion.

Randomness characterization through Bayesian model selection has some clear and natural advantages, as already pointed out in [16], but, unfortunately, it has an important drawback: the number of all possible models for a given length i, given by B_{2^i}, grows supra-exponentially with i: indeed, for $i = 1$, we have two possible models, for $i = 2$, we have 15 possible models, for $i = 3$, we have instead 4140 possible models, while, for $i = 4$, we have 10,480,142,147 models. Thus, even if we are able to acquire data for the evaluation of these many models, it becomes computationally impractical to estimate the posterior for all of them using Equation (2). There is an elegant strategy to overcome this difficulty: one can derive bounds similar to those provided by the Borel-normality criterion, by comparing the log-likelihood ratio between the maximally random model and the maximally biased one. This yields the following bound for the frequencies of occurrence [17]:

$$\sqrt{\sum_{j \leq j'=1}^{2^i-1} \left(\frac{N_i^j(\ell)}{|\ell|_i} - \frac{1}{2^i}\right)\left(\frac{N_i^{j'}(\ell)}{|\ell|_i} - \frac{1}{2^i}\right)} < \sqrt{\frac{i^2}{n^2 \psi_1\left(\frac{1}{2} + \frac{n}{i2^i}\right)} \ln\left(\frac{2^{-n}\Gamma^{2^i}\left(\frac{1}{2}\right)\Gamma\left(\frac{1}{2^{1-i}} + \frac{n}{i}\right)}{\Gamma\left(\frac{1}{2^{1-i}}\right)\Gamma^{2^i}\left(\frac{1}{2} + \frac{n}{i2^i}\right)}\right)}, \tag{4}$$

where ψ_1 is the polygamma function of order 1. Note that, unlike Calude's bound given by Equation (1), this new Borel-type bound couples all frequencies, and, moreover, results in highly restrictive bounds.

3. Ideal Random Number Generation

While intuition dictates that quantum random number generators (QRNG) should be superior to their classical counterparts, such a comparison was carried out in [11] and very recently in [20], with rather disappointing results. For the classical case, the authors used three Pseudo-Random Number Generators (PRNG): the generators included in the software packages Mathematica and Maple, and the digits of π expressed in base 2. For QRNG, they used two devices: (i) Quantis, developed by IDQ [9], a quantum random number generator interfaced with a common computer, and (ii) an experiment from a quantum optics group in Vienna. The latter experiment consists of a very weak light source, attenuated to the single photon level, a beam splitter, and two single photon detectors. Leaving aside the question of which QRNG performs better, the real surprise was that the PRNGs come out in this test with a superior performance, by far, as compared to their quantum counterparts. This result appears to be at odds with the natural randomness associated with quantum phenomena. Why is it that the inherent quantum randomness does not translate into better performance with respect to classical systems? Does randomness, as discussed in this paper, have no impact on the performance of the generators? Is this a fundamental or a technical problem?

In order to explore this apparent paradox, we will discuss the different technical and design difficulties associated with quantum random number generation using light. These days, it is straightforward to detect single photons using avalanche photo-diodes (APD), devices capable of detecting up to a few million single photons per second with $> 60\%$ detection efficiencies employing relatively simple electronics. With this simple design in mind, we only need a single photon source, a beam splitter (BS), and a couple of single-photon detection devices in order to set up a QRNG device. This minimalistic design is sketched in Figure 1.

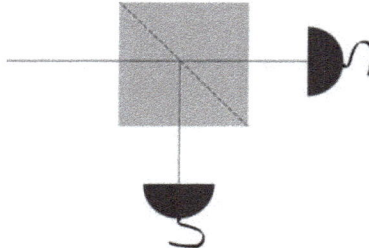

Figure 1. Ideal experimental setup for a naïve QRNG (quantum random number generators) based on an individual photon source and a beam splitter (BS). Neglecting the possible losses, a photon will activate only one of the two detectors, therefore producing a random bit per photon.

4. Experimental Challenges in Random Number Generation

Suppose now that we have a single-photon source and we want to generate a sequence of bits to be tested against the bounds given by Equation (1). Let us start by first focusing on the so-called Borel level, the word length i, which can take a maximum value of $i_{max} = \log_2(\log_2(n))$. In Table 1, we report how i_{max} grows with n, up to a value of $i_{max} = 6$. In order to achieve a Borel level $i_{max} = 6$, we will require a dataset of length 10^{18} events. Assuming that our single-photon generator and detectors can cope with around a million of events per second, we would then require on the order of 600 years to generate a sequence of that length! It turns out that $i_{max} = 5$ is a more realistic value, since it leads to a required dataset ℓ of size 4.3×10^9 events, which can be realistically produced in a couple of hours.

Table 1. Necessary data lengths for maximum Borel level $i_{max} = \log_2(\log_2(n))$. The double exponential relation grows so quickly that it is not possible get to level 6.

Maximum Borel Level $i_{max} = \log_2(\log_2(n))$	Data Length n
1	4
2	16
3	256
4	65,536
5	4,294,967,296
6	18,446,744,073,709,551,616

The bound on the right-hand-side of Equation (1) implies that the frequencies of occurrence $\frac{N_i^j(\ell)}{|\ell|_i}$ for substrings of length $i \leq i_{max}$ cannot deviate from the ideal random one, $1/2^i$, by more than 8.6×10^{-5}, which constitutes an extremely tight tolerance. Hence, in practice, any part of the naïve experimental setup that gives rise to some bias will unfortunately make the dataset ℓ unable to pass the Calude criterion. The first component that we must be wary about is the BS. A regular BS usually has an error figure in the region of 1%, which is very high with respect to the stringent tolerance ℓ would need in order fulfill Borel normality. Is it plausible to correct this using a Polarizing Beam Splitter (PBS), instead of the BS, with an active control through feedback of the state of polarization so as to compensate for any bias in the PBS? In what follows, we investigate this question through a simple experiment. The state of polarization of a single photon entering the PBS can be written as

$$|\psi\rangle = a\,|V\rangle + e^{i\phi}b\,|H\rangle, \tag{5}$$

where $|V\rangle$ and $|H\rangle$ refer to the vertical and horizontal polarization components, respectively. We can approach the state in Equation (5) by transmitting the laser beam trough a half wave plate (HWP) so as to achieve arbitrary rotation of the linear polarization. Assuming a perfect, unbiased BS, we would need an incoming polarization state with $a = b = 1/\sqrt{2}$ so that the resulting sequence of bits is unbiased. If, on the other hand, the PBS exhibits biases (e.g., due to manufacturing error), we can adjust the orientation of the above-mentioned half wave plate so as to adjust precisely the value of our coefficients a and b to compensate for the PBS bias.

Our experimental setup, shown in Figure 2, can be regarded as the minimal realistic device for the implementation of a QRNG. The main questions which we wish to address are: (i) how good are the sequences of bits generated by such a device? In addition, (ii) do they pass the Borel-normality criterion?

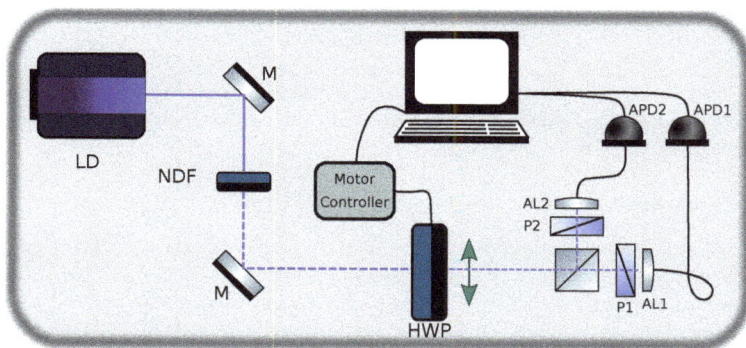

Figure 2. Experimental setup. The relative angle of the half wave plate is controlled in order to reduce bias.

The input state is prepared using the beam from a laser diode (LD). The beam is transmitted through a set of neutral density filters (NDF) with a combined optical density 7.3 for attenuation to a level compatible with the maximum recommended count rate of our single photon detectors. The beam is then transmitted through a half wave plate (HWP) mounted on a motorized rotation stage so as to control its orientation angle relative to the PBS axes. The PBS splits the beam into two spatial modes according to the H and V polarizations, each of which is coupled with the help of an aspheric lens (AL1 and AL2) into a multimode fiber leading to an avalanche photodiode (APD1 and APD2). We include a polariser (P1 and P2), with an extinction ratio, defined as the ratio of the maximum to the minimum transmission of a linearly polarized input, of 100,000:1 prior to each of the aspheric lenses (AL1 and AL2) for a reduction of the non-polarized intensity reaching the detectors.

Suppose now that we prepare our system so that the average relative power $P_i/(P_0 + P_1)$ of each APD detector $i = 0, 1$ starts ideally at $1/2$. In Figure 3, we show how this average relative power evolves with time (see curve labelled "without feedback"). Note that, even though the system starts in a perfectly balanced state, it rapidly deviates from this condition. The slow change of this curve can be attributed to thermal drift while the oscillatory component with a period of approximately half an hour is related to the air conditioning system in the laboratory. These effects can be effectively compensated by rotating the HWP. After some study of the response function of our experimental setup, a correction every minute with a proportional controller was sufficient to correct for all these effects leading to a steady response (see curve labelled "with feedback") [21].

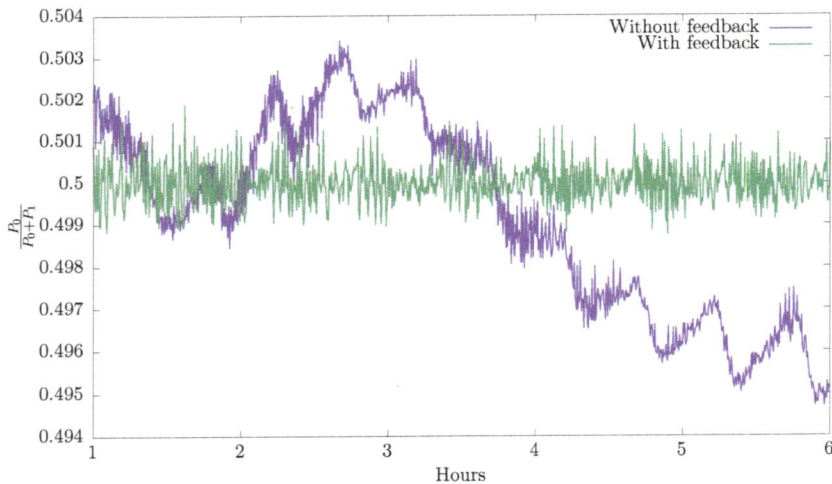

Figure 3. Evolution in time of the normalized power. The device starts in a perfect balanced state but quickly deviates from this condition. By using a feedback mechanism, we can obtain stability of the normalized power within an error of 0.004.

5. First Battery of Results

We have used the experimental setup described in the previous section to generate a sequence of 4,294,967,296 bits, allowing us to test the Borel–Normality criterion up to level $i_{max} = 5$. The results of this analysis are depicted in Figure 4. In the plot, bars represent the deviations from the ideal value for all the strings at each Borel level. For instance, in the first part of the analysis (purple bars), there are only two bars corresponding to the frequency of occurrences of substrings "1" and "0". As our initial setup is very fine-tuned and stable, the bars have practically zero height, with value 5×10^{-6}. The green bars represent the second part of the analysis, or Borel level two, corresponding to frequencies of occurrences of symbols $\{00, 01, 10, 11\}$, and so on. In the same figure, the horizontal lines represent the bound given by the right-hand-side of Equation (1). Our first battery of results are a clear disappointment: only the first set for substrings of length one clearly passes the test, while, for higher lengths, our QRNG fails miserably to pass Calude's criterion.

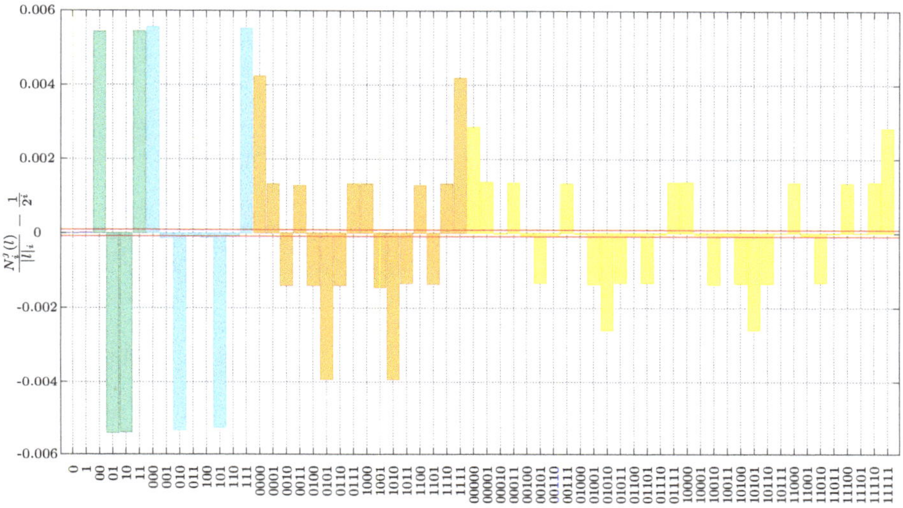

Figure 4. Results from Borel analysis. The first two boxes correspond to the deviations from the mean at the first Borel level; these exhibit the same height but opposite signs. The next four bars (green) represent the deviations for level two, i.e., "00", "01", "10", "11". The blue and yellow boxes represent the deviations for level three and four, respectively. The red lines correspond to Borel's bound, which turns out to be much smaller than the deviations. Only the first level passes the test.

Furthermore, a closer look at the green bars shows that events 00 and 11 appear more frequently than expected, by about 0.005%, (while events 01 and 10 appear less frequently than expected by the same margin). This effect reveals a correlation between "equal events", that is, the same digit appearing twice. In terms of our experiment, this means that it is more probable to observe an event in a detector once a previous event has already been recorded. Other parts of our test validate this: for Borel level three, the yellow bars indicate that events with alternate zeroes and ones (010 and 101) appear less frequently than expected, also by about 0.005%. At Borel levels four and five, the larger deviations appear for events 0101 and 1010, clearly in accordance with the previous results. This indicates that certain parts of our experimental setup are introducing unwanted correlations between bits, which results in the magnitude of some of the deviations to be 50 times larger than expected. Our experimental effort clearly does not suffice for our sequences to pass the Borel–Normality criterion. How is this possible?

6. APD Effects on Introducing Correlations

The two main effects in the behavior of our APDs which can introduce undesired correlations in the resulting sequences of bits are called after-pulsing and dead time [22,23]. The first effect, roughly speaking, corresponds to a false detection event due to the residual effects of an avalanche triggered by a previous event, while the dead time is the time period after each event during which the system is not able to record a subsequent incoming optical signal. In this case, we have a typical dead time of 22 ns, a maximum after-pulsing probability of 1%, and a dark counts rate of 100 counts/s.

While the device exhibits a linear behaviour up to 5×10^6 counts/s, the detection rates used in our experiments are an order of magnitude lower.

The mechanism by which dead time introduces correlations in our data, particularly in experimental arrangements with two or more APDs as in our case, is as follows: suppose that we have an event in one of our detectors. During its dead time, it will have zero probability of recording another incoming event in that detector, while the other detector still exhibits a non-zero probability of recording an event. On the other hand, the way after-pulsing introduces correlations in our data is by increasing the probability of observing consecutively the same event, resulting in the observation of an excess of the events $\{00\}$ and $\{11\}$ in Figure 4.

These two effects can somewhat be corrected either by re-designing our experimental setup and/or modifying the software. Instead of following this route to generate maximally random sequences of bits, let us pursue a rather simple solution as discussed below.

7. Random Number Generation Using Time Measurements

We now follow a method introduced in [3]. Suppose that $\rho(x)$ is the probability density function of a continuous random random variable X on an interval $x \in (a, b)$. Let us further assume that its real value x is represented up to a given precision so that we assign a parity to x according to the parity of its least significant digit. Next, we divide the interval (a, b) into an even number $2L$ of bins and introduce

$$x_i = a + \frac{i(b - a)}{2L}, \qquad i = 0, \ldots, 2L.$$ (6)

Suppose that $2L$ and the precision has been chosen so that x_i is even for i even and odd for i odd. It follows that

$$1 = \int_a^b dx \rho(x) = \sum_{i=0}^{2L-1} \int_{x_i}^{x_{i+1}} dx \rho(x) \equiv \mathcal{N}_{\text{even}} + \mathcal{N}_{\text{odd}},$$ (7)

with

$$\mathcal{N}_{\text{even}} \equiv \sum_{i=0}^{L-1} \int_{x_{2i}}^{x_{2i+1}} dx \rho(x), \qquad \mathcal{N}_{\text{odd}} \equiv \sum_{i=0}^{L-1} \int_{x_{2i+1}}^{x_{2i+2}} dx \rho(x).$$ (8)

Approximating the integrals by the left sum rule, we can write that

$$\mathcal{N}_{\text{even}} \sim \frac{b - a}{2L} \sum_{i=0}^{L-1} \rho(x_{2i}), \qquad \mathcal{N}_{\text{odd}} \sim \frac{b - a}{2L} \sum_{i=0}^{L-1} \rho(x_{2i+1}),$$ (9)

which implies that, roughly, the probability that the least significant digit is odd can be expressed as

$$\mathcal{N}_{\text{odd}} \sim \frac{1}{2} + \frac{1}{2} \sum_{i=0}^{L-1} \rho'(x_{2i}) \left(\frac{b - a}{2L}\right)^2,$$ (10)

where the bias term can be fine-tuned by increasing either the number of bins or through a smooth density $\rho(x)$, or both.

This method can be very easily implemented in the lab as follows. Suppose that the random variable X is the time difference between two consecutive photon arrivals to the detector. In our case, these times are of the order of 500 ns to 10 μs. A typical sequence of these time differences look like:

$$592\ 342\,ps,$$
$$595\ 634\,ps,$$
$$593\ 645\,ps,$$
$$592\ 342\,ps,$$
$$595\ 634\,ps.$$

$$\vdots$$

We can then look at the parity of the least significant digit and assign, for instance, a 0-bit to even parity and 1-bit to odd parity, thus generating a dataset ℓ of n bits. In Figure 5, we show the results of testing such a sequence using the Borel-normality bounds and Bayesian bounds, given by Equations (1) and (4), respectively, up to Borel level $i_{\max} = 5$. The colored bins correspond to the deviations from the ideal value of relative frequencies for all the possible subsequences. These are ordered using its length $i = 1$ (purple bins), $i = 2$ (green bins), $i = 3$ (blue bins), $i = 4$ (orange bins), and $i_{\max} = 5$ (yellow bins). The solid red line corresponds to the Borel bound, 8.6×10^{-5}. In the same graph, the green line is the Bayesian bound given the right-hand-side of Equation (4) that depends on i and therefore it is not a constant, as is the case for the Borel bound. Finally, the height of the various background colored boxes correspond to the values given by the left-hand-side expression of the Bayesian bound.

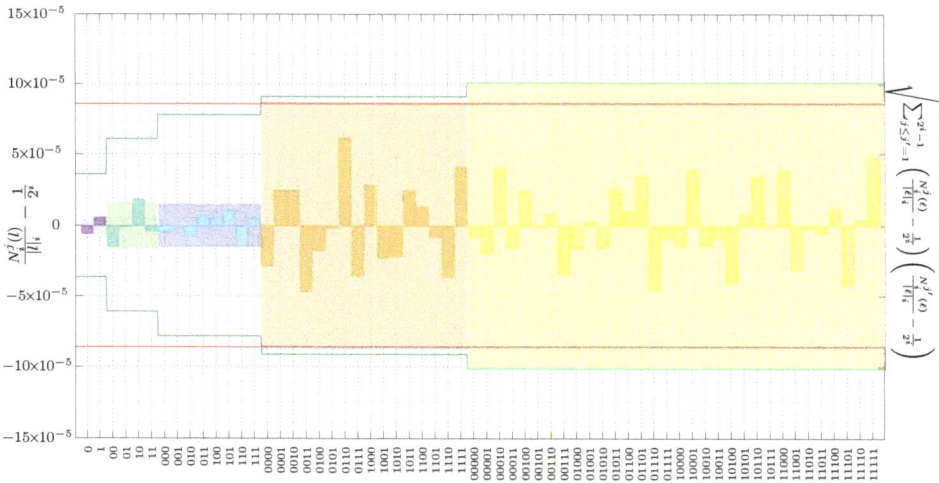

Figure 5. Results for generation using the least significant bits of time tags. In this case, the deviations are very small, so this generation scheme is excellent. The solid red lines represent the maximum deviations allowed by the Calude test, while the solid green lines correspond to the bounds given the Bayesian approach.

As we can see, this extremely simple QRNG passes the Borel-normality criterion up to $i = 5$, and nearly passes the Bayesian criterion (passes it for $i \leq 4$ and slightly exceeds the bound for $i = 5$). Notice that, while the previous experimental setup required an accurate balance between zeroes and ones, in the present case we already have very small deviations at Borel level $i = 1$, less than 10^{-5}, showing the convenience of this method. For $i = 2$ (green bins), the deviations are much larger, almost three times the value for $i = 1$, but nevertheless they pass the test again by a considerable margin. These results show the lack of correlations between consecutive events, which is the main drawback of the previous approach. It is important to note that, in the results at Borel level $i = 4$, there is a substantial increase in the deviations compared with $i = 1, 2, 3$. While this increase may indicate the presence of some as yet unidentified correlations, these are of an insufficient magnitude to reach the bounds. On the other hand, this experimental setup fails to pass some of the requirements of the Bayesian scheme. The deviations derived from the Bayesian criterion are shown in Table 2, and also in Figure 5. As we can see, all Borel levels pass the Bayesian test, except for the last one, albeit by a small margin.

Table 2. Comparison between the left-hand-side and the right-hand-side of the Borel-type bounds given by Equation (4), applied to the sequence of bits generated in our lab. As we can observe, the Bayesian bounds are satisfied for the first four Borel levels, but not the last one by a slight margin.

Borel Level	LHS of Equation (4)	RHS of Equation (4)	LHS < RHS
1	5.719×10^{-6}	3.62956×10^{-5}	Yes
2	1.7129×10^{-5}	6.08097×10^{-5}	Yes
3	1.54974×10^{-5}	7.82572×10^{-5}	Yes
4	8.74186×10^{-5}	9.11726×10^{-5}	Yes
5	1.01138×10^{-4}	1.01069×10^{-4}	No

We can also look at the value of the posterior distribution given by Equation (2) for the maximally random model. The value of the posterior for the four word lengths is reported in Table 3. For the first three Borel levels, the posterior distribution of the maximally random model α_{sym} is very close to one, indicating that, given the dataset, this is the most likely model to have generated such data. For Borel level $i = 4$, we are only able to analyse those models which are in the vicinity, in parameter space, to the maximally random model. These models correspond to partitioning $\Omega^{(5)}$ into two subsets, resulting in 32,767 models, giving a total of 32,768, including the maximally random one. In this case, it turns out that the most likely model is not the maximally random one. Actually, using the value of the posterior probability, this model is ranked in the position 9240 out of all the explored models, and therefore the sequence of bits fails to pass the Bayesian criterion already at Borel level $i = 4$. Note that $i = 5$ was not included in Table 3 because we lack the computational power to address this Borel level.

Table 3. Value of the posterior distribution $P\left(\mathcal{M}_{\text{sym}}^{(i)} \middle| \ell\right)$ given by Equation (2) for the maximally random model. Note that the prior distribution for each Borel model is a flat distribution along all the models tested. This means, for instance, that, at level $i = 3$, while we do not have any observational bias to choose among any particular model, that is $P(\mathcal{M}_{\alpha}^{(i)}) = \frac{1}{4140}$, after observing the data, the maximally random model is the most plausible.

Borel Level i	Number of Models Analysed	$P\left(\mathcal{M}_{\text{sym}}^{(i)} \middle| \ell\right)$
1	$B_{2^i} = 2$	0.999984
2	$B_{2^i} = 15$	0.999634
3	$B_{2^i} = 4140$	0.995476
4	32,768 models considered out of $B_{2^i} = 10,480,142,147$	9.2179×10^{-42}

8. Conclusions

A vast amount of literature exists which claims that QRNGs are superior when compared to their classical counterparts, based on purely theoretical arguments. Indeed, randomness in Quantum Mechanics is usually justified by the unpredictability of individual measurement outcomes given some initial conditions. More concretely, quantum unpredictability is based on no-go theorems, such as Bell's theorem, that simply tells us that, given some initial conditions, it is impossible to predict the outcome of a single measurement. However, in the present review, we have shown that QRNGs actually perform rather poorly in tests of randomness as compared to classical PRNGs. The reason is fairly simple: unpredictability has nothing to do with bias, and while experimental devices based on Quantum Mechanics may produce a truly unpredictable random signal, they also tend, more often than not, to introduce correlations. In particular, for QNRGs based on optical devices, we have been able to account for two, perhaps amongst the many, effects that introduce bias in our data. While in our own experimental work involving a QNRG we have failed to obtain sequences which obey the Borel and Bayesian criteria, we were able to show that extracting sequences from the least significant digits of times of arrival represents a promising strategy.

Author Contributions: J.G.H. and A.B.U. conceived and designed the experiments; A.C.M. and A.S. performed the experiments; I.P.C. and R.D.H.R. developed the method based on Bayesian Inference; A.C.M, A.S. and R.D.H.R. analysed the data. All authors contributed to writing the paper.

Funding: Financial support of UNAM-DGAPA-PAPIIT IA103417 and IN109417 is acknowledged.

Conflicts of Interest: The authors declare no conflict of interest.

References

1. Metropolis, N.; Ulam, S. The Monte Carlo method. *J. Am. Stat. Assoc.* **1949**, *44*, 335–341. [CrossRef] [PubMed]
2. Von Neumann, J. Various techniques used in connection with random digits. *J. Res. Nat. Bur. Stand. Appl. Math. Ser.* **1951**, *12*, 36–38.
3. Isida, M.; Ikeda, H. Random number generator. *Ann. Inst. Stat. Math.* **1956**, *8*, 119–126. [CrossRef]
4. Stojanovski, T.; Kocarev, L. Chaos-based random number generators-part I: Analysis [cryptography]. *IEEE Trans. Circuits Syst. I Fundam. Theory Appl.* **2001**, *48*, 281–288. [CrossRef]
5. Einstein, A.; Podolsky, B.; Rosen, N. Can quantum-mechanical description of physical reality be considered complete? *Phys. Rev.* **1935**, *47*, 777–780. [CrossRef]
6. Bell, J.S. *Speakable and Unspeakable in Quantum Mechanics: Collected Papers on Quantum Philosophy*; Cambridge University Press: Cambridge, UK, 2004.
7. Pironio, S.; Acín, A.; Massar, S.; de La Giroday, A.B.; Matsukevich, D.N.; Maunz, P.; Olmschenk, S.; Hayes, D.; Luo, L.; Manning, T.A.; et al. Random numbers certified by Bell's theorem. *Nature* **2010**, *464*, 1021–1024. [CrossRef] [PubMed]
8. Acín, A.; Masanes, L. Certified randomness in quantum physics. *Nature* **2016**, *540*, 213–219. [CrossRef] [PubMed]
9. Quantis Random Number Generator. Available online: https://www.idquantique.com/random-number-generation/products/quantis-random-number-generator (accessed on 9 October 2018).
10. Rukhin, A.; Soto, J.; Nechvatal, J.; Smid, M.; Barker, E. *A Statistical Test Suite for Random and Pseudorandom Number Generators for Cryptographic Applications*; Technical Report; Booz-Allen and Hamilton Inc.: Mclean, VA, USA, 2001.
11. Calude, C.S.; Dinneen, M.J.; Dumitrescu, M.; Svozil, K. Experimental evidence of quantum randomness incomputability. *Phys. Rev. A* **2010**, *82*, 022102. [CrossRef]
12. Pareschi, F.; Rovatti, R.; Setti, G. Second-level NIST randomness tests for improving test reliability. In Proceedings of the IEEE International Symposium on Circuits and Systems (ISCAS 2007), New Orleans, LA, USA, 27–30 May 2007; pp. 1437–1440.
13. Wasserstein, R.L.; Lazar, N.A. The ASA's statement on p-values: Context, process, and purpose. *Am. Stat.* **2016**, *70*, 129–133. [CrossRef]
14. Calude, C.S. Borel normality and algorithmic randomness. *Dev. Lang. Theory* **1993**, *355*, 113–129.
15. Calude, C.S. *Information and Randomness: An Algorithmic Perspective*, 2nd ed.; Springer Publishing Company: New York, NY, USA, 2010.
16. Díaz Hernández Rojas, R.; Solís, A.; Angulo Martínez, A.M.; U'ren, A.B.; Hirsch, J.G.; Marsili, M.; Pérez Castillo, I. Improving randomness characterization through Bayesian model selection. *Sci. Rep.* **2017**, *7*, 3096. [CrossRef] [PubMed]
17. Díaz Hernández Rojas, R. Mejora en la Caracterización de la Aleatoriedad Usando Selección Bayesiana de Modelos. Master's Thesis, Posgrado en Ciencias Físicas UNAM, Mexico, Mexico, 2017. (In Spanish)
18. Cover, T.M.; Thomas, J.A. *Elements of Information Theory*; John Wiley & Sons: Hoboken, NJ, USA, 2012.

19. Pemmaraju, S.; Skiena, S.S. *Computational Discrete Mathematics: Combinatorics and Graph Theory with Mathematica*; Cambridge University Press: Cambridge, UK, 2003.

20. Abbott, A.A.; Calude, C.S.; Dinneen, M.J.; Huang, N. Experimentally Probing the Incomputability of Quantum Randomness. *arXiv* **2018**, arXiv:1806.08762.

21. Martínez Becerril, A.C. Un Generador Cuántico de Números Al Azar: De un PBS a Etiquetas Temporales. Bachelor's Thesis, Facultad de Ciencias, UNAM, Mexico, Mexico, 2017. (In Spanish)

22. Horoshko, D.B.; Chizhevsky, V.N.; Kilin, S.Y. Afterpulsing model based on the quasi-continuous distribution of deep levels in single-photon avalanche diodes. *J. Mod. Opt.* **2017**, *64*, 191–195. [CrossRef]

23. Kang, Y.; Lu, H.X.; Lo, Y.H.; Bethune, D.S.; Risk, W.P. Dark count probability and quantum efficiency of avalanche photodiodes for single-photon detection. *Appl. Phys. Lett.* **2003**, *83*, 2955–2957. [CrossRef]

entropy

MDPI

Article

Probabilities and Epistemic Operations in the Logics of Quantum Computation

Maria Luisa Dalla Chiara [1,*], **Hector Freytes** [2], **Roberto Giuntini** [2], **Roberto Leporini** [3] and **Giuseppe Sergioli** [2]

1. Dipartimento di Lettere e Filosofia, Università di Firenze, Via Bolognese 52, I-50139 Firenze, Italy
2. Dipartimento di Pedagogia, Psicologia, Filosofia, Università di Cagliari, Via Is Mirrionis 1, I-09123 Cagliari, Italy; hfreytes@gmail.com (H.F.); giuntini@unica.it (R.G.); giuseppe.sergioli@gmail.com (G.S.)
3. Dipartimento di Ingegneria Gestionale, dell'Informazione e della Produzione, Università di Bergamo, viale Marconi 5, I-24044 Dalmine (BG), Italy; roberto.leporini@unibg.it
* Correspondence: dallachiara@unifi.it

Received: 28 August 2018; Accepted: 28 October 2018; Published: 31 October 2018

Abstract: Quantum computation theory has inspired new forms of quantum logic, called *quantum computational logics*, where formulas are supposed to denote pieces of quantum information, while logical connectives are interpreted as special examples of quantum logical gates. The most natural semantics for these logics is a form of *holistic semantics*, where meanings behave in a contextual way. In this framework, the concept of *quantum probability* can assume different forms. We distinguish an absolute concept of probability, based on the idea of *quantum truth*, from a relative concept of probability (a form of *transition-probability*, connected with the notion of fidelity between quantum states). Quantum information has brought about some intriguing epistemic situations. A typical example is represented by teleportation-experiments. In some previous works we have studied a quantum version of the epistemic operations "to know", "to believe", "to understand". In this article, we investigate another epistemic operation (which is informally used in a number of interesting quantum situations): the operation "being probabilistically informed".

Keywords: quantum logics; quantum probability; holistic semantics; epistemic operations

---

1. Introduction

Quantum information and quantum computation have inspired new developments of some basic concepts of the quantum theoretic formalism, which for a long time had been regarded as mysterious and potentially paradoxical. In this framework the concept of *quantum probability* has been investigated according to new perspectives, giving rise to possible applications to fields that are far apart from microphysics (cognitive and social sciences, semantics of natural languages and of the languages of art, see, for instance, [1–3]).

As is well known, the basic idea of quantum computation theory is that information can be stored and transmitted by quantum physical objects. Accordingly, pieces of quantum information can be identified with states of some special quantum systems that are storing the information in question. In the simplest case a piece of quantum information corresponds to a pure state of a single particle: a qubit (or qubit-state), the quantum counterpart of the classical concept of bit. Mathematically a qubit can be represented as a quantum superposition (living in the two-dimensional Hilbert space $\mathbb{C}^2$), whose form is

$$|\psi\rangle = c_0|0\rangle + c_1|1\rangle,$$

where $|0\rangle$ and $|1\rangle$ are the two elements of the canonical orthonormal basis of the space, representing in this framework the two classical truth-values 1 (Truth) and 0 (Falsity). From an intuitive point of view, any qubit

$$c_0|0\rangle + c_1|1\rangle$$

can be regarded as un uncertain information that might be true with probability $|c_1|^2$ and might be false with probability $|c_0|^2$. More generally, a piece of quantum information corresponds to a complex knowledge that can be mathematically represented as a pure or mixed state of a composite quantum system: a density operator ρ of a finite-dimensional Hilbert space whose standard form is

$$\mathcal{H}^{(n)} = \underbrace{\mathbb{C}^2 \otimes \ldots \otimes \mathbb{C}^2}_{n-times} \quad \text{(the } n - \text{fold tensor product of } \mathbb{C}^2\text{)}.$$

Quantum information is processed by (quantum logical) gates: special examples of unitary quantum operations that transform pure and mixed states in a reversible way. Any finite sequence of gates (defined on a space $\mathcal{H}^{(n)}$) gives rise to a quantum circuit: when applied to a given input ρ_{in}, the circuit under consideration transforms ρ_{in} into an output ρ_{out}. This represents a mathematical description for a physical process that might be performed by a quantum computer.

The theory of quantum circuits has inspired a natural logical abstraction, giving rise to the development of new forms of quantum logic that have been termed *quantum computational logics*. In these logics, formulas are supposed to denote pieces of quantum information, while logical connectives are interpreted as special examples of gates. Consequently, all formulas of quantum computational languages turn out to have a typical dynamic character, representing possible computation-actions. The most natural semantics for *quantum computational logics* is a form of *holistic semantics*, where *quantum entanglement* (often described as "mysterious") can be used as a powerful logical resource. Against the classical compositionality-principle, meanings of well-formed expressions of a quantum computational language behave in a holistic and contextual way: the meaning of a global expression determines the meanings of its well-formed parts, and not the other way around. Furthermore, meanings are generally context-dependent: under one and the same interpretation of the language an expression may receive different meanings in different contexts (as, in fact, happens in our current use of natural languages and in many forms of informal reasoning).

An important "character" of the quantum computational semantics is the concept of *quantum probability*, which can assume different forms. A basic notion of *quantum probability* (which plays an important logical role) is connected with the idea of truth. In any quantum computational space $\mathcal{H}^{(n)}$ the concept of truth can be naturally represented as a special projection operator, indicated by $P_1^{(n)}$. For instance, in the case of the space $\mathcal{H}^{(1)} = \mathbb{C}^2$ the truth-concept $P_1^{(1)}$ is identified with the projection $P_{|1\rangle}$ that projects over the closed subspace determined by the bit $|1\rangle$ (corresponding to the classical truth-value 1). In this way, truth is dealt with as a special example of a mathematical representative for a possible physical event. Accordingly, one can naturally apply the basic probabilistic rule of quantum theory, based on the concept of Born-probability. For any qubit $|\psi\rangle = c_0|0\rangle + c_1|1\rangle$, the probability that the quantum information $|\psi\rangle$ satisfies the *truth* can be defined as follows:

$$\mathrm{p}_1(|\psi\rangle) := \|P_{|1\rangle}|\psi\rangle\|^2 = |c_1|^2 = \mathrm{tr}(P_{|\psi\rangle} P_1^{(1)}),$$

where tr is the *trace functional* and $\|P_{|1\rangle}|\psi\rangle\|$ is the length of the vector obtained by projecting $|\psi\rangle$ over the closed subspace determined by $|1\rangle$. In the next Section we will see how the probability-function p_1 can be generalized to all pieces of quantum information, living in any space $\mathcal{H}^{(n)}$.

Interestingly enough, the contextual properties of the holistic quantum computational semantics allow us to understand and to justify (at least to a certain extent) some strange uses of the concept of probability that sometimes occur in the framework of intuitive ways of reasoning (for a general discussion of this problem see, for instance, [1]). For instance, assigning to a conjunction $\alpha \wedge \beta$ a probability-value greater than the probabilities of both members (α, β) is not necessarily "irrational" or "antiscientific". For, it might happen that the meanings of the three sentences $\alpha \wedge \beta$, α, β refer, in fact, to different contexts.

Once fixed the truth-concept $P_1^{(n)}$ (for any space $\mathcal{H}^{(n)}$), the probability-function p_1 represents a kind of absolute concept of probability: any piece of quantum information has a well-determined probability-value of being true. It is interesting to investigate another concept of probability that represents a form of relative probability. Suppose that the pure state

$$|\psi\rangle = \sum_i c_i |\varphi_i\rangle$$

(where every $|\varphi_i\rangle$ is an element of the canonical orthonormal basis of the space $\mathcal{H}^{(n)}$) represents the information of an epistemic agent at a given time (say, at the beginning of an experiment or of a computation). According to the quantum theoretic formalism, the information $|\psi\rangle$ allows us to assign probability-values to other pieces of information, that may represent possible outcomes of a measurement or of a computation. For instance, the probability that the information $|\psi\rangle$ assigns to the outcome $|\varphi_i\rangle$ is $|c_i|^2$. Thus, whenever we have the information $|\psi\rangle$ we might have the information $|\varphi_i\rangle$ with probability $|c_i|^2$. Accordingly, we can write:

$$p_{|\psi\rangle}(|\varphi_i\rangle) = |c_i|^2.$$

We will see how this concept of relative probability (which can be generalized to mixed states) is strongly connected with the concept of fidelity between quantum states.

Quantum information has brought about some intriguing epistemic situations. A typical example is represented by teleportation-experiments, where an agent (say, Alice) transmits qubit-states to a far agent (say, Bob), by using some special quantum non-locality phenomena, which may appear *prima facie* strange and mysterious. These puzzling (non-classical) situations have inspired new ideas in the field of epistemic logics. See, for instance, [4].

As is well known, many standard approaches to epistemic logics have been developed in the framework of a possible world semantics, where the basic epistemic operators ("to know", "to believe") are dealt with as special examples of modal operators. Is it possible to represent epistemic operators as particular examples of quantum operations in the framework of a quantum computational semantics? This question admits a positive answer. In some previous works we have studied a quantum version of the epistemic operators "to know", "to believe", "to understand", whose semantic properties depend on the notion of *quantum truth* and on the probability-function p_1. We have seen, in particular, how quantum knowledge generally gives rise to forms of "reversibility-breaking" that can be compared with what happens in the case of quantum measurements. In this article we will investigate another epistemic concept: the operation "being probabilistically informed", which plays a significant role in a number of interesting quantum situations (for instance, in the case of quantum teleportation-experiments).

2. Pieces of Quantum Information and Quantum Probabilities

We will first recall the basic "mathematical characters" that play an important role in quantum computation. The "mathematical stage" where any piece of quantum information is usually supposed to live is an n-fold tensor product of the Hilbert space $\mathbb{C}^2$:

$$\mathcal{H}^{(n)} = \underbrace{\mathbb{C}^2 \otimes \ldots \otimes \mathbb{C}^2}_{n-times}.$$

Any piece of quantum information is a special mathematical object that lives in a particular quantum computational space $\mathcal{H}^{(n)}$, representing a possible state of a physical system that is storing the information in question. For some applications it may be useful to take, as a basic quantum computational space, a many-dimensional space $\mathbb{C}^d$ (with $d > 2$). In this way, qubits are generalized to qudits. See, for instance, [5]. Important examples are represented by quregisters, qubits, registers, bits and mixtures of quregisters. In the quantum computation-literature the terms "qubit" (or "quantum bit") and "quregister" (or "quantum register") are sometimes used in an ambiguous way. In many cases the expression "qubit" refers to a quantum system (say, an electron or a photon) whose possible pure states live in the space $\mathbb{C}^2$. In some other cases, instead, "qubit" simply means a possible pure state of such a system. A similar ambiguity regards the term "quregister". In this article, we will always use the terms "qubit" and "quregister" in the sense of possible pure states of quantum systems that can store pieces of quantum information.

Definition 1 (Quregisters and registers).

- A quregister *(or* quregister-state*) is a unit vector of a space* $\mathcal{H}^{(n)}$.
- A qubit *(or* qubit-state*) is a quregister of the space* $\mathbb{C}^2$.
- A register *(or* register-state*) is an element*

$$|x_1, \ldots, x_n\rangle = |x_1\rangle \otimes \ldots \otimes |x_n\rangle$$

of the canonical orthonormal basis of the space $\mathcal{H}^{(n)}$ *(where* $x_i \in \{0, 1\}$*).*
- A bit *is a register of the space* $\mathbb{C}^2$.

Any quregister $|\psi\rangle$ of the space $\mathcal{H}^{(n)}$ can be represented as a quantum superposition of registers that belong to the canonical basis of the space:

$$|\psi\rangle = \sum_i c_i |x_{i_1}, \ldots, x_{i_n}\rangle,$$

where c_i are complex numbers (called amplitudes) such that $\sum_i |c_i|^2 = 1$.

Quregisters are pure states representing maximal pieces of information (about the quantum systems under investigation) that cannot be consistently extended to a richer knowledge. More generally, a piece of quantum information may correspond to a non-maximal knowledge: a mixed state (or mixture of quregisters), that is mathematically represented as a density operator ρ of a space $\mathcal{H}^{(n)}$. Of course, any quregister $|\psi\rangle$ corresponds to a special case of a density operator: the projection $P_{|\psi\rangle}$ that projects over the closed subspace determined by $|\psi\rangle$. We will indicate by $\mathfrak{D}(\mathcal{H}^{(n)})$ the set of all density operators of the space $\mathcal{H}^{(n)}$, while $\mathfrak{D} = \bigcup_n \left\{ \mathfrak{D}(\mathcal{H}^{(n)}) \right\}$ will represent the set of all possible pieces of quantum information.

A piece of quantum information is generally stored by a composite system S consisting of some subsystems $S_1, \ldots S_r$, where each part S_i may be, in turn, a composite system. According to the quantum theoretic formalism any possible (pure or mixed) state ρ of S determines the state ρ_i of each subsystem S_i; this state is called the reduced state of ρ with respect to the i-th subsystem. Of course, some composite systems S might be decomposed into parts in different ways; thus, the mathematical formalism shall take into account all possible decomposition-choices. Consider a Hilbert space $\mathcal{H}^{(n)}$ that can be decomposed as

$$\mathcal{H}^{(n)} = \mathcal{H}^{(m_1)} \otimes \ldots \otimes \mathcal{H}^{(m_r)},$$

where $m_1 + \ldots + m_r = n$. Accordingly, any density operator ρ of $\mathcal{H}^{(n)}$ can be regarded as a possible state of a composite system

$$S = S_1 + \ldots + S_r,$$

where $\mathcal{H}^{(m_i)}$ is the Hilbert space associated to the subsystem S_i. Consider now a particular subsystem of S:

$$S_{i_1} + \ldots + S_{i_k}$$

(with $1 \leq i_1, \ldots, i_k \leq r$). We will indicate by

$$Red^{(i_1, \ldots, i_k)}_{[m_1, \ldots, m_r]}$$

the reduced state of ρ with respect to the subsystem $S_{i_1} + \ldots + S_{i_k}$ and with respect to the decomposition $\mathcal{H}^{(n)} = \mathcal{H}^{(m_1)} \otimes \ldots \otimes \mathcal{H}^{(m_r)}$. By simplicity we will omit the subscript $[m_1, \ldots, m_r]$ in the case where the decomposition of $\mathcal{H}^{(n)}$ is obvious.

The mathematical representation of composite systems (via tensor products) has brought about some deep changes in the relationships between parts and whole in the quantum world. As is well known, in perfect harmony with the semantics of classical logic, classical physical systems satisfy a physical compositionality-principle: the states of the subsystems of a given system S determine the state of S and vice versa (from the parts to the whole and from the whole to the parts). In quantum theory the compositionality-principle is strongly violated: the state of a composite system determines the state of all its parts, but generally not the other way around. The mysterious *quantum entanglement* (which has for a long time regarded as potentially paradoxical) is connected with the failure of the compositionality-principle.

What exactly is entanglement? For the sake of simplicity, in this article we will only consider the case of entangled pure states.

Definition 2 (Entangled pure state). *Consider a composite quantum system*

$$S = S_1 + \ldots + S_r,$$

and let $\mathcal{H}_S$, $\mathcal{H}_{S_1}, \ldots, \mathcal{H}_{S_r}$ be the Hilbert spaces associated to the systems $S, S_1, \ldots, S_r$ (respectively). A pure state $|\psi\rangle$ of S is called entangled iff $|\psi\rangle$ cannot be represented as a factorized state

$$|\psi_1\rangle \otimes \ldots \otimes |\psi_r\rangle,$$

where $|\psi_1\rangle, \ldots |\psi_r\rangle$ belong to the spaces $\mathcal{H}_{S_1}, \ldots \mathcal{H}_{S_1}$, respectively.

Important examples of entangled pure states are the so called Bell-states, that live in the space $\mathcal{H}^{(2)} = \mathbb{C}^2 \otimes \mathbb{C}^2$.

Example 1. *A typical Bell-state is the following:*

$$|\psi\rangle = \frac{1}{\sqrt{2}} |0,0\rangle + \frac{1}{\sqrt{2}} |1,1\rangle.$$

Apparently, $|\psi\rangle$ describes the state of a two-particle system ($S = S_1 + S_2$), assigning probability-value $\frac{1}{2}$ to the two following possibilities:

- *both subsystems are in the state $|0\rangle$;*
- *both subsystems are in the state $|1\rangle$.*

One can show that the two reduced states of $|\psi\rangle$ (with respect to the first subsystem S_1 and with respect to the second subsystem S_2) are one and the same mixed state:

$$Red^{(1)}_{[1,1]}(|\psi\rangle) = Red^{(2)}_{[1,1]}(|\psi\rangle) = \frac{1}{2} \mathbb{1}^{(1)}$$

(where $\mathbb{1}^{(1)}$ is the identity operator of the space $\mathcal{H}^{(1)} = \mathbb{C}^2$).
Since $|\psi\rangle$ is a pure state, while $\frac{1}{2}\mathbb{1}^{(1)}$ is a proper mixture, we obtain:

$$P_{|\psi\rangle} \neq Red^{(1)}_{[1,1]}(|\psi\rangle) \otimes Red^{(2)}_{[1,1]}(|\psi\rangle).$$

Thus, the states of the two parts of S turn out to be indistinguishable and entangled *in the context* $|\psi\rangle$.

Let us now turn to the concept of *quantum probability*, a basic "character" of the quantum theoretic formalism that can assume different forms. We will first consider the concept of truth-probability that is naturally connected with the idea of quantum truth. In Section 4 we will see how this concept will play an important role in the semantics of *quantum computational logics*. As noticed in the Introduction, in any space $\mathcal{H}^{(n)}$ the concept of truth can be identified with a special projection operator indicated by $P_1^{(n)}$. In this way, truth is dealt with as a mathematical representative of a possible physical event.

In order to define the truth-concept $P_1^{(n)}$ (of $\mathcal{H}^{(n)}$), we will first distinguish the true registers from the false registers of the space.

Definition 3 (True and false registers). *Let $|x_1, \ldots, x_n\rangle$ be a register of $\mathcal{H}^{(n)}$.*

- $|x_1, \ldots, x_n\rangle$ *is called* true *iff its last bit $|x_n\rangle$ is $|1\rangle$;*
- $|x_1, \ldots, x_n\rangle$ *is called* false *iff its last bit $|x_n\rangle$ is $|0\rangle$.*

Thus, the *truth-value* of a register is determined by its last bit. As we will see in the next Section, this choice turns out to be natural and useful in the theory of some important quantum logical gates, where the last bit of an input-register represents the *target* that is transformed by the gate in question into the final truth-value of the output.

On this basis one can now define the concepts of truth and falsity of a space $\mathcal{H}^{(n)}$.

Definition 4 (Truth and falsity).

- *The* truth *of the space $\mathcal{H}^{(n)}$ is the projection $P_1^{(n)}$ that projects over the closed subspace spanned by the set of all true registers.*
- *The* falsity *of the space $\mathcal{H}^{(n)}$ is the projection $P_0^{(n)}$ that projects over the closed subspace spanned by the set of all false registers.*

As a particular case, we obtain that the *truth* $P_1^{(1)}$ of the space $\mathbb{C}^2$ is the projection $P_{|1\rangle}$ (which projects over the closed subspace determined by the bit $|1\rangle$).

Now, we can naturally apply the basic probabilistic rule of quantum theory: the Born-rule, which determines for any state ρ and for any physical event represented by a projection operator P (of a Hilbert space $\mathcal{H}$), the probability that a quantum system in state ρ verifies the event represented by P. According to this rule we have:

$$Prob_\rho(P) = \mathbf{tr}(\rho P)$$

(where $Prob_\rho(P)$ represents the probability that the state ρ assigns to the event P).

By applying the Born-rule to the particular case of the truth-concept $P_1^{(n)}$ we obtain for any state ρ of $\mathcal{H}^{(n)}$:

$$Prob_\rho(P_1^{(n)}) = \mathbf{tr}(\rho P_1^{(n)}).$$

From an intuitive point of view, $Prob_\rho(P_1^{(n)})$ represents the probability that the quantum information ρ is *true*. Since $P_1^{(n)}$ is a constant projection operator (in the space $\mathcal{H}^{(n)}$) we can briefly write $\mathrm{p}_1(\rho)$, instead of $Prob_\rho(P_1^{(n)})$. Once chosen the truth-concept $P_1^{(n)}$ (in any space $\mathcal{H}^{(n)}$), every piece of quantum information turns out to have a well-determined probability-value of being *true*. An important property of the function p_1 is asserted by the following Lemma.

Lemma 1. *For any $\rho \in \mathfrak{D}(\mathcal{H}^{(n)})$,*

$$\mathrm{p}_1(\rho) = \mathbf{tr}(Red_{[n-1,1]}^{(2)}(\rho)P_1^{(1)}).$$

As observed in the Introduction, it is interesting to consider also a different notion of quantum probability, that represents a kind of relative probability. Let us first refer to the case of quregisters (living in a given space $\mathcal{H}^{(n)}$) and suppose that $|\psi\rangle = \sum_i c_i |x_{i_1}, \ldots, x_{i_n}\rangle$ represents the information that an epistemic agent has at a given time. The quregister $|\psi\rangle$ allows us to assign probability-values to all registers $|x_{i_1}, \ldots, x_{i_n}\rangle$ (which might represent possible outcomes of a measurement or of a computation). For any $|x_{i_1}, \ldots, x_{i_n}\rangle$, the probability that $|\psi\rangle$ assigns to the outcome $|x_{i_1}, \ldots, x_{i_n}\rangle$ is $|c_i|^2$. Apparently, one is dealing with a kind of *relative probability*: when we have the information $|\psi\rangle$, we might have the information $|x_{i_1}, \ldots, x_{i_n}\rangle$ with probability $|c_i|^2$.

In the general case, this idea of relative probability turns out to be strongly connected with the notion of fidelity: a concept (introduced by Uhlmann and by Jozsa), that represents a generalization of the notion of of transition-probability for pure states (see [6,7]).

Definition 5 (Fidelity). *Let $|\psi\rangle$ and $|\varphi\rangle$ be two quregisters of $\mathcal{H}^{(n)}$. The fidelity between $|\psi\rangle$ and $|\varphi\rangle$ is defined as follows:*

$$F(|\psi\rangle, |\varphi\rangle) := |\langle \psi | \varphi \rangle|^2$$

(where $\langle \psi | \varphi \rangle$ is the inner product of $|\psi\rangle$ and $|\varphi\rangle$).

The following theorem sums up some basic properties of the notion of fidelity.

Theorem 1.

1. $F(|\psi\rangle, |\varphi\rangle) \in [0,1]$.
2. $F(|\psi\rangle, |\varphi\rangle) = F(|\varphi\rangle, |\psi\rangle)$.
3. $F(|\psi\rangle, |\varphi\rangle) = \|P_{|\psi\rangle} |\varphi\rangle\|^2 = \|P_{|\varphi\rangle} |\psi\rangle\|^2$.
4. $F(|\psi\rangle, |\varphi\rangle) = \mathrm{tr}(P_{|\psi\rangle} P_{|\varphi\rangle}) = \mathrm{tr}(P_{|\varphi\rangle} P_{|\psi\rangle})$.
5. $F(|\psi\rangle, |\varphi\rangle) = 1$ iff $P_{|\psi\rangle} = P_{|\varphi\rangle}$.
6. $F(|\psi\rangle, |\varphi\rangle) = 0$ iff $|\psi\rangle \perp |\varphi\rangle$.

From an intuitive point of view one can say that the real number $F(|\psi\rangle, |\varphi\rangle)$ measures "how close" are the two pure states $|\psi\rangle$ and $|\varphi\rangle$ in the Hilbert space under consideration.

Since $F(|\psi\rangle, |\varphi\rangle) = \mathrm{tr}(P_{|\psi\rangle} P_{|\varphi\rangle}) = \mathrm{tr}(P_{|\varphi\rangle} P_{|\psi\rangle})$, recalling the Born-rule, the number $F(|\psi\rangle, |\varphi\rangle)$ can be naturally interpreted as a relative probability: when we have the information $|\psi\rangle$, we might have the information $|\varphi\rangle$ with probability $F(|\psi\rangle, |\varphi\rangle)$, or vice versa. Accordingly we can write:

$$\mathsf{p}_{|\psi\rangle}(|\varphi\rangle) = F(|\psi\rangle, |\varphi\rangle) = F(|\varphi\rangle, |\psi\rangle) = \mathsf{p}_{|\varphi\rangle}(|\psi\rangle),$$

stressing that $F(|\psi\rangle, |\varphi\rangle)$ represents the probability of the information $|\varphi\rangle$ under the condition $|\psi\rangle$, or vice versa. In this sense one can say that the notion of fidelity represents a special concept of conditional probability that, unlike the general case, does satisfy the symmetry-property. For a general discussion of conditional probabilities in quantum theory see [8].

The concept of fidelity (which represents a form of transition-probability in the case of pure states) can be generalized to density operators.

Definition 6 (Fidelity for density operators). *Let ρ and σ be two density operators of $\mathcal{H}^{(n)}$.*

$$F(\rho, \sigma) := \mathrm{tr}(\sqrt{\sqrt{\rho}\sigma\sqrt{\rho}})^2.$$

Interestingly enough, Jozsa has proved that this notion of fidelity between density operators coincides with the notion of transition-probability between density operators (via purification of mixtures), investigated by Uhlmann (see [7]).

The concept of fidelity for density operators represents a good generalization of the concept of fidelity for pure states, as stated by the following Lemma.

Lemma 2.

$$F(P_{|\psi\rangle}, P_{|\varphi\rangle}) = |\langle \psi | \varphi \rangle|^2.$$

The following Theorem sums up the basic properties of the concept of fidelity for density operators.

Theorem 2.

1. $F(\rho, \sigma) \in [0, 1]$.
2. $F(\rho, \sigma) = 1$ *iff* $\rho = \sigma$.
3. $F(\rho, \sigma) = F(\sigma, \rho)$.
4. $F(U\rho U^\dagger, U\sigma U^\dagger) = F(\rho, \sigma)$, *for any unitary operator* U *of* $\mathcal{H}^{(n)}$ (*where* $U^\dagger$ *is the adjoint of* U). *Thus, fidelity is preserved by unitary operators.*
5. $F(\rho, \sigma) = \text{tr}(\rho\sigma)$, *if either* ρ *or* σ *is a pure state.*

On this basis, it seems reasonable to assume that the number $F(\rho, \sigma)$ represents a form of relative probability for quantum states (which may be either pure or mixed). Accordingly, we will write (like in the case of pure states):

$$p_\rho(\sigma) = F(\rho, \sigma).$$

Is it possible to define fidelity for density operators that belong to different spaces? Suppose that $\rho \in \mathfrak{D}(\mathcal{H}^{(n)})$, $\sigma \in \mathfrak{D}(\mathcal{H}^{(m)})$ and $n > m$. The intuitive idea is that ρ and σ can be *compared* in the framework of the smaller space $\mathcal{H}^{(m)}$, by referring to a special reduced state of ρ that represents a part of ρ living in $\mathcal{H}^{(m)}$. The choice of comparing pieces of information that live in different spaces in the framework of the smaller space seems to be quite natural. In fact, both in the case of human and of artificial intelligence it often happens that agents endowed with a richer knowledge communicate with less informed agents by "reducing" their information-level to the level of the "more ignorant" agents. Accordingly, the concept of generalized fidelity can be defined as follows.

Definition 7 (Generalized fidelity). *Let* $\rho \in \mathfrak{D}(\mathcal{H}^{(n)})$ *and* $\sigma \in \mathfrak{D}(\mathcal{H}^{(m)})$.

$$F_g(\rho, \sigma) := \begin{cases} F(\rho, \sigma), & \text{if } n = m; \\ F(Red^{(2)}_{[n-m,m]}(\rho), \sigma), & \text{if } n > m; \\ F(\rho, Red^{(2)}_{[m-n,n]}(\sigma)), & \text{if } n < m. \end{cases}$$

The map F_g is clearly symmetric.
Thus, we can put:

$$p_\rho(\sigma) := F_g(\rho, \sigma).$$

We obtain:

$$p_\rho(\sigma) := F_g(\rho, \sigma) = F_g(\sigma, \rho) = p_\sigma(\rho).$$

In this way, for any choice of ρ, the probability-function p_ρ turns out to be defined on the set $\mathfrak{D}$ of all possible pieces of quantum information.

It is interesting to compare the probability functions p_1 and p_ρ. As we have seen, once chosen the truth-concept in any space $\mathcal{H}^{(n)}$, the function p_1 represents a kind of absolute concept of probability: every piece of quantum information ρ (living in whatever space) has a well-determined probability-value of being true. The probability-function p_ρ represents, instead, a relative concept that depends on the choice of ρ. From an intuitive point of view, ρ can be regarded as an information (available for a given epistemic agent) which describes a situation that is essentially characterized by some uncertain and vague features. On the basis of his/her information our agent is able to valuate probabilistically other possible alternative situations, using the p_ρ-function.

Interestingly enough, the truth-probability function p_1 can be represented as a special case of the relative probability p_ρ.

Theorem 3. *For any $\rho \in \mathfrak{D}(\mathcal{H}^{(n)})$,*

$$\mathsf{p}_1(\rho) = \mathsf{p}_\rho(P_1^{(1)}).$$

Proof. Let $\rho \in \mathfrak{D}(\mathcal{H}^{(n)})$. By definition of p_1 we have:

$$\mathsf{p}_1(\rho) = \mathbf{tr}(\rho P_1^{(n)})$$

Suppose that $n = 1$. Then,

$$\mathsf{p}_1(\rho) = \mathbf{tr}(\rho P_1^{(1)}) = F_g(\rho, P_1^{(1)}) = \mathsf{p}_\rho(P_1^{(1)}).$$

Suppose that $n > 1$. Then, by Lemma 1,

$$\mathsf{p}_1(\rho) = \mathbf{tr}(Red_{[n-1,1]}^{(2)}(\rho)P_1^{(1)}) = F_g(\rho, P_1^{(1)}) = \mathsf{p}_\rho(P_1^{(1)}).$$

□

Both the truth-probability function p_1 and the relative probability-function p_ρ determine a preorder relation on the set $\mathfrak{D}$ of all possible pieces of quantum information.

Definition 8 (The truth-preorder). *For any $\sigma_1, \sigma_2 \in \mathfrak{D}$,*

$$\sigma_1 \preceq \sigma_2 \;\; \textit{iff} \;\; \mathsf{p}_1(\sigma_1) \le \mathsf{p}_1(\sigma_2).$$

Definition 9 (The relative preorder). *Consider a density operator ρ. For any $\sigma_1, \sigma_2 \in \mathfrak{D}$,*

$$\sigma_1 \preceq_\rho \sigma_2 \;\; \textit{iff} \;\; \mathsf{p}_\rho(\sigma_1) \le \mathsf{p}_\rho(\sigma_2).$$

One can easily check that both relations $\preceq$ and $\preceq_\rho$ are reflexive, transitive and generally non-antisymmetric.

3. Quantum Logical Circuits

The basic idea of the theory of quantum computers is that computations can be performed by some quantum objects that evolve in time. Recalling that (according to Schrödinger's equation) the time-evolution of quantum systems is mathematically described by unitary operators, it is natural to assume that quantum information is processed by *quantum logical gates* (briefly, gates): special examples of unitary operators that transform (in a reversible way) the pure states of the quantum systems that store the information in question. Any gate $G^{(n)}$ (defined on the space $\mathcal{H}^{(n)}$) can be canonically extended to a unitary operation $^{\mathfrak{D}}G^{(n)}$ (defined on the set $\mathfrak{D}(\mathcal{H}^{(n)})$ of all density operators of $\mathcal{H}^{(n)}$) according to the rule:

$$\forall \rho \in \mathfrak{D}(\mathcal{H}^{(n)}): \;\; ^{\mathfrak{D}}G^{(n)}\rho = G^{(n)}\rho\, G^{(n)\dagger}$$

(where $G^{(n)\dagger}$ is the adjoint of $G^{(n)}$). For the sake of simplicity, we will call gate either a unitary operator $G^{(n)}$ or the corresponding unitary operation $^{\mathfrak{D}}G^{(n)}$.

We will now recall the definitions of some basic gates that play an important role both from the computational and from the logical point of view. We will first consider some gates, called "semiclassical", that cannot "create" superpositions from register-inputs. Two important examples are represented by the negation-gate and by the Toffoli-gate.

Definition 10 (The negation-gate on the space $\mathcal{H}^{(1)}$). *The negation-gate on $\mathcal{H}^{(1)}$ is the linear operator* $\text{NOT}^{(1)}$ *that satisfies the following condition for every element* $|x\rangle$ *of the canonical basis:*

$$\text{NOT}^{(1)}|x\rangle := |1 - x\rangle.$$

The operator $\text{NOT}^{(1)}$ represents a natural quantum generalization of the classical negation. We have:

$$\text{NOT}^{(1)}|0\rangle = |1\rangle; \ \ \text{NOT}^{(1)}|1\rangle = |0\rangle.$$

The negation-gate can be naturally generalized to higher-dimensional spaces. For any $\mathcal{H}^{(n)}$ (with $n > 1$), the operator $\text{NOT}^{(n)}$ is defined for every element $|x_1, \ldots, x_n\rangle$ of the canonical basis as follows:

$$\text{NOT}^{(n)}|x_1, \ldots, x_n\rangle := |x_1, \ldots, x_{n-1}\rangle \otimes \text{NOT}^{(1)}|x_n\rangle.$$

Apparently, $\text{NOT}^{(n)}$ always acts on the last bit of any register of $\mathcal{H}^{(n)}$.
The smallest space where the Toffoli-gate can be defined is the space $\mathcal{H}^{(3)}$.

Definition 11 (The Toffoli-gate on the space $\mathcal{H}^{(3)}$). *The Toffoli-gate on $\mathcal{H}^{(3)}$ is the linear operator* $\text{T}^{(1,1,1)}$ *that satisfies the following condition for every element* $|x, y, z\rangle$ *of the canonical basis:*

$$\text{T}^{(1,1,1,)}|x, y, z\rangle := \begin{cases} |x, y, x \sqcap y\rangle, \ \text{if } z = 0; \\ |x, y, (x \sqcap y)'\rangle, \ \text{if } z = 1 \end{cases}$$

(where $\sqcap$ and ' are the infimum and the complement of the two-valued Boolean algebra based on the set $\{0, 1\}$).

Thus, the Toffoli-gate leaves unchanged the first two bits $|x\rangle$ and $|y\rangle$ (which represent the *control-bits*); while the third bit $|z\rangle$ (representing the *target-bit*) is transformed into

- the bit corresponding to the Boolean conjunction of the two control-bits, when $z = 0$;
- the bit corresponding to the Boolean negation of the conjunction of the two control-bits, when $z = 1$.

Like the negation-gate, the Toffoli-gate also can be generalized to higher-dimensional spaces. For any $m, n \geq 1$, the Toffoli-gate $\text{T}^{(m,n,1)}$ on the space $\mathcal{H}^{(m+n+1)}$ is defined as follows:

$$\text{T}^{(m,n,1)}|x_1, \ldots, x_m, y_1, \ldots, y_n, z\rangle :=$$

$$|x_1, \ldots, x_{m-1}, y_{n-1}, y_1, \ldots, y_{n-2}\rangle \otimes \text{T}^{(1,1,1)}|x_m, y_n, z\rangle.$$

The Toffoli-gate has a special logical interest, since it allows us to define a quantum logical conjunction that behaves as a reversible operation. For any choice of two natural numbers m, n (such that $m, n \geq 1$) the reversible conjunction $\text{AND}^{(m,n)}$ is dealt with as a holistic monadic operator that acts on global pieces of quantum information represented by quregisters of the space $\mathcal{H}^{(m+n)}$. Accordingly, any quregister of $\mathcal{H}^{(m+n)}$ can be regarded as a holistic description of two possible members of the conjunction $\text{AND}^{(m,n)}$, which live in the space $\mathcal{H}^{(m)}$ and $\mathcal{H}^{(n)}$, respectively.

Definition 12 (The conjunction on the space $\mathcal{H}^{(m+n)}$). *For any quregister $|\psi\rangle$ of $\mathcal{H}^{(m+n)}$,*

$$\text{AND}^{(m,n)}|\psi\rangle := \text{T}^{(m,n,1)}(|\psi\rangle \otimes |0\rangle)$$

(where the bit $|0\rangle$ plays the role of an ancilla).

In the particular case of mixed states $\rho \in \mathfrak{D}(\mathcal{H}^{(m+n)})$ we will write:

$$^{\mathfrak{D}}\text{AND}^{(m,n)}(\rho) \ \text{for} \ ^{\mathfrak{D}}\text{T}^{(m,n,1)}(\rho \otimes P_0^{(1)}),$$

where $^{\mathcal{D}}\mathrm{T}^{(m,n,1)}$ is the unitary quantum operation that corresponds to the unitary operator $\mathrm{T}^{(m,n,1)}$.

As a special case consider a register $|x,y\rangle$ of the space $\mathcal{H}^{(2)}$. We obtain:

- $\mathrm{AND}^{(1,1)}|x,y\rangle = \mathrm{T}^{(1,1,1)}|x,y,0\rangle = |1,1,1\rangle$, if $x = y = 1$.
- $\mathrm{AND}^{(1,1)}|x,y\rangle = \mathrm{T}^{(1,1,1)}|x,y,0\rangle = |x,y,0\rangle$, if $x = 0$ or $y = 0$.

Thus, $\mathrm{AND}^{(1,1)}$ represents a "good" quantum generalization of classical conjunction. At the same time, this particular form of quantum conjunction gives rise to a characteristic holistic behavior, which is deeply rooted in the holistic features of the quantum formalism (as shown by the following example).

Example 2. *Consider the quregister represented by the following (entangled) Bell-state:*

$$|\psi\rangle = \frac{1}{\sqrt{2}}|1,1\rangle + \frac{1}{\sqrt{2}}|0,0\rangle.$$

We have:

- $\mathrm{AND}^{(1,1)}|\psi\rangle = \mathrm{T}^{(1,1,1)}(|\psi\rangle \otimes |0\rangle) = \frac{1}{\sqrt{2}}|0,0,0\rangle + \frac{1}{\sqrt{2}}|1,1,1\rangle$;
- $^{\mathcal{D}}\mathrm{AND}^{(1,1)}(P_{|\psi\rangle}) = {}^{\mathcal{D}}\mathrm{T}^{(1,1,1)}(P_{|\psi\rangle} \otimes P_0^{(1)}) = P_{\frac{1}{\sqrt{2}}|0,0,0\rangle + \frac{1}{\sqrt{2}}|1,1,1\rangle}$.

Hence, $\mathrm{AND}^{(1,1)}|\psi\rangle$ *and* $^{\mathcal{D}}\mathrm{AND}^{(1,1)}(P_{|\psi\rangle})$ *represent a pure state of the space* $\mathcal{H}^{(3)}$. *At the same time, the two reduced states of* $P_{|\psi\rangle}$ *turn out to be one and the same proper mixture (of the space* $\mathcal{H}^{(1)}$*):*

$$Red^{(1)}(P_{|\psi\rangle}) = Red^{(2)}(P_{|\psi\rangle}) = \frac{1}{2}\mathrm{I}^{(1)}.$$

Consequently, the conjunction over the global state $P_{|\psi\rangle}$ *cannot be represented as the conjunction of the states of the two separate parts:*

$$\mathrm{AND}^{(1,1)}(P_{|\psi\rangle}) \neq \mathrm{AND}^{(1,1)}(Red^{(1)}(P_{|\psi\rangle}) \otimes Red^{(2)}(P_{|\psi\rangle})).$$

This gives rise to a clear violation of the compositionality-principle.

As semiclassical gates, the negation-gate and the Toffoli-gate are unable to "create" superpositions from register-inputs. Of course, quantum computer theory cannot help using also "genuine quantum gates" that can transform classical inputs (registers) into genuine superpositions (which are responsible for the characteristic parallel structures of quantum computations). An important example is represented by theHadamard-gate.

Definition 13 (The Hadamard-gate on the space $\mathcal{H}^{(1)}$). *The Hadamard-gate on $\mathcal{H}^{(1)}$ is the linear operator $\sqrt{\mathrm{I}}^{(1)}$ that satisfies the following conditions:*

$$\sqrt{\mathrm{I}}^{(1)}|0\rangle = \frac{1}{\sqrt{2}}|0\rangle + \frac{1}{\sqrt{2}}|1\rangle; \quad \sqrt{\mathrm{I}}^{(1)}|1\rangle = \frac{1}{\sqrt{2}}|0\rangle - \frac{1}{\sqrt{2}}|1\rangle.$$

Thus, the Hadamard-gate transforms both bits into two distinct genuine superpositions that might be either true or false with probability $\frac{1}{2}$.

Like the negation and the Toffoli-gate, the Hadamard-gate also can be generalized to higher-dimensional spaces. For any $\mathcal{H}^{(n)}$ (with $n > 1$), the operator $\sqrt{\mathrm{I}}^{(n)}$ is defined for every element $|x_1,\ldots,x_n\rangle$ of the canonical basis as follows:

$$\sqrt{\mathrm{I}}^{(n)}|x_1,\ldots,x_n\rangle := |x_1,\ldots,x_{n-1}\rangle \otimes \sqrt{\mathrm{I}}^{(1)}|x_n\rangle.$$

Quantum computations are performed by quantum circuits. Mathematically any quantum circuit can be described as a finite sequence of gates, all defined on one and the same quantum computational space $\mathcal{H}^{(n)}$. Since gates may be either unitary operators or unitary quantum operations, we will write:

$$\mathcal{C} = (\mathsf{G}_1^{(n)}, \ldots, \mathsf{G}_t^{(n)}) \quad \text{or} \quad \mathcal{C}^{\mathfrak{D}} = (^{\mathfrak{D}}\mathsf{G}_1^{(n)}, \ldots, {}^{\mathfrak{D}}\mathsf{G}_t^{(n)}).$$

Example 3. *An interesting example is represented by the following quantum circuit (also called "Mach–Zehnder circuit"):*

$$\mathcal{C} = (\sqrt{\mathtt{I}}^{(1)}, \mathtt{NOT}^{(1)}, \sqrt{\mathtt{I}}^{(1)}).$$

As is well known, this circuit can be implemented by a Mach–Zehnder interferometer *(Figure 1), where the Hadamard-gate is physically realized by a beam-splitter (BS), while the negation-gate is realized by a pair of mirrors (M).*

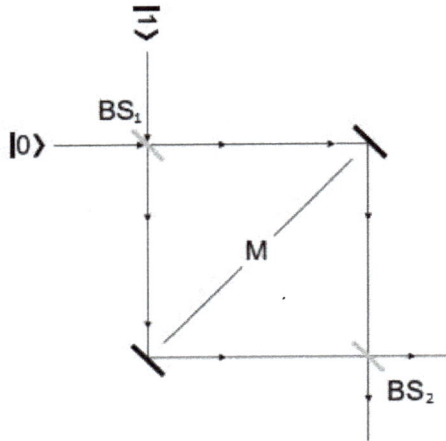

Figure 1. The Mach–Zehnder interferometer.

It is worth while recalling how the properties of the Mach–Zehnder circuit and of the Mach–Zehnder interferometers have been an important object of discussion in many foundational debates about quantum theory. The main intuitive "strangeness" is represented by the following mathematical result (which is confirmed by the experimental evidence) :

$$\sqrt{\mathtt{I}}^{(1)} \, \mathtt{NOT}^{(1)} \, \sqrt{\mathtt{I}}^{(1)} |0\rangle = |0\rangle; \quad \sqrt{\mathtt{I}}^{(1)} \, \mathtt{NOT}^{(1)} \, \sqrt{\mathtt{I}}^{(1)} |1\rangle = -|1\rangle$$

(where the bit $|0\rangle$ is supposed to describe a photon-beam moving along the x-direction, while the bit $|1\rangle$ describes a photon-beam moving along the y-direction). Such result seems to contradict any classical physical expectation, according to which a beam that has entered into the interferometer along the x-direction, after having crossed the second beam-splitter should be detected with probability-value $\frac{1}{2}$ either along the x-direction or along the y-direction.

4. A Logical Abstraction: Quantum Computational Logics

The theory of quantum circuits has inspired a natural logical abstraction, suggesting the development of new forms of quantum logic, that have been called *quantum computational logics*.

As is well known, the prototypical example of quantum logic is the logic created by Birkhoff and von Neumann in their celebrated article "The logic of quantum mechanics", and usually called Birkhoff and von Neumann's quantum logic (see [9]). The basic aim of the original quantum logical approach to quantum theory was the development of an abstract analysis of the relationships between the states of a quantum system S and the quantum events that may occur to S, which can be mathematically represented as projections P of the Hilbert space $\mathcal{H}_S$ associated to S. In the semantics of Birkhoff and von Neumann's quantum logic, the formulas of the quantum logical language are supposed to denote quantum events: projections P (of a given space $\mathcal{H}_S$) to which every state of S assigns a well-determined probability-value. At the same time, the basic logical connectives (negation, conjunction, disjunction) are interpreted as special (generally irreversible) algebraic operations that can be defined on the set of all projections P of $\mathcal{H}_S$.

In the logical community, Birkhoff and von Neumann's quantum logic has been often regarded as a very peculiar and somewhat strange form of non-classical logic, for which some important metalogical questions (like axiomatizability) are still open problems. An axiomatizable version of quantum logic can be obtained by means of a convenient weakening of Birkhoff and von Neumann's quantum logic. This logic (often called abstract quantum logic) can be semantically characterized by referring to the class of all orthomodular lattices (which contains, as special cases, the orthomodular lattices based on the set of all projections of a Hilbert space, see, for instance, [10]).

The semantics of *quantum computational logics* has been inspired by quite different intuitive ideas, which can be briefly sketched as follows:

- any formula α of a quantum computational language is supposed to denote a piece of quantum information ρ, living in a Hilbert space whose dimension depends on the linguistic complexity of α;
- logical connectives are interpreted as particular gates (for a more detailed exposition see [11–13]).

Consequently, unlike the case of traditional quantum logics, the formulas of *quantum computational logics* turn out to have a characteristic dynamic feature, representing possible computation-actions.

In this article we will refer to a minimal (quantum computational) sentential language $\mathcal{L}$, whose alphabet contains:

1. atomic formulas, including two special formulas **t** and **f** that denote the *truth* and the *falsity*, respectively;
2. the following logical connectives:

   - the negation $\neg$, corresponding to the gate negation;
   - the Toffoli-connective $\top$, corresponding to the Toffoli-gate;
   - the Hadamard-connective $\sqrt{id}$, corresponding to the Hadamard-gate.

These connectives simulate, at a syntactical level, the behavior of the corresponding gates. While the negation and the Hadamard-connective are 1-ary connectives (which act on single formulas), the Toffoli-connective is a ternary connective: if α, β are formulas and $\mathbf{q}$ is an atomic formula, then $\top(\alpha, \beta, \mathbf{q})$ is a formula. On this basis, recalling the definition of the quantum computational conjunction $\text{AND}^{(m,n)}$, a binary conjunction-connective $\wedge$ can be defined in terms of the Toffoli-connective:

$$\alpha \wedge \beta := \top(\alpha, \beta, \mathbf{f})$$

(where the false formula **f** plays the role of a *syntactical ancilla*).

We obtain, in this way, an appropriate logical language for the description of a class of quantum circuits. For instance, the Mach–Zehnder circuit:

$$\mathcal{C} = (\sqrt{\mathbb{I}}^{(1)}, \text{NOT}^{(1)}, \sqrt{\mathbb{I}}^{(1)})$$

can be naturally described by the following "Mach–Zehnder formula":

$$\sqrt{id} \, \neg \, \sqrt{id} \, \mathbf{q}.$$

A syntactical notion that plays an important semantic role is the concept of atomic complexity of a formula. As we will see, this concept provides a link between the language and the Hilbert-space environment, where the meanings quantum computational formulas are supposed to live.

Definition 14 (Atomic complexity). *The atomic complexity $At(\alpha)$ of a formula α is the number of occurrences of atomic formulas in α.*

Example 4. *Consider the (contradictory) formula*

$$\alpha = \mathbf{q} \wedge \neg \mathbf{q} = \mathsf{T}(\mathbf{q}, \neg\mathbf{q}, \mathbf{f}).$$

We have: $At(\alpha) = 3$.

For any formula α consider the space $\mathcal{H}^{(At(\alpha))}$ (which is determined by the atomic complexity of α). This space is called *the semantic space of α*, where any piece of quantum information representing a possible *meaning* of α shall live. We will briefly write: $\mathcal{H}^\alpha$, instead of $\mathcal{H}^{(At(\alpha))}$.

Any formula α can be decomposed into its parts, giving rise to a syntactical configuration called the *syntactical tree* of α. Let us first consider a particular example.

Example 5. *Consider again the formula*

$$\alpha = \mathbf{q} \wedge \neg \mathbf{q} = \mathsf{T}(\mathbf{q}, \neg\mathbf{q}, \mathbf{f}).$$

The syntactical tree of α is the following sequence of levels, where each level is a particular sequence of subformulas of α:

$$Level_3^\alpha = (\mathbf{q}, \mathbf{q}, \mathbf{f})$$
$$Level_2^\alpha = (\mathbf{q}, \neg\mathbf{q}, \mathbf{f})$$
$$Level_1^\alpha = (\mathsf{T}(\mathbf{q}, \neg\mathbf{q}, \mathbf{f}))$$

In the general case, the levels of the syntactical tree of a given formula α are determined in the following way:

- the bottom level $Level_1^\alpha$ is (α);
- the top level $Level_h^\alpha$ is the sequence of atomic formulas occurring in α;
- $Level_{i+1}$ (where $1 \leq i < h$) is obtained by dropping the *principal connective* in all molecular formulas occurring at $Level_i$ and by repeating all atomic formulas that occur at $Level_i$.

The syntactical tree of any formula α uniquely determines a quantum circuit: a sequence of gates all defined on the semantic space of α. Such sequence is called the *gate tree of α*. For instance, the gate tree of the formula $\alpha = \mathsf{T}(\mathbf{q}, \neg\mathbf{q}, \mathbf{f})$ is the following circuit:

$$(\mathrm{I}^{(1)} \otimes \mathrm{NOT}^{(1)} \otimes \mathrm{I}^{(1)}, \mathsf{T}^{(1,1,1)}).$$

In fact, the second level of the syntactical tree of α has been obtained from the third level by repeating the first occurrence of $\mathbf{q}$, by negating the second occurrence of $\mathbf{q}$ and by repeating $\mathbf{f}$. The first level has been obtained from the second level by applying the Toffoli-connective to the three sentences occurring at the second level.

While the atomic complexity of α corresponds to the width (i.e., the number of wires) of the circuit described by α, the number of levels of the syntactical tree of α (called the height of α) corresponds to the depth (i.e., the number of computational steps) of the circuit in question.

We will now briefly sum up the basic concepts of the *holistic quantum computational semantics*. The notion of holistic model of the language $\mathcal{L}$ is based on the weaker notion of holistic map: a map

Hol that assigns to each level of the syntactical tree of any formula α a global meaning, represented by a density operator living in the semantic space of α. We have:

$$\text{Hol} : \text{Level}_i^{\alpha} \;\mapsto\; \rho \in \mathfrak{D}(\mathcal{H}^{\alpha}).$$

On this basis, the meaning that Hol assigns to α is identified with the meaning that Hol assigns to the bottom level of the syntactical tree of α:

$$\text{Hol}(\alpha) = \text{Hol}(\text{Level}_1^{\alpha}).$$

Given a formula α, any holistic map Hol determines the *contextual meaning* with respect to the context Hol(α) of any occurrence of a subformula β in α. Suppose that

$$\text{Level}_i^{\alpha} = (\beta_{i_1}, \ldots, \beta_{i_r}).$$

The *contextual meaning* of β_{i_j} with respect to the context Hol(α) can be naturally defined using the notion of *reduced state*:

$$\text{Hol}^{\alpha}(\beta_{i_j}) := Red^{(j)}(\text{Hol}(\text{Level}_i(\alpha))).$$

Remark 1. *Notice how our definition of contextual meaning of a quantum computational formula brings about an interesting connection between the notion of contextuality in linguistic frameworks and the concept of physical contextuality, that plays an important role in many quantum-theoretic problems. As an example, we might recall the debates about the foundational consequences of Kochen and Specker's theorems and about the logical possibility of deterministic completions of quantum theory via a non-contextual hidden- variable theory.*

The concept of holistic model of the language $\mathcal{L}$ can be now defined as a holistic map that satisfies some natural logical restrictions.

Definition 15 (Holistic model). *A holistic model of the language $\mathcal{L}$ is a holistic map Hol that satisfies the following conditions:*

1. *Hol preserves the logical form of all formulas. Thus, the meaning of each Level_i^{α} (different from the top level) of the syntactical tree of α is obtained by applying the corresponding gate G_i^{α} (of the gate tree of α) to the meaning of $\text{Level}_{i+1}^{\alpha}$:*
$$\text{Hol}(\text{Level}_i^{\alpha}) = \mathsf{G}_i^{\alpha}(\text{Level}_{i+1}^{\alpha}).$$

2. *Hol assigns the same contextual meaning to different occurrences (in the syntactical tree of α) of one and the same subformula of α.*

3. *The contextual meanings assigned by Hol to the true formula $\mathbf{t}$ and to the false formula $\mathbf{f}$ are: $P_1^{(1)}$ (the truth) and $P_0^{(1)}$ (the falsity), respectively.*

Suppose that the meaning assigned by a model Hol to a formula α is a pure state, whose form is:

$$c_1|\psi_1\rangle + \ldots + c_n|\psi_n\rangle \quad (\text{where} \quad c_i \neq 0).$$

From an intuitive point of view Hol(α) can be regarded as a vague, ambiguous idea that alludes to other ideas (represented by the pieces of information $|\psi_1\rangle, \ldots, |\psi_n\rangle$) that are, in a sense, all *co-existent*. Notice that any meaning Hol(α) represents a kind of autonomous semantic context that is not necessarily correlated with the meanings of other formulas. Generally we have:

$$\text{Hol}^{\alpha}(\gamma) \neq \text{Hol}^{\beta}(\gamma).$$

Hence, one and the same formula may receive different contextual meanings in different contexts. As, in fact, happens in the case of our normal use of natural languages.

The following Lemma (which might appear *prima facie* obvious) asserts a highly non-trivial property that plays an important role in the development of the holistic quantum computational semantics (for a proof of Lemma 2 see [13]).

Lemma 3. *Consider a formula γ and let η be a subformula of γ. For any model* Hol *and for any formula β there exists a model* *Hol *such that:*

$$\text{*Hol}^{\gamma \wedge \beta}(\eta) = \text{Hol}^{\gamma}(\eta).$$

The concepts of truth, validity and logical consequence (in the framework of the holistic quantum computational semantics) can be now defined as follows.

Definition 16 (Truth, validity and logical consequence).

1. $\vDash_{\text{Hol}} \alpha$ (α *is* true *in a model* Hol) *iff* $p_1(\text{Hol}(\alpha)) = 1$.
2. $\vDash \alpha$ (α *is* valid) *iff for any model* Hol, $\vDash_{\text{Hol}} \alpha$.
3. $\alpha \vDash \beta$ (β *is a* logical consequence *of α) iff for any formula γ such that α and β are subformulas of γ and for any model* Hol,
    $$p_1(\text{Hol}^{\gamma}(\alpha)) \leq p_1(\text{Hol}^{\gamma}(\beta)).$$

Apparently, both truth and logical consequence are, in this semantics, probabilistic concepts, based on the probability-function p_1. In spite of the strong contextual features of the holistic quantum computational semantics, one can prove that this holistic notion of *logical consequence* satisfies the transitivity-property.

Theorem 4.
$$\alpha \vDash \beta \text{ and } \beta \vDash \delta \implies \alpha \vDash \delta.$$

Proof. Assume the hypothesis and suppose, by contradiction, that there exists a model Hol and a formula γ, where α and δ occur as subformulas, such that: $p_1(\text{Hol}^{\gamma}(\alpha)) \not\leq p_1(\text{Hol}^{\gamma}(\delta))$. Consider the formula $\gamma \wedge \beta$. By Lemma 3 there exists a model *Hol such that for any η that is a subformula of γ: $\text{*Hol}^{\gamma \wedge \beta}(\eta) = \text{Hol}^{\gamma}(\eta)$. Thus, we have: $\text{*Hol}^{\gamma \wedge \beta}(\alpha) = \text{Hol}^{\gamma}(\alpha)$ and $\text{*Hol}^{\gamma \wedge \beta}(\delta) = \text{Hol}^{\gamma}(\delta)$. Since we have assumed (by contradiction) that $p_1(\text{Hol}^{\gamma}(\alpha)) \not\leq p_1(\text{Hol}^{\gamma}(\delta))$, we obtain: $p_1(\text{*Hol}^{\gamma \wedge \beta}(\alpha)) \not\leq p_1(\text{*Hol}^{\gamma \wedge \beta}(\delta))$, against the hypothesis (and the transitivity of $\leq$), which imply: $p_1(\text{*Hol}^{\gamma \wedge \beta}(\alpha)) \leq p_1(\text{*Hol}^{\gamma \wedge \beta}(\beta)); p_1(\text{*Hol}^{\gamma \wedge \beta}(\beta)) \leq p_1(\text{*Hol}^{\gamma \wedge \beta}(\delta)); \text{*Hol}^{\gamma \wedge \beta}(\alpha) \leq p_1(\text{*Hol}^{\gamma \wedge \beta}(\delta))$. □

The logic that is semantically characterized by the concept of logical consequence defined in Definition 16 has been called *holistic quantum computational logic* (**HQCL**). One is dealing with a very *weak* form of logic, where many important *logical arguments* (valid either in classical logic or in Birkhoff and von Neumann's quantum logic) may be violated.

The following two Theorems sum up some important logical arguments that are either valid or possibly violated in the framework of the logic **HQCL** (proofs can be found in [13]).

Theorem 5.

1. $\alpha \vDash \alpha$
2. $\alpha \wedge \beta \vDash \alpha; \ \alpha \wedge \beta \vDash \beta$
3. $\alpha \vDash \beta \implies \alpha \wedge \delta \vDash \beta$
4. $\alpha \vDash \neg\neg\alpha; \ \neg\neg\alpha \vDash \alpha$
5. $\alpha \vDash \beta \implies \neg\beta \vDash \neg\alpha$

6. $\mathbf{f} \vDash \beta; \quad \beta \vDash \mathbf{t}$
7. $\sqrt{id}\sqrt{id}\alpha \vDash \alpha; \alpha \vDash \sqrt{id}\sqrt{id}\alpha$
8. $\sqrt{id}(\alpha \wedge \beta) \vDash \sqrt{id}\mathbf{f}; \quad \sqrt{id}\mathbf{f} \vDash \sqrt{id}(\alpha \wedge \beta)$

Theorem 6.

1. $\nvDash \neg(\alpha \wedge \neg\alpha)$
2. $\alpha \nvDash \alpha \wedge \alpha$
3. $\alpha \wedge \beta \nvDash \beta \wedge \alpha$
4. $\alpha \wedge (\beta \wedge \gamma) \nvDash (\alpha \wedge \beta) \wedge \gamma$

Thus, conjunctions are generally non-idempotent, non-commutative and non-associative. Such violations can be explained by recalling the contextual behavior of quantum meanings. It may happen that:

$$p_1(\text{Hol}^\delta(\alpha)) \neq p_1(\text{Hol}^\delta(\alpha \wedge \alpha))$$

$$p_1(\text{Hol}^\delta(\alpha \wedge \beta)) \neq p_1(\text{Hol}^\ell(\beta \wedge \alpha))$$

$$p_1(\text{Hol}^\delta(\alpha \wedge (\beta \wedge \gamma))) \neq p_1(\text{Hol}^\delta((\alpha \wedge \beta) \wedge \gamma)).$$

All this seems to be strongly in agreement with a number of informal ways of reasoning where conjunctions are frequently used as non-idempotent, non-commutative and non-associative logical operations. As is well known, the semantics of natural languages is essentially *holistic and contextual*. We need only think how children learn their mother-language, showing an extraordinary capacity of understanding and using correctly the contextual meanings of expressions that occur in different contexts. And it often happens that the meaning of a global expression is grasped and used in a clear and correct way, while the meanings of its parts appear more vague and ambiguous.

Different forms of *holistic quantum computational logics* (also in a first-order version) can be applied to investigate semantic phenomena where holism, contextuality and ambiguity play an important role, as happens not only in the case of natural languages but also in the languages of art (say, poetry or music, for instance, see [2]). Of course the holistic quantum computational semantics does not forbid compositional semantic situations, which can be described as special cases of the holistic semantics. Interestingly enough, conjunctions are always commutative and associative, but generally non-idempotent in the framework of the compositional fragment of the holistic quantum semantics.

5. Quantum Epistemic Operations and Quantum Probabilities

Quantum information has brought about some intriguing epistemic situations, that have inspired new ideas in the field of epistemic logics. Is it possible to represent epistemic operators as special examples of operations in a Hilbert space environment? This question admits a positive answer.

A characteristic feature of the quantum computational approach to epistemic logic is the use of the notion of *truth-perspective*: each epistemic agent (say, Alice, Bob, ...) is supposed to be associated to a particular truth-perspective that represents his/her idea of truth. Truth-perspective changes may give rise to some interesting relativistic-like epistemic effects: if Alice and Bob have different truth-perspectives, Alice might *see* a kind of deformation in Bob's logical behavior (see [14]).

From the mathematical point of view we assume that the choice of a given truth-perspective is determined by the choice of a particular orthonormal basis of the Hilbert space $\mathbb{C}^2$. In Section 2 we have seen how the truth-concept $P_1^{(n)}$ (of $\mathcal{H}^{(n)}$) has been defined by referring to the canonical basis of the space. But, of course, the choice of a particular basis of a given Hilbert space is a matter of convention. Consider the space $\mathbb{C}^2$. Any orthonormal basis of this space can be described as determined by the application of a unitary operator $\mathfrak{T}$ to the elements of the canonical basis $\{|0\rangle, |1\rangle\}$. We can think that the operator $\mathfrak{T}$ gives rise to a change of *truth-perspective*. While the classical truth-values *Truth* and *Falsity* have been identified with the two bits $|1\rangle$ and $|0\rangle$, assuming a different basis corresponds to a different idea of *Truth* and *Falsity*. Since any basis-change in $\mathbb{C}^2$ is determined by the choice of a particular unitary operator, we can identify a *truth-perspective* with a unitary operator $\mathfrak{T}$ of $\mathbb{C}^2$. We will write:

$$|1_{\mathfrak{T}}\rangle = \mathfrak{T}|1\rangle; \quad |0_{\mathfrak{T}}\rangle = \mathfrak{T}|0\rangle,$$

and we will assume that $|1_{\mathfrak{T}}\rangle$ and $|0_{\mathfrak{T}}\rangle$ represent, respectively, the truth-values *Truth* and *Falsity* of the truth-perspective $\mathfrak{T}$. The *canonical truth-perspective* is, of course, determined by the identity operator $I^{(1)}$. We will indicate by $\mathbf{B}_{\mathfrak{T}}^{(1)}$ the orthonormal basis determined by $\mathfrak{T}$; while $\mathbf{B}_I^{(1)}$ will represent the canonical basis. From a physical point of view, we can suppose that each truth-perspective is associated to an apparatus that allows one to measure a given observable.

Any unitary operator $\mathfrak{T}$ of $\mathcal{H}^{(1)}$ (representing a truth-perspective) can be naturally extended to a unitary operator $\mathfrak{T}^{(n)}$ of $\mathcal{H}^{(n)}$ (for any $n > 1$):

$$\mathfrak{T}^{(n)}|x_1, \ldots, x_n\rangle = \mathfrak{T}|x_1\rangle \otimes \ldots \otimes \mathfrak{T}|x_n\rangle.$$

Accordingly, any choice of a unitary operator $\mathfrak{T}$ of $\mathcal{H}^{(1)}$ determines an orthonormal basis $\mathbf{B}_{\mathfrak{T}}^{(n)}$ for $\mathcal{H}^{(n)}$ such that:

$$\mathbf{B}_{\mathfrak{T}}^{(n)} = \left\{ \mathfrak{T}^{(n)}|x_1, \ldots, x_n\rangle : |x_1, \ldots, x_n\rangle \in \mathbf{B}_I^{(n)} \right\}.$$

Instead of $\mathfrak{T}^{(n)}|x_1, \ldots, x_n\rangle$ we will also write: $|x_{1_{\mathfrak{T}}}, \ldots, x_{n_{\mathfrak{T}}}\rangle$. The elements of $\mathbf{B}_{\mathfrak{T}}^{(1)}$ will be called the $\mathfrak{T}$-*bits* of $\mathcal{H}^{(1)}$; while the elements of $\mathbf{B}_{\mathfrak{T}}^{(n)}$ will represent the $\mathfrak{T}$-*registers* of $\mathcal{H}^{(n)}$.

The notions of *truth*, *falsity* and *truth-probability* can be now generalized to any truth-perspective $\mathfrak{T}$.

Definition 17 ($\mathfrak{T}$-true and $\mathfrak{T}$-false registers).

- $|x_{1_{\mathfrak{T}}}, \ldots, x_{n_{\mathfrak{T}}}\rangle$ *is a* $\mathfrak{T}$-true register *iff* $|x_{n_{\mathfrak{T}}}\rangle = |1_{\mathfrak{T}}\rangle$;
- $|x_{1_{\mathfrak{T}}}, \ldots, x_{n_{\mathfrak{T}}}\rangle$ *is a* $\mathfrak{T}$-false register *iff* $|x_{n_{\mathfrak{T}}}\rangle = |0_{\mathfrak{T}}\rangle$.

Definition 18 ($\mathfrak{T}$-truth and $\mathfrak{T}$-falsity).

- *The* $\mathfrak{T}$-truth *of* $\mathcal{H}^{(n)}$ *is the projection operator* $^{\mathfrak{T}}P_1^{(n)}$ *that projects over the closed subspace spanned by the set of all* $\mathfrak{T}$- *true registers;*
- *the* $\mathfrak{T}$-falsity *of* $\mathcal{H}^{(n)}$ *is the projection operator* $^{\mathfrak{T}}P_0^{(n)}$ *that projects over the closed subspace spanned by the set of all* $\mathfrak{T}$- *false registers.*

Definition 19 ($\mathfrak{T}$-probability). *For any* $\rho \in \mathfrak{D}(\mathcal{H}^{(n)})$,

$$\mathrm{p}_1^{\mathfrak{T}}(\rho) := \mathrm{tr}(\rho \; ^{\mathfrak{T}}P_1^{(n)}).$$

It is worth while noticing that, unlike the probability function p_1, the relative probability function p_ρ cannot be reasonably generalized to different truth-perspectives: as we have seen, the definition of *fidelity* does not depend on the choice of a particular basis of the space.

One can show that all gates can be canonically transposed from the canonical truth-perspective to any truth-perspective $\mathfrak{T}$. Hence, any quantum circuit $\mathcal{C}$ has a corresponding $\mathfrak{T}$-version $\mathcal{C}^{\mathfrak{T}}$ (for any truth-perspective $\mathfrak{T}$).

In some previous works we have studied a quantum representation of the epistemic operations "to know", "to believe", "to understand", whose properties depend on the notion of $\mathfrak{T}$-truth $^{\mathfrak{T}}P_1^{(n)}$ and on the probability-function $\mathrm{p}_1^{\mathfrak{T}}$ (see [14,15]). We will now consider another epistemic operation, which is informally used in a number of interesting quantum situations: the operation "being probabilistically informed".

Let us first recall the definition of a quantum version of the most important concept of epistemic logics: the concept of knowledge-operation.

Definition 20 (Knowledge-operation). *A knowledge-operation of a Hilbert space $\mathcal{H}^{(n)}$ (with respect to the truth-perspective $\mathfrak{T}$) is a map*

$$\mathfrak{K}_{\mathfrak{T}}^{(n)} : \mathfrak{D}(\mathcal{H}^{(n)}) \mapsto \mathfrak{D}(\mathcal{H}^{(n)}).$$

The following conditions are required:

(1) $\mathfrak{K}^{(n)}$ *is associated with an* epistemic domain $EpD(\mathfrak{K}_{\mathfrak{T}}^{(n)})$ *that is a subset of* $\mathfrak{D}(\mathcal{H}^{(n)})$;
(2) $p_1^{\mathfrak{T}}(\mathfrak{K}_{\mathfrak{T}}^{(n)}\rho) \leq p_1^{\mathfrak{T}}(\rho)$, *for any* $\rho \in EpD(\mathfrak{K}_{\mathfrak{T}}^{(n)})$.

As expected, the intuitive interpretation of $\mathfrak{K}_{\mathfrak{T}}^{(n)}\rho$ is the following: the piece of information ρ is known by a given agent whose truth-perspective is $\mathfrak{T}$. The knowledge described by $\mathfrak{K}_{\mathfrak{T}}^{(n)}$ is limited by a given epistemic domain, which is intended to represent the information accessible to our agent, relatively to the space $\mathcal{H}^{(n)}$ (the epistemic domain of $\mathfrak{K}_{\mathfrak{T}}^{(n)}$ should not be confused with the domain of $\mathfrak{K}_{\mathfrak{T}}^{(n)}$, which coincides with the set of all density operators of the space: $\mathfrak{K}_{\mathfrak{T}}^{(n)}\rho$ is defined, even if ρ does not belong to the epistemic domain of $\mathfrak{K}_{\mathfrak{T}}^{(n)}$).Whenever ρ belongs to the epistemic domain of $\mathfrak{K}_{\mathfrak{T}}^{(n)}$, it seems reasonable to assume that the probability-values of ρ and $\mathfrak{K}_{\mathfrak{T}}^{(n)}\rho$ are correlated: the probability of the quantum information asserting that "ρ is known" should always be less than or equal to the probability of ρ. Hence, in particular, we have:

$$p_1^{\mathfrak{T}}(\mathfrak{K}_{\mathfrak{T}}^{(n)}\rho) = 1 \implies p_1^{\mathfrak{T}}(\rho) = 1.$$

But generally, not the other way around. In other words, pieces of quantum information that are certainly known are certainly true (with respect to the truth-perspective in question). This condition is clearly in agreement with a general principle of standard epistemic logics, according to which "knowledge implies truth, but generally not the other way around".

A knowledge-operation $\mathfrak{K}_{\mathfrak{T}}^{(n)}$ is called *non-trivial* iff for at least one density operator $\rho \in EpD(\mathfrak{K}_{\mathfrak{T}}^{(n)})$, $p_1^{\mathfrak{T}}(\mathfrak{K}_{\mathfrak{T}}^{(n)}\rho) < p_1^{\mathfrak{T}}(\rho)$.

Can knowledge-operations be always represented as (reversible) gates? This question has a negative answer, as proved by the following theorem (for a proof of this theorem see [15]).

Theorem 7. *Non-trivial knowledge-operations cannot be generally represented as unitary quantum operations.*

Apparently, the "act of knowing" gives rise to a characteristic reversibility-breaking, which is quite similar to what happens in the case of quantum measurements.

We will now introduce the operation "being probabilistically informed" (indicated by $\mathfrak{I}_{\mathfrak{T}}^{(n)}$), which arises in some interesting situations when an epistemic agent (say, Alice) has a given probabilistic information.

Example 6. *As an example, we can refer to a teleportation-experiment, where, at the initial time, Alice has a probabilistic information, represented by the genuine qubit*

$$|\psi\rangle = c_0|0\rangle + c_1|1\rangle \quad (with \ c_0, c_1 \neq 1),$$

that shall be teleported to the "far" Bob. While Alice is probabilistically informed *about the piece of quantum information $|\psi\rangle$ (with respect to the canonical truth-perspective), one cannot say that "Alice certainly knows $|\psi\rangle$". Since $P_{|\psi\rangle}$ is supposed to belong to Alice's epistemic domain, we would obtain :*

$$p_1(P_{|\psi\rangle}) = 1$$

(against the hypothesis that $|\psi\rangle$ is a genuine qubit).

The intuitive interpretation of $\mathfrak{I}_{\mathfrak{T}}^{(n)}\rho$ is the following: a given epistemic agent, with truth-perspective $\mathfrak{T}$, is probabilistically informed about ρ (in the framework of the space $\mathcal{H}^{(n)}$). As happens in the case of the knowledge-operation $\mathfrak{K}_{\mathfrak{T}}^{(n)}$, the operation $\mathfrak{I}_{\mathfrak{T}}^{(n)}$ is associated to an information-domain $InfD(\mathfrak{I}_{\mathfrak{T}}^{(n)}) \subseteq \mathfrak{D}(\mathcal{H}^{(n)})$, which represents the set of pieces of information that the agent under consideration is able to valuate probabilistically in the domain $\mathfrak{D}(\mathcal{H}^{(n)})$. Unlike the case of knowledge-operations (which shall satisfy the strong condition (2) of Definition 20) we admit the following possibility (which occurs, for instance, in the case of the teleportation-example):

- $\rho \in InfD(\mathfrak{I}_{\mathfrak{T}}^{(n)})$
- $\mathsf{p}_1^{\mathfrak{T}}(\mathfrak{I}_{\mathfrak{T}}^{(n)}\rho) = 1$
- $\mathsf{p}_1^{\mathfrak{T}}(\rho) < 1$

Definition 21 (Probabilistic information). *A probabilistic information-operation of a Hilbert space $\mathcal{H}^{(n)}$ (with respect to the truth-perspective $\mathfrak{T}$) is a map*

$$\mathfrak{I}_{\mathfrak{T}}^{(n)} : \mathfrak{D}(\mathcal{H}^{(n)}) \mapsto \mathfrak{D}(\mathcal{H}^{(n)}).$$

The following conditions are required:

(1) *$\mathfrak{I}_{\mathfrak{T}}^{(n)}$ is associated with an information-domain $InfD(\mathfrak{I}_{\mathfrak{T}}^{(n)})$ that is a subset of $\mathfrak{D}(\mathcal{H}^{(n)})$;*
(2) *$\rho \in Inf(\mathfrak{I}_{\mathfrak{T}}^{(n)}) \ \Rightarrow \ \mathsf{p}_1^{\mathfrak{T}}(\mathfrak{I}_{\mathfrak{T}}^{(n)}\rho) = 1$.*

It is worth while noticing that the domains $EpD(\mathfrak{K}_{\mathfrak{T}}^{(n)})$ and $InfD(\mathfrak{I}_{\mathfrak{T}}^{(n)})$ are not generally closed under the corresponding operations $\mathfrak{K}_{\mathfrak{T}}^{(n)}$ and $\mathfrak{I}_{\mathfrak{T}}^{(n)}$. Hence, the phenomenon of "epistemic self-consciousness" is here avoided: Alice might know something (or might be probabilistically informed about something) without knowing of knowing it (without being informed of being informed about it).

Every epistemic agent can be naturally associated to an epistemic situation, which is characterized by the choice of a truth-perspective, of a knowledge-operation and of a probabilistic information-operation (in any space $\mathcal{H}^{(n)}$).

Definition 22 (Epistemic situation of an agent). *Let i represent an epistemic agent. An epistemic situation for i is a system*

$$\mathfrak{EpSit}_i = (\mathfrak{T}_i, \ EpD_i, \ InfD_i, \ \mathfrak{K}_i, \ \mathfrak{I}_i),$$

where:

(1) *$\mathfrak{T}_i$ represents the truth-perspective of i.*
(2) *EpD_i is a map that assigns to any $n \geq 1$ a set $EpD_i^{(n)} \subseteq \mathfrak{D}(\mathcal{H}^{(n)})$ that represents the information accessible to i in the information-environment $\mathfrak{D}(\mathcal{H}^{(n)})$.*
(3) *$InfD_i$ is a map that assigns to any $n \geq 1$ a set $InfD_i^{(n)} \subseteq EpD_i^{(n)}$ that represents the information that i is able to valuate probabilistically in the information-environment $\mathfrak{D}(\mathcal{H}^{(n)})$.*
(4) *$\mathfrak{K}_i$ is a map that assigns to any $n \geq 1$ a knowledge-operation $\mathfrak{K}_{\mathfrak{T}_i}^{(n)}$ (defined on $\mathcal{H}^{(n)}$), which describes the knowledge of i with respect to the information-environment $\mathfrak{D}(\mathcal{H}^{(n)})$. The epistemic domain associated to the operation $\mathfrak{K}_{\mathfrak{T}_i}^{(n)}$ is the set $EpD_i^{(n)}$.*
(5) *$\mathfrak{I}_i$ is a map that assigns to any $n \geq 1$ an information-operation $\mathfrak{I}_{\mathfrak{T}_i}^{(n)}$ (defined on $\mathcal{H}^{(n)}$), which describes the probabilistic information of i with respect to the information-environment $\mathfrak{D}(\mathcal{H}^{(n)})$. The information-domain associated to the operation $\mathfrak{I}_{\mathfrak{T}_i}^{(n)}$ is the set $InfD_i^{(n)}$.*
(6) *$\forall \rho \in \mathfrak{D}(\mathcal{H}^{(n)}) : \mathsf{p}_1^{\mathfrak{T}_i}(\mathfrak{K}_{\mathfrak{T}_i}^{(n)}\rho) \ \leq \ \mathsf{p}_1^{\mathfrak{T}_i}(\mathfrak{I}_{\mathfrak{T}_i}^{(n)}\rho).$*

The probability of knowing a given information is less than or equal to the probability of having a probabilistic valuation about it. But, generally, not the other way around.

On this basis one can develop a *quantum epistemic semantics* for a first-order language that can be express formulas like:

- *Ka*α (*a* knows α);
- *Ia*α (*a* is probabilistically informed about α).

This semantics allows us to represent and to justify a number of significant features of our normal use of epistemic notions in the framework of natural language. Interestingly enough, the unpleasant phenomenon of logical omniscience is here avoided. Due to the limits of epistemic domains, Alice might know a given sentence without knowing all its logical consequences. Furthermore, knowledge and probabilistic information are not generally closed under logical conjunction, in accordance with what happens in the case of concrete memories both of human and of artificial intelligence (see [14]).

In conclusion, we have seen how the holistic quantum computational semantics provides some useful abstract tools that can be naturally applied to a formal analysis of concepts and problems in fields that may be far apart from microphysics. Some interesting examples concern the use of crucial epistemic concepts (like knowledge, information, belief, understanding) both in the case of rigorous scientific arguments and in some informal ways of reasoning. Other examples regard the role of ambiguity, vagueness, uncertainty and contextuality either in scientific theories or in our normal use of natural languages or in the languages of art (say, poetry or music).

According to some traditional philosophical views, ambiguity and holism represent characteristic features of human thought that cannot be adequately analyzed in the framework of scientific theories, whose semantics is supposed to be essentially "sharp" and "analytical". Interestingly enough, quantum logics (in their different versions) have provided a significant bridge that might fill a gap between humanistic and scientific disciplines.

Author Contributions: Funding acquisition, H.F., R.G. and G.S.; Writing—original draft, M.L.D.C., H.F., R.G., R.L. and G.S.

Funding: This work has been partially supported by Regione Autonoma della Sardegna in the framework of the project "Time-logical evolution of correlated microscopic systems" (CRP 55, L.R. 7/2007, 2015), by Fondazione Banco di Sardegna & Regione Autonoma della Sardegna in the framework of the project "Science and its logics, the representation's dilemma", cup: F72F16003220002, and by Fondazione Banco di Sardegna of the project "Strategies and Technologies for Scientific Education and Dissemination", cup: F71I17000330002.

Conflicts of Interest: The authors declare no conflict of interest.

References

1. Aerts, D.; Sassoli de Bianchi, M.; Sozzo, S.; Veloz, T. Modeling Human Decision-Making: An Overview of the Brussels Quantum Approach. *Found. Sci.* **2018**, 1–28. [CrossRef]
2. Dalla Chiara, M.L.; Giuntini, R.; Luciani, A.R.; Negri, E. *From Quantum Information to Musical Semantics*; College Publications: London, UK, 2012.
3. Haven, E.; Khrennikov, A. *Quantum Social Science*; Cambridge University Press: Cambridge, UK, 2013.
4. Beltrametti, E.; Dalla Chiara, M.L.; Giuntini, R.; Sergioli, G. Quantum teleportation and quantum epistemic semantics. *Math. Slovaca* **2012**, *62*, 1121–1144. [CrossRef]
5. Dalla Chiara, M.L.; Giuntini, R.; Leporini, R.; Sergioli, G. A many-valued approach to quantum computational logics. *Fuzzy Sets Syst.* **2018**, *335*, 94–111. [CrossRef]
6. Jozsa, R. Fidelity for mixed quantum states. *J. Mod. Opt.* **1994**, *41*, 2315–2323. [CrossRef]
7. Uhlmann, A. The "transition probability" in the state space of a ∗-algebra. *Rep. Math. Phys.* **1976**, *9*, 273–279. [CrossRef]
8. Beltrametti, E.; Cassinelli, G. *The Logic of Quantum Mechanics*; Encyclopedia of Mathematics and its Applications 15; Addison-Wesley: Cambridge, MA, USA, 1981.
9. Birkhoff, G.; von Neumann, J. The logic of quantum mechanics. *Ann. Math.* **1936**, *37*, 823–843. [CrossRef]
10. Dalla Chiara, M.L.; Giuntini, R.; Greechie, R. *Reasoning in Quantum Theory: Sharp and Unsharp Quantum Logics*; Kluwer: Dordrecht, The Netherlands, 2004.

11. Dalla Chiara, M.L.; Giuntini, R.; Leporini, R. Logics from Quantum Computation. *Int. J. Quantum Inf.* **2005**, *3*, 293–337. [CrossRef]
12. Dalla Chiara, M.L.; Giuntini, R.; Ledda, A.; Leporini, R.; Sergioli, G. Entanglement as a Semantic Resource. *Found. Phys.* **2011**, *40*, 1494–1518. [CrossRef]
13. Dalla Chiara, M.L.; Giuntini, R.; Leporini, R.; Sergioli, G. Holistic Logical Arguments in Quantum Computation. *Math. Slovaca* **2016**, *66*, 313–334.
14. Dalla Chiara, M.L.; Giuntini, R.; Leporini, R.; Sergioli, G. A First-order Epistemic Quantum Computational Semantics with Relativistic-like Epistemic Effects. *Fuzzy Sets Syst.* **2016**, *298*, 69–90. [CrossRef]
15. Beltrametti, E.; Dalla Chiara, M.L.; Giuntini, R.; Leporini, R.; Sergioli, G. Epistemic Quantum Computational Structures in a Hilbert space Environment. *Fundam. Inf.* **2012**, *115*, 1–14.

entropy

MDPI

Article

Enhancing Extractable Quantum Entropy in Vacuum-Based Quantum Random Number Generator

Xiaomin Guo [1,2], **Ripeng Liu** [1,2], **Pu Li** [1,2], **Chen Cheng** [1,2], **Mingchuan Wu** [1,2] and **Yanqiang Guo** [1,2,*]

[1] Key Laboratory of Advanced Transducers and Intelligent Control System, Ministry of Education, Taiyuan 030024, China; guoxiaomin@tyut.edu.cn (X.G.); liuripeng0944@163.com (R.L.); lipu8603@126.com (P.L.); chengchen248@163.com (C.C.); wumingchuan1@163.com (M.W.)

[2] College of Physics and Optoelectronics, Taiyuan University of Technology, Taiyuan 030024, China

* Correspondence: guoyanqiang@tyut.edu.cn; Tel.: +86-351-6018-249

Received: 9 July 2018; Accepted: 22 October 2018; Published: 24 October 2018

Abstract: Information-theoretically provable unique true random numbers, which cannot be correlated or controlled by an attacker, can be generated based on quantum measurement of vacuum state and universal-hashing randomness extraction. Quantum entropy in the measurements decides the quality and security of the random number generator (RNG). At the same time, it directly determines the extraction ratio of true randomness from the raw data, in other words, it obviously affects quantum random bits generating rate. In this work, we commit to enhancing quantum entropy content in the vacuum noise based quantum RNG. We have taken into account main factors in this proposal to establish the theoretical model of quantum entropy content, including the effects of classical noise, the optimum dynamical analog–digital convertor (ADC) range, the local gain and the electronic gain of the homodyne system. We demonstrate that by amplifying the vacuum quantum noise, abundant quantum entropy is extractable in the step of post-processing even classical noise excursion, which may be deliberately induced by an eavesdropper, is large. Based on the discussion and the fact that the bandwidth of quantum vacuum noise is infinite, we propose large dynamical range and moderate TIA gain to pursue higher local oscillator (LO) amplification of vacuum quadrature and broader detection bandwidth in homodyne system. High true randomness extraction ratio together with high sampling rate is attainable. Experimentally, an extraction ratio of true randomness of 85.3% is achieved by finite enhancement of the laser power of the LO when classical noise excursions of the raw data is obvious.

Keywords: quantum random number; vacuum state; maximization of quantum conditional min-entropy

1. Introduction

Randomness is one vital ingredient in modern information science, in the regime of both classical and quantum [1,2], since encryption is founded upon the trust in random numbers [3–5]. The demand for true and unique randomness in these applications has triggered various proposals for producing random numbers based on the measurements of quantum observables, which offer the verifiability and ultimate in randomness. In the past two decades, there has been tremendous development for various types of quantum RNG [6–15]. Among these proposals, random number generation based on homodyne measurement of quantum vacuum state is especially appealing in practice since highly efficient photodiodes working at room temperature can be applied [11]. Vacuum state is a pure quantum state with the lowest energy and independent of any external physical quantities. It cannot be correlated or controlled by an attacker, therefore unique random numbers can be yielded by measuring the quadrature amplitude of the vacuum state [16,17]. All the components in this scheme, including laser source, beam splitter and photo detectors have been integrated on a single chip recently [18]. Meanwhile, bit conversion and post-processing are easy to be implemented in virtual "hardware" inside the field-programmable gate array (FPGA). Chip-size integration of the QRNG is expectable. Several dedicated researches have been developed to enhance the generation rate of random bits in this proposal, such as schemes based on optimization of the digitization algorithm [19], implementation of fast randomness extraction in the post-processing [20], application of squeezing vacuum state to increase entropy in raw data [21] and optimization of ADC parameters to improve the quantum entropy in the raw data [22]. In this paper, considering the effects of the classical noise, we discuss the role of homodyne gain in enhancing quantum entropy in the vacuum-based quantum RNG working in the optimum dynamical ADC range scenario. Conditional min-entropy is applied to critically assess the quantum entropy in the quantum RNG. It is the key input parameter of randomness extractor and determines the extraction ratio of true randomness from the raw random sequence, thereby affects the generation speed of quantum RNG significantly.

2. Quantum Entropy Evaluation and Enhancing in Vacuum-Based Quantum RNG

Entropy is defined relative to one's knowledge of an experiment's output prior to observation. The larger the amount of the entropy, the greater the uncertainty in predicting the value of an observation. Among types of entropy, min-entropy is a very conservative measure. In cryptography, the unpredictability of secret values is essential. The min-entropy measure the probability that a secret is guessed correctly in the first trial. For mathematically determining min-entropy of a secret, the first thing is to precisely identify the distribution that the secret was generated from [23].

Quadrature fluctuation of optical quantum vacuum state, the nature initial state of optical field at room temperature, is the noise source for the random bits generation in this scheme. According to Born's rule, the measurement outcome of a pure quantum state can be intrinsically random. A single measurement of the quadrature of the vacuum state is completely random and multiple repeated measurements satisfy the Gaussian distribution statistically, so we can extract random bits from the measurement results. Based on homodyne measurement, the microscopic fluctuations of quadrature of the vacuum state are detected, amplified and transferred into an electric signal

$$V_{\text{vac}} \propto \left\langle i_-^2 \right\rangle - \left\langle i_- \right\rangle^2 \propto 4\alpha^2 [\delta X(t)^2 \cos(\theta)^2 + \delta Y(t)^2 \sin(\theta)^2] \tag{1}$$

i_- is the difference current from the two detectors. Measured quantum quadrature of vacuum state in any local phase is amplified by the factor $\alpha^2 = g_{TIA}\alpha_L^2$, which includes the amplification effects from LO gain and electronics gain in the system [24]. Without regard to classical noise, the electric signals (voltage or current) obey a Gaussian distribution:

$$P(V_{\text{vac}}) = \frac{1}{\sqrt{\pi}\alpha} \exp(-\frac{V_{\text{vac}}^2}{\alpha^2}). \tag{2}$$

The coefficient α has to be calibrated to rescale histogram of the associated marginal distribution in optical homodyne tomography (OHT) [25]. In this scheme of quantum random numbers generation, α is associated with the quantum entropy contained in the measured data and it is the critical parameter for digitization of the measured analogue signal.

When classical noise is taken into account, such as electronic noise and local noise resulted from imperfect balancing in balanced homodyne detection (BHD), the observed probability distribution of the electric signal is in the form of a convolution of the scaled vacuum state marginal distribution and the classical noise histogram

$$P_{\text{obs}}(V) = \frac{1}{\alpha} \int P(\frac{V'}{\alpha}) P_{\text{cl}}(V - V') dV'. \tag{3}$$

without loss of generality, the broadband electric noise and the LO noise distribution can be assumed to be Gaussian:

$$P_{cl}(V_{cl}) = \frac{1}{\sqrt{\pi B}} \exp(-\frac{V_{cl}^2}{B}).$$ (4)

The vacuum noise and the classical noise as two variables with normal distribution, are independent with each other, thus their sum is also normally distributed with a total variance equal to the sum of the two variances.

According to Equations (2)–(4), the homodyne measurement of the vacuum state yields a signal distribution as follows

$$P_{obs}(V) = \frac{1}{\sqrt{\pi}\sqrt{\alpha^2 + B}} \exp(-\frac{V^2}{\alpha^2 + B}),$$ (5)

with the measurement variance of

$$\sigma_{obs}^2 = \sigma_{quan}^2 + \sigma_{cl}^2 = (\alpha^2 + B)/2,$$ (6)

where factor 2 is added to renormalize the distribution. Then the quantum and classical noise ratio (*QCNR*) in the homodyne measurement system is defined as

$$QCNR = 10Lg(\sigma_{quan}^2/\sigma_{cl}^2).$$ (7)

The *QCNR* related to the signal-to-noise ratio of homodyne detection, is defined as the ratio between the mean square noise of the measured vacuum state and the electronic noise, that is, the quantity

$$S = (\alpha^2 + B)/B = \sigma_{obs}^2/\sigma_{cl}^2,$$ (8)

or the clearance between the shot noise power spectrum and electronic noise power in dB units, $10Log_{10}(S)$ dB, reads on spectrum analyzer. In other words, when the homodyne detection system works in linear region, the *QCNR* of the raw data can be indicated from the clearance shown by spectrum analyzer.

In our proposal, as a continuous-variable, the measurement output consisting of scaled quadrature of the vacuum state and the classical noise is discretized by an *n*-bit ADC with a dynamical range $[-R + \delta/2, R - 3\delta/2]$. The sampled signals are binned over 2^n bins with width of $\delta = R/2^{n-1}$ and are assigned a corresponding bit combination with length of *n*. 3-bit ADC binning is shown in Figure 1a as an example.

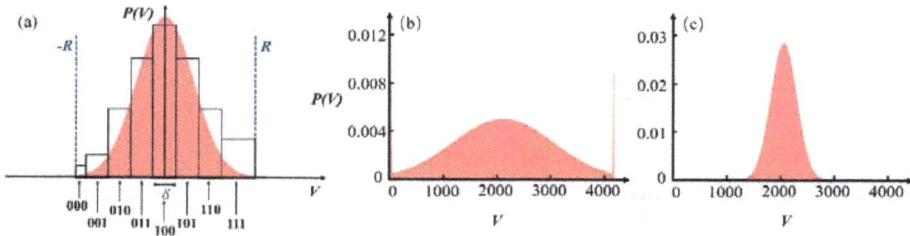

Figure 1. (a) Model of 3-bit analog-digital converter (ADC); (b) Numerical simulations of acquisition conditions for a Gaussian signal when dynamical ADC range is chosen too small; (c) too big.

In order to design an entropy source that provides an adequate amount of entropy per output bit string, the developer must be able to accurately estimate the amount of entropy that can be provided by sampling its noise source. The behavior of the other components included in the entropy must also be known clearly since the behavior of the other components may affect the assessment of the entropy. In our system, the randomness or the entropy in the measurements could derive from multiple factors, such as the quantum fluctuation, classical influences on it and even malicious attack from the third part [5]. Especially and strictly, quantum conditional min-entropy is used to evaluate the maximal amount of randomness extractable from the total entropy of the system [26]. Firstly, the min-entropy for the Gaussian distribution is defined as

$$H_{\min}(X) = -\text{Log}_2(\max_{V \in \{0,1\}^n} \text{Prob}[X = V]). \tag{9}$$

In this scheme, the min-entropy of the probability distribution of quadrature measurements can be accurately predicted from the probability density function of the quantum signal. The maximum probability in (9) can be acquired based on the probability distribution discretized by the bins

$$P_{\text{bin}}(V_i) = \begin{cases} \int_{-\infty}^{-R+\delta/2} P_{obs}(V)dV, i = i_L, \\ \int_{V_i-\delta/2}^{V_i+\delta/2} P_{obs}(V)dV, i_L < i < i_M, \\ \int_{R-3\delta/2}^{+\infty} P_{obs}(V)dV, i = i_M. \end{cases} \tag{10}$$

Each bin is labelled by an integer $i \in \{i_L, ..., i_M\}$, with $i_L = -2^{n-1}$, the least significant bits (LSB) bin, $i_M = 2^{n-1} - 1$, the most significant bit (MSB) bin and $V_i = i \times \delta$.

Secondly, some restrictions must be taken into account in analog-digital conversion process. Those samples go off-scale, that is, points in saturation will be recorded as extrema values as depicted in Figure 1b. So, underestimating the range will induce too many blocks of zeros and ones. Conversely, overestimating the signal range will lead to undue unused bins (Figure 1c). In either situation, some bit combinations are too frequent to be considered random. It is necessary to adjust the amplitude of the analogue signal and the ADC dynamical range in order to employ the full n-bit sampling properly whenever possible.

Further, considering the influence of classical noise on the measurement outcome, ADC dynamical range should be optimized over the classical noise shifted quantum signal probability distribution. In application scenario, inevitable classical noise excursion in the measurement system will result in nonzero mean in the measured signal probability distribution. On the other hand, eavesdropper may induce a deliberate offset over the sampling period. In a word, a noticeable classical noise excursion, Δ, need to be considered in the optimization of the sampling dynamical range.

Taking into account all these factors offered above, we rewrite the discretized probability distribution as,

$$P_{\text{bin}}(V_i|V_{cl}) = \begin{cases} \int_{-\infty}^{-R+\delta/2-\Delta} P(V_i|V_{cl})dV, i = i_L, \\ \int_{V_i-\delta/2-\Delta}^{V_i+\delta/2-\Delta} P(V_i|V_{cl})dV, i_L < i < i_M, \\ \int_{R-3\delta/2-\Delta}^{+\infty} P(V_i|V_{cl})dV, i = i_M. \end{cases} \tag{11}$$

where,

$$P(V_i|V_{cl}) = \frac{1}{\sqrt{\pi}\alpha} \exp(-\frac{(V - V_{cl})^2}{\alpha^2}) \tag{12}$$

is the probability density distribution of the quantum signal given full knowledge of the classical noise V_{cl}, where $V_{cl} \in [V_{cl,\min}, V_{cl,\max}]$ with an excursion of Δ. Finally, the quantum conditional min-entropy is expressed as

$$H_{\min}(V|V_{cl}) = -\text{Log}_2\Big[\text{Max}(\frac{1}{2}\{1 + \text{Erf}\big[\frac{-2(V_{cl,\min}+R+\Delta)+\delta}{2\alpha}\big]\},$$
$$\text{Erf}(\frac{\delta}{2\alpha}), \frac{1}{2}\{1 + \text{Erf}\big[\frac{2(V_{cl,\max}-R+\Delta)+3\delta}{2\alpha}\big]\})\Big]. \tag{13}$$

In the best-case scenario of ADC sampling range, the measurement outcome probability in the center bin is equal to the higher one of the first and the last bins. In this way, the quantum conditional min-entropy is information theoretically provably estimated and the amount of quantum-based randomness in the total noise signal is rigorously evaluates. In applications with the requirement of information security, a random sequence is demanded to be truly unpredictable and have maximum entropy [27].

At the same time, the conditional min-entropy sets the lower bound of extractable randomness from the raw measurements and quantifies the least amount of randomness possessed by each sample or $P = H_{\min}(X)/n$ bit per raw bit. Quantum randomness can be distilled from raw data by applying information theoretically provable Toeplitz-hash extractor. As discussed above, the key point is to find out the QCNR and derive the probability distribution of the quantum signal. The higher the QCNR, the more true randomness can be extracted from the raw measurement. Only when QCNR is high enough, both the quality and the security of the random number generator are guaranteed. Fulfilling the condition of optimal dynamical sampling range R, minimum-entropy of the quantum signal for growing clearance is theoretically analyzed. Proceeding from the directly measurable quantity, homodyne clearance, corresponding QCNR is derived from Equation (8). Then quantum noise variances are expressed as multiples of the σ_{cl}. For different clearance, probabilities of middle bin and the LSB/MSB are compared and the optimal sampling range R is decided based on Equation (11). Finally, based on Equation (13), the quantum conditional min-entropy in optimal sampling range scenario as a function of different classical noise excursion is analyzed.

The classical noise excursions in our raw data have been collected from multiple measurements, which range from almost 3 to 29 times of classical noise standard deviation σ_{cl}. In application scenario, much larger DC offset may be induced deliberately by the eavesdropper. In Figure 2, we show the quantum conditional min-entropy, $H_{\min}(V|V_{cl})$, as a function of homodyne detection clearance for three different classical noise excursions under the precondition of optimal sampling range. $\Delta = 3\sigma_{cl}$ is the smallest classical noise excursion among our multiple measurements, $\Delta = 40\sigma_{cl}$, a larger classical noise excursion for comparison and $\Delta = 17.2\sigma_{cl}$ is the excursion in the raw data from which we extract true random numbers. As shown in Figure 2, the extractable random bits are robust against the decline of QCNR while the classical excursion is subtle. Whereas if classical noise excursion is evidence, one can achieve high secure randomness only when clearance is high enough.

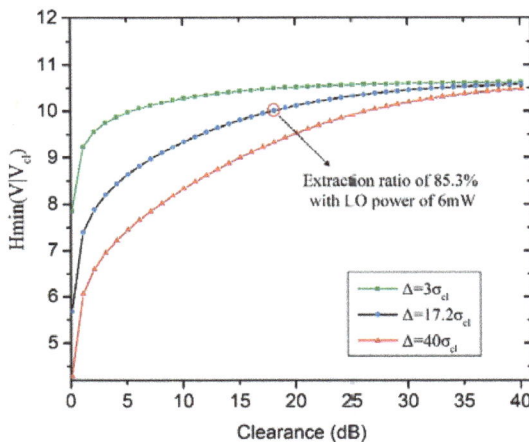

Figure 2. Optimized $H_{\min}(V|V_{cl})$ as a function of homodyne detection clearance among different classical excursions. The theoretical value circled in red corresponding to the highest extraction ratio of true randomness in our experiment.

The clearance relies on the total gain in homodyne detection system (also α in Equation (1)), including the LO amplification and the electrical gain. In quantum state measurements and reconstructions, the clearance needed between shot noise and classical noise is dependent on the amount of squeezing and entanglement one wishes to measure. Empirically, the homodyne system should satisfy the condition that the measured shot noise is 10 dB higher than the classical noise among the analysis frequency range [28,29]. High TIA gain and moderate dynamical range are required so that shot noise is the dominant spectral feature among the detection frequency range. In this scheme of quantum RNG, however, high QCNR, but also large detection bandwidths, are pursued, since the cut-off frequency of the homodyne detector upper bound the sampling frequency in random numbers generation process [30].

On the other hand, the classical effects, which blur the distribution and cause classical entropy in the raw bit sequence, include imperfect balancing of LO, non-unit quantum efficiency and electronic noise of the detectors [31–34]. The non-unit detector efficiency can almost completely overcome by using special fabricated diodes and the quantum efficiencies of more than 99% have been reported [35]. The detrimental electronic noise depends on numerous components in the circuit part as expressed by

$$V_{EL,noise} = R\sqrt{(4KT/R_{PD} + I_{PD,dk}^2 + 4KT/R_r + I_{TIA,c}^2) + (V_{TIA,v}/R)^2} \tag{14}$$

One term is from the photodiode (PD) and comprise of thermal noise and dark current noise of PD, both of which are usually negligible thanks to its big shunt resistance R_{PD} and low dark current $I_{PD,dk}$ [36]. The other term is from the TIA circuit including thermal noise $4KT/R_r$, input noise current $I_{TIA,c}$ and input noise voltage $V_{TIA,v}$ of the operational amplifier. The electrical gain of TIA amplifies quantum fluctuations as well as the electronic fluctuations, so the electronic noise included in the homodyne raw measurements comes mainly from the amplified TIA circuit noise. LO effectively acts as a noise-less amplifier for the quantum fluctuations of the vacuum state and the electrical noise is independent of the LO. In fact, the optical fluctuations seen by the detector can be made much larger than the electronic fluctuations by increasing the laser intensity of LO beam to enhance the QCNR signally [37].

At the same time, the gain of a typical op-amp is inversely proportional to frequency and characterized by its gain–bandwidth product (GBWP). As a trade-off, lower electrical gain put up with higher op-amp bandwidth. In fact, theoretically, vacuum quadrature fluctuates with unlimited bandwidth in the frequency domain. The random number generation rate in this scheme is ultimately limited by the bandwidth of the homodyne detector. Increased bandwidth of op-amp allows higher sampling rate.

3. Experiment and Results

Experimentally, we dedicate to enhance quantum entropy in quantum RNG by enhancing the laser power of LO beam to noise-independently amplify quadrature fluctuation of vacuum state on the premise of optimizing ADC sampling range. An extraction ratio of true randomness of 85.3% is achieved by finite enhancement of the LO power when classical noise excursions of the raw data is obvious and the extracted random sequences passed the NIST (National Institute of Standards and Technology), Diehard and the TestU01 tests.

The experimental setup is depicted in Figure 3. A 1550 nm laser diode (LD) is driven by constant current with thermoelectric temperature control with a maximal out power of 15 mW. A half-wave plate and a polarizing beamsplitter (PBS2) were combined to serve as accurate 50/50 beamsplitting. Single-mode continuous-wave laser beam from the laser incident into one port of the beamsplitter and acts as the LO, while the other port was blocked to ensure that only the vacuum state could enter in. The vacuum field and the LO interfere on the symmetric beamsplitter to form two output beams with balanced power. The outputs are simultaneously detected by balanced homodyne detector (PDB480C, Thorlabs Inc., Newton, MA, USA) to cancel the excess noise in LO while amplify the quadrature amplitude of the vacuum state, which fluctuates randomly and is independent of any external physical quantities.

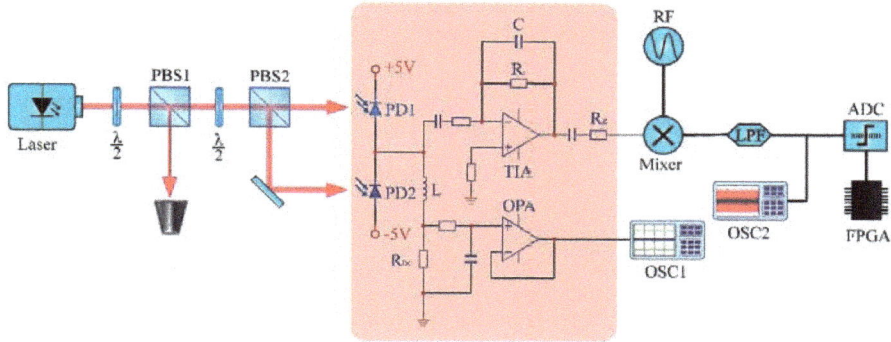

Figure 3. Schematic of the experiment for the quantum random number generator based on homodyne measurements of the quadrature amplitudes of the vacuum state.

Classical noise in the photocurrents is rejected effectively over the whole detection band while the clearance has dependence on frequency as shown in Figure 4. We filtered out a part of the vacuum spectrum, where the clearance is almost consistent, to extract true randomness based on a certain quantum conditional min-entropy and analyze the effect of LO intensity on the conditional min-entropy. The shot noise limited signal from the homodyne detector is mixed down with a 200 MHz carrier (HP8648A) and then passes through a low-pass-filter (LPF, with 50 MHz cut-off frequency (BLP50+, Mini-Circuits Corp., Brooklyn, NY, USA), that is, we actually use 100 MHz vacuum sideband frequency spectrum centered at 200 MHz to act as the random noise resource.

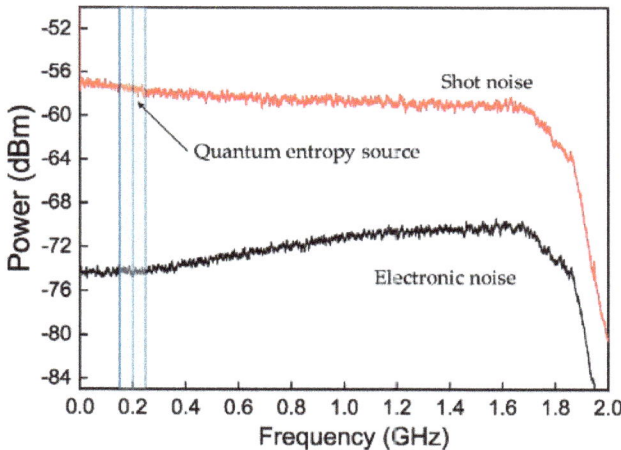

Figure 4. Amplified vacuum noise power spectral when local oscillator (LO) power is 6 mW. 100 MHz vacuum sideband centered at 200 MHz is filtered out as the entropy source of quantum RNG.

In OHT, BHD system is established and locked to every relative phase to measure the marginal distributions of electromagnetic field quadrature for completely reconstruction of quantum states [25]. While the random numbers generation scheme discussed here focus on a marginal distribution of vacuum state in any one phase thanks to the space rotational invariance of its distributions in the phase space, that is no active modulation or phase (or polarization) stabilization is required.

We present the *QCNR* as a function of the LO power arriving at the PD. The electrical noise variance is relatively consistent for certain TIA gain. The clearance depends only on the LO power. The noise power is given by

$$P_{dBm} = 10 \lg \left(\frac{4e^2 (P/hv)\eta BR^2}{Z \times 1 \, mW} \right), \tag{15}$$

where e is the electron charge, $\eta = 0.9$ is the quantum efficiency of the photodiode (Hamamatsu G8376), $B = 100$ KHz the resolution bandwidth, $R = 16 \times 10^3$ V/A the transimpedance gain of the photo detector and $Z = 50 \, \Omega$ the load impedance [38]. For each power value the distribution of the random data was analyzed in time domain in the form of histogram to calculate the *QCNR*. *QCNR* as a function of the LO power figured out from the measured clearance levels is plotted with open circles in Figure 5. The LO power received by each PD is gradually increased from 300 µW to 6 mW by rotating the HWP before PBS1. Here we interpolate between the experimental points to obtain the dependence of *QCNR* on LO power. It is shown as the black dashed line in Figure 5. The experimental results are given by red open circles and can be fitted well by the theoretical curve with a transimpedance gain of 13.1×10^3 V/A. The experimental results are about 2 dB lower than the theoretically excepted *QCNR*, which is due to uncertainties in determining the transimpedance of the detector and the transmission losses in the LPF.

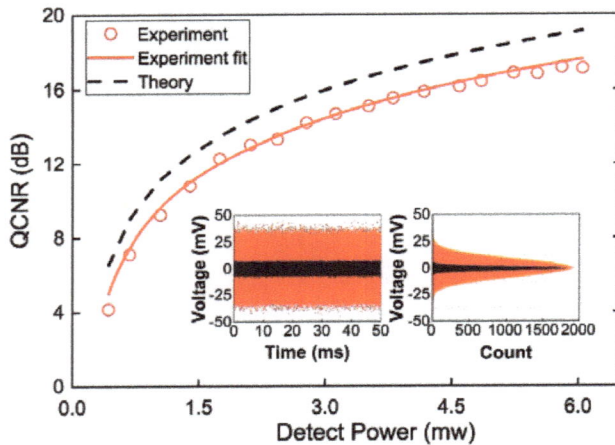

Figure 5. *QCNR* as a function of the LO power. Inset: Resulting histograms of the vacuum (red) and electronic (black) noise obtained at a LO power of 6 mW.

We increase the LO power up to 6 mW to achieve the largest *QCNR* of 17.8 dB in our system, limited by the maximal output power of the laser. The signal is sampled with a rate of 100 MHz, upper limit of twice the LPF band for the sampling rate to avoid temporal correlation between samples. The resolution is 12 bits and the dynamical range is optimized according to the histogram of the time series acquired with reasonably larger sampling range. The amplitude acquisition scale of oscilloscope (SDA806Zi-A, LeCroy, New York, NY, USA) is continuously adjustable. By choosing the analog-digital conversion range appropriately and tuning the LO intensity finely, the amount of off-scale points can be controlled within allowed statistical deviation. The number of saturated points is easy to restrain on-line from the oscilloscope. The distributions of the random data in time domain and in histogram are shown as insets of Figure 5. The measured total variance of the raw data and electrical noise variance are 154.43 mV^2 and 5.89 mV^2, respectively. The classical noise excursions of the raw data are about 17.2 times of the classical noise standard deviation σ_{cl}. Then the probability distribution of the quantum signal is derived and the conditional min-entropy in the quantum signal is worked out to be 10.13 bit per sample, as circled in red in Figure 2.

Finally, information-theoretically provable post-processing scheme, Toeplitz-hashing extractor, is constructed on an FPGA to extract true randomness from the raw data and uniform the Gaussian biased binary stream [39]. A binary Toeplitz matrix of $m \times n$ is constructed with a seed of $m + n - 1$ random bits (the seed can be reused since the Toeplitz-hashing extractor is a strong extractor). m final random bits are extracted by multiplying the matrix and n raw bits, where $m/n \leq P$ and $P = H_{min}(X)/n$. We employ 4096×3520 Toeplitz Hash extractor to distil 10.13 bits/sample. The extraction ratio of 85.3% is the highest as ever reported. We recorded the data with the size of 1 G bits to undergo random test. 1000 sequences with each one 1 M bits are applied to the NIST test and significant level is set as $\alpha = 0.01$. The NIST test is successful if final P-values of all sequences are larger than α with a proportion within the range of $(1 - \alpha) \pm 3\sqrt{(1 - \alpha)\alpha/n} = 0.99 \pm 0.00944$ for 15 test suits [40]. P-value shown in the Figure 6 are the worst cases of our test outcomes.

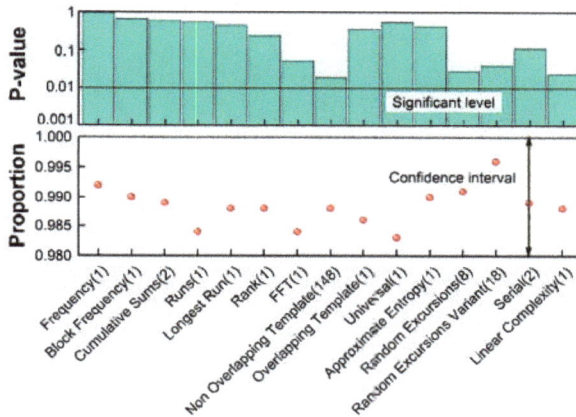

Figure 6. Results of the NIST statistical test suite for a 10^9-bit sequence.

Results of the Diehard statistical test suite for the same data file is shown in Figure 7. Kolmogorov-Smirnov (KS) test is used to obtain a final *p*-value to measure the uniformity of the multiple *p*-values. The test is considered successful if all the final *p*-values lies in the range from 0.01 to 0.99 [41].

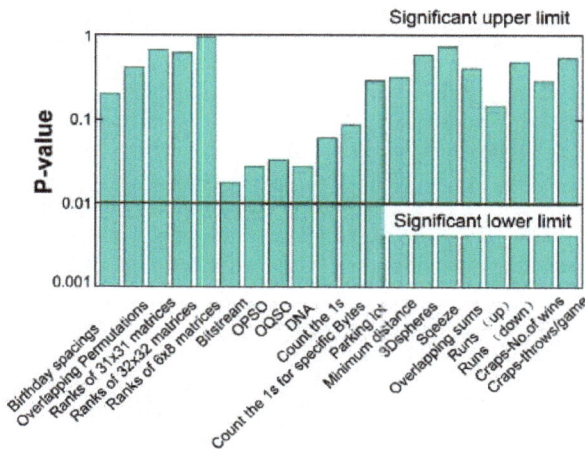

Figure 7. Results of the Diehard statistical test suite for a 10^9-bit sequence.

Constrained by the computational power of crush of TestU01, small crush test is performed with a data size of 8 G bits [42]. The random numbers can pass all the statistical tests successfully. The *p*-value from a failing test converges to 0 or 1. Where the test has multiple *p*-values, the worst case is tabled in Figure 8. All the test items are passed successfully.

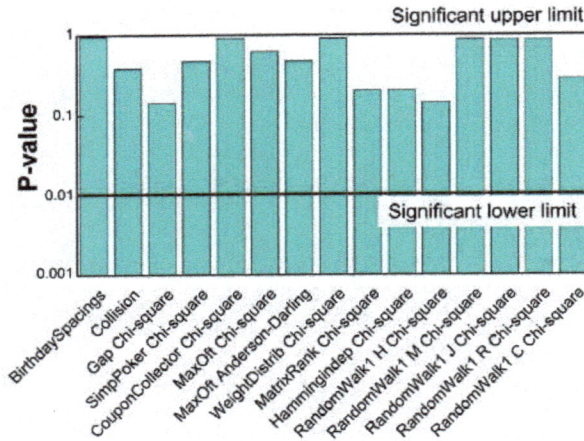

Figure 8. Results of the TestU01 statistical test suite for a 5×10^9-bit sequence.

On the other hand, we reduce the LO power in the homodyne system to 400 µW and correspondingly, the clearance declines to 4.06 dB. The time series of the system outcomes are collected and statistically analyzed. Classical noise excursion in the Gaussian distribution is about 19.3 times of the classical noise standard deviation. Based on theoretical calculation, the min-entropy is worked out to be 7.73 bits/sample. The hash extraction results with maximum extraction ratio of 0.63 can pass the NIST, Diehard and TestU01 tests finally.

4. Conclusions

To summary, in this work, we discussed the role of LO power plays in random number generation based on quantum detection of vacuum state. When classical noise excursion in the system is trivial, LO power in the homodyne system affect the quantum entropy in raw data insignificantly. Nevertheless, in realistic scenario, the mean of the measured signal distribution is normally nonzero, even much larger noise excursion may be induced deliberately by the eavesdropper. In this case, enough real randomness is attainable only when QCNR is high enough. With the LO power enhanced, the vacuum quadrature fluctuations are amplified independent of the electrical noise and the quantum entropy content in the raw data is enhanced effectively. Thus, we propose large dynamical range and moderate TIA gain to pursue higher LO amplification of vacuum quadrature and larger detection bandwidth in homodyne system for higher sampling rate in random numbers generation. Higher hash extraction ratio along with higher sampling rate will enhance the real random number generation rate effectively. More importantly, the quantum RNG system is more robust against to the third part attack.

5. Discussion

The central mathematical concept in true RNG is entropy, which is the assessment standard of the security and quality of a RNG. There are many types of entropy. In recent years, min-entropy, a very conservative evaluation, is applied to lower bound the entropy content in quantum RNG and as the indicator for extraction ratio of universal hash extractor. In our work and some ever works [15,19], quantum conditional min-entropy are deduced to impose stricter removal of side signal. Min-entropy is estimated by using the most common value estimate. However, the most common value estimate is more appropriate for IID (independent identically distribution). For non-IID distribution, the estimate may provide an overestimation. The NIST Special Publication 800-90 series of Recommendations provides guidance on the construction and validation of random bit generators (RBGs) in the form of deterministic random bit generators, in which pseudorandom bits are generated by using an unknown seed, or in the form of non-deterministic random bit generators that can be used for cryptographic applications. Entropy source validation is necessary in order to obtain assurance that all relevant requirements of this Recommendation are met.

As discussed above, the raw noise-source output in our proposal is biased, Toeplitz hash extractor (conditioning component) is used in the design to reduce that bias to an appropriately level before the RNG exports any bits. For non-IID data, a list of estimators is proposed and the minimum of all the estimates is taken as the entropy assessment of the entropy source for the entropy source validation for the Recommendation. We apply our raw bit strings to the test suit on line [43]. The Test result is shown in Figure 9. Because the size of the sample space in our work is 2^{12}, we take the lower 8 bits to meet the applicability of the test. The resulting min-entropy is taken from the minimum of all the estimates as 5.818 per 8 bits. The restart tests are passed. Although the ratio of 72.7% is lower than the evaluation of quantum conditional min-entropy, the quality of our entropy source is validated.

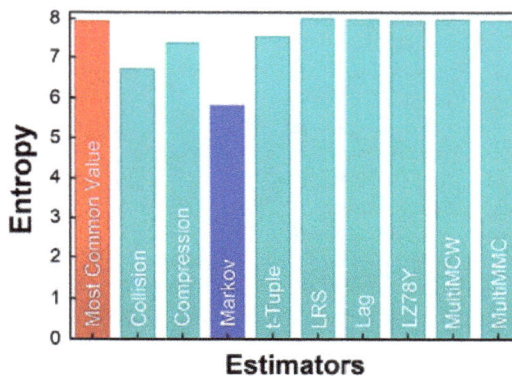

Figure 9. Entropy estimates NIST 800-90B for a 1.6×10^6-bit sequence.

Author Contributions: X.G. and Y.G. designed the study and wrote the paper; X.G. carried out calculations and analyzed the data; R.L., C.C. and M.W. contributed to the experiment; X.G. and P.L. performed the random number tests. All authors discussed the results at all stages. All authors have read and approved the final manuscript.

Funding: This research was funded by the National Natural Science Foundation of China (NSFC) (Grants Nos. 61505136, 61875147, 61775158, 61671316), Shanxi Scholarship Council of China (SXSCC) (Grants No. 2017-040) and Natural Science Foundation of Shanxi Province (Grants No. 201701D221116).

Acknowledgments: We would like to thank Tiancai Zhang for fruitful discussions.

Conflicts of Interest: The authors declare no conflict of interest.

References

1. Korzh, B.; Lim, C.C.W.; Houlmann, R.; Gisin, N.; Li, M.J.; Nolan, D. Provably secure and practical quantum key distribution over 307 km of optical fibre. *Nat. Photonics* **2014**, *9*, 163–168. [CrossRef]
2. Ferguson, N.; Schneier, B.; Kohno, T. *Cryptography Engineering: Design Principles and Practical Applications*; John Wiley & Sons: Hoboken, NJ, USA, 2010.
3. Stefanov, A.; Gisin, N.; Guinnard, O.; Guinnard, L.; Zbinden, H. Optical quantum random number generator. *J. Mod. Opt.* **2000**, *47*, 595–598. [CrossRef]

4. Gisin, N.; Ribordy, G.; Tittel, W.; Zbinden, H. Quantum cryptography. *Rev. Mod. Phys.* **2002**, *74*, 145–195. [CrossRef]

5. Toffoli, T. Entropy? honest! *Entropy* **2016**, *18*, 247. [CrossRef]

6. Rarity, J.; Owens, P.; Tapster, P. Quantum random-number generation and key sharing. *J. Mod. Opt.* **1994**, *41*, 2435–2444. [CrossRef]

7. Guo, H.; Tang, W.Z.; Liu, Y.; Wei, W. Truly random number generation based on measurement of phase noise of a laser. *Phys. Rev. E* **2010**, *81*, 051137. [CrossRef] [PubMed]

8. Ma, H.Q.; Xie, Y.; Wu, L.A. Random number generation based on the time of arrival of single photons. *Appl. Opt.* **2005**, *44*, 7760–7763. [CrossRef] [PubMed]

9. Yan, Q.R.; Zhao, B.S.; Liao, Q.H.; Zhou, N.R. Multi-bit quantum random number generation by measuring positions of arrival photons. *Rev. Sci. Instrum.* **2014**, *85*, 615–621. [CrossRef] [PubMed]

10. Ren, M.; Wu, E.; Liang, Y.; Jian, Y.; Wu, G.; Zeng, H.P. Quantum random-number generator based on a photon-number-resolving detector. *Phys. Rev. A* **2011**, *83*, 1293–1304. [CrossRef]

11. Gabriel, C.; Wittmann, C.; Sych, D.; Dong, R.F.; Mauerer, W.; Andersen, U.L. A generator for unique quantum random numbers based on vacuum states. *Nat. Photonics* **2010**, *4*, 711–715. [CrossRef]

12. Qi, B.; Chi, Y.M.; Lo, H.-K.; Qian, L. High-speed quantum random number generation by measuring phase noise of a single-mode laser. *Opt. Lett.* **2010**, *35*, 312–314. [CrossRef] [PubMed]

13. Xu, F.H.; Qi, B.; Ma, X.F.; Xu, H.; Zheng, H.X.; Lo, H.K. Ultrafast quantum random number generation based on quantum phase fluctuations. *Opt. Express* **2012**, *20*, 12366. [CrossRef] [PubMed]

14. Marangon, D.G.; Vallone, G.; Villoresi, P. Source-device-independent ultrafast quantum random number generation. *Phys. Rev. Lett.* **2017**, *118*, 060503. [CrossRef] [PubMed]

15. Cao, Z.; Zhou, H.; Ma, X.F. Loss-tolerant measurement-device-independent quantum random number generation. *New J. Phys.* **2015**, *17*, 125011. [CrossRef]

16. Sych, D.; Leuchs, G. Quantum uniqueness. *Found. Phys.* **2015**, *45*, 1613–1619. [CrossRef]

17. Fiorentino, M.; Santori, C.; Spillane, S.M.; Beausoleil, R.G.; Munro, W.J. Secure self-calibrating quantum random-bit generator. *Phys. Rev. A* **2006**, *75*, 723–727. [CrossRef]

18. Abellan, C.; Amaya, W.; Domenech, D.; Muñoz, P.; Capmany, J.; Longhi, S. Quantum entropy source on an InP photonic integrated circuit for random number generation. *Optica* **2016**, *3*, 989–994. [CrossRef]

19. Symul, T.; Assad, S. M.; Lam, P.K. Real time demonstration of high bitrate quantum random number generation with coherent laser light. *Appl. Phys. Lett.* **2011**, *98*, 231103. [CrossRef]

20. Shi, Y.C.; Chng, B.; Kurtsiefer, C. Random numbers from vacuum fluctuations. *Appl. Phys. Lett.* **2016**, *109*, 041101. [CrossRef]

21. Zhu, Y.Y.; He, G.Q.; Zeng, G.H. Unbiased quantum random number generation based on squeezed vacuum state. *Int. J. Quantum Inf.* **2012**, *10*, 1250012. [CrossRef]

22. Haw, J.Y.; Assad, S.M.; Lance, A.M.; Ng, N.H. Y.; Sharma, V.; Lam, P.K. Maximization of extractable randomness in a quantum random-number generator. *Phys. Rev. Appl.* **2015**, *3*, 054004. [CrossRef]

23. Turan, M.S.; Barker, E.; Kelsey, J.; McKay, K.A.; Baish, M.L.; Boyle, M. NIST Draft Special Publication 800-90 B: Recommenda-tion for the Entropy Sources Used for Random Bit Generation. Available online: https://csrc.nist.gov/csrc/media/publications/sp/800-90b/draft/documents/sp800-90b_second_draft.pdf (accessed on January 2018).

24. Kumar, R.; Barrios, E.; MacRae, A.; Gairns, E.; Huntington, E.H.; Lvovsky, A.I. Versatile wideband balanced detector for quantum optical homodyne tomography. *Opt. Commun.* **2012**, *285*, 5259–5267. [CrossRef]

25. Lvovsky, A.I.; Raymer, M.G. Continuous-variable optical quantum state tomography. *Rev. Mod. Phys.* **2005**, *81*, 299–332. [CrossRef]

26. Konig, R.; Renner, R.; Schaffner, C. The operational meaning of min- and max-entropy. *IEEE Trans. Inform. Theory* **2009**, *55*, 4337–4347. [CrossRef]

27. Stipčević, M. Quantum random number generators and their applications in cryptography. *Adv. Photon Count. Tech.* **2012**, 837504.

28. Vahlbruch, H.; Mehmet, M.; Chelkowski, S.; Hage, B.; Franzen, A.; Lastzka, N. Observation of squeezed light with 10-db quantum-noise reduction. *Phys. Rev. Lett.* **2008**, *100*, 033602. [CrossRef] [PubMed]

29. Olivares, S.; Paris, M.G.A. Bayesian estimation in homodyne interferometry. *J. Phys. B At. Mol. Opt. Phys.* **2012**, *42*, 55506–55512. [CrossRef]

30. Shen, Y.; Tian, L.; Zou, H.X. Practical quantum random number generator based on measuring the shot noise of vacuum states. *Phys. Rev. A* **2010**, *81*, 063814. [CrossRef]
31. Mcclelland, D.E.; Mckenzie, K.; Gray, M.B.; Ping, K.L. Technical limitations to homodyne detection at audio frequencies. *Appl. Opt.* **2007**, *46*, 3389–3395.
32. Gramdi, S.; Zavatta, A.; Bellini, M.; Paris, M.G.A. Experimental quantum tomography of a homodyne detector. *New J. Phys.* **2017**, *19*, 053051.
33. Combes, J.; Wiseman, H. Quantum feedback for rapid state preparation in the presence of control imperfections. *J. Phys. B At. Mol. Opt. Phys.* **2011**, *44*, 154008 [CrossRef]
34. Chrzanowski, H.M.; Assad, S.M.; Bernu, J.; Hage, B.; Lund, A.P.; Ralph, T.C. Reconstruction of photon number conditioned states using phase randomized homodyne measurements. *J. Phys. B At. Mol. Opt. Phys.* **2013**, *46*, 104009. [CrossRef]
35. Oshima, T.; Okuno, T.; Arai, N.; Suzuki, N.; Ohira, S.; Fujita, S. Vertical solar-blind deep-ultraviolet schottky photodetectors based on beta-Ga_2O_3 substrates. *Appl. Phys. Express* **2008**, *1*, 011202. [CrossRef]
36. Graeme, J. *Photodiode Amplifiers: OP AMP Solutions*; McGraw-Hill: New York, NY, USA, 1995.
37. Jin, X.L.; Su, J.; Zheng, Y.H.; Chen, C.; Wang, W.Z.; Peng, K.C. Balanced homodyne detection with high common mode rejection ratio based on parameter compensation of two arbitrary photodiodes. *Opt. Express* **2015**, *23*, 23859. [CrossRef] [PubMed]
38. Gray, M.B.; Shaddock, D.A.; Harb, C.C.; Bachor, H.-A. Photodetector designs for low-noise, broadband and high-power applications. *Rev. Sci. Instrum.* **1998**, *69*, 3755–3762. [CrossRef]
39. Carter, J.L.; Wegman, M.N. Universal classes of hash functions (Extended Abstract). *J. Comput. Syst. Sci.* **1977**, *18*, 106–112.
40. Rukhin, A.; Soto, J.; Nechvatal, J.; Miles, S.; Barker, E.; Leigh, S. *A Statistical Test Suite for Random and Pseudorandom Number Generators for Cryptographic Applications*; National Institute of Standards and Technology: Gaithersburg, MD, USA, 2001.
41. Marsaglia, G. DIEHARD Battery of Tests of Randomness. 1995.
42. L'Ecuyer, P.; Simard, R. TestU01: A C library for empirical testing of random number generators. *ACM Trans. Math. Softw.* **2007**, *33*, 22. [CrossRef]
43. Turan, M.S.; Barker, E.; Kelsey, J.; McKay, K.A.; Baish, M.L.; Boyle, M. "The SP800-90B_Entropy Assessment Python Package". 2008. Available online: https://github.com/usnistgov/SP800-90B_EntropyAssessment (accessed on 3 August 2018).

Article

Entanglement of Three-Qubit Random Pure States

Marco Enríquez [1], Francisco Delgado [1,*] and Karol Życzkowski [2,3]

1   Escuela de Ingeniería y Ciencias, Tecnológico de Monterrey, Atizapán 52926, Mexico; menriquezf@itesm.mx
2   Faculty of Physics, Astronomy and Applied Computer Science, Jagiellonian University, ul. Łojasiewicza 11, 30-348 Kraków, Poland; karol.zyczkowski@uj.edu.pl
3   Center for Theoretical Physics, Polish Academy of Sciences, Al. Lotników 32/46, 02-668 Warsaw, Poland
*   Correspondence: fdelgado@itesm.mx; Tel.: +52-55-5864-5670

Received: 29 August 2018; Accepted: 26 September 2018; Published: 29 September 2018

Abstract: We study entanglement properties of generic three-qubit pure states. First, we obtain the distributions of both the coefficients and the only phase in the five-term decomposition of Acín et al. for an ensemble of random pure states generated by the Haar measure on $U(8)$. Furthermore, we analyze the probability distributions of two sets of polynomial invariants. One of these sets allows us to classify three-qubit pure states into four classes. Entanglement in each class is characterized using the minimal Rényi-Ingarden-Urbanik entropy. Besides, the fidelity of a three-qubit random state with the closest state in each entanglement class is investigated. We also present a characterization of these classes in terms of the corresponding entanglement polytope. The entanglement classes related to stochastic local operations and classical communication (SLOCC) are analyzed as well from this geometric perspective. The numerical findings suggest some conjectures relating some of those invariants with entanglement properties to be ground in future analytical work.

Keywords: quantum entanglement; three-qubit random states; entanglement classes; entanglement polytope; anisotropic invariants

1. Introduction

Entanglement is possibly the most interesting and complex issue in Quantum Mechanics. Due to this phenomenon it is not possible to describe properties of individual subsystems, even though the entire system is known to be in a concrete pure quantum state. Quantification of entanglement is still a challenge for any quantum system consisting of more than two parts [1,2]. The difficulty of the problem grows quickly with the growing number of subsystems and it becomes intractable in the asymptotic limit [3]. Several measures of quantum entanglement were proposed [4], but even in the case of pure states of a multipartite quantum system, it is not possible to identify the single state which can be called the most entangled, as the degree of entanglement depends on the measure used [5].

On the other hand, entanglement in bipartite systems is already well understood. In the case of pure states, a key tool in describing entanglement properties is the Schmidt decomposition as any entanglement measure is a function of the Schmidt coefficients [2]. Dealing with three-party pure states, the problem becomes more intricate as the corresponding state is represented by a tensor rather than a matrix, so one cannot rely on the Schmidt decomposition related to the singular value decomposition of a matrix. Nevertheless, several decompositions for three-qubit states have been studied in literature [6–8]. More recently, a canonical form for symmetric three-qubit states has been proposed, showing that in this case the number of entanglement parameters can be reduced from five to three [9].

Early studies on correlation in composite quantum systems revealed that for three or more parties there exist quantum states with different forms of entanglement [8], as the states from one entanglement class cannot be converted by local operations to any states of the other class. As the number of parties increases, the number of entanglement classes grows quickly [10]. Since local operations cannot generate entanglement, one usually assumes that a faithful measure of quantum entanglement should be invariant under local unitary operations and should not grow under arbitrary local operations.

For a given class of operations there exist invariants which are constant along every orbit of equivalent states [11,12]. A full set of invariants determines a given orbit of locally equivalent states. However, such sets of invariants are established only for systems consisting of few parties of a small dimension including the simplest multipartite case of three-qubit systems [13–15].

An interesting question arises: To what extent single-particle properties can provide information about the global entanglement [16]? The issue is related to the so-called quantum marginal problem: Given a set of reduced density matrices one asks whether they might appear as partial trace of a given state of a composed system [17]. Necessary conditions for such a "compatibility problem" were provided in [18] for the two-qubit system and then developed by Klyachko [19] for the general case. These conditions can be expressed as a set of linear inequalities concerning the eigenvalues of the density matrix corresponding to the entire system and eigenvalues of the reduced matrices. Interestingly, for multipartite systems the compatibility problem is related to the entanglement characterization [20]. For instance, eigenvalues of three one-qubit reduced matrices of any three-qubit pure state belong to the entanglement polytope and some of its parts correspond to certain classes of quantum entanglement [21].

Not knowing a particular quantum state corresponding to a physical system it is interesting to ask, what are properties of a typical state? More formally, one defines an ensemble of pure quantum states induced by the unitary invariant Fubini-Study measure [2] and computes mean values of various quantities averaging over the unitary group with respect to the Haar measure. Such random quantum states are physically interesting as they arise during time-evolution of quantum systems corresponding to classically chaotic systems [22,23] and are relevant for problems of quantum information processing [24,25].

Research on non-local properties of generic multipartite states has been intensive in recent years. This includes entanglement in two qudit systems [26–28], pairwise entanglement in multi-qubit systems [29–31], entropic relations and entanglement [32], correlations and fidelities in qutrits system [33], a characterization of entanglement through negativities and tangles in several qubits systems and its relation to the emergence of the bulk geometry [34]. More recently, genuine entanglement for typical states for a system composed out of three subsystems with d levels each was studied with help of the geometric measure of entanglement [35], while for generic four-qubit Alsina analyzed the distribution of the hyperdeterminant [36].

The aim of this work is to extend the analysis of entanglement properties of generic states of three-qubit systems. We focus our attention on the five-term decomposition of an arbitrary pure state [15] as it allows one to construct a set of polynomial invariants and to identify the classes of entanglement. We generated an ensemble of pure quantum states induced by the Haar measure on the unitary group $U(8)$ corresponding to the system composed of three qubits and investigated the distribution of various entanglement measures and local invariants.

The paper is organized as follows. In Section 2 we review the five-term decomposition of a three-qubit state and study statistical properties of the coefficients in such a representation of a generic state. In Section 3, we investigate properties of the three qubits invariants, I_k and J_k [15] as well as two newly discovered anisotropic invariants [37]. We obtain their probability distributions, either exact or approximate, and compare them with accurate numerical approximations. The fourth section presents an analysis for the entanglement classes defined in terms of the latter invariants. As a comparative element, we use the Rényi and the minimal Rényi-Ingarden-Urbanik (RIU) entropies [35] to analyze possible meanings for such classes. Another measure, the maximum overlap with respect to a selected entanglement class, allows us to identify for an arbitrary three-qubit state the closest state in each class resembling it. In Section 5 we discuss a characterization of quantum entanglement through the corresponding entanglement polytope and we show how entanglement classes can be distinguished from a geometrical viewpoint. The last section presents concluding remarks, a list of open questions with suggestions concerning the future work.

2. The Canonical Five-Term Decomposition

A three-qubit state in the Hilbert space $\mathcal{H}^{\otimes 3}$ involves eight terms, thus, it can be written as

$$|\psi\rangle = t^{ijk}|ijk\rangle, \quad t^{ijk}\bar{t}_{ijk} = 1, \quad t^{ijk} \in \mathbb{C}, \tag{1}$$

where we have used the repeated scripts notation. It is known [15] that through local unitaries, the number of terms in $|\psi\rangle$ can be reduced from eight to five. First, we define the two square matrices T_0 and T_1 whose entries are given by $(T_i)_{jk} = t^{ijk}$, with $i, j, k = 0, 1$. A local unitary transformation $U \otimes 1_2 \otimes 1_3$ acting on the first qubit produces

$$T_0' = u_{00}T_0 + u_{01}T_1, \quad T_1' = -\bar{u}_{01}T_0 + \bar{u}_{00}T_1. \tag{2}$$

The matrix U is taken such that $\det(T_0') = 0$. On the other hand, the transformation $1_2 \otimes V \otimes W$ changes the matrices T_i according to VT_iW. We choose V and W so that T_0' can be diagonalized via the singular value decomposition (SVD). Explicitly, at the end of this procedure we arrive at

$$T_0'' = \begin{pmatrix} \lambda_0 & 0 \\ 0 & 0 \end{pmatrix}, \quad T_1'' = \begin{pmatrix} \tilde{\lambda}_1 & \tilde{\lambda}_2 \\ \tilde{\lambda}_3 & \tilde{\lambda}_4 \end{pmatrix}. \tag{3}$$

In addition, the phase of the coefficients $\tilde{\lambda}_2, \tilde{\lambda}_3$ and $\tilde{\lambda}_4$ can be absorbed into $\tilde{\lambda}_1$ to yield the decomposition

$$|\psi\rangle = \lambda_0|000\rangle + \lambda_1 e^{i\phi}|100\rangle + \lambda_2|101\rangle + \lambda_3|110\rangle + \lambda_4|111\rangle, \tag{4}$$

where $\lambda_i, \phi \in \mathbb{R}$. Besides $\sum \lambda_i^2 = 1$. According to [15], the only phase ϕ should be restricted to $0 < \phi < \pi$ to assure the uniqueness of the decomposition.

Distribution of the Coefficients

We take an ensemble of 10^6 random states in $\mathcal{H}^{\otimes 3}$ distributed according to the unitary invariant measure on the group $U(8)$ and then first reduce them into the five-term representation (4), then we track each coefficient λ_k to compute numerically its probability distributions as well as the distribution of the phase ϕ. The result is shown in Figure 1 depicting the value of each component λ_k versus their relative normalized density on $\mathcal{H}^{\otimes 3}$. Note that the state (1) depends on 14 real parameters, say $\mathbf{p} = (p_1, \dots, p_{14})$ where each p_μ is the real or imaginary part of t_{ijk}. The unitary invariance implies that after the action of the transformation $U \otimes V \otimes W$ on the state $|\psi\rangle$ the distribution of the coefficients λ_i's and the phase ϕ fulfils $\mathcal{P}(\mathbf{p}) = J \times \mathcal{P}(\lambda)$, where $\lambda = (\lambda_0, \lambda_1, \lambda_2, \lambda_3, \lambda_4, \phi)$ and J is the Jacobian of the transformation. The evaluation of this 14×14 determinant becomes cumbersome and one has to rely on numerical methods to compute the marginal distributions $P(\lambda_k)$ of the coefficients of the state (4) as well as the phase ϕ. The data presented in Figure 1b suggest that the phase ϕ is distributed uniformly on the entire range, $P(\phi) = 1/\pi$ for $\phi \in [0, \pi]$. As the beta distribution has been used to model the behavior of random variables limited to finite length intervals in several contexts [24,35,38], we propose the following distribution $P_i(\lambda_i) = c \lambda_i^a (1 - \lambda_i)^b$, to fit the distributions of the coefficients λ_j. The numerical fits are depicted as solid lines in Figure 1a and the values of the best fitting parameters are reported in Table 1. Results presented suggest that the coefficients λ_1, λ_2 and λ_3 are distributed according to the same probability distribution. Hence, we conjecture that out of the six real parameters in Equation (4), only four are required to characterize entanglement in three-qubit random states, say $\{\lambda_0, \lambda_1, \lambda_4, \phi\}$. Interestingly, the coefficients λ_0 and λ_4 are related with the invariant J_4 connected with the three-qubit genuine entanglement (for the definition see subsequent section).

As generic three-qubit states are typically strongly entangled [35], this analysis illustrates how each coefficient λ_j of a given state is linked with the degree of its entanglement. Note particularly how low values of λ_1, λ_2 and λ_3 are more representative for entangled states in contrast to λ_4, the distribution of which appears to be balanced. Furthermore, the higher values of the coefficient λ_0 correspond to the states with larger entanglement. This is particularly interesting as in the decomposition of Carteret et al. this coefficient yields the maximum overlap with the closest separable state [7].

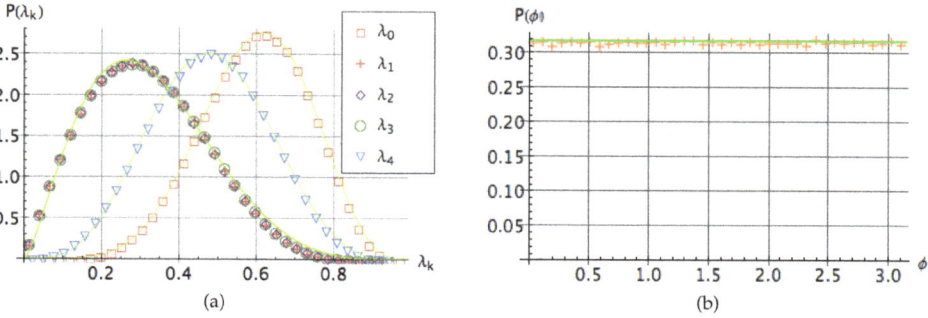

Figure 1. Probability distribution of the Acín parameters in the state (4): (**a**) the coefficients $\lambda_k, k = 0, 1, ..., 4$ and (**b**) the phase ϕ for a set of 10^6 three-qubit random states on $\mathcal{H}_2^{\otimes 3}$. Solid lines represent the best numerical fit in all the cases, the parameters of which are listed in Table 1.

Table 1. Best numerical fit parameters of the distributions $P_i(\lambda_i) = c\,\lambda_i^a(1 - \lambda_i)^b$ for $i = 0, 1, 2, 3, 4$.

i	a	b	c
0	3.74	6.05	1856.85
1	67.76	4.25	1.52
2	68.40	4.27	1.53
3	66.75	4.24	1.52
4	795.16	4.37	3.96

3. Three-Qubits Polynomial Invariants

Local unitary (LU) transformations performed on individual subsystems define orbits of locally equivalent multipartite states. Local invariants can be understood as coordinates in the space of orbits of locally equivalent states. Any complete set of local invariants allows one to distinguish between different orbits of locally equivalent states and thus to describe the degree of quantum entanglement [7]. For pure states of a three-qubit system, the space of orbits has six dimensions and it is possible to find six algebraically independent invariants [39].

In this section we will analyze the distributions $P(I_k)$ and $P(J_k)$ on $\mathcal{H}^{\otimes 3}$ for the corresponding three-qubit invariants (under local operations) I_k [13] and J_k [40], with $k = 1, ..., 5$. These polynomial invariants set representative classes on $\mathcal{H}^{\otimes 3}$ and cannot be directly used as the measures of genuine entanglement.

Distribution of the Invariants

We first consider the set of five invariants used in [40]

$$I_2 = \mathrm{tr}(\rho_A^2), \quad I_3 = \mathrm{tr}(\rho_B^2), \quad I_4 = \mathrm{tr}(\rho_C^2),$$
$$I_5''' = \mathrm{tr}[(\rho_A \otimes \rho_B)\rho_{AB}], \quad I_6 = |\mathrm{Hdet}(T)|^2 \tag{5}$$

where ρ_i stands for the reduced density matrix of the i-th system, ρ_{ij} is the reduced density matrix when the partial trace respect the system k is performed while i, j, k is a permutation of A, B, C. The last invariant is related to the hyperdeterminant Hdet of the tensor coefficients $T = (t^{ijk})$ representing the state (1).

The invariants are labeled according to the notation used by Sudbery [13]. Note that the squared norm of the state (1) is in itself a polynomial invariant usually denoted as I_1. In Figure 2a–c we show the probability distribution of the above set of invariants over an ensemble of 10^6 random states. Moreover, as for $k = 2, 3, 4$ the quantity I_k is related with the linear entropy, $S_k = 1 - I_k$, the corresponding distributions show that the entanglement of each qubit with the other two is the same no matter which partial trace is performed. On the other hand, the invariants I_k in terms of the coefficients t^{ijk} are written as [13]:

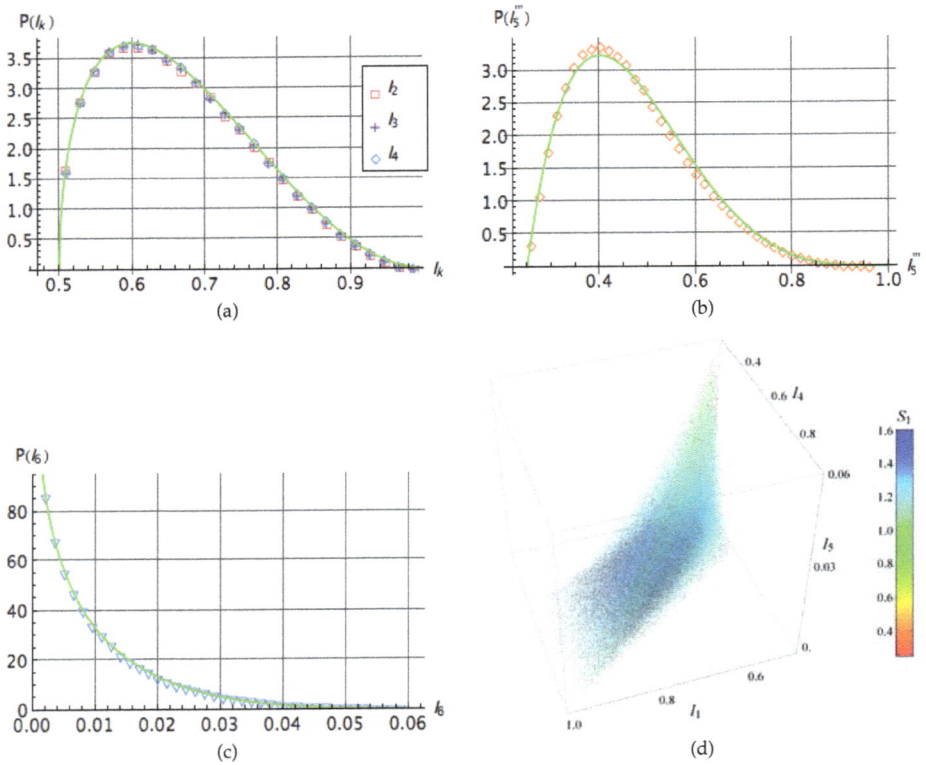

Figure 2. (a–c) Probability distribution for the polynomial invariants $I_i, i = 1, ..., 5$ for a set of 10^6 three-qubit random states. Solid line in panel (a) stands for the distribution (9), while in panels (b,c) the best numerical distributions are depicted by green curves. In panel (d) a dispersion plot comparing I_1, I_4 and I_5 is shown. In addition, each dot has been colored as function of its S_1 Rényi entropy [41] calculated after the five terms reduction.

$$I_2 = t^{i_1 j_1 k_1} \bar{t}_{i_2 j_1 k_1} t^{i_2 j_2 k_2} \bar{t}_{i_1 j_2 k_2}, \quad I_3 = t^{i_1 j_1 k_1} \bar{t}_{i_1 j_2 k_1} t^{i_2 j_2 k_2} \bar{t}_{i_2 j_1 k_2}, \quad I_4 = t^{i_1 j_1 k_1} \bar{t}_{i_1 j_1 k_2} t^{i_2 j_2 k_2} \bar{t}_{i_2 j_2 k_1},$$

$$I_5''' = t^{i_1 j_1 k_1} \bar{t}_{i_1 j_2 k_2} t^{i_2 j_2 k_2} \bar{t}_{i_2 j_3 k_1} t^{i_3 j_3 k_3} \bar{t}_{i_3 j_1 k_3} \tag{6}$$

$$I_6 = 4|\epsilon_{i_1, j_1} \epsilon_{i_2 j_2} \epsilon_{k_1 \ell_1} \epsilon_{k_2 \ell_2} \epsilon_{i_3 k_3} \epsilon_{j_3 \ell_3} t^{i_1 i_2 i_3} t^{j_1 j_2 j_3} t^{k_1 k_2 k_3} t^{\ell_1 \ell_2 \ell_3}|^2$$

where the convention of summation over repeated indexes is used and $\epsilon_{k,\ell}$ stands for the Levi-Civita tensor of order two. Since the coefficients can be regarded as a column of a random unitary matrix, we can compute the average value of each invariant by evaluating integrals of polynomial functions over the unitary group with respect to unique normalized Haar measure. Using symbolic integration [42] we obtain $\langle I_k \rangle = 2/3$ for $k = 2, 3, 4$. This result is consistent with the mean purity of a single qubit traced out from a 2×4 system reported in [43]. Moreover, $\langle I_5''' \rangle = 7/15$ and $\langle I_5'''^2 \rangle = 133/572$. In order to compute the mean value of I_6, we use the second moment of the three-tangle τ reported in [35] with the fact $\tau^2 = 16I_6$ to get $\langle I_6 \rangle = 1/110$. On the other hand, to compute the distributions of the invariants $P(I_k)$ for $k = 2, 3, 4$, we first note that the joint density of eigenvalues ϑ_1 and ϑ_2 of a single qubit traced out of a system of a three-qubit system is given in Equation (6) of [43] with $N = 2$ and $K = 4$. This reads

$$\mathcal{P}(\vartheta_1, \vartheta_2) = 210 \, \delta(1 - \vartheta_1 - \vartheta_2)(\vartheta_1 - \vartheta_2)^2 \vartheta_1^2 \vartheta_2^2 \tag{7}$$

where δ stands for the Dirac delta. As each I_k is nothing other than the purity of a single qubit reduced density matrix, we can compute the probability distribution by performing the following integral

$$P(I_k) = 210 \int_0^1 \int_0^1 d\vartheta_1 d\vartheta_2 \mathcal{P}(\vartheta_1, \vartheta_2) \delta(I_k - \vartheta_1^2 - \vartheta_2^2), \tag{8}$$

this yields

$$P(I_k) = \frac{105}{2}(1 - I_k)^2 (2I_k - 1)^{1/2}, \quad 1/2 \leq I_k \leq 1, \quad k = 2, 3, 4. \tag{9}$$

This probability distribution is depicted in Figure 2. In addition, we approximate the distribution $P(I_5''')$ by the following beta distribution

$$P_{F_5}(I_5''') = \frac{\Gamma(a + b + 2)}{3^{a+b+1}4^{a+1}\Gamma(a+1)\Gamma(b+1)}(1 - I_5''')^a (4I_5''' - 1)^b, \tag{10}$$

requiring the first two moments of this distribution coincide with the exact two moments of $P(I_5''')$ reported above. We found $a = 21,989/5691$ and $b = 5554/5691$. On the other hand, the distribution of the square of the three tangle was approximated in [35] by a Beta distribution. Thus, making a variable change in this result we may approximate $P(I_6)$ by

$$P_{F_6}(I_6) = \frac{2}{\sqrt{I_6}} \text{Beta}(31/17, 62/17, 4\sqrt{I_6}), \quad 0 \leq I_6 \leq 1/16. \tag{11}$$

As the distributions of the invariants I_2, I_3 and I_4 are the same, we only need three invariants to characterize the entanglement in the set of three-qubit random states, say (I_2, I_5''', I_6). In Figure 2d we show a dispersion plot whose three axes correspond to such invariants and their colors correspond to their S_1 Rényi entropy calculated after of the five terms reduction [35] (which will be properly presented in the next section) in agreement with the side color scale.

We also consider the set of invariants proposed by Acín et al. [15]. These invariants allow to identify different entanglement classes (which will be discussed in the next section) and can be written in terms of the six parameters of the five-term decomposition as

$$J_1 = |\lambda_1\lambda_4 e^{i\varphi} - \lambda_2\lambda_3|^2, \quad J_2 = \mu_0\mu_2, \quad J_3 = \mu_0\mu_3,$$
$$J_4 = \mu_0\mu_4, \quad J_5 = \mu_0(J_1 + \mu_2\mu_3 - \mu_1\mu_4), \tag{12}$$

where $\mu_i = \lambda_i^2$. For this analysis, the same set of 10^6 random states was considered but they are now used to obtain the corresponding values of them through their expressions in terms of the five-term coefficients [15]. All these invariants can be calculated departing from the set of λ_i. The outcomes are shown in the Figures 2 and 3 in their respective ranges. Note in the Figure 3a–c how for J_1, J_2 and J_3

the distribution is biased on low values of these invariants, denoting a possible relation with higher entanglement. For the quantity J_4, related to the hyperdeterminant, the distribution peaks around of $\frac{1}{16}$, denoting that separability as well as genuine entanglement are absent in the most of states in $\mathcal{H}^{\otimes 3}$. A similar feature is observed for J_5 but varying sharply for negative and positive values. On the other hand, the invariants J_k's can be expressed in terms of the quantities I_k's [15]

$$J_1 = \frac{1}{4}(1 + I_2 - I_3 - I_4 - 2\sqrt{I_6}), \quad J_2 = \frac{1}{4}(1 - I_2 + I_3 - I_4 - 2\sqrt{I_6}),$$
$$J_3 = \frac{1}{4}(1 - I_2 - I_3 + I_4 - 2\sqrt{I_6}), \quad J_4 = \sqrt{I_6},$$
$$J_5 = \frac{1}{4}(3 - 3I_2 - 3I_3 - I_4 + 4I_5 - 2\sqrt{I_6}).$$

Such expressions are useful to compute some averages. For instance, as $\langle\sqrt{I_5}\rangle = \langle\tau\rangle/4$ it is immediate to compute $\langle J_4 \rangle = 1/12$. From the above definitions we can calculate directly $\langle J_k \rangle = 1/24$, for $k = 1, 2, 3$ and $\langle J_5 \rangle = 1/120$. We approximate the probability distributions $P(J_k)$ with $k = 1, 2, 3$ by a distribution $P_{F_k}(J_k) \sim J_k^a(1 - 4J_k)^b$, where the parameters in this case are determined numerically to yield the best fit. In addition, making use of the approximation (11) for the distribution of the invariant I_6. One can obtain the following approximation for the distribution of the variable J_4

$$P_{F4}(J_4) = 4\text{Beta}(31/17, 62/17; 4J_4), \quad 0 \le J_4 \le 1/4 \tag{13}$$

On the other hand, as the distributions for J_1, J_2 and J_4 are uniform among them, we may characterize the entanglement using only the invariants J_1, J_4 and J_5. In Figure 3d we depict a scatter plot using these invariants as coordinates, similarly as in Figure 2d for I_k.

Another interesting invariant is the one obtained by Kempe [44]

$$I_5 = 3\text{tr}(\rho_A \otimes \rho_B)\rho_{AB} - \text{tr}\rho_A^3 - \text{tr}\rho_B^3 = t^{i_1 j_1 k_1} t^{i_2 j_2 k_2} t^{i_3 j_3 k_3} \bar{t}_{i_1 j_2 k_3} \bar{t}_{i_2 j_3 k_1} \bar{t}_{i_3 j_1 k_2}, \tag{14}$$

which distinguishes locally indistinguishable states. In terms of the Acín parameters, it reads

$$\begin{aligned} I_5 &= 1 - 3\lambda_4^2 - 3\lambda_3^2 + 3\lambda_3^4 + 3\lambda_4^4 + 3\lambda_1^2\lambda_3^2 + 6\lambda_3^2\lambda_4^2 \\ &\quad + \left(\lambda_1^2 \left(3 - 6\lambda_3^2 \right) - 3 \left(\lambda_3^2 - 1 \right) \left(2\lambda_3^2 + 2\lambda_4^2 - 1 \right) \right) \lambda_2^2 \\ &\quad + 6\lambda_1\lambda_3\lambda_4 \left(\lambda_1^2 + \lambda_2^2 + \lambda_3^2 + \lambda_4^2 \right) \lambda_2 \cos\phi + \left(3 - 6\lambda_3^2 \right) \lambda_2^4. \end{aligned} \tag{15}$$

Note that the form (14) of the Kempe Invariant I_5 is manifestly permutation symmetric. Although this quantity cannot be considered as a legitimate measure of entanglement, Osterloh has pointed out [45] that different values of I_5 allow to distinguish between different local orbits of three qubit pure states. Integrating Equation (14) using symbolic integration on the Haar measure, we found that $\langle I_5 \rangle = 2/5$ and $\langle I_5^2 \rangle = 499/2860$. In Figure 4a, we show the probability distribution of the invariant I_5, which can be approximated by the distribution

$$P_{F_{I_5}}(\kappa) = \frac{9^{a+1}\Gamma(a + b + 2)}{7^{a+b+1}\Gamma(a+1)\Gamma(b+1)}(1 - \kappa)^a(9\kappa - 2)^b, \quad 2/9 \le \kappa \le 1 \tag{16}$$

where $a = 90/23$ and $b = 283/621$ are settled by the condition that the first two moments of $P_{F_{I_5}}(I_5)$ correspond with the first two moments of $P(I_5)$ provided above. We remark that sextic invariant I_5''' can be written in terms of the Kempe invariant and the quadratic and quartic invariants [13].

Recently, an alternative set of invariants characterizing a three-qubit pure state $|\psi\rangle$ was proposed by Cheng and Hall [37]. To define them, consider a two-qubit reduced density matrix $\rho^{kl} =$

$\text{Tr}_m|\psi_{klm}\rangle\langle\psi_{klm}|$ where indices k,l,m denote three subsystems A,B,C and $m \neq k,l$. Any such a matrix of order four can be written in its Bloch representation,

$$\rho^{kl} = \frac{1}{4}\left(\mathbf{1}^k \otimes \mathbf{1}^l + \mathbf{K}\cdot\vec{\sigma}^k \otimes \mathbf{1}^l + \mathbf{1}^k \otimes \mathbf{L}\cdot\vec{\sigma}^l + \sum_{i,j=1}^{3} T_{i,j}^{k,l}\sigma_i^k \otimes \sigma_j^l\right) \tag{17}$$

where $\vec{\sigma}^k = (\sigma_1^k, \sigma_2^k, \sigma_3^k)$, while $\mathbf{K}$ and $\mathbf{L}$ denote the Bloch vectors for parts k and l respectively. Entries of the correlation matrices of the reduced states read $T_{n,m}^{k,l} = \langle\sigma_n^k \otimes \sigma_m^l\rangle = \text{Tr}(\sigma_n^k \otimes \sigma_m^l\rho^{kl})$, while the superscripts denote two out of three subsystems A,B,C as required to determine a two-qubit partial trace. Let s_j^{kl} denote the eigenvalues of the symmetric matrix $S^{kl} = T^{kl}\left(T^{kl}\right)^\dagger$ and the average value read, $s_{\text{iso}}^{kl} = (s_1^{kl} + s_2^{kl} + s_3^{kl})/3$. The invariants are constructed in terms of the pairwise anisotropic strengths δs_j^{AB}, δs_j^{AC} and δs_j^{BC} with $j = 1,2,3$ which read $\delta s_j^{kl} = s_j^{kl} - s_{\text{iso}}^{kl}$, with $k,l = A,B,C$. It was shown [37] that the pairwise anisotropic strengths fulfil the relations

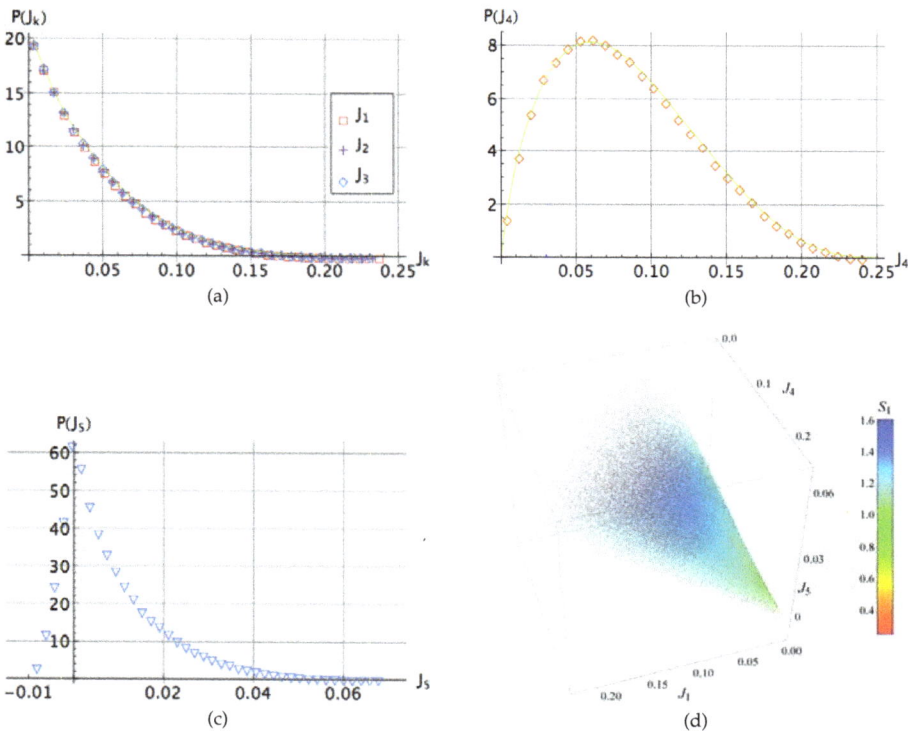

Figure 3. (a–c) Probability distribution for the polynomial invariants $J_i, i = 1,...,5$ for a set of 10^6 three-qubit random states. In all graphics the numerical best fit distribution is depicted as the green line. In Figure (d) we show a dispersion plot comparing J_1, J_4 and J_5. In addition, each dot has been colored as a function of its S_1 Rényi entropy [41] calculated after the five term reduction in agreement with the side color scale.

$$\delta s_j = \delta s_j^{AB} = \delta s_j^{AC} = \delta s_j^{BC}, \quad j = 1,2,3 \tag{18}$$

and they are also invariant under local transformations as well as any permutation of the parties. Hence, the anisotropic strength and the anisotropic volume can be defined as

$$s_{\text{ani}}^2 = \sum_i (\delta s_i)^2, \qquad V_{\text{ani}} = \prod_j \delta s_i. \tag{19}$$

Note that for a given three-qubit pure state $|\psi\rangle$ the above invariants can be related with parameters entering the five-term form (4) —see Supplementary Material in [37].

In Figure 4b–d we show the probability distributions of the pairwise anisotropic strengths as well as the probability distribution of the invariants s_{ani} and V_{ani} for an ensemble of 10^6 three-qubit random states. We approximate numerically the distribution of the quantities δs_1, δs_2 and s_{ani} by respective beta distributions:

$$P_F(\delta s_1) = c_1 (a_1 - \delta s_1)^{\beta_1} (\delta s_1 - a_2)^{\beta_2}, \quad P_F(\delta s_2) = c_2 (a_3 - \delta s_2)^{\beta_3} (a_4 + \delta s_2)^{\beta_4},$$

$$P_F(s_{\text{ani}}) = c_3 (a_5 - s_{\text{ani}})^{\beta_5} s_{\text{ani}}^{\beta_6}, \tag{20}$$

while the positive part of the distribution of V can be approximated by an exponential distribution,

$$P_F(V_{\text{ani}}) = c_4 e^{-b V_{\text{ani}}}. \tag{21}$$

The fitting parameters read $c_i = (472.7, 1299, 135.6, 54.9)$ for $i = 1, \dots, 4$; $a_j = (0.66, 0.01, 0.11, 0.33, 0.72)$ for $j = 1, \dots, 5$; $\beta_i = (2.5, 2.04, 1.88, 1.92, 2.26, 1.63)$ for $i = 1, \dots, 6$ and $b \approx 61.6$. Interestingly, the distribution of the negative quantity δs_3 displays a singular peak, while the distribution of V_{ani} attains its maximum at anisotropic volume close to zero and exhibits an exponential decay.

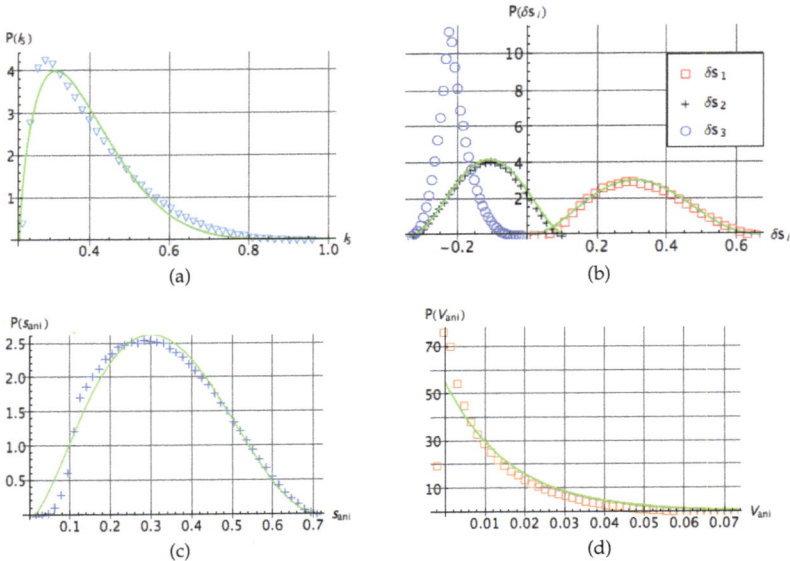

Figure 4. Probability distribution of: (a) the Kempe invariant I_5 (the green line stands for the probability distribution (16)). (b) The pairwise anisotropic strengths δs_j with $j = 1, 2, 3$. (c) The invariant s_{ani} and (d) The invariant V_{ani}. Solid lines in all cases correspond to the best numerical fit.

4. Three-Qubits Entanglement Classes

A state classification has been presented in [15] based on the minimal number of product states in (4). Acín et al. reported some entanglement classes which are presented in Table 2. The conditions for such class states are expressed in terms of the invariants J_k. Thus, in this section we consider the invariant classes introduced there, departing from the coefficients of the five-term representation in $\mathcal{H}^{\otimes 3}$. These classes barely describe some families around some characteristic states in this space. The first aim is to analyze how those classes represent the entanglement of each state included there, mainly based on the entanglement distribution knowledge on $\mathcal{H}^{\otimes 3}$ [35]. Note that in some classes the direct imposition of the conditions on the invariants leaves some product states that differ from those reported by Acín, that is to say, to obtain such product states an additional LU transformation is required. Such cases are remarked with $\star$ in Table 2.

Table 2. Acín entanglement classes introduced in [15]. Besides $\Delta_J \equiv (J_4 + J_5)^2 - 4(J_1 + J_4)(J_2 + J_4)(J_3 + J_4)$. Basis elements marked with $\star$ are not directly obtained, instead they have additional relabellings. Besides, the fourth column shows the identification of each class with subsets of the entanglement polytope. The point G stands for $(1/2, 1/2, 1/2)$. Details are presented in Section 5.

Class	Conditions	States	Entanglement Polytope
1	$J_i = 0$	$\lvert 000 \rangle$	point $\mathcal{O} = (0,0,0)$
2a	All $J_i = 0$ apart from J_1	$\lvert 000 \rangle, \lvert 011 \rangle^\star$	lines $\overline{\mathcal{O}A}$, $\overline{\mathcal{O}B}$ and $\overline{\mathcal{O}C}$
2b	All $J_i = 0$ apart from J_4	$\lvert 000 \rangle, \lvert 111 \rangle$	line $\overline{\mathcal{O}G}$
3a	$\begin{aligned} J_1 J_2 + J_1 J_3 + J_2 J_3 \\ \sqrt{J_1 J_2 J_3} = J_5/2, J_4 = 0 \end{aligned} =$	$\lvert 000 \rangle, \lvert 101 \rangle, \lvert 110 \rangle$	$\triangle_2 \mathcal{O}AB, \triangle_2 \mathcal{O}AC, \triangle_2 \mathcal{O}BC, \triangle_2 ABC$
3b	$J_1 = J_2 = J_5 = 0$	$\lvert 000 \rangle, \lvert 110 \rangle, \lvert 111 \rangle$	$\triangle_2 ABG, \triangle_2 ACG, \triangle_2 BCG$
4a	$J_4 = 0, \sqrt{J_1 J_2 J_3} = J_5/2$	$\lvert 000 \rangle, \lvert 100 \rangle, \lvert 101 \rangle, \lvert 110 \rangle$	$\triangle_3 \mathcal{O}ABC$
4b	$J_2 = J_5 = 0$	$\lvert 000 \rangle, \lvert 100 \rangle, \lvert 110 \rangle, \lvert 111 \rangle$	
4c	$\begin{aligned} J_1 J_4 + J_1 J_2 + J_1 J_3 + J_2 J_3 \\ \sqrt{J_1 J_2 J_3} = J_5/2 \end{aligned} =$	$\lvert 000 \rangle, \lvert 101 \rangle, \lvert 110 \rangle, \lvert 111 \rangle$	
4d	$\Delta_J = 0, \sqrt{J_1 J_2 J_3} = \lvert J_5 \rvert /2$	$\lvert 000 \rangle, \lvert 010 \rangle, \lvert 100 \rangle, \lvert 111 \rangle^\star$	

4.1. The Minimal Decomposition Entropy

We characterize the entanglement degree of the classes in Table 2 using the minimal Rényi-Ingarden-Urbanik (RIU) entropy, also known as minimal decomposition entropy [35]. For the state (1) this is defined as

$$ S_q^{\text{RIU}}(\psi) := \min_{U_{\text{loc}}} S_q \left[p(U_{\text{loc}} \lvert \psi \rangle) \right], \tag{22} $$

where $p(\cdot)$ stands for the probability vector related to the state (1) and the minimum is taken on all local transformations $U_{\text{loc}} = U_1 \otimes U_2 \otimes U_3$. Note that S_q is the q-order Rényi entropy [41]. Depending on the parameter q the quantity (22) provides information about the state [35]. Thus, for

- $q = 0$: The decomposition entropy is related to the tensor rank of the state $\lvert \psi \rangle$. As a direct consequence of the decomposition (4) we have $S_0^{\text{RIU}}(\psi) \leq 5$.
- $q = 1$: The minimal decomposition entropy $S_1^{\text{RIU}}(\lvert \psi \rangle)$ determines the minimal information gained by the environment after performing a projective von-Neumann measurement of the pure state $\lvert \psi \rangle \langle \psi \rvert$ in an arbitrary product basis [46].
- $q \to \infty$: In such a limiting case, the minimal RIU entropy is associated with the maximal overlap with the closest separable state $\Lambda_{\max} = \max \lvert \langle \psi \lvert \chi_{\text{sep}} \rangle \rvert^2$. Indeed, it can be shown that $S_\infty^{\text{RIU}}(\lvert \psi \rangle) = -\log \Lambda_{\max}$. See [35] for details.

A direct computation shows that for a state in class 1, the minimal RIU entropy vanishes regardless of the value of the parameter q. The corresponding calculation for the other entanglement classes is presented below.

4.1.1. Classes 2

A direct calculation shows that the decomposition of states in class 2b is optimal. That is to say, if the state is given by

$$|\varphi_{2b}\rangle = \cos\alpha|000\rangle + \sin\alpha|111\rangle, \quad 0 < \alpha < \pi/2, \tag{23}$$

the minimal decomposition entropy reads

$$S_1^{\text{RIU}}(\varphi_{2b}) = -\cos^2\alpha\ln(\cos^2\alpha) - \sin^2\alpha\ln(\sin^2\alpha). \tag{24}$$

Our numeric calculations indicate that for the class 2a, the Acín decomposition is optimal as well. The states with the largest minimal decomposition entropy in each class are

$$|\varphi_{2a}^{\max}\rangle = \frac{1}{\sqrt{2}}|000\rangle + \frac{1}{\sqrt{2}}|111\rangle, \quad |\varphi_{2b}^{\max}\rangle = \frac{1}{\sqrt{2}}|100\rangle + \frac{1}{\sqrt{2}}|111\rangle, \tag{25}$$

note the reported basis for class 2b in Table 2 is different due to additional changes commonly reported in the literature. A simple calculation shows the LU equivalence of the two local basis. Note that the state $|\varphi_{2b}^{\max}\rangle$ is bi-separable and it attains the same minimal decomposition entropy as the *GHZ* state.

4.1.2. Classes 3

Any state belonging to class 3a can be parametrized as

$$|\varphi_{3a}\rangle = \sin\theta_1\sin\theta_2|000\rangle + \sin\theta_1\cos\theta_2|101\rangle + \cos\theta_1|110\rangle, \quad 0 < \theta_1,\theta_2 < \pi/2 \tag{26}$$

note such state is LU-equivalent to the symmetric state

$$|\widetilde{\varphi}_{3a}\rangle = \sin\theta_1\sin\theta_2|100\rangle + \sin\theta_1\cos\theta_2|001\rangle + \cos\theta_1|010\rangle, \tag{27}$$

hence, the minimal RIU entropy can be computed using the method described in [35] for symmetric states. In particular, if $\cos\theta_1 = 1/\sqrt{3}$ and $\sin\theta_2 = 1/\sqrt{2}$ we obtain the well-known *W*-state for which $S_1^{\text{RIU}}(W) = \ln 3$, which is the largest value of S_1^{RIU} for this class.

On the other hand, a state in class 3b can be written as

$$|\varphi_{3b}\rangle = \sin\theta_1\sin\theta_2|000\rangle + \sin\theta_1\cos\theta_2|110\rangle + \cos\theta_1|111\rangle, \quad 0 < \theta_1,\theta_2 < \pi/2. \tag{28}$$

No state in class 3b has greater S_1^{RIU} than the *W*-state. For a general state in these classes, the minimal decomposition entropy as a function of parameters θ_1 and θ_2 is depicted in Figure 5. Note that regions of maximal S_1^{RIU} entropy are around the values θ_1,θ_2 for the maximal entropy for such states.

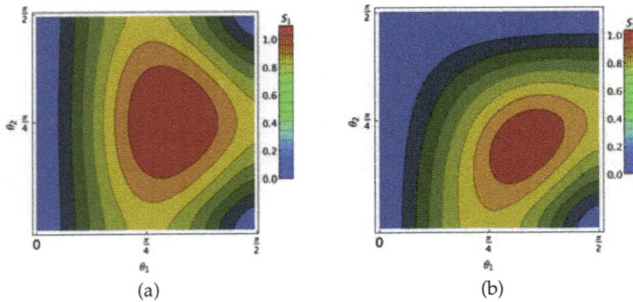

Figure 5. (a) The minimal decomposition entropy level curves as function of the parameters θ_1 and θ_2 for a state in class 3a; (b) Same as (a) for a state in class 3b.

4.1.3. Classes 4

A general state in each one of the classes 4 can be written as

$$|\varphi_{4a}\rangle = \beta_1|000\rangle + e^{i\varphi}\beta_2|100\rangle + \beta_3|101\rangle + \beta_4|110\rangle \tag{29}$$

$$|\varphi_{4b}\rangle = \beta_1|000\rangle + e^{i\varphi}\beta_2|100\rangle + \beta_3|110\rangle + \beta_4|111\rangle \tag{30}$$

$$|\varphi_{4c}\rangle = \beta_1|000\rangle + \beta_2|101\rangle + \beta_3|110\rangle + \beta_4|111\rangle \tag{31}$$

$$|\varphi_{4d}\rangle = \beta_1|000\rangle + \beta_2|010\rangle + \beta_3|100\rangle + \beta_4|111\rangle \tag{32}$$

where $\beta_1 = \sin\theta_1 \sin\theta_2 \sin\theta_3$, $\beta_2 = \sin\theta_1 \sin\theta_2 \cos\theta_3$, $\beta_3 = \sin\theta_1 \cos\theta_2$ and $\beta_4 = \cos\theta_1$. As for class 2b, the basis elements for class 4d reported in Table 2 are not those directly obtained from (4). Class 4d corresponds to the real class (with all components real, thus $e^{i\varphi} = \pm 1$) which allows for the performance of an additional reduction to only four terms. As in the previous case, we get the surfaces of minimal decomposition entropy in terms of parameters θ_1, θ_2 and θ_3 in the Figure 6. Those figures exhibit for each class the behavior for the entropy. There, the frontiers of the regions shown $\theta_1, \theta_2, \theta_3 = 0, \pi/2$ correspond to separable states. In addition, our numerical calculations show that the minimal decomposition entropy is independent of the phase ϕ. We also numerically found that the the largest $S_1^{\text{RIU}}(\psi_{4a}^{\max}) = 1.213$ is attained for a state in class 4a with $\theta_1 = 3\pi/10$, $\theta_2 = 4\pi/15$ and $\theta_3 = 23\pi/60$. Note that this value is smaller than the one reported earlier [35] as the maximal for a random state with five components.

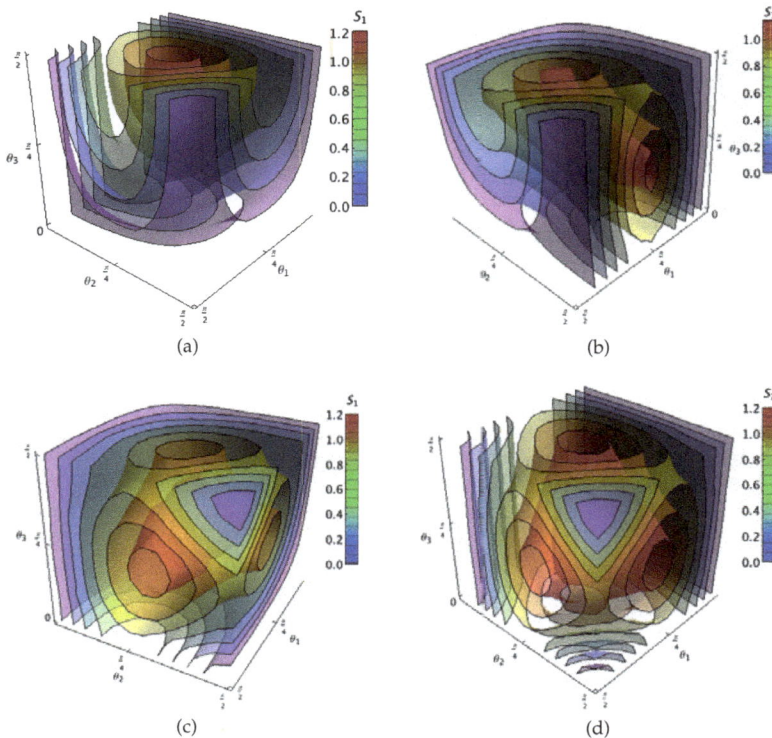

Figure 6. Surfaces of equal entanglement for classes 4 measured with respect the minimal decomposition entropy as function of the parameters θ_1, θ_2 and θ_3 defining each class. Different panels correspond to a state in: **(a)** Class 4a; **(b)** Class 4b; **(c)** Class 4c; **(d)** Class 4d.

4.2. The Maximum Overlap with an Entanglement Class

Given an ensemble of random states, a natural question arises: how many states of such ensemble belong to a particular Acín entanglement class? To tackle this question, observe first that numeric calculations imply $\langle S_0^{\mathrm{RIU}}(\psi) \rangle = \log 5$. Hence a generic three-qubit state has five non trivial components in the decomposition (4). As each class has at most four components, we rather consider the following quantity

$$\Lambda_i(\beta) = \max_{|\varphi\rangle, U_{\mathrm{local}}} \{ |\langle \varphi | U_{\mathrm{local}}^\dagger | \beta \rangle|^2 : |\varphi\rangle \in \text{Class } i \}, \tag{33}$$

where $i = \{1, 2a, 2b, 3a, 3b, 4a, 4b, 4c, 4d\}$ and $U_{\mathrm{local}} = U_1 \otimes U_2 \otimes U_3$. Such quantity provides an information, how much a given state $|\beta\rangle$ on $\mathcal{H}^{\otimes 3}$ differs from the closest state $|\varphi\rangle$ in the Acín entanglement class i [15]. Note that the quantity Λ_i can be interpreted as the maximal fidelity of a given state $|\beta\rangle$ with respect to the closest state belonging to the class i. In particular, if $i = 1$ the results are consistent with $S_\infty^{\mathrm{RIU}}(\beta)$ (see [35]) as this yields the maximum overlap with the closest separable state.

By taking a set of 10^5 random states in $\mathcal{H}^{\otimes 3}$, we get their projection Λ_i on each Acín class, tracking their hyperdeterminant $\mathrm{Hdet}(|\varphi\rangle)$, which is clearly invariant under local transformations. Then we perform a numerical optimization on the three parameters depicting a local transformation on each qubit (nine in total) together with the necessary coefficients depicting an arbitrary state in each class [15]. Finally, we also track the hyperdeterminant of such a state, $\mathrm{Hdet}(|\beta\rangle)$. With this information, we construct the corresponding distribution $\rho(\Lambda_i)$ of each projection i (33).

Numerical results are shown jointly in Figure 7. First, the line plot shows the value of $\rho(\Lambda_i)$ on the left axis versus the value of projection Λ_i on the horizontal axis. Superposed, a dispersion plot of the entire set of states being analyzed is shown in color. Each dot represents a random state located vertically on their projection value Λ_i and horizontally in its hyperdeterminant value $\mathrm{Hdet}(|\beta\rangle)$, which remains invariant under the local optimization procedure. Additionally, each dot is colored in agreement with the hyperdeterminant of the best class element $|\varphi\rangle$ obtained in the optimization. Colors are assigned from red for separable states to green for maximal genuine entanglement. This structure of the plot allows one to compare the closeness between $|\beta\rangle$ and $|\varphi\rangle$ in terms of genuine entanglement. Note the graph corresponding to class 4d has been omitted because it is equivalent to that of class 4c: All coefficients in the class are real, then by exchanging 0 and 1 in all qubits and swapping the qubits 1 and 3 we get the same state with local operations. Thus, the maximal overlap and the hyperdeterminant statistics do not change.

Note particularly how in the Figure 7a the closest class states have $\mathrm{Hdet}(|\varphi\rangle) = 0$ for some random states which have $\mathrm{Hdet}(|\beta\rangle)$ near from the highest value $\frac{1}{4}$ maintaining a closer distance $\Lambda_{4a} \approx 1$. The opposite phenomenon is also observed in Figure 7b,c,e,g where some class states with $\mathrm{Hdet}(|\varphi\rangle) \approx \frac{1}{4}$ (in green) are close to some random states with lower $\mathrm{Hdet}(|\beta\rangle)$ values. On the other hand, in Reference [24] the distribution of the fidelity between two random states has been computed analytically. However, in our case the problem becomes more complicated due to the optimization of the fidelity over all local unitaries.

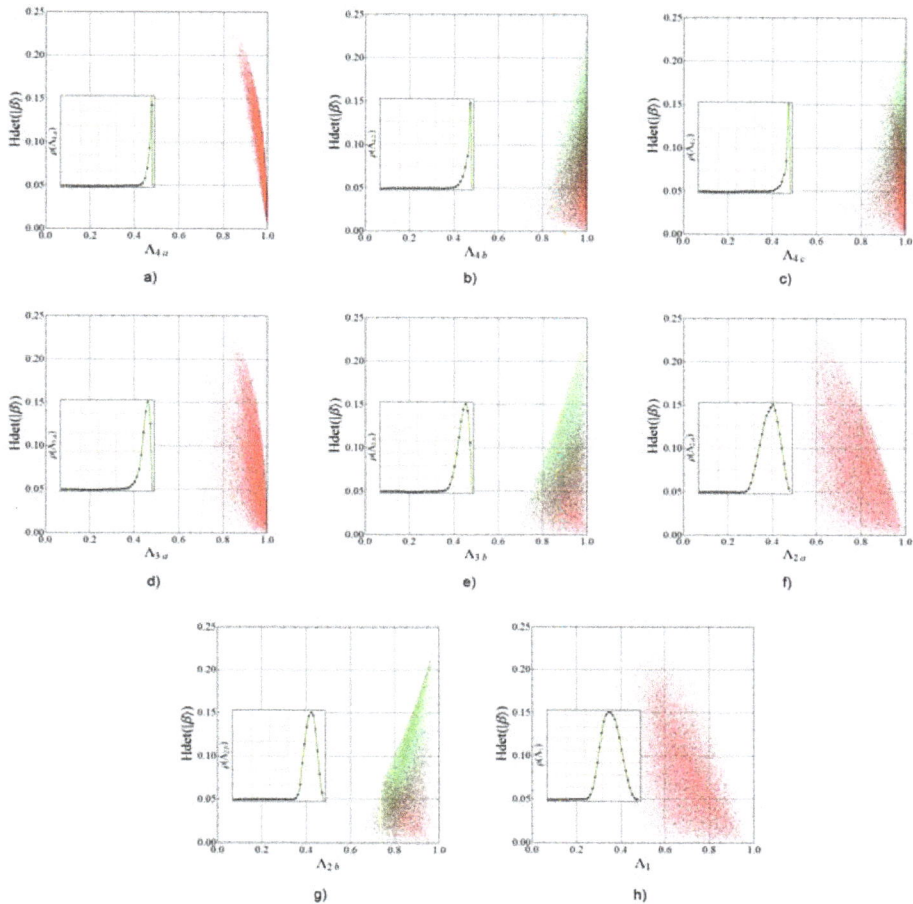

Figure 7. Dispersion graphs showing Hdet($|\beta\rangle$) versus the maximum overlap Λ_i, colored from red (separable) to green (maximal genuine entanglement). Each panel correspond to one of the Acín classes (see Table 2) as follows: (**a**) Class 4a; (**b**) Class 4b; (**c**) Class 4c; (**d**) Class 3a; (**e**) Class 3b; (**f**) Class 2a; (**g**) Class 2b; (**h**) Class 1. Besides, probability distributions of the maximum overlap (33) are shown in the inset of each plot (vertical scale on the left). We have taken an ensemble of 10^5 three-qubit random states. Graphs of classes 4c and 4d are equivalent so this last was omitted (see details in the core text).

5. The Entanglement Polytope of Three Qubits

Let $\lambda_k^{\min}$ denote the smallest eigenvalue of the reduced density matrix of the subsystem of three qubits, where $k = A, B, C$. The following set of compatibility conditions

$$\lambda_A^{\min} \leq \lambda_B^{\min} + \lambda_C^{\min}, \quad \lambda_B^{\min} \leq \lambda_A^{\min} + \lambda_C^{\min}, \quad \lambda_C^{\min} \leq \lambda_A^{\min} + \lambda_B^{\min}. \tag{34}$$

Form particular examples of polygon inequalities obtained by Higuchi et al. for systems of several qubits [17]. The smaller eigenvalue of a one-qubit system is not larger then $1/2$ so that $0 \leq \lambda_k^{\min} \leq 1/2$. Inequalities (34) determine jointly a convex polytope in the three-space $(\lambda_A^{\min}, \lambda_B^{\min}, \lambda_C^{\min})$. Its five vertices represent distinguished three-qubit states: Fully separable states are identified by the point

$SEP = (0,0,0)$ whereas points $A = (1/2, 1/2, 0)$, $B = (1/2, 0, 1/2)$ and $C = (0, 1/2, 1/2)$ stand for bi-separable states. The GHZ-state is located at $GHZ = (1/2, 1/2, 1/2)$. The convex hull of these points is known as the Kirwan polytope [21,47,48]. In addition, the identification of a state belonging to an entanglement classes reported in [20] is summarized in Table 2.

Consider now an ensemble of three-qubit random states. For such states, the probability distribution of the minimal eigenvalue of a single-particle reduced density matrix fulfils $P(\lambda_{\min}) = P(\lambda_A^{\min}) = P(\lambda_B^{\min}) = P(\lambda_C^{\min})$. Using the following relation between the two eigenvalues ϑ_1 and ϑ_2 of a single qubit reduced density matrix

$$\lambda_{\min} = \min(\vartheta_1, \vartheta_2) = \frac{1}{2}(\vartheta_1 + \vartheta_2) - \frac{1}{2}|\vartheta_1 - \vartheta_2|,$$

we can compute the probability distribution of the minimal eigenvalue $\lambda_{\min}$ as

$$P(\lambda_{\min}) = \int_0^1 \int_0^1 d\vartheta_1 d\vartheta_2 \mathcal{P}(\vartheta_1, \vartheta_2) \delta[\lambda_{\min} - (\vartheta_1 + \vartheta_2)/2 + |\vartheta_1 - \vartheta_2|/2)], \tag{35}$$

where $\mathcal{P}(\vartheta_1, \vartheta_2)$ is the joint density (7) and δ stands for the Dirac delta function. Performing the integral, we obtain

$$P(\lambda_{\min}) = 420[\lambda_{\min}(2\lambda_{\min} - 1)(1 - \lambda_{\min})]^2, \quad 0 \leq \lambda_{\min} \leq 1/2. \tag{36}$$

This distribution is depicted in Figure 8. Besides, a direct calculation yields the average value $\langle \lambda_{\min} \rangle = 29/128$. In general, the k-the moment of $\lambda_{\min}$ reads

$$\langle \lambda_{\min}^k \rangle = \frac{105}{2^k} \left[\frac{\Gamma(k+3)}{\Gamma(k+6)} - \frac{\Gamma(k+4)}{\Gamma(k+7)} + \frac{\Gamma(k+5)}{4\Gamma(k+8)} \right]. \tag{37}$$

Note that a given pure state can be identified with a point in the entanglement polytope. Its coordinates are $(\lambda_A^{\min}, \lambda_B^{\min}, \lambda_C^{\min})$. This is shown in Figure 8b for an ensemble of 10^6 three-qubit random states colored according to their joint probability distribution in the polytope. To compute such probability distribution, the space containing the whole polytope $[0, \frac{1}{2}]^{\times 3}$ was divided into 80^3 cubic cells. Then, we state the statistics of random states falling in each cell to get the probability density of those states (by volume unity). Note that the closer the points are to the faces, the lower the value of the distribution. In Figure 8c we depict a transverse cut by the plane containing the vertices S, C and GHZ to depict the distribution of the inner points. This shows that random states are more concentrated near the line joining the vertices SEP and GHZ, which corresponds to class 2a.

On the other hand, two quantum pure states attain the same amount of entanglement if they belong to the same class, that is to say if there is a finite probability of success that they can be converted into each other using stochastic local operations and classical communication, referred to as SLOCC by its acronyms. For the case of three qubits, there exist two SLOCC classes of entanglement: the one containing the GHZ state, which exhibits genuine entanglement and the W class [8].These classes can be distinguished from the entanglement polytope. Numerical calculation shows that around 6% of the states are placed in the upper polytope, so that they belong to the GHZ SLOCC class [21]. As the invariant I_6 discriminates between such classes in panel Figure 8d we show the ensemble of random states colored with respect to this invariant. For states placed near the bi-separable faces I_6 goes to zero, whereas the states landing in the GHZ simplex are characterized by a positive value of this invariant. An equivalent approach can be done dealing with the maximum eigenvalues of the reduced single qubit density matrices. For such a case, the joint probability distribution is known [49] and hence the fraction of random states in the GHZ pyramid was computed in Reference [50] yielding $13/216 \approx 6.02\%$ which is consistent with our numerical calculation.

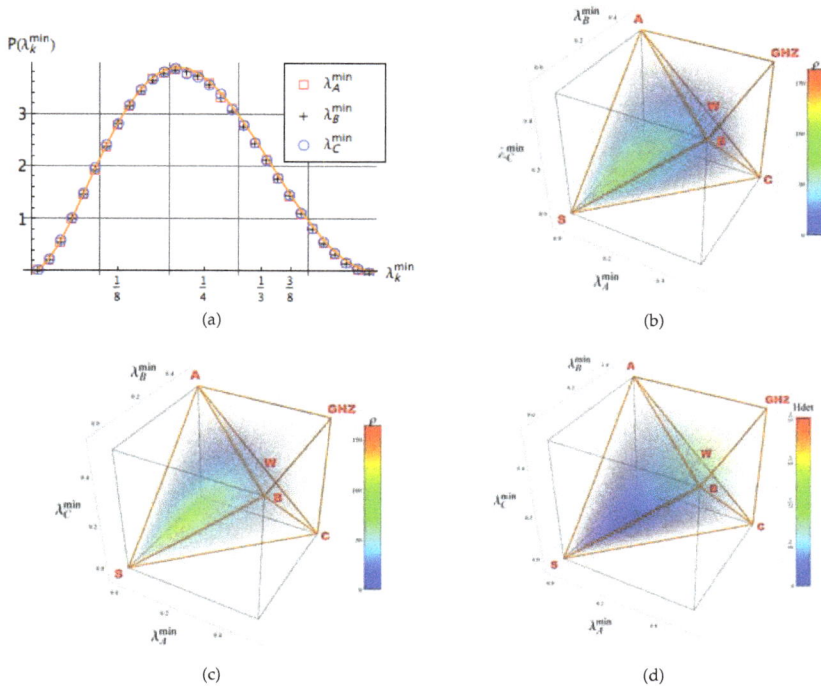

Figure 8. (**a**) Probability distribution of the minimal eigenvalue of a single qubit reduced system (36). (**b**) An ensemble 10^6 of three-qubit random states depicted in the entanglement polytope. The color scale stands for the joint probability distribution. (**c**) Detail of (**b**): A transversal section by the plane which contains the points S, C and GHZ. (**d**) The ensemble of three qubit random states labeled by colour settled according to the value of the invariant I_6.

6. Conclusions and Future Work

We studied various quantities describing a three-qubit pure quantum state and analyzed their probability distributions obtained for an ensemble of random pure states generated by the unitary invariant Haar measure. In particular, we investigated the distribution of the six parameters determining the five-terms decomposition (4) of a three-qubit state. The phase of the complex coefficient occurs to be uniformly distributed. The distributions of the amplitudes λ_0 and λ_4 differ from the distribution describing the remaining three coefficients. Interestingly, these two coefficients can be related with the degree of entanglement as the invariant J_4 depends only on them. In addition, we have also analyzed the probability distributions of two sets of polynomial invariants. The invariants I_1, I_2 and I_3 follow the same distribution. Thus, out of the five independent invariants, only three are necessary to characterize entanglement in three-qubit states. This fact is consistent with the second set of invariants reported by Acín et al. as the distributions of the invariants J_1, J_2 and J_3 do coincide. For each invariant its mean value was computed using symbolic integration with respect to the unitary invariant Haar measure. Moreover, we have also obtained the probability distribution of the anisotropic strength s_{ani} and the anisotropic volume V_{ani} introduced recently in [37]. These invariants are useful in the study of strong monogamy relations, geometric discord and fidelity of remote state preparation and studies of violation of the Bell inequality. In this last context, one could ask for the probability that one of the three pairs violates a Bell inequality. However, these results will be reported elsewhere.

On the other hand, the set of invariants $\{J_k\}$ allows us to identify certain entanglement classes, whose entanglement was described through the minimal decomposition entropy. Moreover, highly entangled states with respect to this measure were identified in each class. Our results imply that the more terms in the decomposition (4) of a three-qubit state, the larger its degree of entanglement measured by the minimal decomposition entropy.

The numerical outcomes provide us with several insights about possible meanings of the entanglement invariants. First, there is an apparent underlying statistical equivalence between coefficients λ_1, λ_2 and λ_3 (and their low values suggest a closer position of the states respecting genuine entanglement states in terms of the RIU entropy statistics for the overall 3-qubits random states). The same aspects seems true for I_2, I_3, I_4 and J_1, J_2, J_3 invariants. Together, larger values for λ_0 and low values for I_5''' and J_4 seem related with the presence of genuine entanglement (this affirmation is based on the fact that larger values of RIU entropy are statistically more common for the overall 3-qubit random states).

Other outcomes relative to the type a in the Acín classes exhibit separable states. There, the growing number of the class $(1, 2, ..., 4)$ reflects the inclusion of most of the random states for three qubits (see Figure 7a,d,f,h). In this sense, the use of RIU entropy as an exemplary measure of quantum entanglement allows us to provide a classification of three qubit states and to describe their hierarchy. The invariants with respect to local transformations are useful to identify certain types of entangled structures in the entire system. As shown in Figure 7, the states displaying genuine entanglement appear closer from other states in the classes with no genuine entanglement. Although smooth measures of entanglement depend on the state in a continuous way, a small variation of a state can lead to a considerable change of its entanglement. This feature was observed in larger systems [51]. In such a scenario, the current analysis in the quest of understanding the hierarchy of entanglement, could set directions to transform states from maximally entangled into separable ones. By using the $SU(2)$ decomposition procedure, [52] has been clear about the existence of basic $U(1) \times SU(2)$ operations among entangled pairs, showing how the entanglement phenomena can be generated in a structured way form basic operations then transiting from separable to genuine entangled states. This suggests that programmed local operations combined with entangling operations between two previous entangled pairs can be realized in order to connect such state types. Thus, basic separable states could be transformed into maximal entangled states as $|GHZ\rangle$ and $|W\rangle$ only with a series of such operations. In a more ambitious task, those single types of operations could suggest they could be responsible for the transit from certain classes to others among the hierarchies of entanglement. In such a process, the track in the change of the invariants values could provide a strong road-map for such transit.

Finally, we have analyzed the probability distribution of the maximal fidelity of a random state with respect to the closest representative of each entanglement class. The highest maximal fidelity is obtained for classes 4a–d listed in Table 2. This can be seen from the fact that the distributions of five coefficients in the decomposition (4) are highly non-trivial, as these quantities carry some information concerning the degree of entanglement. Our study comprises several ways to analyze the entanglement in a three-qubit system showing the fact that entanglement can be characterized from different approaches, each one providing different aspects of non-locality. Therefore, we hope the results of this work will shed some light on the matter.

Author Contributions: M.E. and F.D. performed the numerical calculations. M.E. obtained the analytical results. K.Ż. proposed the main idea of Section 5. All authors contributed equally to write the paper.

Funding: This research received no external funding.

Acknowledgments: The support of Escuela de Ingeniería y Ciencias of Tecnológico de Monterrey as well as the support of CONACyT are gratefully acknowledged. K.Ż. acknowledges support by Narodowe Centrum Nauki under the grant number DEC-2015/18/A/ST2/00274. We are obliged to Michael J.W. Hall for drawing our attention to the reference [37].

Conflicts of Interest: The authors declare no conflict of interest.

References

1. Walter, M.; Gross, D.; Eisert, J. Multi-partite entanglement. *arXiv* **2016**, arXiv:1612.02437.
2. Bengtsson, I.; Życzkowski, K. *Geometry of Quantum States: An Introduction to Quantum Entanglement*, 2nd ed.; Cambridge University Press: Cambridge, UK, 2017.
3. Gurvits, L. Classical complexity and quantum entanglement. *J. Comput. Syst. Sci.* **2004**, *69*, 448–484. [CrossRef]
4. Horodecki, R.; Horodecki, P.; Horodecki, M.; Horodecki, K. Quantum entanglement. *Rev. Mod. Phys.* **2009**, *81*, 865–942. [CrossRef]
5. Enríquez, M.; Wintrowicz, I.; Życzkowski, K. Maximally entangled multipartite states: A brief survey. *J. Phys. Conf. Ser.* **2016**, *698*, 012003. [CrossRef]
6. Higuchi, A.; Sudbery, A. How entangled can two couples get? *Phys. Lett. A* **2000**, *273*, 213–217. [CrossRef]
7. Carteret, H.A.; Higuchi, A.; Sudbery, A. Multipartite generalisation of the Schmidt decomposition. *J. Math. Phys.* **2000**, *41*, 7932. [CrossRef]
8. Dür, W.; Vidal, G.; Cirac, J.I. Three qubits can be entangled in two inequivalent ways. *Phys. Rev. A* **2000**, *62*, 062314. [CrossRef]
9. Meill, A.; Meyer, D.A. Symmetric three-qubit-state invariants. *Phys. Rev. A* **2017**, *96*, 062310. [CrossRef]
10. Verstraete, F.; Dehaene, J.; De Moor, B.; Verschelde, H. Four qubits can be entangled in nine different ways. *Phys. Rev. A* **2002**, *65*, 052112. [CrossRef]
11. Albeverio, S.; Fei, S. A note on invariants and entanglements. *J. Opt. B* **2011**, *3*, 223. [CrossRef]
12. Grassl, M.; Rötteler, M.; Beth, T. Computing local invariants of qubit systems. *Phys. Rev. A* **1998**, *58*, 1833. [CrossRef]
13. Sudbery, A. On local invariants of pure three-qubit states. *J. Phys. A Math. Gen.* **2001**, *34*, 643–652. [CrossRef]
14. Holweck, F.; Luque, J.; Thibon, J. Entanglement of four qubit systems: A geometric atlas with polynomial compass I (the finite world). *J. Math. Phys.* **2014**, *55*, 012202. [CrossRef]
15. Acín, A.; Andrianov, A.; Jané, E.; Tarrach, R. Three-qubit pure-state canonical forms. *J. Phys. A* **2001**, *34*, 6725–6739. [CrossRef]
16. Sawicki, A.; Walter, M.; Kuś, M. When is a pure state of three qubits determined by its single-particle reduced density matrices? *J. Phys. A* **2013**, *46*, 055304. [CrossRef]
17. Higuchi, A.; Sudbery, A.; Szulc, J. One-qubit reduced states of a pure many-qubit state: Polygon inequalities. *Phys. Rev. Lett.* **2003**, *90*, 107902. [CrossRef] [PubMed]
18. Bravyi, S. Requirements for compatibility between local and multipartite quantum states. *Quantum Inf. Comp.* **2004**, *4*, 12–26.
19. Klyachko, A. Quantum marginal problem and representations of the symmetric group. *arXiv* **2004**, arXiv:quant-ph/0409113.
20. Han, Y.J.; Zhang, Y.S.; Guo, G.C. Compatible conditions, entanglement, and invariants. *Phys. Rev. A* **2004**, *70*, 042309. [CrossRef]
21. Walter, M.; Doran, B.; Gross, D.; Christandl, M. Entanglement polytopes: Multiparticle entanglement from single-particle information. *Science* **2013**, *340*, 1205–1208. [CrossRef] [PubMed]
22. Kuś, M.; Mostowski, J.; Haake, F. Universality of eigenvector statistics of kicked tops of different symmetries. *J. Phys. A Math. Gen.* **1988**, *21*, L1073–L1077. [CrossRef]
23. Haake, F. *Quantum Signatures of Chaos*, 2nd ed.; Springer Verlag: Berlin, Germany, 2001.
24. Życzkowski, K.; Sommers, H.-J. Average fidelity between random quantum states. *Phys. Rev. A* **2005**, *71*, 032313. [CrossRef]
25. Giraud, O.; Žnidarič, M.; Georgeot, B. Quantum circuit for three-qubit random states. *Phys. Rev. A* **2009**, *80*, 042309. [CrossRef]
26. Kendon, V.M.; Życzkowski, K.; Munro, W.J. Bounds on entanglement in qudit subsystems. *Phys. Rev. A* **2002**, *66*, 062310. [CrossRef]
27. Cappellini, V.; Sommers, H.-J.; Życzkowski, K. Distribution of G concurrence of random pure states. *Phys. Rev. A* **2006**, *74*, 062322. [CrossRef]
28. Kumar, S.; Pandey, A. Entanglement in random pure states: Spectral density and average von Neumann entropy. *J. Phys. A* **2011**, *44*, 445301. [CrossRef]

29. Vivo, P.; Pato, M.P.; Oshanin, G. Random pure states: Quantifying bipartite entanglement beyond the linear statistics. *Phys. Rev. E* **2016**, *93*, 052106. [CrossRef] [PubMed]

30. Kendon, V.; Nemoto, V.K.; Munro, W. Typical entanglement in multiple-qubit systems. *J. Mod. Opt.* **2002**, *49*, 1709–1716. [CrossRef]

31. Facchi, P.; Florio, G.; Pascazio, S. Probability-density-function characterization of multipartite entanglement. *Phys. Rev. A* **2006**, *74*, 042331. [CrossRef]

32. Korzekwa, K.; Lostaglio, M.; Jennings, D.; Rudolph, T. Quantum and classical entropic uncertainty relations. *Phys. Rev. A* **2014**, *89*, 042122. [CrossRef]

33. Fannes, M. Multi-state correlations and fidelities. *Int. J. Geom. Methods Mod. Phys.* **2012**, *9*, 1260021. [CrossRef]

34. Rangamani, M.; Rota, M. Entanglement structures in qubit systems. *J. Phys. A* **2015**, *48*, 385301. [CrossRef]

35. Enríquez, M.; Puchała, Z.; Życzkowski, K. Minimal Rényi-Ingarden-Urbanik entropy of multipartite quantum states. *Entropy* **2015**, *17*, 5063–5084. [CrossRef]

36. Alsina, D. Multipartite Entanglement and Quantum Algorithms. Ph.D. Thesis, Universitat de Barcelona, Barcelona, Spain, 2017.

37. Cheng, S.; Hall, M.J.W. Anisotropic invariance and the distribution of quantum correlations. *Phys. Rev. Lett.* **2017**, *118*, 010401. [CrossRef] [PubMed]

38. Grendar, M. Entropy and effective support size. *Entropy* **2006**, *8*, 169–174. [CrossRef]

39. Carteret, H.A.; Linden, N.; Popescu, S.; Sudbery, A. Multi-particle entanglement. *Found. Phys.* **1999**, *29*, 527–552. [CrossRef]

40. Acín, A.; Andrianov, A.; Costa, L.; Jané, E.; Latorre, J.; Tarrach, R. Generalized Schmidt decomposition and classification of three-quantum-bit states. *J. Phys. Lett.* **2000**, *85*, 1560–1563. [CrossRef] [PubMed]

41. Rényi, A. On measures of information and entropy. In Proceedings of the Fourth Berkeley Symposium on Mathematics, Statistics and Probability, Berkeley, CA, USA, 20 June–30 July 1960.

42. Puchała, Z.; Miszczak, J.A. Symbolic integration with respect to the Haar measure on the unitary groups. *Bull. Pol. Acad. Sci. Tech. Sci.* **2017**, *65*, 21. [CrossRef]

43. Życzkowski, K.; Sommers, H.-J. Induced measures in the space of mixed quantum states. *J. Phys. A* **2001**, *34*, 7111–7125. [CrossRef]

44. Kempe, J. Multiparticle entanglement and its applications to cryptography. *Phys. Rev. A* **1999**, *60*, 910–916. [CrossRef]

45. Osterloh, A. Classification of qubit entanglement: $SL(2, \mathbb{C})$ versus $SU(2)$ invariance. *Appl. Phys. B* **2010**, *98*, 609–616. [CrossRef]

46. Maziero, J. Understanding von Neumann entropy. *Rev. Bras. Ensino Fís.* **2015**, *37*, 1314. [CrossRef]

47. Sawicki, A.; Oszmaniec, M.; Kuś, M. Critical sets of the total variance can detect all stochastic local operations and classical communication classes of multiparticle entanglement. *Phys. Rev. A* **2012**, *86*, 040304. [CrossRef]

48. Sawicki, A.; Oszmaniec, M.; Kuś, M. Convexity of momentum map, Morse index, and quantum entanglement. *Rev. Math. Phys.* **2014**, *26*, 1450004. [CrossRef]

49. Christandl, M.; Doran, B.; Kousidis, S.; Walter, M. Eigenvalue distributions of reduced density matrices. *Commun. Math. Phys.* **2014**, *332*, 1–52. [CrossRef]

50. Zhao, Y.Y.; Grassl, M.; Zeng, B.; Xiang, G.Y.; Zhang, C.; Li, C.F.; Guo, G.C. Experimental detection of entanglement polytopes via local filters. *Quantum Inf.* **2017**, *3*, 11. [CrossRef]

51. Delgado, F. Assembling large entangled states in the Rényi-Ingarden-Urbanik entropy measure under the $SU(2)$-dynamics decomposition for systems built from two-level subsystems. In Proceedings of the 6th Annual International Conference on Physics, Athens, Greece, 23–26 July 2018.

52. Delgado, F. $SU(2)$ decomposition for the Quantum Information Dynamics in 2d-Partite Two-Level Quantum Systems. *Entropy* **2018**, *20*, 610. [CrossRef]

![entropy logo] **entropy**

MDPI

Article

New Entropic Inequalities and Hidden Correlations in Quantum Suprematism Picture of Qudit States [†]

Margarita A. Man'ko [1,*] and Vladimir I. Man'ko [1,2,3]

[1] Lebedev Physical Institute, Russian Academy of Sciences, Leninskii Prospect 53, Moscow 119991, Russia
[2] Moscow Institute of Physics and Technology (State University), Institutskii per. 9, Dolgoprudnyi, Moscow Region 141700, Russia; manko@sci.lebedev.ru
[3] Department of Physics, Tomsk State University, Lenin Avenue 36, Tomsk 634050, Russia
[*] Correspondence: mmanko@sci.lebedev.ru
[†] We dedicate this article to the memory of Professor E. C. G. Sudarshan, the Great Scientist and our friend and colleague with whom we were happy to collaborate during many years.

Received: 17 August 2018; Accepted: 7 September 2018; Published: 11 September 2018

![check for updates]

Abstract: We study an analog of Bayes' formula and the nonnegativity property of mutual information for systems with one random variable. For single-qudit states, we present new entropic inequalities in the form of the subadditivity and condition corresponding to hidden correlations in quantum systems. We present qubit states in the quantum suprematism picture, where these states are identified with three probability distributions, describing the states of three classical coins, and illustrate the states by Triada of Malevich's squares with areas satisfying the quantum constraints. We consider arbitrary quantum states belonging to N-dimensional Hilbert space as $(N^2 - 1)$ fair probability distributions describing the states of $(N^2 - 1)$ classical coins. We illustrate the geometrical properties of the qudit states by a set of Triadas of Malevich's squares. We obtain new entropic inequalities for matrix elements of an arbitrary density $N \times N$-matrix of qudit systems using the constructed maps of the density matrix on a set of the probability distributions. In addition, to construct the bijective map of the qudit state onto the set of probabilities describing the positions of classical coins, we show that there exists a bijective map of any quantum observable onto the set of dihotomic classical random variables with statistics determined by the above classical probabilities. Finally, we discuss the physical meaning and possibility to check derived inequalities in the experiments with superconducting circuits based on Josephson junction devices.

Keywords: entropy; correlations; qubits; probability representation; Bayes' formula

1. Introduction

Generic states of quantum systems are identified with the density matrices [1,2] or the density operators $\hat{\rho}$ acting in a Hilbert space. The pure states of quantum systems are identified with the state vectors $\mid \psi \rangle$ belonging to the Hilbert space [3] and complex wave functions [4,5] $\psi(x) = \langle x \mid \psi \rangle$, where x is an observable, e.g., the continuous position of a particle. The physical meaning of the wave function $\psi(x)$ is related to measuring the observable x; in the state $\mid \psi \rangle$, the measurement of the position of a particle yields the probability density $|\langle x \mid \psi \rangle|^2 = |\psi(x)|^2$, which does not contain information on the phase of the complex wave function.

For spin-s systems with discrete observables like spin projections $m = -s, -s + 1, \ldots, s - 1, s$; $s = 0, 1/2, 1, 3/2, 2, \ldots$, the state vectors belong to the Hilbert space of finite dimension $N = 2s + 1$, and the complex wave function $\psi(m) = \langle m \mid \psi \rangle$ determines the probability distribution $|\langle m \mid \psi \rangle|^2 = |\psi(m)|^2$ associated with the state $\mid \psi \rangle$. The phase of the wave function is not determined by the probability distribution; in view of this fact, information on the state $\mid \psi \rangle$, contained in the probability density $|\psi(x)|^2$ or in the probability distribution $|\psi(m)|^2$, is not sufficient to describe the particle's pure state or the spin-s pure state.

The aim of this paper is to consider the old problem of looking for a such formulation of quantum mechanics, where the system states can be identified with fair probability distributions of measurable observables only. Such a possibility is based on quantum tomography methods of measuring [6] quantum states, using the formalism of reconstructing [7,8] the state Wigner function by means of Radon transform [9].

Wigner introduced the Wigner function [10] $W(q, p)$ of the position q and momentum p that is similar to the probability density $f(q, p)$ describing the classical particle state in the presence of fluctuations. The Wigner function can take negative values and, due to this circumstance, it is called the quasidistribution function. The Wigner function is related to the density matrix $\rho(x, x')$ of the quantum particle state by an invertible Fourier transform and contains the same information on the state as the density matrix.

There exist other analogous quasidistributions like the Husimi–Kano Q-function [11,12] and Glauber–Sudarshan function [13,14], which are functions on the phase space. The suggestion to identify quantum states with fair probability densities was presented in [15], where the probability density, called symplectic tomogram, was used. An analogous approach was elaborated for spin states in [16,17], where the spin tomograms, being fair probability distributions of spin projections m on an arbitrary direction in the space given by a unit vector $\vec{n}$, were shown to determine the density matrix $\rho(m, m')$.

In this paper, on the example of qubits, we show the bijective map of density operators of spin-1/2 states onto the probability distributions. Since for probability distributions the notion of Shannon entropy [18], relative entropy, and Tsallis entropy [19] is the standard tool to characterize the statistical properties of the systems, we obtain, in view of the map introduced, some new relations like entropic equalities and inequalities for quantum spin states. Other kinds of entropies also exist like Rényi entropy [20], non-Shannonian and generalized (c, d) entropies; see, e.g., [21,22]. In this paper, we consider new relations connected with Shannon and Tsallis entropies. In addition, we discuss new geometric interpretation of spin-1/2 (qubit) states in terms of the Triada of Malevich's squares [23,24] and its relation to the Bloch sphere geometry of these states. Employing the identification of qubit states with probability distributions, we present the construction of quantum observables (Hermitian 2×2-matrices) in terms of sets of classical-like variables and provide the bijective map of the qubit states (density matrices) and observables onto classical-coin probability distributions and classical observables associated with these coins.

We present the evolution equations for the density matrices of qubit states in the form of kinetic equations for probability distributions determining the qubit states. We formulate the superposition principle of qubit state vectors as a new addition rule for the probabilities determining the states. In addition, we express the Born rule for calculating the probability $|\langle \psi_1 | \psi_2 \rangle|^2 = w_{12}$ as a function of probabilities determining the pure states $| \psi_1 \rangle$ and $| \psi_2 \rangle$. Then, we extend the probability representation of qubit states and express the matrix elements of an arbitrary density $N \times N$-matrix in terms of classical-coin probability distributions. We consider in detail examples of qutrit (spin-1), identifying the qutrit state with a set of Triadas of Malevich's squares. We present new relations of areas of Malevich's squares and the possibility of checking these relations in the experiments with superconducting circuits.

The other goal of this work is to study within the probability representation of quantum states [15–17,25–28] (reviewed in [29]) the triangle geometry of qudit states and discuss Bayes' formula for systems without subsystems and correlations (called the hidden correlations) in such systems. It is worth noting that the classical probability distributions were discussed within the framework of state vectors for spin-1/2 systems by Khrennikov [30–32] and the superposition principle for spin-1/2 states was expressed as the nonlinear superposition of classical probability distributions in [33,34]. Malevich's squares and the approach called the suprematism in art are described in [35].

This paper is organized as follows.

We present the notion of random variables in Section 2 and study Bayes' formula for systems with one random variable in Section 3. We discuss qubit states in Section 4 and consider classical-coin random variables for qubit systems in Section 5. Then, we review the notion of quantum suprematism in Section 6 and study qutrit states in the probability representation in Section 7. We devote Section 8 to the superposition principle for the probabilities, demonstrating this principle on the example of qutrits. Within the framework of the probability representation, we formulate the superposition principle for qudit states in Section 9. Finally, in Section 10, we provide the conclusions and perspectives.

2. Random Variables and Probabilities

In probability theory, the notion of random variables and probability distributions were discussed using rigorous approaches presented, for example, in [36–39]. We employ here the following empiric approach. We define the relation of random variables to sets of integer numbers following [40,41]. Given a set of N different events, these events are associated with integers $j = 1, 2, \ldots, N$. We call relative frequencies $P(j)$ of the realization of these random events in a series of experiments "the probabilities of the events" where $0 \leq P(j) \leq 1$. The function $P(j)$ is the probability distribution; it is normalized $\sum_{j=1}^{N} P(j) = 1$.

The properties of the events are characterized by some functions $f(j)$, which we call observables. In this approach, random variables are mapped onto the integers $j = 1, 2, \ldots, N$. The physical meaning of the events can be different; for example, in the casino roulette, the event is the appearance of some integer number j which is chosen from a set of integer numbers located between 1 and N. The event may be also considered as positions "UP" and "DOWN" of two coins; in this case, the integer number j is mapped onto a pair (a, b) of integer numbers labeling the position of each coin. In both cases, the relative frequencies of the events can be associated with the integer j, but the interpretation of this random variable is different. In the case of casino roulette, we say about one random variable, and in the case of two coins, we have two random variables associated with labeling positions of two coins by other two integer numbers (a, b). An analogous approach to random events can be employed in quantum mechanics.

We extend the above approach to classical probabilities using in this case the identification of random events with the integers $1 \leq j, j' \leq N$ labeling the matrix elements of the density matrix $\rho_{jj'}$ determining the states, e.g., of qudit with spin s, where $N = 2s + 1$, or of the N-level atom. The physical observables are given by the Hermitian matrices $f_{jj'}$, where indices of rows and columns are identified with the random variables $1 \leq j, j' \leq N$. It is important that we can interpret the described above association of integers j analogously to the case of classical casino roulette and the case of two classical coins considering numerically the same density matrices $\rho_{jj'}$ either as the density matrices of noncomposite (nondivisible) systems (an analog of the casino roulette) or as the density matrices of bipartite systems (an analog of the states of two coins).

In the next section, we consider Bayes' formula, in view of the approach under discussion, using it for one random variable and applying the map of integer numbers $1, 2, \ldots, N$ onto pairs of random numbers.

3. Bayes' Formula for the Probability Distribution of One Random Variable

In this section, we discuss the application of Bayes' formula available for probability distributions of several random variables to the case of the probability distribution of one random variable.

First, we recall Bayes' formula and the notion of conditional probability distribution for statistics of two random variables. Given the function $1 \geq P(j, k) \geq 0$, where $j = 1, 2, \ldots, n_1$, $k = 1, 2, \ldots, n_2$, and $n_1 n_2 = N$, with the normalization condition

$$\sum_{j=1}^{n_1} \sum_{k=1}^{n_2} P(j, k) = 1. \tag{1}$$

This function is identified with the probability distribution of two random variables j and k. The marginal probability distributions

$$P_1(j) = \sum_{k=1}^{n_2} P(j,k), \qquad P_2(k) = \sum_{j=1}^{n_1} P(j,k) \tag{2}$$

determine the statistical properties of each random variable.

The conditional probability distribution of the first random variable j for given k is presented by the formula; see [36],

$$P(j \mid k) = \frac{P(j,k)}{P_2(k)}, \tag{3}$$

which means that

$$P(j,k) = P_2(k)P(j \mid k). \tag{4}$$

For the case of joint probability distributions of two random variables describing the statistics of the bipartite system, these relations correspond to Bayes' formula connecting marginal probability distributions and conditional probability distributions of these random variables. In Appendix A, we present an example of application of the above formulas for a particular case $N = 4$. In view of the example from Appendix A, we are in the position to formulate the rule for introducing Bayes' formula for the probability distribution $\mathcal{P}(n)$; $N = n_1, n_2$ of one random variable. We apply the map of integers n onto pairs of integers j and k, such that $j = 1, 2, \ldots, n_1$ and $k = 1, 2, \ldots, n_2$. Then, for marginal probability distributions and conditional probability distributions, we use the known expression for joint probability distribution of two variables and define these distributions, in view of the invertible map of integers $1, 2, \ldots, N \leftrightarrow (1,1), (2,1), \ldots, (n_1,1)(1,2), (2,2), \ldots, (n_1,2), \ldots, (n_1,n_2)$, where $N = n_1 n_2$. This map can be described by the functions discussed in [42–44].

Following [44], we determine the functions $y(x_1, x_2)$, $x_1(y)$, and $x_2(y)$, where $1 \leq x_1 \leq X_1$, $1 \leq x_2 \leq X_2$, and $1 \leq y \leq N = X_1 X_2$, as

$$y(x_1, x_2) = x_1 + (x_2 - 1)X_1, \tag{5}$$

$$x_1(y) = y \bmod X_1, \quad 1 \leq y \leq N, \tag{6}$$

$$x_2(y) - 1 = \frac{y - x_1(y)}{X_1} \bmod X_2, \quad 1 \leq y \leq N. \tag{7}$$

We use these functions for representing the probability distribution of one random variable as a joint probability distribution of two random variables. To do this, we introduce in Equations (5)–(7) the following notation: $y \equiv n$, $x_1 \equiv j$, $x_2 \equiv k$, $X_1 \equiv n_1$, $X_2 \equiv n_2$, $N = n_1 n_2 = X_1 X_2$, $n = 1, 2, \ldots, N$, $P(j,k) \equiv y(x_1, x_2)$, and $f(y) = \mathcal{P}(n)$. In the case of $N = 4$ and $n_1 = n_2 = 2$, the map introduced just provides the relations $\mathcal{P}(1) = P(1,1)$, $\mathcal{P}(2) = P(2,1)$, $\mathcal{P}(3) = P(1,2)$, and $\mathcal{P}(4) = P(2,2)$ discussed above. Nevertheless, the functions introduced describe the invertible map of the probability distribution of one random variable $\mathcal{P}(n)$; $n = 1, 2, \ldots, N$ onto the joint probability distribution $P(j,k)$ of two random variables $j = 1, 2, \ldots, n_1$ and $k = 1, 2, \ldots, n_2$, with $N = n_1 n_2$, for arbitrary integers n_1 and n_2. In our new notation, $n = n(j,k)$, $j = j(n)$, and $k = k(n)$. Taking into account this discussion, we introduce Bayes' formula for the probability distribution $\mathcal{P}(n)$ of one random variable; it reads

$$P(j(n) \mid k(n)) = \frac{\mathcal{P}(n(j,k))}{\sum_{j=1}^{n} \mathcal{P}(n(j,k))}, \qquad n = n_1 n_2, \tag{8}$$

where functions $n(j,k)$, $j(n)$, and $k(n)$ are constructed in [44].

The relation of the joint probability distribution $P(j,k)$ to the marginal probability distributions corresponds to the presence of correlations in the system with two random variables. Since we introduced an analog of two random variables and their marginal and conditional probability distributions, the relation of these distributions reflect correlations, which we called [45] the hidden correlations for systems without subsystems. Such correlations exist for both classical and quantum systems.

Bayes' formula can also be considered for the probability distribution of one random variable $\mathcal{P}(n)$, if the integer $n = 1, 2, \ldots, N$ and $N = n_1 n_2 n_3$, where n_1, n_2, and n_3 are integers of the joint probability distribution of three random variables $P(j, k, l)$, with $j = 1, 2, \ldots, n_1$, $k = 1, 2, \ldots, n_2$, and $l = 1, 2, \ldots, n_3$. To do this, we use an analogous invertible map [44] of functions $y(x_1, x_2, x_3)$, $x_1(y)$, $x_2(y)$, and $x_3(y)$, taking integer values $y = 1, 2, \ldots, N = X_1 X_2 X_3; 1 \leq x_i \leq X_i; i = 1, 2, 3$, defined by the relations

$$y(x_1, x_2, x_3) = x_1 + (x_2 - 1)X_1 + (x_3 - 1)X_1 X_2, \tag{9}$$

$$x_1(y) = y \bmod X_1, \tag{10}$$

$$x_2(y) - 1 = \frac{y - x_1(y)}{X_1} \bmod X_2, \tag{11}$$

$$x_3(y) - 1 = \frac{y - x_1(y) - x_2(y) X_1}{X_1 X_2} \bmod X_3. \tag{12}$$

After substitution $x_1 \equiv j$, $x_2 \equiv k$, $x_3 \equiv l$, and $y \equiv n$, we arrive at an analog of Bayes' formula for one random variable

$$P(j(n) \mid k(n), l(n)) = \frac{P(j(n), k(n), l(n))}{\sum_{j=1}^{n_1} P(n(j, k, l))}. \tag{13}$$

To illustrate this formula, we consider the example of $N = 8 = 2 \cdot 2 \cdot 2$, i.e., $n_1 = n_2 = n_3 = 2$; the map of integers reads $1 \leftrightarrow (1, 1, 1)$, $2 \leftrightarrow (2, 1, 1)$, $3 \leftrightarrow (1, 2, 1)$, $4 \leftrightarrow (2, 2, 1)$, $5 \leftrightarrow (1, 1, 2)$, $6 \leftrightarrow (2, 1, 2)$, $7 \leftrightarrow (1, 2, 2)$, $8 \leftrightarrow (2, 2, 2)$. This means that the probability distribution $\mathcal{P}(n)$ takes the values $\mathcal{P}(1) \equiv P(1, 1, 1)$, $\mathcal{P}(2) \equiv P(2, 1, 1)$, $\mathcal{P}(3) \equiv P(1, 2, 1)$, $\mathcal{P}(4) \equiv P(2, 2, 1)$, $\mathcal{P}(5) \equiv P(1, 1, 2)$, $\mathcal{P}(6) \equiv P(2, 1, 2)$, $\mathcal{P}(7) \equiv P(1, 2, 2)$, and $\mathcal{P}(8) \equiv P(2, 2, 2)$. The joint probability distribution $P(j, k, l)$ has the values given by numbers $\mathcal{P}(n)$, and Bayes' formula obtained provides, e.g., the conditional probability

$$P(j(n) = 1 \mid k(n) = 1, l(n) = 1) = \frac{P(1, 1, 1)}{P(1, 1, 1) + P(2, 1, 1)} = \frac{\mathcal{P}(1)}{\mathcal{P}(1) + \mathcal{P}(2)}.$$

In the quantum case, the map of integers discussed provides a tool to consider the density matrix of qudit state $\rho_{nn'}$, where $n, n' = 1, 2, \ldots, N$, as the density matrix of a multipartite system. For example, at $N = 4$, the ququart density matrix can be interpreted as the density matrix of two two-level atoms, using the map discussed. In fact, if $n, n' = 1, 2, 3, 4$, we consider the density matrix as $\rho_{nn'} \equiv \rho_{jk, j'k'}$, where $j, j', k, k' = 1, 2$. Formally, we obtain the density matrix of the two-qubit system, which has the same numerical matrix elements that the 4×4-matrix $\rho_{nn'}$; this means that all numerical properties of the density matrix of two-qubit state and ququart state are identical.

This fact provides the possibility to consider formal entanglement properties of ququart system. For example, if we consider the pure state, $(\rho^2)_{nn'} = \rho_{nn'}$, then the properties of linear entropy $S = 1 - \mathrm{Tr}(\rho^2(1))$, with $(\rho(1))_{jj'} = \sum_{k=1}^{2} \rho_{jk, j'k}$, where the indices j, k, j' are determined by the numbers $n, n' = 1, 2, 3, 4$ according to the discussed map, characterize the entanglement degree in the bipartite system.

For the four-level atom, one has the same numerical characteristics. From the viewpoint of the matrix properties, the ququart state with the density matrix, having only different from zero matrix elements $\rho_{11} = \rho_{14} = \rho_{41} = \rho_{44} = 1/2$, provides the linear entropy $S = 1/2$ corresponding to maximum entangled state of two qubits. The interpretation of this phenomenon for systems without subsystems is the presence of hidden correlations in the degrees of freedom of such systems, formally analogous to quantum correlations associated with the entanglement phenomenon, e.g., in bipartite systems of two qubits.

4. Probability Representation of Spin-1/2 States

We start our introduction of the probability representation of quantum system states with the consideration of spin-1/2 systems. These systems realize qubits and their states, as well as they are realized by two-level atom systems. In standard formulation of quantum mechanics, the spin-1/2 pure states are described by Pauli spinors, which are complex vectors $\mid \psi \rangle$ with two components, i.e., $\mid \psi \rangle = \begin{pmatrix} \varphi_1 \\ \varphi_2 \end{pmatrix}$. The vectors $\mid \psi \rangle$ belong to the two-dimensional Hilbert space $\mathcal{H}$ with the scalar product

$$\langle \psi^{(1)} \mid \psi^{(2)} \rangle = \varphi_1^{(1)*} \varphi_1^{(2)} + \varphi_2^{(1)*} \varphi_2^{(2)}. \tag{14}$$

The state vectors are normalized $\langle \psi \mid \psi \rangle = 1$, and $|\varphi_1|^2 + |\varphi_2|^2 = 1$. The density operators [1,2] of the pure states $\rho_\psi = \mid \psi \rangle \langle \psi \mid$ in matrix form read $\rho_\psi = \begin{pmatrix} \varphi_1^* \varphi_1 & \varphi_1^* \varphi_2 \\ \varphi_2^* \varphi_1 & \varphi_2^* \varphi_2 \end{pmatrix}$. This matrix has the properties of Hermiticity $\rho_\psi^\dagger = \rho_\psi$ and nonnegativity $\rho_\psi \geq 0$, as well as it has the unit trace $\mathrm{Tr}\, \rho_\psi = 1$.

The physical meaning of the state-vector components φ_1 and φ_2 and matrix elements of the density matrix $\rho(\psi)$ is determined by the relation of these values to operators of physical observables associated with spin projection operators $\hbar \sigma_x/2$, $\hbar \sigma_y/2$, and $\hbar \sigma_z/2$ onto the axes x, y, and z, respectively. Here, the Pauli matrices σ_x, σ_y, and σ_z are

$$\sigma_x = \begin{pmatrix} 1 & 1 \\ 1 & 0 \end{pmatrix}, \qquad \sigma_y = \begin{pmatrix} 0 & -i \\ i & 0 \end{pmatrix}, \qquad \sigma_z = \begin{pmatrix} 1 & 0 \\ 0 & -1 \end{pmatrix}, \tag{15}$$

and $\hbar$ is the Planck constant. In this paper, we use dimensionless units and assume $\hbar = 1$. Three normalized eigenstates of the matrices $\sigma_x/2$, $\sigma_y/2$, and $\sigma_z/2$ with eigenvalues $+1/2$ have the form

$$\mid \psi_x \rangle = \frac{1}{\sqrt{2}} \begin{pmatrix} 1 \\ 1 \end{pmatrix}, \qquad \mid \psi_y \rangle = \frac{1}{\sqrt{2}} \begin{pmatrix} 1 \\ i \end{pmatrix}, \qquad \mid \psi_z \rangle = \begin{pmatrix} 1 \\ 0 \end{pmatrix}. \tag{16}$$

The vectors are identified with spin-1/2 states, in which the spin projections on the axes x, y, and z are equal to $+1/2$. The corresponding density matrices read

$$\mid \psi_x \rangle \langle \psi_x \mid = \begin{pmatrix} 1/2 & 1/2 \\ 1/2 & 1/2 \end{pmatrix}, \quad \mid \psi_y \rangle \langle \psi_y \mid = \begin{pmatrix} 1/2 & -i/2 \\ i/2 & 1/2 \end{pmatrix}, \quad \mid \psi_z \rangle \langle \psi_z \mid = \begin{pmatrix} 1 & 0 \\ 0 & 0 \end{pmatrix}. \tag{17}$$

Any density matrix describing mixed state of the spin-1/2 system $\rho = \begin{pmatrix} \rho_{11} & \rho_{12} \\ \rho_{21} & \rho_{22} \end{pmatrix}$, such that $\rho^\dagger = \rho$, $\mathrm{Tr}\, \rho = 1$, and $\rho \geq 0$ (i.e., the matrix has nonnegative eigenvalues), is determined by three real parameters. The physical meaning of these parameters can be clarified, if one considers the probabilities to obtain in the state with the density matrix ρ the spin projections $+1/2$ on the axes x, y, z, which we denote as p_1, p_2, and p_3, respectively. The probabilities p_1, p_2, and p_3 play a fundamental role in describing the spin-1/2 states and, as we show, they determine the density matrix of this system. These probabilities are given by the Born rule as follows:

$$p_1 = \mathrm{Tr}\, (\rho \mid \psi_x \rangle \langle \psi_x \mid), \quad p_2 = \mathrm{Tr}\, (\rho \mid \psi_y \rangle \langle \psi_y \mid), \quad p_3 = \mathrm{Tr}\, (\rho \mid \psi_z \rangle \langle \psi_z \mid). \tag{18}$$

The spin tomogram $w(m|\vec{n})$ introduced in [16,17], being equal to the conditional probability of spin projection $m = \pm 1/2$ onto the direction given by the unit vector $\vec{n}$, is expressed in terms of the probability vector $\vec{p} = (p_1, p_2, p_3)$ [23], i.e.,

$$w(m|\vec{n}) = (1/2) + m(\vec{p} - \vec{p}_0)\vec{n}, \qquad \vec{p}_0 = (1/2, 1/2, 1/2). \tag{19}$$

In view of relations (18) employed as the equations for matrix elements of ρ, it is not difficult to rewrite the density matrix ρ in the form where its matrix elements are expressed in terms of the probabilities p_1, p_2, and p_3 [23,24,46,47]; we have

$$\rho = \begin{pmatrix} p_3 & p_1 - (1/2) - i(p_2 - 1/2) \\ p_1 - (1/2) + i(p_2 - 1/2) & 1 - p_3 \end{pmatrix}. \tag{20}$$

The standard parameters of the Bloch sphere of qibit states x_1, x_2, and x_3 are connected with the probabilities through the bijective map $x_k = 2p_k - 1; k = 1, 2, 3$.

If the spin-1/2 state is the pure state, its density matrix satisfies the constraint $\rho^2 = \rho$ that provides the condition for probabilities

$$(p_1 - 1/2)^2 + (p_2 - 1/2)^2 = p_3(1 - p_3). \tag{21}$$

In this case, the Pauli spinor of the pure state $|\psi\rangle$ can also be expressed in terms of the three probabilities satisfying condition (21), i.e.,

$$|\psi\rangle = \begin{pmatrix} \sqrt{p_3} \\ \dfrac{p_1 - 1/2}{\sqrt{p_3}} + i\dfrac{p_2 - 1/2}{\sqrt{p_3}} \end{pmatrix}. \tag{22}$$

For mixed states, the nonnegativity condition of the density matrix (nonnegativity condition for its eigenvalues) yields the inequality for the probabilities p_1, p_2, and p_3; it reads

$$(p_1 - 1/2)^2 + (p_2 - 1/2)^2 + (p_3 - 1/2)^2 \leq 1/4. \tag{23}$$

As we see, all information on the spin-1/2 state density matrix (and its Pauli spinor describing the pure state $|\psi\rangle$) is identified with three probabilities $0 \leq p_1, p_2, p_3 \leq 1$ satisfying inequality (23).

This observation provides the possibility to consider again very old problem of quantum mechanics, namely: Is it possible to formulate the notion of quantum states employing only ingredients of classical probability theory of systems with fluctuations, such as the probability distributions?

We observed that for spin-1/2 systems it is enough to have three probability distributions given by the probability vectors $\vec{P}_1 = \begin{pmatrix} p_1 \\ 1 - p_1 \end{pmatrix}$, $\vec{P}_2 = \begin{pmatrix} p_2 \\ 1 - p_2 \end{pmatrix}$, and $\vec{P}_3 = \begin{pmatrix} p_3 \\ 1 - p_3 \end{pmatrix}$. Inequality (23) is the only one quantum condition which should be respected by the probabilities. Thus, instead of vectors $|\psi\rangle$ and density matrices ρ, we can introduce the notion of spin-1/2 states, employing the set of three probability distributions or identify the state with the vector $\vec{P} = \begin{pmatrix} p_1 \\ p_2 \\ p_3 \end{pmatrix}$. This means that all quantum phenomena like, e.g., quantum interference, can be described in terms of the probabilities. Here, it worth noting that the interference of classical probabilities was discussed in [30–32]. For example, the superposition principle of quantum states, expressed in terms of normalized and orthogonal state vectors $|\psi_1\rangle$ and $|\psi_2\rangle$ by the equality

$$|\psi\rangle = \sqrt{\Pi_3}\,|\psi_1\rangle + \sqrt{1 - \Pi_3}\,e^{i\xi}\,|\psi_2\rangle, \tag{24}$$

where $|\psi\rangle$ is again the state vector, can be formulated as "superposition" of probabilities.

We present the result in the form of a nonlinear addition of two vectors

$$\vec{\mathcal{P}}^{(1)} \oplus \vec{\mathcal{P}}^{(2)} = \vec{\mathcal{P}}^{(3)}, \tag{25}$$

where components of the vectors are the probabilities $p_1^{(k)}$, $p_2^{(k)}$, and $p_3^{(k)}$; $k = 1, 2, 3$, satisfying equality (21) for each value of k. The notation of addition $\oplus$ is also associated with the probability vector $\vec{\Pi} = \begin{pmatrix} \Pi_1 \\ \Pi_2 \\ \Pi_3 \end{pmatrix}$, where the components $0 \leq \Pi_1, \Pi_2, \Pi_3 \leq 1$ satisfy equality (21). These three probabilities are related to the superposition parameters as follows:

$$\cos\xi = \frac{\Pi_1 - 1/2}{\sqrt{\Pi_3(1 - \Pi_3)}}, \qquad \sin\xi = \frac{\Pi_2 - 1/2}{\sqrt{\Pi_3(1 - \Pi_3)}}. \tag{26}$$

The three components of the vector $\vec{P}^{(3)}$ are functions of the three probability vectors $\vec{P}_1$, $\vec{P}_2$, and $\vec{\Pi}$; they read [23,24,48]

$$p_3^{(3)} = \Pi_3 p_3^{(1)} + (1 - \Pi_3)p_3^{(2)} + 2\sqrt{p_3^{(2)}p_3^{(2)}}\,(\Pi_1 - 1/2), \tag{27}$$

$$p_1^{(3)} - 1/2 = \Pi_3(p_1^{(1)} - 1/2) + (p_1^{(2)} - 1/2)(1 - \Pi_3)$$

$$+ \left[(\Pi_1 - 1/2)(p_1^{(1)} - 1/2) + (\Pi_2 - 1/2)(p_2^{(1)} - 1/2)\right]\sqrt{p_3^{(2)}/p_3^{(1)}}$$

$$+ \left[(\Pi_1 - 1/2)(p_1^{(2)} - 1/2) - (\Pi_2 - 1/2)(p_2^{(2)} - 1/2)\right]\sqrt{p_3^{(1)}/p_3^{(2)}}, \tag{28}$$

and the third one $p_2^{(3)}$ is determined in view of Equation (21).

As an example of the superposition of vectors $\dfrac{1}{\sqrt{2}}\begin{pmatrix} 1 \\ 1 \end{pmatrix} = \mid \psi_1 \rangle$ and $\dfrac{1}{\sqrt{2}}\begin{pmatrix} 1 \\ -1 \end{pmatrix} = \mid \psi_2 \rangle$ given by (24) and described by the probabilities $(1, 1/2, 1/2)$ and $(0, 1/2, 1/2)$, respectively, we obtain $p_3^{(3)} = \Pi_1$ and $p_1^{(3)} = \Pi_3$.

One can check that the obtained numbers $p_1^{(3)}$, $p_2^{(3)}$, and $p_3^{(3)}$ are nonnegative; they satisfy equality (21), which determines the probabilities $p_2^{(3)}$ and $(1 - p_2^{(3)})$. The unitary evolution of the probabilities p_1, p_2, and p_3 is described by the following transform of the matrix ρ (20)

$$\rho \to u\rho u^{\dagger}, \tag{29}$$

where the unitary 2×2 matrix u is such that $u^{\dagger} = u^{-1}$ and

$$u(t) = e^{-iHt} = \begin{pmatrix} u_{11}(t) & u_{12}(t) \\ u_{21}(t) & u_{22}(t) \end{pmatrix}, \tag{30}$$

with the Hamiltonian $H = \begin{pmatrix} H_{11} & H_{12} \\ H_{21} & H_{22} \end{pmatrix}$. From this evolution, which corresponds to the von Neumann equation

$$\frac{\partial\rho}{\partial t} + i[H, \rho] = 0, \tag{31}$$

the evolution formula for the probabilities follows; it reads

$$\begin{pmatrix} p_3(t) \\ p_1(t) - (1/2) - i(p_2(t) - 1/2) \\ p_1(t) - (1/2) + i(p_2(t) - 1/2) \\ 1 - p_3(t) \end{pmatrix} = [u(t) \otimes u^*(t)] \begin{pmatrix} p_3 \\ p_1 - (1/2) - i(p_2 - 1/2) \\ p_1 - (1/2) + i(p_2 - 1/2) \\ 1 - p_3 \end{pmatrix}. \tag{32}$$

This result is the solution of Equation (31) written as the kinetic equation for probabilities $p_1(t)$, $p_2(t)$, and $p_3(t)$. Equation (32) describes the temporal evolution of the initial probabilities p_1, p_2, and p_3, which convert at time t to probabilities $p_1(t)$, $p_2(t)$, and $p_3(t)$ satisfying relation (23).

5. Quantum Observables and Classical-Coin Random Variables for Qubit Systems

The quantum observable for spin-1/2 system is described by the Hermitian matrices $A = \begin{pmatrix} A_{11} & A_{12} \\ A_{21} & A_{22} \end{pmatrix}$. It is possible [34] to consider quantum statistical properties of this observable in the probability representation associating the matrix elements A_{jk} with classical-like random variables. Introducing the notation $A_{12} = x - iy$, $A_{11} = z_1$, and $A_{22} = z_2$, we can rewrite the mean value of the observable $\langle A \rangle$ in the form

$$\langle A \rangle = \operatorname{Tr} \rho A = \langle x \rangle_{cl} + \langle y \rangle_{cl} + \langle z \rangle_{cl}, \tag{33}$$

where

$$\langle x \rangle_{cl} = p_1 x + (1 - p_1)(-x), \quad \langle y \rangle_{cl} = p_2 y + (1 - p_2)(-y), \quad \langle z \rangle_{cl} = p_3 z_1 + (1 - p_3) z_2. \tag{34}$$

This form shows that the mean values of the observable A calculated using the standard formalism of quantum mechanics provide the connection with classical-like means of three dichotomic random variables x_{cl}, y_{cl}, and z_{cl}, employing the values $(x, -x)$, $(y, -y)$, and (z_1, z_2), respectively.

The probability distributions for these values are given as $\vec{P}_1 = (p_1, 1 - p_1)$, $\vec{P}_2 = (p_2, 1 - p_2)$, and $\vec{P}_3 = (p_3, 1 - p_3)$, which determine the density matrix ρ of the qubit state. The quantum observable in the quantum suprematism representation has a classical analog.

We point out that the highest moments of the observable A given as $\langle A^n \rangle = \operatorname{Tr} \rho A^n$, $n = 2, 3, \ldots$ are also expressed in terms of classical random variables x_{cl}, y_{cl}, and z_{cl}, and the expressions reflect quantum correlations of classical-like random observables due to the nonnegativity condition of the density matrix (23). One can also associate the matrix elements of arbitrary qudit observables with artificial-coin random variables.

6. Quantum Suprematism Representation

The relations described in the previous section can be illustrated in the quantum suprematism picture [23,24,48], where the probabilities p_1, p_2, and p_3 determine the Triada of Malevich's squares. We construct a triangle with vertices A_1, A_2, and A_3, which are located on three simplexes—sides of equilateral triangle with the side length equal to $\sqrt{2}$, and three squares (black, red, and white) determined by the sides of the triangle (Figure 1).

We call these squares the Triada of Malevich's squares following [23,24,34,46,47]. The areas of the squares S_{A_1,A_2}, S_{A_2,A_3}, and S_{A_3,A_1} are

$$\begin{aligned} S_{A_1,A_2} &= 2 + 2p_2^2 - 4p_2 - 2p_3 + 2p_3^2 + 2p_2 p_3, \\ S_{A_2,A_3} &= 2 + 2p_3^2 - 4p_3 - 2p_1 + 2p_1^2 + 2p_3 p_1, \\ S_{A_3,A_1} &= 2 + 2p_1^2 - 4p_1 - 2p_2 + 2p_2^2 + 2p_1 p_2. \end{aligned} \tag{35}$$

The sum of the areas of Malevich's squares, being the function of probabilities p_1, p_2, and p_3, reads

$$S = 2 \left[3 + 2 \left(p_1^2 + p_2^2 + p_3^2 \right) - 3 \left(p_1 + p_2 + p_3 \right) + p_1 p_2 + p_2 p_3 + p_3 p_1 \right]. \tag{36}$$

The map of the Bloch sphere parameters x_k; $k = 1, 2, 3$, onto the probabilities $p_k = (x_k + 1)/2$ can be used to express the area S in terms of the parameters x_k.

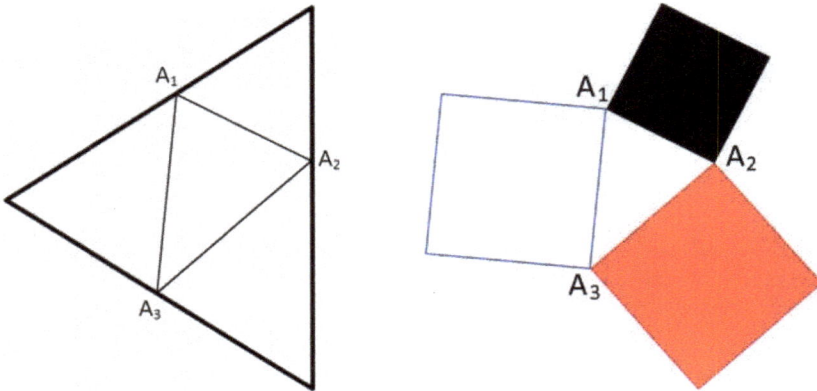

Figure 1. The equilateral triangle with a side length equal to $\sqrt{2}$. Each of three sides is simplex corresponding to the coin probabilities p_k and $(1 - p_k)$ satisfying the relation $p_k + (1 - p_k) = 1$; $k = 1, 2, 3$. Here, the points on simplexes with probabilities p_1, p_2, and p_3 determine the triangle $A_1 A_2 A_3$ (**on the left**). The three squares (black, red, and white) called Triada of Malevich's squares, which are constructed using the sides of the triangle $A_1 A_2 A_3$. The squares are in the one-to-one correspondence with the density matrix of qubit (spin-1/2) states (**on the right**).

For quantum states, the sum satisfies the inequality $S \leq 3$ [47]. For classical-coin states, the sum can take maximum value $S_{cl} = 6$. In view of this fact, the quantization condition (23) provides the possibility to clarify the difference of classical and quantum properties of the systems, which states are illustrated in the quantum suprematism representation by Triadas of Malevich's squares.

The quantum states of spin-1/2 systems are described by spin tomogram $w(m \mid \vec{n})$ [16], and the spin tomogram determines the density matrix of the spin state [17]. The spin tomogram can be expressed by the probabilities p_1, p_2, and p_3; the tomogram is the probability to obtain the spin projection m onto the direction in space determined by the unit vector $\vec{n} = (\sin\theta\cos\varphi, \sin\theta\sin\varphi, \cos\theta)$.

For spin-1/2, the spin projection m takes two values $\pm 1/2$, and the tomogram (19) provides

$$w(m = +1/2 \mid \vec{n}) = \vec{n}(\vec{p} - \vec{p}_0)) + 1/2, \text{ where } \vec{p} = \begin{pmatrix} p_1 \\ p_2 \\ p_3 \end{pmatrix} \text{ and } \vec{p}_0 = \begin{pmatrix} 1/2 \\ 1/2 \\ 1/2 \end{pmatrix}. \text{ The tomogram}$$

$w(m \mid \vec{n})$ can be interpreted as the conditional probability distribution and, in view of this fact, one can obtain a new entropic inequality associated with this distribution. For example, the relative Tsallis entropy for two distributions $w(m \mid \vec{n}_1)$ and $w(m \mid \vec{n}_2)$ satisfies the inequality, which yields the new condition for probabilities,

$$(1-q)^{-1} \left\{ [\vec{n}_1(\vec{p} - \vec{p}_0) + 1/2]^q [\vec{n}_2(\vec{p} - \vec{p}_0) + 1/2]^{1-q} \right.$$
$$\left. + [(1/2) - \vec{n}_1(\vec{p} - \vec{p}_0)]^q [(1/2) - \vec{n}_2(\vec{p} - \vec{p}_0)]^{1-q} - 1 \right\} \geq 0. \tag{37}$$

In the limit $q \to 1$, this inequality provides the nonnegativity condition for the von Neumann relative entropy

$$[\vec{n}_1 (\vec{p} - \vec{p}_0) + 1/2] \ln \left\{ [\vec{n}_1 (\vec{p} - \vec{p}_0) + 1/2] [\vec{n}_2(\vec{p} - \vec{p}_0) + 1/2]^{-1} \right\}$$
$$+ [(1/2) - \vec{n}_1(\vec{p} - \vec{p}_0)] \ln \left\{ [(1/2) - \vec{n}_1(\vec{p} - \vec{p}_0)] [(1/2) - \vec{n}_2(\vec{p} - \vec{p}_0)]^{-1} - 1 \right\} \geq 0. \tag{38}$$

The obtained new entropic inequalities for probabilities p_1, p_2, and p_3 can be checked in the experiments with superconducting qubits, as well as the maximum value $S = 3$ for the sum of areas of Malevich's squares.

7. Qutrit States in the Probability Representation

One can extend the consideration of suprematism representation to the case of any qudits, for example, qutrit states with the density matrix $\rho = \begin{pmatrix} \rho_{11} & \rho_{12} & \rho_{13} \\ \rho_{21} & \rho_{22} & \rho_{23} \\ \rho_{31} & \rho_{32} & \rho_{33} \end{pmatrix}$. Using the tool [23,24,47] of embedding this matrix into 4×4-matrix $R = \begin{pmatrix} \rho & 0 \\ 0 & 0 \end{pmatrix}$, one can obtain three qubit-state density matrices applying the partial tracing procedure. These three-qubit density matrices are

$$\rho(1) = \begin{pmatrix} \rho_{11} + \rho_{22} & \rho_{13} \\ \rho_{31} & \rho_{33} \end{pmatrix}, \quad \rho(2) = \begin{pmatrix} \rho_{11} + \rho_{33} & \rho_{12} \\ \rho_{21} & \rho_{22} \end{pmatrix}, \quad \rho(3) = \begin{pmatrix} \rho_{22} & \rho_{23} \\ \rho_{32} & \rho_{11} + \rho_{33} \end{pmatrix}. \quad (39)$$

Since these matrices can be expressed in terms of probabilities

$$\rho(k) = \begin{pmatrix} p_3^{(k)} & p_1^{(k)} - (1/2) - i(p_2^{(k)} - 1/2) \\ p_1^{(k)} - (1/2) + i(p_2^{(k)} - 1/2) & 1 - p_3^{(k)} \end{pmatrix}; \quad k = 1,2,3, \quad (40)$$

the qutrit density matrix elements can also be expressed in terms of these probabilities $p_{1,2,3}^{(k)}$. In fact, we arrive at

$$\rho = \begin{pmatrix} p_3^{(1)} + p_3^{(2)} - 1 & p_1^{(2)} - (1/2) - i(p_2^{(2)} - 1/2) & p_1^{(1)} - (1/2) - i(p_2^{(1)} - 1/2) \\ p_1^{(2)} - (1/2) + i(p_2^{(2)} - 1/2) & 1 - p_3^{(2)} & p_1^{(3)} - (1/2) - i(p_2^{(3)} - 1/2) \\ p_1^{(1)} - (1/2) + i(p_2^{(1)} - 1/2) & p_1^{(3)} - (1/2) + i(p_2^{(3)} - 1/2) & 1 - p_3^{(1)} \end{pmatrix}; \quad (41)$$

in other notation, Equation (41) is given in Appendix B. The probabilities $p_{1,2,3}^{(k)}$ satisfy inequality (23) for $k = 1, 2, 3$ and also the nonnegativity condition for $\det \rho \geq 0$. The expression of matrix elements of the qutrit density matrix in terms of the probability distributions follows a new entropic inequality for the matrix elements:

$$\frac{1}{1-q} \left\{ \left[\tfrac{1}{2} (\rho_{13} + \rho_{31} + 1) \right]^q \left[\tfrac{1}{2} (\rho_{13} + \rho_{31} + 1) \right]^{1-q} \right.$$
$$\left. + \left[\tfrac{1}{2} (-\rho_{13} - \rho_{31} + 1) \right]^q \left[\tfrac{1}{2} (1 - \rho_{23} - \rho_{32}) \right]^{1-q} - 1 \right\} \geq 0. \quad (42)$$

This new inequality for the qutrit state comes from applying the nonnegativity condition of the Tsallis entropy expressed in terms the qubit-state tomogram by Equations (37) and (38) to the qutrit-state density matrix.

Now, we derive the other inequality for the probabilities determining the qutrit state (41). First, we construct the qubit density matrices following [49]

$$\rho(2) = \begin{pmatrix} p_3^{(2)} & p_1^{(2)} - (1/2) - i(p_2^{(2)} - 1/2) \\ p_1^{(2)} - (1/2) + i(p_2^{(2)} - 1/2) & 1 - p_3^{(2)} \end{pmatrix}, \quad (43)$$

$$\rho(2) = \begin{pmatrix} p_3^{(1)} & p_1^{(1)} - (1/2) - i(p_2^{(1)} - 1/2) \\ p_1^{(1)} - (1/2) + i(p_2^{(1)} - 1/2) & 1 - p_3^{(1)} \end{pmatrix}. \quad (44)$$

The subadditivity condition provides the inequality for the probabilities $p_{1,2,3}^{(k)}$, $k = 1, 2, 3$; it reads

$$- \operatorname{Tr} \rho(1) \ln \rho(1) - \operatorname{Tr} \rho(2) \ln \rho(2) \geq -\operatorname{Tr} (\rho \ln \rho), \tag{45}$$

where ρ, being given by (41), is determined by the probabilities.

In addition, the probabilities determining the qutrit states satisfy the Tsallis entropic inequality for the distance between the states

$$\frac{1}{1-q} \left(\operatorname{Tr} \rho(1)^q \rho(2)^{1-q} - 1 \right) \geq 0. \tag{46}$$

Inequalities (42), (45), and (46) are compatible with the nonnegativity condition of the qutrit density matrix.

Since the qubit density matrices (43) and (44) are obtained from the same qutrit density matrix, the distance between these qubit states characterizes the hidden correlations between the artificial qubits associated with the qutrit density matrix (41). The inequality can be checked in the experiments where the tomography of qutrit states is performed, e.g., in the experiments with superconducting circuits based on Josephson junction devices [50,51].

Since the qutrit states are described by the probabilities determining the states of three artificial qubits, the density matrix (39) can be mapped onto the set of three Triadas of Malevich's squares.

8. Pure Qutrit States and Their Superposition in the Probability Representation

In [52,53], the superposition principle of quantum states was formulated as a nonlinear addition rule of the pure-state density operators. Namely, given two density operators $\hat{\rho}_1 = | \psi_1 \rangle \langle \psi_1 |$ and $\hat{\rho}_2 = | \psi_2 \rangle \langle \psi_2 |$ satisfying the conditions $\hat{\rho}_1^2 = \hat{\rho}_1$, $\hat{\rho}_2^2 = \hat{\rho}_2$, $\operatorname{Tr} \hat{\rho}_1 = \operatorname{Tr} \hat{\rho}_2 = 1$, and $\hat{\rho}_1 \hat{\rho}_2 = 0$. Then, for arbitrary real numbers $0 \leq \lambda_1, \lambda_2 \leq 1$; $\lambda_1 + \lambda_2 = 1$ and the density operator $\hat{\rho}_0 = | \psi_0 \rangle \langle \psi_0 |$; $\hat{\rho}_0^2 = \hat{\rho}_0$, the state with the density operator $\hat{\rho}_\psi$ of the form

$$\hat{\rho}_\psi = \lambda_1 \hat{\rho}_1 + \lambda_2 \hat{\rho}_2 + \sqrt{\lambda_1 \lambda_2} \frac{\hat{\rho}_1 \hat{\rho}_0 \hat{\rho}_2 + \hat{\rho}_2 \hat{\rho}_0 \hat{\rho}_1}{\sqrt{\operatorname{Tr} (\hat{\rho}_1 \hat{\rho}_0 \hat{\rho}_2 \hat{\rho}_0)}} \tag{47}$$

satisfies the conditions $\hat{\rho}_\psi^\dagger = \hat{\rho}_\psi$, $\hat{\rho}_\psi^2 = \hat{\rho}_\psi$, and $\operatorname{Tr} \hat{\rho}_\psi = 1$.

The nonlinear addition rule (47) corresponds to the interference formula of two orthogonal pure states $| \psi_1 \rangle$ and $| \psi_2 \rangle$ of the form $| \psi \rangle = \sqrt{\lambda_1} | \psi_1 \rangle + e^{i\varphi} \sqrt{\lambda_2} | \psi_2 \rangle$, where the phase φ is coded by an artificial density operator $\hat{\rho}_0$.

One can use generic Equation (47) to formulate the superposition rule of qudit states expressed in terms of the probabilities; we obtain such a formula for two qutrit states. For this, we introduce three probability vectors, i.e., three probability distributions $\vec{\Pi}_1 = (\Pi_1, 1 - \Pi_1)$, $\vec{\Pi}_2 = (\Pi_2, 1 - \Pi_2)$, and $\vec{\Pi}_3 = (\Pi_3, 1 - \Pi_3)$, where $0 \leq \Pi_1, \Pi_2, \Pi_3 \leq 1$, and the phase φ is defined by the relations (26),

$$\cos \varphi = \frac{\Pi_1 - 1/2}{\sqrt{\Pi_3 (1 - \Pi_3)}}, \qquad \sin \varphi = \frac{\Pi_2 - 1/2}{\sqrt{\Pi_3 (1 - \Pi_3)}}. \tag{48}$$

This means that we arrive at the condition $(\Pi_1 - 1/2)^2 + (\Pi_2 - 1/2)^2 + (\Pi_3 - 1/2)^2 = 1/4$ equivalent to the condition (21). To obtain the superposition rule for qutrit states in the probability representation, we employ the expression of density matrix (41), introduce three 8-vectors $\vec{\mathcal{P}}_1$, $\vec{\mathcal{P}}_2$, and $\vec{\mathcal{P}}_0$ of the form (see Appendix B):

$$\begin{aligned}
\vec{\mathcal{P}}_1 &= \left(p_1^{(I)}, p_2^{(I)}, p_3^{(I)}, p_4^{(I)}, p_5^{(I)}, p_6^{(I)}, p_7^{(I)}, p_8^{(I)} \right), \\
\vec{\mathcal{P}}_2 &= \left(p_1^{(II)}, p_2^{(II)}, p_3^{(II)}, p_4^{(II)}, p_5^{(II)}, p_6^{(II)}, p_7^{(II)}, p_8^{(II)} \right), \\
\vec{\mathcal{P}}_0 &= \left(p_1^{(III)}, p_2^{(III)}, p_3^{(III)}, p_4^{(III)}, p_5^{(III)}, p_6^{(III)}, p_7^{(III)}, p_8^{(III)} \right),
\end{aligned} \tag{49}$$

and identify three qutrit states $\hat{\rho}_1$, $\hat{\rho}_2$, and $\hat{\rho}_3$ with these vectors $\vec{\mathcal{P}}_1$, $\vec{\mathcal{P}}_2$, and $\vec{\mathcal{P}}_0$, since the density matrices are determined by their components. Formula (47) provides the dependence of the probability vector $\vec{\mathcal{P}}_\psi$, which determines the state $\hat{\rho}_\psi$, on the probabilities $\vec{\mathcal{P}}_1$, $\vec{\mathcal{P}}_2$, $\vec{\mathcal{P}}_0$, and $\vec{\Pi}_{1,2,3}$. The density matrix (41) of the qutrit pure state satisfies the condition $\rho^2 = \rho$, which provides the formula for the probabilities following from the equality of the matrix elements $\sum_k \rho_{jk}\rho_{km} = \rho_{jm}$ written in terms of probabilities.

Thus, we formulated the result for qutrits. We rewrite Equation (47) expressing the matrices $\hat{\rho}_\psi$, $\hat{\rho}_1$, $\hat{\rho}_2$, and $\hat{\rho}_0$ in terms of probabilities. The equality of matrix elements provides the expression of the probability vectors $\vec{\mathcal{P}}_\psi$ as functions of the probability vectors $\vec{\mathcal{P}}_1$, $\vec{\mathcal{P}}_2$, $\vec{\mathcal{P}}_0$, and $\Pi_{1,2,3}$. The approach can be extended to other qudit states.

9. Probability Representation of the Density $N \times N$-Matrix of the Qudit State

We generalize Equation (41) and write the matrix element ρ_{jk} for an arbitrary qudit state (N-level atom state) in the form [48]:

$$\rho_{jk} = p_1^{(jk)} - (1/2) - i(p_2^{(jk)} - 1/2), \quad k > j,$$
$$\rho_{jj} = 1 - p_3^{(jj)}, \quad j \geq 2, \tag{50}$$
$$\rho_{22} = 1 - \sum_{j=2}^{N} \rho_{jj},$$

where $p_1^{(jk)}$ and $p_2^{(jk)}$ are the probabilities for artificial qubits (spin-1/2 states) to have spin projections $m = +1/2$ onto the x and y axes, respectively. The diagonal matrix elements ρ_{jj} depend on probabilities $p_3^{(jj)}$, which are probabilities of artificial spin projections $m = +1/2$ on the z-axis. Thus, all the matrix elements of the qudit density matrix are expressed in terms of probabilities for $(N^2 - 1)$ classical coins to have positions "UP" or "DOWN." The new inequality (42) is also valid for the qudit density matrix (50) as well as the entropic inequalities for the matrix elements where $1 \to j$ and $3 \to k$.

The superposition principle of pure states given by relation (47) for the density matrices provides the connection of the probabilities determining two orthogonal pure states $\hat{\rho}_1$ and $\hat{\rho}_2$ and the density matrix $\hat{\rho}_0$ with the probabilities determining the pure superposition state $\hat{\rho}_\psi$. In spite of the fact that the formulas are cumbersome, their existence demonstrates that such quantum phenomenon as interference of quantum states for arbitrary N-level atoms can be formulated as nonlinear superposition of classical probabilities (cf. [30–32]). Thus, for any qudit state, one has relationships connecting the probability distributions determining the qudit states in view of generic formula (47).

10. Conclusions

To conclude, we point out the main results of our study.

We introduced the notion of hidden correlations for systems without subsystems, using the explicitly written functions (5)–(7) and (9)–(12) providing the invertible map of the integers in both classical and quantum domains. This approach provides the possibility to write Bayes' formula and introduce the conditional probability distribution for given probability distribution of one random variable (22). Using the probability description of qubit states, we presented the solution of the von Neumann equation for the two-level atom as the transform of probabilities determining the state density matrix (35).

We obtained new inequality for the qubit-state probabilities, which can be checked experimentally as the nonnegativity condition of classical relative Tsallis entropy—Equation (38). The properties of generalized entropies discussed and employed in [21,22] can be also studied in view of the probability distributions determining the matrix elements of the density matrix. We will do this in future publications.

The tomographic reconstruction of the density matrix, e.g., in the experiments with superconductive circuits [50,51], provides the possibility to find the probabilities that should satisfy the inequalities.

In view of a generic form of the superposition principle formulated in terms of density matrices (47), we presented the general approach to get the addition rule for classical probabilities determining the qutrit pure state. In this sense, we considered two related problems. One problem is to reformulate the standard description of quantum mechanics by means of wave functions and density matrices, using the Hilbert space formalism, in terms of classical probabilities; this can be done within the framework of quantum tomography [15–17,29]. The inverse approach is to consider classical probabilities associated with quantumlike objects in Hilbert spaces [30–32]. In this paper, we concentrate on the presentation of the first problem and show that the quantum formalism of Hilbert spaces can be mapped bijectively onto the classical-like formalism of the probability theory and geometry of simplexes. New results obtained in this approach are the nonlinear addition rules of probabilities, giving the probabilities. Such rules correspond either to Born rules of quantum mechanics or the superposition principle of pure states of qudits.

In addition, we obtained an explicit form of the unitary evolution for probabilities determining the probabilities in terms of their transform by means of unitary matrices for two-level atoms.

To illustrate the map of Bloch sphere parametrization of qubit states onto the probability representation of the states, where the probabilities satisfy the quantum constraint inequalities, we employed the geometric representation of the probabilities, in view of the suprematism picture of Triadas of Malevich' squares. The approach developed and its properties in the case of generic qudit states will be elaborated in the future publication.

In addition, we point out that the tomographic-probability approach was applied in signal analysis [54,55] due to the descriptions of the signals by an analog of the Wigner function proposed by Ville [56]. The probability properties considered above can be used in the signal theory as well.

Author Contributions: Both authors contributed equally to the conception, design, and methodology of this study. Both equally to the analysis of the results and the conclusions. M.A.M. wrote the first draft of the manuscript. Both authors contributed equally to final writing of the manuscript.

Funding: Some results presented in this paper were reported at the 19th Växjö Conference on Quantum Foundations "Towards Ultimate Quantum Theory" (Linnaeus University, Växjö, Sweden, June 11–14, 2018), and we thank the Organizers and especially Professor Khrennikov for invitation and kind hospitality. The work of V.I.M. was partially performed at the Moscow Institute of Physics and Technology, where V.I.M. was supported by the Russian Science Foundation under Project No. 16-11-00084. Also, V.I.M. acknowledges the partial support of the Tomsk State University Competitiveness Improvement Program.

Conflicts of Interest: The authors declare no conflict of interest.

Appendix A

We consider a simple example of $N = 4$ and $n_1 = n_2 = 2$. We have the distribution $P(j,k)$ of four numbers $1 \geq P(1,1), P(1,2), P(2,1), P(2,2) \geq 0$. In view of Equations (2) and (3), we have the conditional probabilities:

$$P(1 \mid 1) = \frac{P(1,1)}{P(1,1) + P(2,1)}, \qquad P(2 \mid 1) = \frac{P(2,1)}{P(1,1) + P(2,1)},$$

$$P(1 \mid 2) = \frac{P(1,2)}{P(1,2) + P(2,2)}, \qquad P(2 \mid 2) = \frac{P(2,2)}{P(1,2) + P(2,2)}.$$

Numbers $P(1 \mid 1)$, $P(2 \mid 1)$, $P(1 \mid 2)$, and $P(2 \mid 2)$ are nonnegative and normalized, $\sum_{j=1}^{2} P(j \mid k) = 1$.

Now, we consider the example of the same four nonnegative numbers, which we denote as $\mathcal{P}(n)$, with $n = 1, 2, 3, 4$, and define the numbers as $\mathcal{P}(1) = P(1,1)$, $\mathcal{P}(2) = P(2,1)$, $\mathcal{P}(3) = P(1,2)$, and $\mathcal{P}(4) = P(2,2)$. The set of numbers $\mathcal{P}(n)$ can be interpreted as the probability distribution of one random variable n. Since Bayes' formula is determined through only the numbers $P(j,k)$, we can apply it using the invertible map of integers $1 \leftrightarrow 1,1$, $2 \leftrightarrow 2,1$, $3 \leftrightarrow 1,2$, $4 \leftrightarrow 2,2$ and introducing two "marginal probability distributions" $P_1(1) = P(1,1) + P(1,2) = \mathcal{P}(1) + \mathcal{P}(3)$, $P_1(2) = P(2,1) + P(2,2) = \mathcal{P}(2) + \mathcal{P}(4)$ and $P_2(1) = P(1,1) + P(2,1) = \mathcal{P}(1) + \mathcal{P}(2)$, $P_2(2) = P(1,2) + P(2,2) = \mathcal{P}(3) + \mathcal{P}(4)$. After this, we can introduce the "conditional probability distributions" in view of the definition given by the same relations (3) and (4) but rewritten in terms of numbers $\mathcal{P}(n)$, with $n = 1,2,3,4$. We define the conditional probabilities

$$P(1 \mid 1) = \frac{\mathcal{P}(1)}{\mathcal{P}(1) + \mathcal{P}(3)}, \qquad P(2 \mid 1) = \frac{\mathcal{P}(2)}{\mathcal{P}(1) + \mathcal{P}(3)},$$

$$P(1 \mid 2) = \frac{\mathcal{P}(3)}{\mathcal{P}(2) + \mathcal{P}(4)}, \qquad P(2 \mid 2) = \frac{\mathcal{P}(4)}{\mathcal{P}(2) + \mathcal{P}(4)}.$$

Appendix B

To express the qutrit density matrix ρ in the probability representation, we employ the 8-vector $\vec{\mathcal{P}} = (p_1, p_2, p_3, p_4, p_5, px_6, p_7, p_8)$ and obtain

$$\rho = \begin{pmatrix} p_3 + p_6 - 1 & p_4 - (1/2) - i(p_5 - 1/2) & p_1 - (1/2) - i(p_2 - 1/2) \\ p_4 - (1/2) + i(p_5 - (1/2)) & 1 - p_6 & p_7 - (1/2) - i(p_8 - (1/2)) \\ p_1 - (1/2) + i(p_2 - 1/2) & p_7 - (1/2) + i(p_8 - 1/2) & 1 - p_3 \end{pmatrix}. \quad \text{(A1)}$$

We have the probability $p_9 = 1 - p_6$, and nine probability distributions, corresponding to classical-coin positions, read $\vec{p}_j = (p_j, 1 - p_j)$; $j = 1, 2, \ldots, 8$ and $\vec{p}_9 = (p_9, 1 - p_9) = (1 - p_6, p_6)$. Thus, the density matrix of qutrit state formally corresponds to the probability distributions describing positions "UP" and "DOWN" of nine classical coins. In our notation, the ninth coin probability distribution is completely correlated with the sixth coin position, i.e., it is determined by the probability distribution of the sixth coin. The other correlations are described by nonnegativity conditions for the density-matrix eigenvalues. The density matrices ρ_1, ρ_2, and ρ_0 have the form (A1) with the replacement $p_j \to p_j^{(I)}, p_j^{(II)}, p_j^{(III)}$; $j = 1, 2, \ldots, 8$.

The matrix ρ_0 is determined by the probability vector $\vec{\mathcal{P}}_0 = \left\{ p_j^{(0)} \right\}$; $j = 1, 2, \ldots, 8$. All of the 8-vectors $\vec{\mathcal{P}}_0$, $\vec{\mathcal{P}}_1$, and $\vec{\mathcal{P}}_2$ satisfy the condition that the corresponding density matrices satisfy the equality $\text{Tr}\,\hat{\rho}_1^2 = \text{Tr}\,\hat{\rho}_2^2 = \text{Tr}\,\hat{\rho}_0^2 = 1$ and the property $\text{Tr}\,\hat{\rho}_\psi^2 = 1$ follows relation (47).

References

1. Landau, L. Das Dämpfungsproblem in der Wellenmechanik. *Z. Phys.* **1927**, *45*, 430–441. (In German) [CrossRef]
2. Von Neumann, J. Wahrscheinlichkeitstheoretischer Aufbau der Quantenmechanik. *Gött. Nach.* **1927**, *1*, 245–272. (In German)
3. Dirac, P.A.M. *The Principles of Quantum Mechanics*; Clarendon Press: Oxford, UK, 1981; ISBN 9780198520115.
4. Schrödinger, E. Quantisierung als Eigenwertproblem (Erste Mitteilung). *Ann. Phys.* **1926**, *79*, 361–376. (In German) [CrossRef]
5. Schrödinger, E. Quantisierung als Eigenwertproblem (Zweite Mitteilung). *Ann. Phys.* **1926**, *80*, 489–527. (In German) [CrossRef]
6. Smithey, D.T.; Beck, M.; Raymer, M.G.; Faridani, A. Measurement of the Wigner distribution and the density matrix of a light mode using optical homodyne tomography: Application to squeezed states and the vacuum. *Phys. Rev. Lett.* **1993**, *70*, 1244–1247. [CrossRef] [PubMed]
7. Bertrand, J.; Bertrand, P. A tomographic approach to Wigner's function. *Found. Phys.* **1987**, *17*, 397–405. [CrossRef]
8. Vogel, K.; Risken, H. Determination of quasiprobability distributions in terms of probability distributions for the rotated quadrature phase. *Phys. Rev. A* **1989**, *40*, 2847–2849. [CrossRef]

9. Radon, J. Über die Bestimmung von Funktionen durch ihre Integralwerte langs gewisser Mannigfaltigkeiten. *Berichte Sachsische Akademie der Wissenschaften Leipzig* **1917**, *29*, 262–277. (In German)

10. Wigner, E. On the quantum correction for thermodynamic equilibrium. *Phys. Rev.* **1932**, *40*, 749–759. [CrossRef]

11. Husimi, K. Some formal properties of the density matrix. *Proc. Phys. Math. Soc. Jpn.* **1940**, *22*, 264–314.

12. Kano, Y. A new phase-space distribution function in the statistical theory of the electromagnetic field. *J. Math. Phys.* **1965**, *6*, 1913–1915. [CrossRef]

13. Glauber, R.J. Coherent and incoherent states of the radiation field. *Phys. Rev.* **1963**, *131*, 2766–2788. [CrossRef]

14. Sudarshan, E.C.G. Equivalence of semiclassical and quantum mechanical descriptions of statistical light beams. *Phys. Rev. Lett.* **1963**, *10*, 277–279. [CrossRef]

15. Mancini, S.; Man'ko, V.I.; Tombesi, P. Symplectic tomography as classical approach to quantum systems. *Phys. Lett. A* **1996**, *213*, 1–6. [CrossRef]

16. Dodonov, V.V.; Man'ko, V.I. Positive distribution description for spin states. *Phys. Lett. A* **1997**, *229*, 335–339. [CrossRef]

17. Man'ko, V.I.; Man'ko, O.V. Spin state tomography. *J. Exp. Theor. Phys.* **1997**, *85*, 430–434. [CrossRef]

18. Shannon, C.E. A mathematical theory of communication. *Bell Syst. Tech. J.* **1948**, *27*, 379–423. [CrossRef]

19. Tsallis, C. Nonextensive statistical mechanics and thermodynamics: Historical background and present status. In *Nonextensive Statistical Mechanics and Its Applications*; Abe, S., Okamoto, Y., Eds.; Lecture Notes in Physics; Springer: Berlin, Germany, 2001; Volume 560, pp. 3–98.

20. Rényi, A. On measures of entropy and information. In *Proceedings of the Fourth Berkeley Symposium on Mathematical Statistics and Probability*; University of California Press: Berkeley, CA, USA, 1961; Volume 1, pp. 547–561.

21. Hanel, R.; Thumer, S. Generalized (c,d) entropy and aging random walks. *Entropy* **2013**, *15*, 5324–5337. [CrossRef]

22. Jisba, P.; Korbel, J. Maximum entropy principle in statistical interference: Case for non-Shannonian entropies. **2018**, arXiv:1808.01172 [cond-mat.stat-mech].

23. Chernega, V.N.; Man'ko, O.V.; Man'ko, V.I. Triangle geometry for qutrit states in the probability representation. *J. Russ. Laser Res.* **2017**, *38*, 416–425. [CrossRef]

24. Chernega, V.N.; Man'ko, O.V.; Man'ko, V.I. Quantum suprematism picture of Malevich's squares triada for spin states and the parametric oscillator evolution in the probability representation of quantum mechanics. *J. Phys. Conf. Ser.* **2018**, in press, arXiv:1712.01927.

25. Man'ko, V.I.; Marmo, G.; Ventriglia, F.; Vitale, P. Metric on the space of quantum states from relative entropy. Tomographic reconstruction. *J. Phys. A* **2017**, *50*, 335302. [CrossRef]

26. Filippov, S.N.; Man'ko, V.I. Symmetric informationally complete positive operator valued measure and probability representation of quantum mechanics. *J. Russ. Laser Res.* **2010**, *31*, 211–231. [CrossRef]

27. Castaños, O.; López-Peña, R.; Man'ko, M.A.; Man'ko, V.I. Squeeze tomography of quantum states. *J. Phys. A* **2004**, *37*, 8529–8544. [CrossRef]

28. Man'ko, M.A.; Man'ko, V.I. Properties of nonnegative hermitian matrices and new entropic inequalities for noncomposite quantum systems. *Entropy* **2015**, *17*, 2876–2894. [CrossRef]

29. Asorey, M.; Ibort, A.; Marmo, G.; Ventriglia, F. Quantum tomography twenty years later. *Phys. Scr.* **2015**, *90*, 074031. [CrossRef]

30. Khrennikov, A. Quantum-like representation algorithm: Transformation of probabilistic data into vectors on Bloch's sphere. *arXiv* **2008**, arXiv:0803.1391v1.

31. Khrennikov, A. The principle of supplementarity: A contextual probabilistic viewpoint to complementarity, the interference of probabilities and incompatibility of variables in quantum mechanics. *Found. Phys.* **2005**, *35*, 1655–1693. [CrossRef]

32. Khrennikov, A. Interference of probabilities and number field structure of quantum models. *Ann. Phys.* **2003**, *12*, 575–585. [CrossRef]

33. Chernega, V.N.; Man'ko, O.V.; Man'ko, V.I. God plays coins or superposition principle for classical probabilities in quantum suprematism representation of qubit states. *J. Russ. Laser Res.* **2018**, *39*, 128–139. [CrossRef]

34. Chernega, V.N.; Man'ko, O.V.; Man'ko, V.I. Probability representation of quantum observables and quantum states. *J. Russ. Laser Res.* **2017**, *38*, 324–333. [CrossRef]

35. Shatskikh, A. *Black Square: Malevich and the Origin of Suprematism*; Yale University Press: New Haven, CT, USA, 2012.

36. Kolmogorov, A.N. *Foundation of the Theory of Probability*; Chelsea: New York, NY, USA, 1956.

37. Shiryaev Albert, N. *Probability-1*; Springer: Berlin, Germany, 2016.

38. Holevo, A.S. *Probabilistic and Statistical Aspects of Quantum Theory*; North Holland: Amsterdam, The Netherlands, 1982.

39. Khrennikov, A. *Probability and Randomness. Quantum versus Classical*; World Scientific: Singapore, 2016.

40. Man'ko, M.A.; Man'ko, V.I. Inequalities for nonnegative numbers and information properties of qudit tomograms. *J. Russ. Laser Res.* **2013**, *34*, 203–218. [CrossRef]

41. Man'ko, M.A.; Man'ko, V.I. Hidden quantum correlations in single qudit systems. *J. Russ. Laser Res.* **2015**, *36*, 301–311. [CrossRef]

42. De Pasquale, A. Bipartite entanglement of large quantum systems. *arXiv* **2012**, arXiv:1206.6749.

43. De Pasquale, A.; Facchi, P.; Giovannetti, V.; Parisi, G.; Pascazio, S.; Scardicchio, A. Statistical distribution of the local purity in a large quantum system. *J. Phys. A Math. Theor.* **2012**, *45*, 015308. [CrossRef]

44. Man'ko, V.I.; Seilov, Z. The partition formalism and new entropic-information inequalities for real numbers on an example of Clebsch-Gordan coefficients. *J. Russ. Laser Res.* **2017**, *38*, 50–60. [CrossRef]

45. Man'ko, M.A.; Man'ko, V.I. Hidden correlations and entanglement in single-qudit states. *J. Russ. Laser Res.* **2018**, *38*, 301–311. [CrossRef]

46. Chernega, V.N.; Man'ko, O.V.; Man'ko, V.I. Triangle geometry of the qubit state in the probability representation expressed in terms of the Triada of Malevich's Squares. *J. Russ. Laser Res.* **2017**, *38*, 141–149. [CrossRef]

47. Lopez-Saldivar, J.A.; Castaños, O.; Nahmad-Achar, E.; López-Peña, R.; Man'ko, M.A.; Man'ko, V.I. Geometry and entanglement of two-qubit states in the quantum probabilistic representation. *Entropy* **2018**, *20*, 630. [CrossRef]

48. Man'ko, M.A.; Man'ko, V.I. From quantum carpets to quantum suprematism—The probability representation of qudit states and hidden correlations. *Phys. Scr.* **2018**, *93*, 084002. [CrossRef]

49. Chernega, V.N.; Man'ko, O.V.; Man'ko, V.I. New inequality for density matrices of single qudit states. *J. Russ. Laser Res.* **2014**, *35*, 457–461. [CrossRef]

50. Devoret, M.H.; Schoelkopf, R.J. Superconducting circuits for quantum information: An outlook. *Science* **2013**, *339*, 1169–1174. [CrossRef] [PubMed]

51. Pashkin, Y.A.; Yamamoto, T.; Astafiev, O.; Nakamura, Y.; Averin, D.V.; Tsai, J.S. Quantum oscillations in two coupled charge qubits. *Nature* **2003**, *421*, 823–826. [CrossRef] [PubMed]

52. Man'ko, V.I.; Marmo, G.; Sudarshan, E.C.G.; Zaccaria, F. Interference and entanglement: An intrinsic approach. *Phys. A Math. Gen.* **2002**, *35*, 7137–7157.

53. Man'ko, V.I.; Marmo, G.; Sudarshan, E.C.G.; Zaccaria, F. Positive maps of density matrix and a tomographic criterion of entanglement. *Phys. Lett. A* **2004**, *327*, 353–364. [CrossRef]

54. Man'ko, M.A.; Man'ko, V.I.; Mendes, R.V. Tomograms and other transforms: A unified view. *J. Phys. A Math. Gen.* **2001**, *34*, 8321–8332. [CrossRef]

55. Mendes, R.V. Non-commutative tomography and signal processing. *Phys. Scr.* **2015**, *90*, 074022. [CrossRef]

56. Ville, J. Theorie et Applications de la Notion de Signal Analytique. *Cables Transm.* **1948**, *24*, 61–74. (In French)

entropy

MDPI

Article

"The Heisenberg Method": Geometry, Algebra, and Probability in Quantum Theory

Arkady Plotnitsky

Literature, Theory, Cultural Studies Program, Purdue University, West Lafayette, IN 47907, USA; plotnits@purdue.edu

Received: 2 June 2018; Accepted: 23 August 2018; Published: 30 August 2018

Abstract: The article reconsiders quantum theory in terms of the following principle, which can be symbolically represented as *QUANTUMNESS* → *PROBABILITY* → *ALGEBRA* and will be referred to as the QPA principle. The principle states that the quantumness of physical phenomena, that is, the specific character of physical phenomena known as quantum, implies that our predictions concerning them are irreducibly probabilistic, even in dealing with quantum phenomena resulting from the elementary individual quantum behavior (such as that of elementary particles), which in turn implies that our theories concerning these phenomena are fundamentally *algebraic*, in contrast to more geometrical classical or relativistic theories, although these theories, too, have an algebraic component to them. It follows that one needs to find an algebraic scheme able make these predictions in a given quantum regime. Heisenberg was first to accomplish this in the case of quantum mechanics, as matrix mechanics, whose matrix character testified to his algebraic method, as Einstein characterized it. The article explores the implications of the Heisenberg method and of the QPA principle for quantum theory, and for the relationships between mathematics and physics there, from a nonrealist or, in terms of this article, "reality-without-realism" or RWR perspective, defining the RWR principle, thus joined to the QPA principle.

Keywords: algebra; causality; geometry; probability; quantum information theory; realism; reality

---

> Perhaps the success of the Heisenberg method points to a purely algebraic method of description of nature, that is, to the elimination of continuous functions from physics. Then, however, we must give up, in principle, the space–time continuum.
>
> —Albert Einstein, "Physics and Reality" (1936)

1. Introduction

This article reconsiders quantum theory, from quantum mechanics to quantum field theory to quantum information theory, primarily focusing on quantum mechanics, in terms of the following principle, which can be symbolically represented as:

$$QUANTUMNESS \rightarrow PROBABILITY \rightarrow ALGEBRA$$

and will be referred to as the QPA principle. This principle states, first, defining the experimental nature of my first implication, *QUANTUMNESS* → *PROBABILITY*, that the quantumness of physical phenomena, that is, the specific character of physical phenomena known as quantum, implies that our predictions concerning them are irreducibly probabilistic or statistical, even in dealing with quantum phenomena resulting from the elementary individual quantum behavior (such as that of elementary particles). This, in turn implies, defining the theoretical character on my second implication, *PROBABILITY* → *ALGEBRA*, that our theories concerning these phenomena, quantum theories, are fundamentally algebraic, in contrast to more geometrical classical or relativistic theories.

Although this distinction is not unconditional, because quantum theories do have geometrical aspects, while, conversely, geometrical theories have algebraic aspects, it is, I shall argue, irreducible; and understanding this distinction, along with the shared aspects of both types of theories, helps us to shed new light on the relationships between algebra and geometry in physics.

Physically, quantum phenomena are defined by the fact that, in considering them, Planck's constant, h, must be taken into account, which allows one to use the classical theory in describing them, but not in predicting them. By "quantum physics" I shall refer to the overall assembly of the available quantum phenomena and theoretical accounts of these phenomena. The terms "classical physics" will be used along parallel lines for classical phenomena, which need not depend on h (or on c, the role of which defines relativistic phenomena). While, however, the role of h is irreducible in quantum phenomena, their specificity as quantum is defined by a broader set of physical features, such as the uncertainty relations (which do contain h), complementarity, and quantum correlations, some of which are not linked to h, at least not expressly. On the other hand, some of these features, although not all of them, are also exhibited by classical phenomena or found in mathematical models different from those of the standard quantum mechanics or quantum field theory. The ultimate distinction between quantum and classical phenomena is the subject of ongoing investigations and debates, on which subject I shall further comment below. From present perspective, h may not pertain to quantum objects or behavior but only to our theories, and enters these theories via the interactions between quantum objects and measuring instruments.

Quantum phenomena also obey the principle of discreteness or the QD principle, which, necessarily coupled to the individuality of quantum phenomena, may indeed represent the "essence" or "quantumness" of quantum theory, according to Bohr [1]. In other words, quantum phenomena are individual and discrete in relation to each other, which, as emphasized by N. Bohr, is not the same as the atomic, Democritean, discreteness of elementary quantum objects themselves, which was initially (following Planck's discovery of quantum physics in 1900) seen as defining quantum physics as quantum [2]. By "elementary" I refer to those quantum objects, also known as "elementary particles," that cannot be considered as composite. Either character, elementary or composite, could be ascertained on the basis of effects such objects have on measuring instruments, keeping in mind that some particles considered elementary can reveal themselves to be composite, as it happened in the case of hadrons that were found to be composed of quarks and gluons.

It follows that the difference between objects and phenomena is irreducible in quantum theory, as opposed to classical theory, specifically classical mechanics, which deals with individual classical objects or simple classical systems. Rigorously speaking, as Kant already realized, this difference exists there as well, but it can be disregarded insofar as we can, at least ideally and in principle, consider such objects by neglecting the interference of observation. This is not possible in quantum physics, at least not explicitly. It is under debate whether it is possible indirectly, inferentially, to establish the independent nature and behavior of quantum objects, similarly to the way in which we can treat, by means of classical mechanics, the nature and behavior of the elemental constituents of the systems considered in classical statistical physics, even though we do not directly observe this behavior. (This approach helped the nineteenth-century physics to confirm the existence of atoms.) The interpretation of quantum phenomena adopted in this article, following Bohr and Heisenberg (in his early work, discussed below [3]), precludes this possibility of attributing any independent properties to quantum objects, although it does not preclude alternative interpretations of quantum phenomena. The QPA principle would still hold for most of these interpretations, possibly, in contrast to the present view, under the assumption of a continuously connected underlying reality. There is, thus, another implication: *QUANTUMNESS* → *DISCRETENESS*. I shall, however, subsume the discreteness of quantum phenomena under quantumness.

I shall discuss the concepts of "geometry," "algebra," and "probability," and the relationships between them in more detail below. Briefly, I understand geometry as the mathematical formalization of spatiality in terms of measurement (while topology as referring to the structure of spatiality apart from measurement), algebra as the mathematical formalization of the relationships between symbols, arithmetic as dealing with numbers, and probability as the mathematical formalization the likelihood of events and expectations concerning them (the corresponding mathematical fields are geometry, topology, algebra, number theory, and probability theory). Geometrical and topological objects always have algebraic components, while algebraic objects need not have a geometrical component.

Now, there is next to nothing geometrical about probability or probability theory. The origin of probability theory coincides with the rise of algebra, in the works of Cardano, Fermat, Descartes, and Pascal. Some form of algebra was necessary for probability theory, as Hacking persuasively argued in explaining why the theory emerged in the seventeenth century rather than earlier [4]. Analytic geometry and calculus were introduced around the same time by, the first by Fermat and Descartes, and the second by Newton and Leibniz (although Fermat was, again, an important precursor, especially as concerns the algebraic aspects of calculus), and these fields, too, were the product of the algebraization of mathematics, a defining feature of the mathematics and physics of modernity, even though geometry continued to dominate both until the nineteenth century.

It is true that probability theory, uses spatialized mathematical concepts, such as that of "probability space," introduced by Kolmogorov as part of his axiomatization of probability theory, just as quantum mechanics uses the concept of "Hilbert space." Kolmogorov's concept follows the concepts of "space" developed in functional analysis and measure theory (which Kolmogorov used to axiomatize probability theory [5]), that of Hilbert space, among them. As I shall argue, however, these concepts are more algebraic than geometrical: They have algebraic structures that geometrical objects possess but are not aimed at representing the physical space, the original and still continuing task of geometry, even though it has, as a mathematical field, developed far beyond its concerns with nature or its relations to physics.

Quantum mechanics reshaped the relationships between the algebra of probability and the algebra of theoretical physics, as against previous uses of probability, for example, in classical statistical physics. There the relationships between them is underlain by a geometrical picture of the behavior of the individual constituents of the systems considered, assumed to follow the laws of classical mechanics. By contrast, as became apparent beginning with Planck's discovery of quantum phenomena, even elementary individual quantum objects and the events they give rise to had to be treated probabilistically. One needed, accordingly, to find a new theory to make correct probabilistic or statistical predictions concerning them, a task that quantum theory pursued from its inception, with mixed results. Heisenberg was able to accomplish this task with quantum mechanics as matrix mechanics, which avoided the deficiencies of "the old quantum theory," as it became called after the introduction of quantum mechanics, and which only predicted the probabilities of what was observed in measuring instruments, as quantum phenomena, without describing the behavior of quantum objects [3]. Heisenberg's use of his matrix variables as operators in linear vector spaces (essentially, infinite-dimensional Hilbert spaces over **C**) defined the algebraic nature of "the Heisenberg method," as Einstein characterized it [6]. This was in contrast to Schrödinger's more geometrical method in his wave mechanics, accompanied by a geometrical conception of quantum-level reality in terms of a continuous vibrational process, a conception never worked out by Schrödinger to accord with the experimentally established discrete features of quantum phenomena. Indeed, the physical and mathematical demands of accounting for these features led Schrödinger to a mathematically equivalent scheme [7]. I shall explain below why Schrödinger's equation may be seen in probabilistically predictive terms, without (geometrically) representing either the propagation of wave or the motion of particles in space and time.

Bohr, in his initial assessment of Heisenberg's discovery in 1925, before Schrödinger's wave version, but after Born and Jordan's paper, which gave Heisenberg's initial scheme its proper matrix form [8], explained Heisenberg's approach as follows [9]:

> In contrast to ordinary mechanics, *the new quantum mechanics does not deal with a space–time description of the motion of atomic particles.* It operates with manifolds of quantities which

replace the harmonic oscillating components of the motion and symbolize the possibilities
of transitions between stationary states in conformity with the correspondence principle
[which requires that quantum and classical predictions coincide in the classical limit].
These quantities satisfy certain relations which take the place of the mechanical equations of
motion and the quantization rules [of the old quantum theory]. (emphasis added)

Thus, in effect combining the QPA principle with another principle, the *reality-without-realism*,
RWR, principle (defined by the fact that "the new quantum mechanics does not deal with a space–time
description of the motion of atomic particles" and explained in detail below), Heisenberg's quantum
mechanics was a radical departure from the preceding history of modern (mathematical-experimental)
physics, from Galileo's mechanics to Einstein's relativity and even to the previous quantum theory.
All these theories were based on such descriptions or representations, either phenomenally visualizable,
usually on geometrical *lines* (referring to the continuous motion of the objects considered), at least in
dealing with elementary individual processes, as in classical mechanics, or beyond visualization but
given a conceptual or mathematical representation, as in relativity when dealing with photons and
velocities close to *c*. Classical statistical physics still relied and even depended on the representational
treatment of the behavior of the elemental constituents of the multiplicities considered, constituents
that were viewed as behaving in accordance with the laws of classical mechanics. The description of
individual quantum objects and the corresponding mathematical representation became partial in
the so-called old quantum theory, such as Bohr's atomic theory, introduced in 1913 and developed
by him and others over the following decade [10,11]. The old (semiclassical) quantum theory only
provided such a representation, in terms of orbits, for stationary states of electrons in atoms, but
not for the discrete transitions, "quantum jumps," between stationary states. The relativistic law
of addition of velocities (defined by the Lorentz transformation) in special relativity, $s = \frac{v+u}{1+(vu/c)^2}$,
for collinear motion (*c* is the speed of light in a vacuum), runs contrary to any intuitive (geometrical)
representation of motion that we can have. This concept of motion is, thus, no longer a mathematical
refinement of a daily concept of motion in the way the classical concept of motion is. Relativity was
the first physical theory that defeated our ability to form a phenomenal conception of an elementary
physical process, and it was a radical change in the history of physics. The reason that this aspect of
relativity worried Einstein less than quantum mechanics was that relativity still offered a conceptual,
as well as mathematical, representation of the behavior of individual systems, which could, moreover,
be handled deterministically rather than probabilistically. Besides, there was plenty of geometry in
the theory, all the more so in general relativity, the geometry grounded in the spacetime continuum,
which Einstein wanted to preserve as part of physical reality at all levels and which was threatened
by the Heisenberg method [6]. Ultimately, photons are quantum objects and are treated by quantum
electrodynamics, which, in the present view, no longer represents the behavior of quantum objects in
the corresponding (high-energy) regimes any more than quantum mechanics.

Heisenberg abandoned a geometrical representation of stationary states retained in Bohr's 1913
theory and, thus, any description or representation, even a mathematical one, of *the behavior of quantum
objects*, in accordance with the RWR principle. Stationary states were only represented algebraically
by energy-level values, considered apart from representing the behavior of electrons themselves
to which these energy levels were assigned. One was no longer thinking, as in classical physics or
relativity, in terms of predicting, even in considering elementary individual processes, the spacetime
behavior of the objects considered due to (continuous) changes in their states, assumed to be definable
independently of the interactions between objects and measuring instruments, and representable
by the corresponding formalism. Instead, following Bohr's thinking concerning quantum jumps in
his 1913 atomic theory, one was thinking in terms of discontinuous transitions between physical
states of quantum objects and the probabilities or statistics of such transitions, as only probabilistic
predictions were possible experimentally [12]. These states are, moreover, only manifested as effects
of the interactions between quantum objects and measuring instruments. While not part of Bohr's
1913 theory, this understanding became central to Bohr's thinking following Heisenberg's discovery

and came to define Bohr's interpretation of quantum phenomena and quantum mechanics. (Although quantum phenomena and quantum mechanics are commonly interpreted jointly, as they were by Bohr or are here, quantum phenomena could be given an interpretation independent of a theory accounting for them.)

Heisenberg's theory may be seen in terms of transition from geometry to algebra in fundamental physics, which was acutely sensed by Einstein, who was hardly welcoming this transition [6]:

> [P]erhaps the success of the Heisenberg method points to a purely algebraic method of description of nature, that is, to the elimination of continuous functions from physics. Then, however, we must give up, in principle, the space–time continuum [at the ultimate level of reality]. It is not unimaginable that human ingenuity will some day find methods which will make it possible to proceed along such a path. At present however, such a program looks like an attempt to breathe in empty space.

Einstein was equally unhappy with the recourse to probability in dealing with elementary individual processes, and he thought it would be avoided by a kind of fundamental theory he envisioned, a continuous geometrical field theory of the type general relativity was. Earlier, he referred to Heisenberg's scheme as a magical trick, "Jacob's pillow," of Göttingen: it was not "the real thing" and "[did] not really bring us any closer to the secret of the 'old one,'" who, Einstein added in his famous pronouncement, "at any rate is ... not playing at dice" [13]. Ultimately, Einstein was more concerned with the absence of realism at the fundamental level. As, however, he must have realized, randomness and the recourse to probability are automatic in this absence.

I shall also argue here, as a bridge to considering quantum information theory, that, while not, technically, quantum-informational, Heisenberg's thinking could be viewed as quantum-informational *in spirit*, and conversely, quantum information theory as Heisenbergian in spirit, and thus both as algebraic in spirit [14,15]. The reason for this view is that the quantum-mechanical situation, as Heisenberg conceived of it, was, in retrospect, defined by:

(a)   certain *already obtained* information, concerning the energy of an electron, derived from spectral lines (due to the emission of radiation by the electron), *observed* in measuring instruments; and

(b)   certain possible future information, concerning the energy of this electron, *to be obtainable* from spectral lines *to be observed* in measuring instruments and predictable (on experimental grounds) in probabilistic or statistical terms by the mathematical formalism of one or another quantum theory.

Heisenberg's aim was to develop such a formalism without assuming that this formalism needed to represent a spatiotemporal process connecting these two sets of information or how each set comes about. Heisenberg's quantum mechanics was about quantum information, albeit not only about it. It was equally about the nature of quantum objects, even though and because this nature was beyond human knowledge and even thought. But then, this is also true about much foundational thinking in quantum information theory, which aims to understand the ultimate nature of reality through the nature of quantum information.

The remainder of this article proceeds as follows: the next section outlines my main concepts. Section 3 addresses algebra and geometry in fundamental physics. Section 4 revisits Heisenberg's discovery of quantum mechanics and Bohr's interpretation of it. Section 5 considers some recent work in quantum information theory.

2. Fundamentals of the QPA/RWR Approach to Quantum Theory

The currently standard version of quantum theory, the only one to be considered in this article, is comprised of three theories, all discovered in quick succession between 1925 and 1928. The first is quantum mechanics for continuous variables in infinite-dimensional Hilbert spaces (QM), the second is quantum theory for discrete variables in finite-dimensional Hilbert spaces (QTFD), and the third is quantum field theory in Hilbert spaces that are tensor products of finite and infinite dimensional Hilbert spaces (QFT), initially introduced in the form of quantum electrodynamics (QED). All these theories are algebraic and probabilistic or statistical and are governed by the QPA principle. QFT, which handles high-energy physics, is comprised of several theories, constituting the standard model of particle physics: quantum electrodynamics (QED), the theory of weak forces, and the theory of strong forces, quantum chromodynamics (QCD). While the first two are unified or (as some prefer to see it) "merged" in the electroweak theory, the unification of all three, known as "grand unification," has not been achieved. More troubling is that QFT and general relativity are inconsistent with each other. This inconsistency is one of the greatest outstanding problems of fundamental physics, which motivated string and M-brane theories, and alternative approaches, including in quantum information theory (e.g., [16]).

The interpretation of quantum phenomena and quantum theory adopted here is defined by a nonrealist or "reality-without-realism" (RWR) view of quantum theory in any of its versions, a view that follows "the Copenhagen spirit of quantum theory" [*Kopenhagener Geist der Quantenheorie*], as Heisenberg called it [17]. This characterization, abbreviated here to "the spirit of Copenhagen," is preferable to the more common "Copenhagen interpretation," because there is no single such interpretation, even in the case of Bohr, who changed his views a few times [18] (here I shall be primarily concerned with the ultimate version of his interpretation). This is an important point. First, there is much confusion concerning this fact by both critics and advocates of Bohr and the spirit of Copenhagen. Secondly, at stake *are interpretations*, those (again, several) in the spirit of Copenhagen amidst still others, and not the ultimate truth of nature, which we do not know and may never know or even imagine and concerning which this article makes no definitive claims. In most of this article, I will be concerned with QM. I will give some attention to QTFD in the context of quantum information theory, which has been primarily concerned with it. QFT will only be mentioned in passing, although the QPA principle and the RWR principle apply there, and historically, QFT, beginning with QED, has been used to support the spirit of Copenhagen all along. I shall now outline the key concepts grounding my argument, in part in order to avoid misunderstandings concerning them, because these concepts or, more accurately, concepts designated by these terms can be defined otherwise.

It is fitting to begin by addressing the concept of concept, first, because, it is rarely adequately considered in physical or philosophical literature, and secondly and more importantly, because the role of concepts is not sufficiently appreciated in the philosophy of physics, especially the analytic philosophy of physics. If, as Wilczek, a leading elementary particle theorist and a Nobel Prize laureate, argues, "the primary goal of fundamental physics is to discover profound concepts that illuminate our understanding of nature," then creative thinking in fundamental physics is defined by concepts and is advanced by the discovery or invention of new concepts [19]. But what is a physical concept, and what is a concept in the first place? Wilczek does not explain it, taking it for granted or assuming some general sense of it presumably shared by his readers. One might safely assume, given the specific concepts that Wilczek invokes, such as that of "elementary particle" associated with that of "symmetry group," that the concepts in question have mathematical components, the presence of which has defined the concepts of all modern, post-Galilean, theoretical physics. I shall also understand a physical theory as an organized assemblage of concepts in the sense about to be defined, an assemblage that relate certain physical objects or phenomena, usually in terms of propositions that are considered to be true, at least with a sufficient practical, even if not fully definitive, justification.

It is the latter aspect that tends to dominate the concepts of theory used in the analytic philosophy of physics. This aspect is of course indispensable: no physical theory, or philosophical argument concerning theoretical physics, can bypass it. I would, nevertheless, argue, following Borel's 1907 critique of the logically based understanding of mathematics, which is, in my view, applicable to the logically based understanding of theoretical physics as well [20]. For Borel, a truly fertile invention in mathematics and theoretical physics alike consists of the discovery of new concepts that enable a new point of view from which to interpret the facts, followed by a search for the necessary proofs by plausible reasoning, and only then, necessarily, bringing logic in. According to Gray, "Borel's criticisms point quite clearly toward a problem that has not gone away in philosophers' treatment of mathematics: a tendency to reduce it to some essence than not only deprives it of purpose but is false to mathematical practice. The logical enterprise, even if it had succeeded, would only have been an account of part of mathematics, its deductive skeleton" [21]. This, I would contend, is true about much of the analytic philosophy of physics, again, indispensable as the propositional and logical aspects of theoretical physics are, in physics also as concerns correspondence with the available experimental evidence.

Bohr clearly understood the significance of these aspects of a physical theory and specifically QM, and used them in addressing Einstein's criticism: "In my opinion, there could be no other way to deem a logically consistent mathematical formalism as inadequate than by demonstrating the departure of its consequences from experience or by proving that its predictions did not exhaust the possibilities of observation, and Einstein's argumentation could be directed to neither of these ends" [22]. Bohr also understood, however, now in agreement with Einstein, that the invention of concepts play a decisive creative role in theoretical physics. Einstein saw "conceptual construction" [*begrrifliche Konstruction*] as essential to and irreducible in physics [23]. He also saw the practice of theoretical physics as that of the invention of new concepts through which one can approach reality, sometimes even to the point of overriding the experimental evidence [24]. Riemann, a major inspiration for Einstein's general relativity, including as concerns conceptual construction, observed already in 1854: "From Euclid to Lagrange this darkness [in our understanding of the nature of geometry] has been dispelled neither by the mathematicians nor the philosophers who have concerned themselves with it. The reason [*Grund*] for this is undoubtedly because the general concept of multiply extended magnitudes, which includes spatial magnitudes, remains completely unexplored. I have therefore first set myself *the task of constructing the concept* of a multiply extended magnitude from general notions of magnitude" [25]. This led Riemann to his concept of manifold [*Maningfaltigkeit*], central to modern geometry and topology [26]. This article, too, while recognizing the indispensability of logical and propositional structures of physical theories, gives concepts and the invention of concepts the defining role the creative practice of theoretical physics.

I shall adopt the following understanding of concepts, in part following Deleuze and Guattari, whose thinking was inspired by Riemann and especially his concept of manifold [26,27]. In this definition, a concept is not merely a generalization from particulars (which is commonly assumed to define concepts) or a general or abstract idea, although a concept may contain such ideas, specifically abstract mathematical ideas in physics or in mathematics itself, where these ideas may become concepts in the present sense. A concept is a multicomponent entity, defined by the *organization* of its components, and some of these components may be concepts in turn. The definition may be very basic, but it reflects an essential character of concepts in any domain, from daily life to the stratosphere of mathematics and science. What is crucial is how this basic architecture is specifically instantiated in a given concept, which is defined by both the nature of the components and their organization, by how they relate to each other in the structure of the concept.

Consider as, an example, the concept of motion, first, as it is used in daily life: it will involve various components, such as a change of place, speed, acceleration, moving bodies, etc., which belong to our phenomenal intuition and are not defined rigorously, especially mathematically, but are still parts of the concept defined by the organization of these components. Now, one can, as both Bohr and Heisenberg did, see classical mechanics as a physical and mathematical refinement of these daily concepts by means of such mathematically defined concepts as coordinates, momentum, angular momentum, energy, and so forth, and thus also that of motion. While the concepts of classical physics are derived from the concepts of daily life, they are both mathematical and subject to an experimental verification. A concept could also be borrowed from a preceding physical theory and modified (or left intact). Every concept and every theory, no matter how innovative, has a history and depends on it. In quantum theory (QM, QFT or QTFD), in RWR-type interpretations, classical concepts are no longer applicable to quantum objects and their behavior, while they still have a limited applicability at the level of quantum phenomena, defined by effects of the interaction between quantum objects and measuring instruments. The concepts of quantum theory still have their history in classical physics, both physically (when applied to measuring instruments) and mathematically, for example, by adopting the concept of the Hamiltonian, while changing the variables from those of functions of real variables of classical mechanics to operator variables (in Hilbert spaces over **C**) in QM or QFT. The standard conservations laws (those of momentum, energy, and angular momentum), too, are preserved, although new conservation laws are added, such as the conservation of probability current in QM and QFT, or the conservation of baryon or lepton number in QFT. According to Heisenberg [28]:

> The concepts of velocity, energy, etc., have been developed from simple experiments with common objects, in which the mechanical behavior of macroscopic bodies can be described by use of such words. The same concepts have then been carried over to the electron, since in certain fundamental experiments electrons show a mechanical behavior like that of the objects of common experience [or classical mechanics]. Since it is known, however, that this similarity exists only in a certain limited region of phenomena, the applicability of the corpuscular theory must be limited in the corresponding way . . .

> As a matter of fact, it is experimentally certain only that light [too] sometimes behaves as if it possessed some of the attributes of a particle [as reflected in the uncertainty relations], but there is no experiment which proves that it possesses all the properties of a particle; similar statements hold for matter [e.g., electrons] and wave motion. The solution of the difficulty is that the two mental pictures [derived from classical physics] which experiments lead us to form—the one of particles, the other of waves—are both incomplete and have only the validity of analogies which are accurate only in limited cases. It is a trite saying that "analogies cannot be pushed too far," but they may be justifiably used to describe things for which our language has no words. Light and matter are both single entities, and the apparent duality arises in the limitation of our language.

In the RWR-type view, quantum objects and behavior are beyond any representation, including mathematical one (which need not depend on language or physical concepts) or, in the view adopted here, are beyond conception. Either view transforms the wave-particle duality into the viewpoint defined by the concept complementarity, moreover, in the way in which, contrary to a common view of complementarity, there is no wave-particle complementarity. As Bohr noted, in referring to complementarity, a word that does not appear to have been used as *a noun* before Bohr (as opposed to the adjective "complementary") and was introduced by Bohr to designate a new concept: "In the last resort an artificial word like 'complementarity' which does not belong to our daily concepts serves only briefly to remind us of the epistemological situation [found in quantum physics], which at least in physics is of an entirely novel character" [29]. Both the epistemological situation in question, essentially that of the RWR-type, and the architecture of the concept of complementarity, which does a great more than merely serving as such a reminder, will be discussed below. My main point at the moment is that complementarity is a new physical concept with several interrelated components. As most innovative concepts, complementarity, when introduced, was not defined by generalization from available entities: it was something entirely new, although it, too, had its history in physics and beyond [30,31]. It then functioned, in part, by generalizing multiple specific entities, such as specific complementary configurations, say, those of the position or the momentum measurement, always mutually exclusive at any given moment of time in the case of quantum phenomena. All physical concepts, including those of classical physics (which are closer to our daily concepts), are *physical* concepts, with mathematical components, ultimately divorced from their daily meaning and should be treated as such, which is not always the case, especially when it comes to complementarity and other concepts introduced by Bohr (e.g., [31]).

These considerations extend to the concept of theory, again, *as understood here*, as other definitions of this concept are possible: a theory is an organized (conceptually, logically, or otherwise) assemblage of concepts, as just defined. Every theory, again, has its history in preceding theories and can change by modifying its concepts or the relationships among them. A viable physical theory must, however, relate, by means of logically consistent and experimentally verifiable propositions (possibly probabilistic or statistical in nature), to the multiplicity of phenomena or objects that are assumed to form the reality considered by this theory. This relation, in modern physics usually by means of mathematical models (defined below), might be representational, and derive its predictive capacity, essential for any physical theory, from this representation, or be merely predictive, possibly only probabilistically or statistically predictive. I refer to both phenomena and objects, because, as Kant realized, they are not the same even in classical mechanics, which deals with individual classical objects or sufficiently small classical systems. However, classical objects, say, planets moving around the Sun, and our phenomenal representation of them could be treated as the same for all practical purposes. This is because our observational interference could, in principle, be neglected or compensated for, thus allowing us to consider this behavior independently, a circumstance that does not appear to be expressly noted by Kant but that is crucial to Bohr, because this is no longer the case in quantum physics [32,33]. Doing so was assumed to be possible, at least in principle, in the case of all classical physical objects, even when they were not or even could not be observed, as in the case of atoms or molecules in the kinetic theory of gases.

Quantum phenomena put this assumption into question. Defined by the effects of the interactions between quantum objects and measuring instruments, quantum *phenomena* are observable in the same way as are classical physical objects and could be treated as classical objects. By contrast, the "uncontrollable" (quantum) nature of these interactions precludes any observation and, in RWR-type interpretations, an inferential reconstitution of the independent behavior of quantum *objects* [34,35]. Nobody has ever observed a moving electron or photon as such, independently, to the degree that the concept of motion, as opposed to a change of a state, ultimately applies to them, or any kind of quantum object. It is only possible to observe traces, such as spots on photographic plates, left by their interactions with measuring instruments. This still allows for a spectrum of assumptions concerning quantum objects and their behavior, beginning with the assumption of the existence of such objects (or what is so idealized), inferred from these traces.

The present interpretation, while assuming this existence on the basis of these effects and their particular character (not found in classical physics), places quantum objects and behavior beyond conception, which I shall term "the strong RWR view," rather than only representation, which I shall term "the weak RWR view." The strong RWR view is a radical position. Not all interpretations in the spirit of Copenhagen go that far. Thus, while there are indications that Bohr, especially in the ultimate version of his interpretation (my primary focus here), might have agreed with this view (e.g., [35]), he never expressly stated so. One could assume the possibility of a mathematical representation of quantum objects and behavior in the absence of a physical conception of them. While Bohr's and the present view exclude this possibility, Heisenberg was open to it in his later thinking (e.g., [36]).

The history of a theory is accompanied by the history of its interpretations. The history of QM, in particular, has been shaped by a seemingly uncontainable proliferation of, sometimes conflicting, interpretations. It is not possible to survey these interpretations here. Each rubric on by now a long list (e.g., the Copenhagen, the many-worlds, consistent-histories, modal, relational, transcendental-pragmatist, and so forth) contains different versions. The literature dealing with each interpretation is immense. Standard reference sources would list and summarize most common rubrics. Although often implicit, an interpretation is essential for establishing the relationships between a theory and the phenomena or objects it considers, essential to any theory. This is customarily done by means of mathematical models.

I define a *mathematical* model *in physics* as a mathematical structure or a set of mathematical structures that enables such relationships (the concept of models in mathematics or mathematical logic is a separate subject, put aside here). As that of theory and other major concepts discussed here, the concept of a mathematical model or model, in the first place, has a long history, which is also a history of diverse definitions, and literature on the subject is extensive as well. It is not my aim to discuss the subject as such or engage with this literature, which would be difficult within the scope of this article. The present concept of a mathematical model, while relatively open, is sufficient to accommodate those models that I shall consider. A more detailed discussion of the present view of mathematical models is given [37,38] and of modeling in general, on the lines of analytic philosophy of science, in [39,40]. One must also keep in mind the difference between a mathematical model *of* a theory (a concept especially important in mathematical logic or the philosophy of mathematics) and a mathematical model *used by* a theory, with which I am concerned here. Mathematical models used in physics may be geometrical (as in general relativity, for example) or algebraic, as in QM and QFT, although geometrical models contain algebraic elements. (The geometrical aspects of algebraic models are a more complex matter considered below.) The relationships between a model and the objects or phenomena considered may be representational. In this case the elements of a model and relations among them would correspond or map the elements of reality and the relations among them and relate the theory to reality by means of this mathematical representation. The predictive capacity of the theory, essential for any theory, would then derive this representation. The mathematical models used in classical mechanics or relativity are examples of such models. Models may, however, also be strictly predictive, without being representational, as are the mathematical models used in QM or QFT, in RWR-type interpretations, the predictions of which are, moreover, probabilistic or statistical, again, even in the case of elementary individual quantum objects and behavior. An interpretation of a given theory is, thus, always an interpretation of how the mathematical model or models used by it relate to the phenomena or objects considered. A theory may, however, involve other interpretive aspects, defined by its concepts. For example, part of Bohr's and, following Bohr, the present interpretation is a particular (interpretive) concept of measuring instruments used in quantum physics. According to this concept, the observable parts of measuring instruments are described my means of classical physics, while these instruments also have quantum parts, through which they interact with quantum objects, an interaction "irreversibly amplified" to what is observed in measuring instruments [41]. Placing quantum objects and behavior beyond representation or even conception is another interpretative feature, in the second case without giving quantum objects and behavior concepts.

Rigorously, a different interpretation of a given mathematical model defines a different theory in the present definition of a theory, because this interpretation may involve specific concepts, such as the ones just mentioned in the case of the interpretation adopted here, which may not be shared by other interpretations, even those in the spirit of Copenhagen. For simplicity, however, I shall speak of the corresponding interpretation of the theory itself containing a given mathematical model, interpreted by this theory, say, of one or another interpretation of QM. Thus, initially, Heisenberg's and Schrödinger's *versions of the formalism* appeared as two different mathematical models, giving the same predictions. They were also accompanied by *two different theories*, initially designated quantum mechanics and wave mechanics, the first strictly algebraic and the second geometrical, by virtue of conceiving quantum-level reality as a continuous wave-like process, each theory, moreover, given different interpretations at the time (actually, interpreting either theory posed major difficulties then). These two models were quickly proven to be mathematically equivalent (there are several proofs, all of which involve additional assumptions and complexities, the most general one given by the Stone-von Neumann theorem), which allowed one to unify the mathematical model of QM, ultimately in terms of its Hilbert-space formalism, with some yet more abstract versions added later. By contrast, the two theories—quantum mechanics (underlying the model of matrix mechanics) and wave mechanics—which were based in two different sets of concepts remained different. While Schrödinger's theory, based in the idea of a wave-like ultimate reality, had receded by the late 1920s, as Schrödinger's wave function received an interpretation as a tool for predicting probabilities, rather than representing any physical process, this theory has never been entirely abandoned.

I now turn to the concept of reality, which I shall approach via Bohr's elaboration, partially cited above, concerning the epistemological situation that the concept of complementarity reflects [29]:

> The renunciation of the ideal of causality in atomic physics which has been forced on us is
> founded logically only on our not being any longer in a position to speak of the autonomous

behavior of a physical object, due to the unavoidable interaction between the object and the measuring instruments which [interaction] in principle cannot be taken into account, if these instruments according to their purpose shall allow the unambiguous use of the concepts necessary for the description of experience. In the last resort an artificial word like "complementarity" which does not belong to our daily concepts serves only briefly to remind us of the epistemological situation here encountered, which at least in physics is of an entirely novel character.

It follows that "our not being any longer in a position to speak of the autonomous behavior of a physical object," demands "a radical revision of our attitude toward the problem of physical reality," ultimately depriving us of realism [42]. "The renunciation of the [classical] ideal of causality," invoked by Bohr, is automatic, because, as I explain below, causality requires realism [42].

I shall now introduce a concept of reality that permits this revision. This concept itself is very general and is, arguably, in accord with most, even if not all (which would be impossible), currently available concepts of reality in realism and nonrealism, which would, respectively, assume this reality to be representable or at least conceivable and to be beyond representation or even conception. By *reality* I refer to that which exists or is assumed to exist, without making any claim concerning the *nature* of this existence, which thus may be placed beyond representation or even conception. I understand existence as a capacity to have effects on the world with which we interact and that, because it exists, has such effects upon itself. To ascertain such effects entails representation of these effects, but not necessarily of how they come about, which implies that a given theory might assume different levels of reality, some allowing for a representation or at least conceptions and others not.

In physics, the primary reality considered is that of matter, including radiation, generally governed by the concept of field, classical or quantum. The idea of matter is still a product of thought, which, however, is customarily assumed to be a product of the material processes in the brain, and thus of matter. Matter is commonly, but not always (although exceptions are rare), assumed to exist independently, and to have existed when we did not exist and to continue to exist when we will no longer exist, which may be seen as defining the independent existence of matter. This view is upheld in the RWR-type interpretations of QM, but in the absence of a representation or even conception of the character of this existence, for example, as either discrete or continuous. Discreteness only pertains to quantum phenomena, observed in measuring instruments, while continuity has no physical significance at all. It is only a feature of the formalism of QM, which, while mathematically continuous, relates to discrete phenomena by predicting the probabilities or statistics of their occurrence.

Physical theories prior to quantum theory have been realist theories, usually *representational* realist theories. Such theories aim to represent the corresponding objects and their behavior by mathematical models, assumed to idealize how nature works, an assumption sometimes referred to as "scientific realism." More exactly, as noted earlier, such a theory is a representation that is then realized by a mathematical model, which mathematically represents the reality considered. Thus, classical mechanics (used in dealing with elemental individual objects and small classical systems), classical statistical mechanics (used in dealing, statistically, with large classical systems), or chaos theory (used in dealing with classical systems that exhibit a highly nonlinear behavior) are all realist theories, as concerns the ultimate reality they consider. While classical statistical mechanics does not represent the overall behavior of the systems considered because their great mechanical complexity prevents such a representation, it assumes that the individual constituents of these systems are represented by classical mechanics. As indicated earlier, the status of these theories as realist could be questioned, on Kantian lines, even in the case of classical mechanics, where the representational idealizations used are more in accord with our phenomenal experience, which is only partially the case in relativity. However, all these cases still allow for viable idealized realist and (classically) causal models.

One could also define another type of realism, which is not representational. This realism encompasses theories that presuppose an independent structure of reality governing the behavior of the ultimate objects these theories consider, while allowing that this architecture cannot be represented, even ideally, either at a given moment in history or perhaps ever, but if so, only due to practical epistemological limitations. In the first eventuality, a theory that is merely predictive may be accepted for lack of a realist alternative, but under the assumption that a future theory will do better, in particular as a representational realist theory. Einstein adopted this view toward QM, which he expected to be eventually replaced by such a theory.

The assumption of realism of either type is abandoned or even precluded in reality-without-realism (RWR) type interpretations of quantum phenomena and QM, beginning with that of Bohr. In such interpretations, the mathematical model of QM, defined by its mathematical formalism, becomes a strictly probabilistically or statistically predictive, rather than *deterministic*, model, even in considering elementary individual quantum objects and processes, which form quantum-level reality, while suspending or even precluding a representation and possibly a conception of this reality, and an assumption that this reality is *causal*, or classically causal. I distinguish "causality" and "determinism." By classical causality I refer, ontologically, to the conception that the state of the system considered is determined at all future moments of time, once it is determined at a given moment of time, and by determinism, epistemologically, to the possibility of predicting the outcomes of such processes ideally exactly. This conception of causality has defined modern classical physics since Descartes, Galileo, and Newton, and philosophy beginning at least with Plato. As will be seen, causality may be defined differently, first, in a relativistic or local sense and, second, in a quantum-theoretical probabilistic sense. The probabilistic or statistical character of quantum predictions must, however, be maintained by realist interpretations of QM or alternative theories (such as Bohmian mechanics), to accord with quantum experiments, where only probabilistic or statistical predictions are possible. This is because the repetition of identically prepared experiments in general leads to different outcomes, and unlike in classical physics, this difference cannot be diminished beyond the limit defined by Planck's constant, h, by improving the capacity of our measuring instruments, as manifested in the uncertainty relations, which would remain valid even if we had perfect instruments.

RWR-type interpretations do assume the concept of *reality*, defined as that which is assumed to exist, without, in contrast to realist theories, making any claims concerning the *character* of this existence, which is what makes this concept of *reality* that of "reality *without* realism" [43,44]. The existence of quantum objects or something that leads to this idealization (it is still an idealization) is inferred from the totality of effects they have on world we observed, specifically on experimental technology, without making claims concerning their independent behavior. Such interpretations place quantum objects and processes either beyond representation, the weak RWR view, or more radically, beyond conception, the strong RWR view, which I adopt here. As I said, Heisenberg at the time of his discovery and Bohr at nearly all stages of his thinking held at least a weak RWR view, with Bohr eventually moving closer to the strong RWR view, while Heisenberg eventually adopted a form of mathematical realism. In 1927, Bohr briefly and ambivalently entertained the idea that independent quantum behavior and thus the ultimate nature of quantum reality could be represented, moreover, causally, by the mathematical formalism of QM, while indeterminism was introduced by measurement [45]. Bohr's ambivalence was due to the fact that one deals with the formalism over **C**, which is difficult to associate with physical representation and the fact that Schrödinger's wave equation in fact applied to the coordinate and not a real space: "The symbolic character of Schrödinger's method appears not only from the circumstance that its simplicity, similarly to that of the matrix theory, depends essentially upon the use of imaginary arithmetic quantities. But above all there can be no question of an immediate connection with our ordinary conceptions because the 'geometrical' problem represented by the wave equation is associated with the so-called co-ordinate space, the number of dimensions of which is equal to the number of degrees of freedom of the system, and, hence, in general greater than the number of dimensions of ordinary space" [46]. In any event, Bohr quickly abandoned the view that independent quantum behavior is represented by QM, under the impact of his exchanges with Einstein. However, championed by both Dirac's and von Neumann's influential books [47,48], this view has persisted and remains common [49].

Although Kant's philosophy may be seen as an important precursor the RWR view, the strong RWR view is manifestly more radical than Kant's view of noumena or things-in-themselves vis-à-vis phenomena or appearances formed in our minds. According to Kant, while noumena are unknowable, they are still in principle conceivable, especially when one's thinking is helped by what he calls "Reason" [*Vernunft*], a higher faculty than "Understanding" [*Verstand*], which only concerns phenomena, although there is no guarantee, even for Reason, that this conception is correct [50]. Even the weak RWR view is still more radical than that of Kant, because, while a conception of quantum objects and behavior is in principle possible, it cannot be unambiguously used in considering quantum phenomena, at least as things stand now. I am not saying that the strong RWR view is physically necessary, but only that it is interpretively possible. There does not appear to be any experimental data that would compel one to prefer either the strong or the weak RWR view, or to definitively claim for either anything beyond its consistency or effectiveness. These views are, however, different philosophically because they reflect different limits that nature allows our thought in reaching its ultimate constitution.

Two qualifications are in order. First, one could, in principle, see the claim concerning merely the existence or reality of something to which a theory can relate without representing it as a form of realism. This use of the term realism is sometimes found in advocating interpretations of QM that are nonrealist in the present sense (e.g., [51–53]), although none of these works adopted the strong RWR view. However, the present definition of realism or, similarly, ontology is more in accord with most understandings of realism, representational or nonrepresentational, including in considering quantum theory in physics and the philosophy of physics. Secondly, the present argument does not aim to deny realism even in the present sense, still a generally preferred view.

It could be assumed that *something* "happens" between observations, as manifested in changes that we observed in the instruments used, such as a discrete change of the energy, a "quantum jump," of an electron in an atom (statistical as any claim concerning such changes may be), if one keeps in mind the provisional nature of such concepts as "happens." According to Heisenberg [54]:

> There is no description of what happens to the system between the initial observation and the next measurement. . . . The demand to "describe what happens" in the quantum-theoretical process between two successive observations is a contradiction in adjecto, since the word "describe" [or "represent"] refers to the use of classical concepts, while these concepts cannot be applied in the space between the observations; they can only be applied at the points of observation.

The same, it follows, must apply to the word "happen" or any word we use, and we must use words and concepts associated to them, even when we try to restrict ourselves to mathematics as much as possible. There can be no physics without language, but quantum physics imposes new limitations on using it. Heisenberg adds later in the same book: "But the problems of language are really serious. We wish to speak in some way about the structure of the atoms and not only about 'facts'—the latter being, for instance, the black spots on a photographic plate or the water droplets in a cloud chamber. But we cannot speak about the atoms in ordinary language" [55].

On the other hand, as Heisenberg noted on an earlier occasion, mathematics is, "fortunately," free from the limitations of daily language and concepts, fortunately because one could take advantage of this freedom in creating QM, doing which might not even have been possible otherwise [56]. Mathematics, especially algebra (geometry is more connected to our phenomenal intuition of spatiality), also allows to circumvent the limits our phenomenal, representational intuition, also involving visualization, sometimes used, including by Bohr, to translate the German word for intuition, *Anschaulichkeit*. Bohr often spoke of quantum objects and behavior as beyond altogether beyond visualization, although ultimately for him both were beyond any representation, including mathematical one (e.g., [57,58]). As free from these limitations of language and ordinary, or philosophical or even physical concepts, mathematics could be assumed to represent quantum-level reality, as Heisenberg eventually came to believe. *Physics and Philosophy* and his other later writings do give mathematics at least some capacity to do so, still in algebraic terms, to the point of defining, following Wigner [59], elementary particles themselves as representations of symmetry groups. However, while crucial in QM or, even more so, in QFT, the role of symmetry need not depend on realism, physical or mathematical, because symmetry groups can be viewed as part of the probabilistically or statistically predictive machinery of QM and QFT. There are ontological and, specifically, geometrical symmetries, for example, those embedded in conservation laws by Noether's theorems, which apply in quantum theory, where, however, they are manifested at the macro level of measuring instruments, described classically. The concept of group is, however, algebraic, even when used, on realist lines, in classical physics or relativity, or of course in geometry or topology. In any event, a form of mathematical (algebraic) realism advocated by Heisenberg in his later works appears to exclude the application of representational language or concepts apart from mathematical ones to the ultimate constitution of reality [60]. On the other hand, as his argument in his paper introducing QM [3] (discussed in Section 4) suggests, at the time of his discovery Heisenberg appears to have seen mathematics' freedom from these limitations, while crucial for QM and even making its invention possible, in terms of its probabilistically predictive rather than representational capacity, a view held by Bohr, a least from 1928 on. For Bohr, again, a mathematical representation, even if not a conception of quantum objects and behavior was "*in principle* excluded," along with a physical one, at least as things stands now [35].

"As things stand now" is an important qualification, equally applicable to the strong RWR view, even though it might appear otherwise, given that this view precludes any conception of the ultimate reality not only now but also ever, by placing it beyond thought altogether. The qualification "as things stand now" still applies because a return to realism is possible, either on experimental or theoretical grounds even for those who hold this view. This return may take place because quantum theory, as currently constituted (QM, QFT, and QTFD), may be replaced by an alternative theory that allows for or requires a realist interpretation, or because RWR-type interpretations, either of the weak or the strong type, may become obsolete, even for those who hold this view, with quantum theory in place in its present form. As things stand now, either RWR view is interpretively possible. It is also possible, however, that the RWR view, in either weak or strong version, will remain part of our future fundamental theories, as the development of QFT theory appears to indicate. QFT has been open to the RWR-type view and the corresponding interpretation from its inception with Dirac until now and was used, specifically by Bohr, in support of this view (e.g., [61,62]). It also conforms to the QPA principle, even though it may appear and in some respects is more geometrical than QM, given the role of certain geometrical concepts, such as symmetries, in quantum theory, especially in QED, such as gauge symmetry, introduced by Weyl, initially in his (failed) attempt to (geometrically) unify general relativity and electromagnetism. In QED or QFT in general, however, the "geometry" of gauge symmetries is symbolic, and their real significance is the invariance under the corresponding gauge groups, say, as applies to the "phases" of electrons ("phase" being a symbolic concept, which physically relates to probabilities), is just part of the algebra of QFT theory which ultimately relates to the probabilities or statistics of experiments. As I said, there are spatial (or temporal) geometrical symmetries, such as those involved, by Noether's theorems, in conservation laws, or still others, which are used in QFT, or QM, but these are only manifested at macro-levels. Feynman's path integrals, which suggest trajectories and thus geometry, can be seen along the same algebraic lines (as part of the algebraic probabilistic machinery of QFT), especially given that these "paths" do not refer to the actual motion of particles, although the subject, admittedly, needs more discussion, which cannot be pursued here.

It is also true that we use visual tools, such as Feynman's diagrams. Enormously helpful as they are, however, Feynman's diagrams are only *diagrams*, heuristic devices: they do not represent the quantum processes to which they refer, even if one holds that these processes are representable. The role of Feynman's diagrams may be said to represent the predictive workings of the *formalism* of QED or QFT and thus to help one to work with this formalism in order to make probabilistic or statistical predictions concerning the outcomes of the experiments these diagrams are connected to, predictions only possible rigorously, numerically, by means of the algebra of QED or QFT. This, however, is quite different from representing quantum behavior itself. In sum, while there are additional nuances as concerns their difference, the formalism of QFT is algebraic in the same way as is that of QM, and the QPA and (interpretively) RWR principles combine analogously.

I now turn to the question of causality. As noted, RWR-type interpretations make the absence of classical causality nearly automatic. This absence is strictly automatic if one adopts the strong RWR view, which places the ultimate nature of reality beyond conception, because the assumption that this nature is classically causal would imply at least a partial conception of this reality. However, even if one adopts the weak RWR view, which only precludes a representation of this reality, classical causality is still difficult to maintain in considering quantum phenomena. This is because to do so one requires a degree of representation, analogous to that found in classical physics, that appears to be prevented, in particular, by the uncertainty relations (which are independent of QM). Schrödinger expressed this difficulty, while disparaging QM, or at least the spirit of Copenhagen, as "the doctrine born of distress," in his cat-paradox paper: "if a classical state does not exist at any moment, it can hardly change causally," where a classical state is defined by the (ideally) exact position and momentum of an object at any moment of time [63]. According to Bohr, who did not share Schrödinger's reservations [64]:

> It is most important to realize that the recourse to probability laws under such circumstances is essentially different in aim from the familiar application of statistical considerations as practical means of accounting for the properties of mechanical systems of great structural complexity. In fact, in quantum physics we are presented not with intricacies of this kind, but with the inability of the classical frame of concepts to comprise the peculiar feature[s] of the elementary [quantum] processes.

While "the classical frame of concepts" might refer to those of classical physics, Bohr might have been here closer to the strong RWR view, because at this stage of his thinking, he argues that all our representational concepts ("object" and "process," among them) are classical, possibly apart from purely mathematical concepts, considered as entirely divorced from any phenomenal representation. Indeed, according to Wittgenstein, we may be unable to conceive of a process that is not causal [65]. Complementarity is different: While it has representational aspects, referring to phenomena observed in measuring instruments and thus to our experience, it does not represent the independent properties and behavior of quantum objects, but is instead designed to deal with a lack of this representation.

The question of causality is, however, a subtle matter, especially given that one can define concepts of causality that are not classical, and it merits a further discussion. First, I shall consider the concepts of indeterminacy, randomness, chance, and probability, again, as I understand them, because they, too, can be defined otherwise. In the present definition, indeterminacy or chance is a more general category, while randomness will refer to a most radical form of indeterminacy, when even a probability is not and cannot be assigned to a possible future event. Indeterminacy (including randomness) and chance may be understood as different from each other as well. These differences are, however, not germane in the present context, and I shall for convenience only refer to indeterminacy. An indeterminate, including random, event may or may not result from some underlying classical causal processes, whether this process is accessible to us or not. The first eventuality defines classical indeterminacy or randomness, conceived as ultimately underlain by a hidden classically causal architecture; the second irreducible indeterminacy and randomness. The ontological validity of an application of the latter cannot be guaranteed: it is impossible to ascertain that an apparently indeterminate or random sequence is in fact indeterminate or random, and there is no mathematical proof that any sequence is [66]. This concept is an assumption that may only be practically justified insofar as an effective theory or interpretation is developed.

As explained, factually, quantum phenomena only preclude determinism, because identically prepared quantum experiments, as concerns the state of measuring instruments, in general lead to different outcomes. Only the statistics of multiple identically prepared experiments are repeatable. It would be difficult, if not impossible, to do science without being able to reproduce at least the statistical data. The lack of classical causality or of realism in the RWR-type interpretations of quantum phenomena and QM are *interpretive inferences* from this situation and additional features such as correlations, the uncertainty relations, or complementarity. Such interpretations, again, do not exclude the possibility of causal or realist interpretations of QM, or alternative causal or realist quantum theories, such as Bohmian mechanics (which is, however, nonlocal), or theories defined by deeper underlying causal dynamics, which makes QM an "emergent" theory, such as A. Khrennikov's "pre-quantum classical statistical field theory" [44,67].

Although sometimes glossed over, the difference between probability and statistics is important in quantum theory. I would like to briefly comment on this difference and on the role of probability and statistics in quantum theory more generally from the RWR-type perspective. My remarks cannot do justice to the subject, extensively considered in literature (e.g., [5,68]). They are only aimed to address those points that are especially relevant for my argument. "Probabilistic" commonly refers to our estimates of the probabilities of either individual or collective events, such as that of a coin toss or of finding a quantum object in a given region of space. "Statistical" refers to our estimates concerning the outcomes of identical or similar experiments, such as that of multiple coin-tosses or repeated identically prepared experiments with quantum objects, or to the average behavior of certain objects or systems. The standard use of the term "quantum statistics" refers to the behavior of large multiplicities of identical quantum objects, such as electrons and photons, which behave differently, in accordance with, respectively, the Fermi-Dirac and the Bose-Einstein statistics, for identical particles with, respectively, half-integer and integer spin. The Bayesian understanding defines probability as a degree of belief concerning a possible occurrence of an individual event on the basis of the relevant information we possess (e.g., [69] or, in a different version [70]). This makes the probabilistic estimates, generally, subjective, although there may be agreement (possibly among a large number of individuals) concerning such estimates. The frequentist understanding, also referred to as "frequentist *statistics*," defines probability in terms of sample data by emphasis on the frequency or proportion of these data, which is considered more objective. In quantum physics, as noted, exact predictions are in general impossible even in dealing with elemental individual processes and events. This situation could, however, be interpreted either on Bayesian lines, under the assumption that a probability could be assigned to individual quantum events, or on frequentist lines, under the assumption that each individual effect is strictly random. A prominent recent example of a nonrealist Bayesian approach is Quantum Bayesianism, QBism, which, however, contains other philosophical dimensions (e.g., [52]). Although most of its argument would apply if one adopts a Bayesian view, this article adopts the frequentist, RWR-type, view, considered in detail in [43,44]. Bohr and Heisenberg appear to have been inclined to a statistical view of the type adopted here [71].

A brief qualification might be in order concerning two different uses of the statistical just mentioned, concerning, respectively, multiple repeated experiment and the average behavior of large system. One can make a *probabilistic* estimate for an event of finding an electron gas occupying less than a given volume, similarly to that of finding a quantum object in a given region of space. (In fact, this is true for events pertaining to classical statistical systems.) My point here, however, is that, unlike in classical mechanics, in QM we are dealing with randomness or probabilities even in considering events associated with elemental individual objects, such as electrons, rather than with large ("statistical") multiplicities, and that such individual events can be interpreted on either Bayesian or statistical lines, in the latter case, under the assumption that such individual events are strictly random. These two interpretations would also distinguish individual events, such as those of finding an electron gas occupying less than a given volume. In the statistical interpretation in the present sense we would need to repeat this experiment many times to establish these statistics, while no probability is, in general, assigned to a given event.

Finally, probability introduces an element of order into situations defined by the role of randomness in them and enables us to handle such situations better. In other words, probability or statistics is about the interplay of indeterminacy or randomness and order. This interplay takes on a unique significance in quantum physics, because of the existence of quantum correlations, such as the EPR or (as they are also known) EPR-Bell correlations, found in the experiments of the Einstein-Podolsky-Rosen (EPR) type and considered, in the case of discrete variables, in Bell's and the Kochen-Specker theorems, and related findings. These correlations are a form of statistical order. They are properly predicted by QM, which is, thus, along with and responding to quantum phenomena themselves, as much about order as about indeterminacy or randomness, and, most crucially, about their unique combination in quantum physics. The correlations themselves are collective, statistical, and as such they would not depend on either interpretation, Bayesian or frequentist, of our predictions concerning the individual events involved. That, in certain circumstances, indeterminate or random individual events form statistically correlated and thus ordered multiplicities is, however, one of the greatest mysteries of quantum physics, which makes it as much about order as about indeterminacy and randomness, a statistically correlated order without an ontologically underlying classical order that merely cannot be accessed epistemologically.

I shall now consider two alternative conceptions of causality important for my argument. Thus, the term "causality" is often used in accordance with the requirements of special relativity, which restricts (classical) causes to those occurring in the backward (past) light cone of the event that is seen as an effect of this cause, while no event can be a cause of any event outside the forward (future) light cone of that event. In other words, no physical causes can propagate faster that the speed of light in a vacuum, *c*, which requirement also implies temporal locality. Technically, this requirement only *restricts* classical causality by a relativistic antecedence postulate, rather than precludes it, and relativity theory itself, special or general, is (locally) a classically causal and indeed deterministic theory. By contrast, while, as a probabilistic or statistical theory of quantum phenomena, QM, at least in RWR-type interpretations, lacks classical causality, its probabilistic or statistical predictions are consistent with both temporal and spatial locality, and hence the relativistic antecedence. The same is true in the case of QTFD or QFT, and QFT in its standard form conforms to special relativity (although there are nonrelativistic versions of QFT). Thus, the compatibility with relativistic or, more generally, locality requirements would be maintained insofar as an already performed experiment determines, probabilistically or (if repeated many times) statistically, a possible outcome of a future experiment, without assuming classical causality. Determinism is, again, precluded on experimental grounds. Whatever *actually happens* is defined by spatially and temporally local factors, although the probabilistic or statistical *predictions*, could concern distant events, sometimes, as in the EPR-type experiments, without previously performing a measurement on the object concerning which one makes a prediction [34,72].

Relativistic causality is, thus, a manifestation of a more general concept or principle, that of locality. This principle states that no instantaneous transmission of physical influences between spatially separated physical systems ("action at a distance") is allowed or that physical systems can only be physically influenced by their immediate environment. It is true that locality is a spatial or spatiotemporal concept (there is a temporal locality, which precludes, for example, retroaction in time and backward in time causality), which makes it geometrical. However, although it is an effect of the ultimate reality (which is, in RWR-type interpretations, beyond representation or even conception), locality manifests itself only classically, in what is observed in measuring instruments, where the geometrical considerations are fully applicable. Events, as we observe them, do happen in space and time: otherwise they could not be observed. In general, some geometrical considerations remain unavoidable in algebraic physical theories, such as QM, in part, again, because these theories must relate, at least in term predictions to phenomena observed in space and time.

Locality of quantum phenomena and QM was at stake in the Bohr-Einstein debate from its inception in the late 1920s, but especially following EPR's paper [34,72]. As Bohr argued in his reply to EPR's paper (which argued that QM is either incomplete or else nonlocal), standard QM avoids nonlocality, at least in the RWR-type interpretations of the theory and quantum phenomena themselves, even though, as I said, under certain circumstances, such as those of the EPR-type experiments, QM can make *predictions* concerning the state of spatially separated systems, while, crucially to Bohr's argument, the physical circumstances of making these predictions and verifying them are local [73]. The question of the locality of QM or quantum phenomena is, however, a matter of much debate and controversy, especially in the wake of the Bell and Kochen-Specker theorems and related findings, as well as numerous experiments dealing with correlations, beginning with, most famously, those by Aspect [74], based on Bohm's version of the EPR experiments. These debates cannot be addressed within the scope of this article, and the literature dealing with these subjects is nearly as extensive as that on interpretations of QM (e.g., [75–78]). As in Bohr's exchange with EPR, the question of the relationships between locality and realism, or a lack thereof, figures centrally in these debates and the findings just mentioned.

Finally, I would like to propose the concept of quantum causality. I shall do so via Bohr's concept of complementarity, which Bohr saw as a generalization of causality. Complementarity is defined by:

(a)   a mutual exclusivity of certain phenomena, entities, or conceptions; and yet
(b)   the possibility of considering each one of them separately at any given point; and
(c)   the necessity of considering all of them at different moments for a comprehensive account of the totality of phenomena that one must consider in quantum physics.

Complementarity may be seen as a reflection of the fact that, in a radical departure from classical physics or relativity, the behavior of quantum objects of the same type, say, electrons, is not governed, individually or collectively, by the same physical law, in all possible contexts, specifically in complementary contexts. Speaking of "*physical* law" in this connection requires caution, because, in Bohr's interpretation, there is no physical law representing this behavior, not even a probabilistic law if one adopts a statistical, rather than a Bayesian, view of the individual quantum behavior. The behavior of quantum objects leads to mutually incompatible observable physical effects in complementary contexts. On the other hand, the mathematical formalism of QM offers correct probabilistic or statistical predictions (no other predictions are possible) of quantum phenomena *in all contexts*.

It follows, especially if one adopts an RWR-type of interpretation, that the nature of both experimental and theoretical physics changes. Experimentally we no longer track, as we do in classical physics or relativity, the independent behavior of the systems considered, track, in effect geometrically, what happens in any event. Instead we define what *will* happen in the experiments we perform, by *how* we experiment with nature by means of our experimental technology, even though and because we can only predict what will happen probabilistically or statistically. Thus, in the double-slit experiment, the two alternative setups of the experiment, whether we, respectively, can or cannot know, even in principle, through which slit each particle, say, an electron, passes, we obtain two different outcomes of the statistical distributions of the traces on the screen (with which each particle collides). Although sometimes associated with the "wave" behavior of quantum objects, the "interference pattern" observed in the second setup is a statistically ordered pattern of discrete traces left by the collisions between the particles and the screen. This is one of the reasons why Bohr avoided speaking of wave-particle complementarity, even though the latter is commonly used to illustrate Bohr's concept. In Bohr's view, quantum objects could not be represented either in terms of particles or in terms of waves, and the pattern of traces in questions were, again, only effects on the interactions between quantum objects and measuring instruments. Or, in effect equivalently to the double-slit experiments, we can set up our apparatus so as to measure and correspondingly predict, again, probabilistically or statistically, either the position or the momentum of a given quantum object, but never both together. Either case requires a separate experiment, incompatible with the other, rather than representing "the arbitrarily picking out of different elements of physical reality at the cost of other such elements [all pertaining to the same quantum object]" within the same physical situation, by tracking either one of its aspects or the other, as we do in classical mechanics [79]. There, this is possible because we can, at least in principle, assign simultaneously both quantities within the same experimental arrangement. In quantum physics, we cannot. Quantum physics, again, changes what experiments do: they *define* what will happen, rather than follow what is bound to happen in accordance with classical causality. It is true that we can sometimes define by an experiment what will happen in classical physics. In this case, however, we can then observe the resulting process without affecting it by observation. This is not the case in quantum physics, because *any new observation* defines a new course of events. Only *some observations* do in classical physics. By the same token, at least in RWR-type interpretations, quantum theory only tells us possible things about the future, never about the past, which is only determined by measurements.

It is this probabilistic or statistical determination (which precludes classical causality but respects locality) of what can happen as a result of our conscious decision concerning which experiment to perform at a given moment in time, that defines what I call "quantum causality" [80]. Whatever is registered as a quantum event (providing the initial data) defines a possible set of, probabilistically or statistically, predictable future events, outcomes of possible future experiments. This definition is in accord with recent views causality in quantum information theory (e.g., [81–83]), except that it is linked to our conscious decision concerning experiments we perform, which is rarely considered. It is, however, this aspect of the situation that brings complementarity into play, because, in complementary situations, such a decision irrevocably rules out the possibility of making any predictions concerning certain other, complementary, events.

With these considerations in mind one can understand Bohr's view of complementarity as a generalization of causality [84]. On the one hand, "our freedom of handling the measuring instruments, characteristic of the very idea of experiment" in all physics, our "free choice" concerning what kind of experiment we want to perform is essential to complementarity [79]. On the other hand, as against classical physics or relativity, implementing our decision concerning what we want to do will allow us to make only certain types of predictions and will exclude the possibility of certain other, *complementary*, types of predictions. Complementarity generalizes causality in the absence of classical causality and, in the first place, realism, because it defines which reality can and cannot be brought about by our decision concerning what experiment to perform.

3. Geometry and Algebra in Physics and Beyond

However one assesses Einstein's skeptical attitude toward the Heisenberg method, his view of it as algebraic was, I argue here, correct, and it helps one to better understand and contextualize Heisenberg's discovery. Before I consider this discovery itself, however, a more proper examination of this characterization is necessary. First, I briefly revisit the basic understanding of geometry and algebra, starting with algebra, which is more straightforward, because geometry involves further complexities, in part by virtue of always containing algebraic components, while algebra can be free from geometry. Most generally, algebra is, as said, the mathematical formalization of the relationships between symbols, which makes it part of all mathematics, at least all modern mathematics (as the ancient Greek geometry only contained arithmetic). Thus, for example, both mathematical logic and calculus are forms of algebra in this general sense, even though, as fields, each contains its specificity. Another, narrower or field specific sense of algebra, that referring to algebraic structures such as groups or associative algebras, is equally crucial to quantum theory and its algebraic character. (As noted, symmetry groups are also crucial to geometry and geometrical physical theories, such as relativity, but there, too, they form part of the algebraic structures associated to geometrical and topological ones.) While the role of symmetry groups in quantum theory only became apparent a few years after Heisenberg's discovery, it was well in place by the time of Einstein's comment on the algebraic nature of the Heisenberg method in 1936. Third, algebra also refers to, and was born from, the study of algebraic equations, which still define a large part of the mathematical discipline of algebra. Forth, finally, the algebra of probability is central, at a fundamental, rather than, as in classical physics, merely practical, level in quantum theory. Heisenberg's algebraic method and, following his work, quantum theory encompasses all these senses and aspects of algebra, some of which are shared with classical physics or relativity.

Geometry, generally defined here as the mathematical formalization of spatiality, especially (although not only) in terms of measurement is a more complex matter, because, on the one hand, this formalized spatiality still connects to our general phenomenal intuition, including visualization, of spatiality, and on the other, the role of mathematical formalization in geometry connects it to algebra. This connection allows one to generalize geometrical or topological objects far beyond anything our phenomenal intuition can access, especially by means of visualization. I would like now to address some of these complexities, as they pertain to my argument concerning the algebraic character of quantum theory. It would not be possible to treat this complex subject more generally.

Thus, we speak of and can rigorously define Hilbert *spaces* and their *geometry*, and in this sense, one could speak, as Dirac was reportedly fond of doing, of geometrical thinking in QM or QFT. I would argue, however, that these spaces and geometries are such primarily by extrapolation or, as it were, metaphorically, while, rigorously, they are essentially forms of *algebra*, as against the more standard forms of geometry, such as Euclidean or even non-Euclidean geometry, or differential geometry, used in general relativity, which are still more closely connected to our phenomenal intuition of spatiality. I would contend that this distinction is warranted and useful even though it is not absolute and there are gray areas. The geometries of these more conventional spaces are defined by certain algebraic properties and relations (beginning with metric), some of which can be used to define more abstract objects. Thus, these properties and relations define the *structure* of Hilbert spaces or, similarly, other mathematical "spaces," such as those of projective geometries, abstract algebraic varieties, the spaces of noncommutative geometry, geometric groups, and so forth, in the absence of certain other, more conventionally spatial elements and structures, geometrical or topological, found in more conventional spatial objects, such as and in particular $\mathbf{R}^3$. The latter is the mathematical space that is the closest to our phenomenal spatiality, even though some of its mathematical (topological and geometrical) properties, beginning with continuity are far beyond our phenomenal intuition. As Weyl argued: "the conceptual world of mathematics is so foreign to what the intuitive continuum presents to us that the demand for coincidence between the two must be dismissed as absurd" [85]. "Coincidence" is, of course, not the same as "relation," which might be unavoidable, at least in that it is difficult to think of continuity or spatially apart from one or other phenomenal intuitions of it. On the other hand, it is entirely possible to mathematically define continuous mathematical objects, such as $\mathbf{R}^3$, algebraically. This is why I speak of the extrapolated or metaphorical character of "space" and "geometry" of such objects as Hilbert spaces. Technically, $\mathbf{R}^3$ is a Hilbert space too, but it need not be considered as such, while Hilbert spaces considered in quantum theory do. This extrapolated or algebraic character arises because of the infinite-dimensional nature of some of those spaces or because they are defined over **C**, which appear irreducible in QM. While the Hilbert spaces involved are finite in the case of discrete variables in QTFD, they can have higher dimensions and are over **C**.

It is difficult to be certain, especially from reported statements, what Dirac exactly had in mind in his appeals to geometrical thinking in quantum theory. If, however, one is to judge by his writings, they appear to suggest that at stake were algebraic properties and relations modeled on those found in geometrical objects, as just explained. Indicatively, notwithstanding his insistence on the role of geometrical thinking in Dirac, Darigold's analysis of this thinking shows precisely the significance of this type of algebra there. Thus, he says: "roughly, Dirac's quantum mechanics could be said to be to ordinary mechanics what noncommutative geometry is to intuitive geometry" [86]. However, noncommutative geometry, the invention of which was in part inspired by the mathematics of QM, is a form of this kind of algebra [87,88].

In what sense, then, apart from being defined by such algebraic structures, may such spaces be seen as spaces, in particular, as relates to our phenomenal intuition, including visualization, from which QM departs, although, in Bohr's or the present view it breaks with any representation of quantum objects and behavior, including a mathematical one? The subject is complex and it is far from sufficiently explored in cognitive psychology and related fields, an extensive research during recent decades notwithstanding, including as concerns cultural or technological (digital technology in particular) factors affecting our spatial thinking. Accordingly, it would be difficult to make any definitive claims. It does appear, however, that, these factors notwithstanding, our three-dimensional phenomenal intuition is shared by us cognitively and even neurologically in shaping our sense of spatiality. Part of this sense appear to be Euclidean, insofar as it corresponds to what is embodied in $\mathbf{R}^3$ (again, a mathematical concept), keeping in mind that the idea of empty space, apart from bodies of one kind or another defining or faming it, is an extrapolation, because we cannot have such a conception phenomenally or, as Leibniz argued against Newton, physically. We can have a mathematical conception of space itself. To what degree our phenomenal spatiality is Euclidean remains an open question, for example, in dealing with the visual perception of extent and perspective (e.g., [89–91]).

It is nearly certain, however, that, when we visualize such algebraically defined spatial objects as Hilbert spaces or even more conventional geometrical spaces or geometries, once the number of dimensions is more than three, or for spaces of any dimensions defined over fields other than $\mathbf{R}$, such as $\mathbf{C}$ or algebraic fields of finite characteristics, we visualize only three- (and even two-) dimensional configurations and supplement them by algebraic structures and intuitions. Feynman instructively explained this process in describing visual intuition in thinking about quantum objects [92]. Obviously, such anecdotal evidence hardly suffices for any definitive claim, but it appears to be in accord with some of the current neurological and cognitive-psychological research, as just mentioned, which suggests the dependence of our spatial intuition, including visualization, on two- and three-dimensional phenomenality.

As noted earlier, according to Bohr and Heisenberg, classical mechanics, a mathematical science of bodies and motion in space and time, is a mathematical refinement of our daily phenomenal thinking concerning space, time, and motion, no longer workable in quantum theory or, again, even in relativity in considering photons or motions with speed close to the speed of light in a vacuum, in which cases, however, the behavior of the objects considered can be properly handled algebraically, in quantum theory in probabilistic or statistical terms [56,93]. Accordingly, one could maintain the difference that I argue for between geometrical character of classical mechanics or (with qualifications) relativity and the algebraic character of quantum theory, especially in RWR-type representations of the latter.

This is not to say that the geometrical or topological character of such objects disappears; quite the contrary, this character remains crucial on at least two counts. First, algebra has a special form that may be called "spatial algebra." Selecting a good term poses some difficulties because such suitable terms as "geometric algebra" and "algebraic geometry" are already in use for designating, respectively, the Clifford algebra over a vector space with a quadratic form, and the study of algebraic varieties, defined as the solutions of the systems of polynomial equations. "Spatial algebra" arises from algebraic structures that *mathematically* define what are conventionally known as geometrical or topological objects, such as projective spaces or topological manifolds, and reflect their proximity to $\mathbf{R}^3$ and mathematical spatial objects that are close to our phenomenal intuition, and the geometry and topology associated with it. This proximity may be left behind in the rigorous mathematical treatment of such objects, beginning with $\mathbf{R}^3$. The same type of spatial algebra also defines objects, from projective and finite (topologically discrete) spaces to the infinite-dimensional spaces, where these connections to phenomenally visualizable spatial objects are only spatial algebraic, unless used by an extrapolation or metaphorically, as in the case of a projective space (a set of lines through the origin of a vector space, such as $\mathbf{R}^2$ in the case of projective plane, with projective curves defined algebraically, as algebraic varieties) or an infinite-dimensional Hilbert space (the points of which are typically square-integrable functions or infinite series). Spatial algebra is the algebraization of spatiality that makes it rigorously mathematical.

At the same time, however, and this is the second count on which mathematical objects defined by spatial algebra retain their connections to geometrical and topological thinking, analogies with $\mathbf{R}^3$ continue to remain useful and even indispensable. Such analogies may be rigorous (and hence algebraic) or phenomenally intuitive or metaphorical. Thus, the analogues of the Pythagorean theorem or parallelogram law in Euclidean geometry, which holds in infinite-dimensional Hilbert spaces over either **R** or **C**, are important, including in applications to physics, such as QM. More generally, our thinking concerning geometrical and topological objects is not entirely translatable into algebra. This was well understood by Hilbert in his axiomatization of Euclidean geometry, even though this axiomatization had a spatial algebraic character, including in establishing an algebraic model (the field of complex numbers) for his system of axioms in order to prove its consistency [94].

One can gain a further insight into this situation by considering a related principle, "Think Geometrically, Prove Algebraically," advanced by Tate, whose thinking bridged number theory and algebraic geometry in highly original and profound ways. The principle was introduced in the book (with Silverman), on "the rational points of elliptic curves." The title-phrase itself combines algebra ("rational points") and "geometry" (curves), and implies that geometry, at least beyond that of **R**^{3} and even there, requires algebra to be mathematically rigorous. According to Silverman and Tate [95]:

> It is also possible to look at polynomial equations and their solutions in rings and fields other than **Z** or **Q** or **R** or **C**. For example, one might look at polynomial with coefficients in the finite field **F**$_p$ with p elements and ask for solutions whose coordinates are also in the field **F**$_p$. You may worry about your geometric intuitions in situations like this. How can one visualize points and curves and directions in **A**^{2} when the points of **A**^{2} are pairs (x, y) with $x, y \in$ **F**$_p$? There are two answers to this question. The first and most reassuring is that you can continue to think of the usual Euclidean plane, i.e., **R**^{2}, and most of your geometric intuitions concerning points and curves will still be true when you switch to coordinates in **F**$_p$. The second and more practical answer is that the affine and projective planes and affine and projective curves are defined *algebraically* in terms of ordered pairs (r, s) or homogeneous triples $[a, b, c]$ without any reference to geometry. So in proving things one can work algebraically using coordinates, without worrying at all about geometrical intuitions. We might summarize this general philosophy as: *Think Geometrically, Prove Algebraically*".

The affine and projective planes and curves can, in principle, be defined without any reference to ordinary language and concepts, which are more difficult and perhaps impossible to avoid in geometry, that is, in the kind of intuitive geometry they refer to, rather than spatial algebra that ultimately defines all geometry rigorously. Rigorously, in these cases, we think algebraically, too, by using spatial algebra, even if with the help of geometrical intuitions. It is true that a mathematician like Tate can develop and use intuition in dealing with discrete geometries as such, say, that of the Fano plane of order 2, which has the smallest number of points and lines (seven each). However, beyond the fact that they occur in the two-dimensional regular plane, the diagrammatic representations of even the Fano plane are still difficult to think of as other than spatial algebra, in this case, combinatorial in character. For the moment, while useful and even indispensable, our Euclidean intuitions are limited even when we deal with algebraic curves in the usual Euclidean plane, let alone in considering something like a Riemann surface as a curve over **C**, or curves in finite geometries, abstract algebraic varieties, Hilbert spaces, the spaces of noncommutative geometry, or geometric groups, a great example of the extension, by a reversal, of spatial algebra to conventionally algebraic objects as it is. This also means, as Tate must have been aware, that mathematical *thinking* concerning geometrical and topological objects cannot be reduced to those naïve intuitions. Silverman and Tate's next example from differential calculus, that of finding a tangent line to a curve, confirms this point. The invention of calculus, an essentially algebraic form of mathematics, was not about proving things algebraically, as the standard of proof then was geometry, which compelled Newton to present his mechanics in terms of geometry in his 1687 *Principia*, in order, as he said, to assure a geometrical demonstration of his findings, also in the direct sense of showing something by means of visualization, rather than in terms of calculus [96]. Calculus was about *thinking* algebraically, as was especially manifested in Leibniz's version, rather than about rigorous proofs.

Einstein was, then, quite correct in characterizing the Heisenberg method as "algebraic." The "geometry" of the Hilbert spaces of quantum theory only confirms the algebraic character of this method, especially manifested in his matrix or operator algebra. Still, a few additional qualifications are necessary for rigorously maintaining Einstein's and the present view of the Heisenberg method as algebraic. First of all, in saying that "we must give up, in principle, the space-time continuum," Einstein must have had in mind the spacetime continuum in representing, by means of the corresponding theory, the ultimate reality considered, and possibly in attributing the space-continuum to this reality. The idea that this reality may ultimately be discrete had been around for quite a while by then: it was, for example, proposed by Riemann as early as 1854, by way of a remarkable phrase "the reality underlying space," thus potentially divorcing on proto-RWR lines reality and realism, at least a form of realism according to which our phenomenal or even mathematical representation of space was continuous [97]. The idea of the discrete nature of the ultimate reality has acquired new currency in view of QM and QFT, advocated by, among others, Heisenberg in the 1930s, and is still around.

One must also keep in mind the complexity of the algebra of QM or QFT, which involves objects that are not, in general, discontinuous, although certain key elements involved are no longer continuous functions, such as those in classical physics. Some continuous functions are retained, because the Hilbert spaces involved are those of such functions, considered as infinite-dimensional vectors in the case of continuous variables such as position and momentum, which variables themselves are represented by operators. These functions are those of complex (rather than, as in classical physics, real) variables and the vector spaces that they comprise or associated objects, such as operator algebras, have special properties, such as noncommutativity. Indeed, given that it deals with Hilbert spaces, QM or QFT involves mathematical objects whose continuity is denser than that of regular continua such as the (real number) spacetime continuum of classical physics or relativity. In contrast to these theories, however, the continuous and differential mathematics used in quantum theory, along with the discontinuous algebraic one, relates, in terms of probabilistic predictions, to the physical discontinuity defining quantum phenomena, which are discrete in relation to the observable spacetime continuum and to each other, while, at least in non-realist, RWR-type, interpretations, quantum objects and their behavior are not given any physical or mathematical representation—continuous or discontinuous (or mixed). Born and Jordan developed a differential calculus, "symbolic differentiation," as they called it, for matrices used in quantum mechanics [8,98]. So did Dirac in his first paper on quantum mechanics [99]. In the style of Leibniz, this differentiation was defined algebraically by using the noncommutation rules. This quantum differentiation enables one to retain the differential equations of classical mechanics and their accompanying machinery, such as and in particular the Poisson bracket, while using new quantum variables. The quantum-mechanical analogue of the Poisson bracket is the expression $\frac{2\pi i}{h}(pq - qp)$, as Dirac was first to realize, with far-reaching implications for quantum mechanics. Dirac's starting point, again, in the style of Leibniz, was the quantum-mechanical analogue of the rule for the differential of the product of two functions, which may be seen as a linear operator and which may be suitably algebraically quantized [99].

4. From Geometry to Algebra, with Heisenberg

For nearly a century since the publication of von Neumann's seminal *The Mathematical Foundation of Quantum Mechanics* [48], the mathematical models of quantum theory—QM, QTFD, and QFT—have been commonly defined in terms of the Hilbert-space formalism, which remains dominant, even though there are other versions, such as C*-algebras and, more recently, the one based in category theory. Von Neumann's aim was to give a proper mathematical grounding to the quantum-mechanical formalism, already developed by Heisenberg, Born, Jordan, Dirac, and others. By contrast, Heisenberg's aim was to *find* a successful theory accounting for the behavior of the electrons in the atoms, which he accomplished with his new mathematical model, admittedly preliminary, but quickly developed into a more rigorous model, matrix mechanics, by Born and Jordan [8]. Dirac offered the most general version of the formalism before von Neumann, who thought that Dirac's version lacked a proper mathematical rigor, in part because of Dirac's use of delta function, not considered mathematically legitimate. Eventually, in the 1940s, delta function was given a proper definition, as the so-called "distribution," a functional, by Schwartz, which legitimatized Dirac's formalism mathematically.

As stated from the outset, Heisenberg abandoned the project of representing the behavior of electrons in atoms. It is true that he only thought that such a representation was unlikely to be achieved at the time, as opposed to arguing, as Bohr did in the 1930s, that such a representation and even an analysis, if not conception, of quantum behavior was "*in principle* excluded" [35]. Still, Heisenberg's was an audacious and radical move, which decisively shaped Bohr's subsequent thinking, although Heisenberg's thinking was in turn influenced by Bohr's 1913 atomic theory. Bohr's theory only partially abandoned the geometrical representation of quantum behavior in the case of "quantum jumps"—transitions between the stationary states of electrons, conceived in terms of electrons orbiting the nuclei. While Bohr's theory, as developed by him and others had major successes, by the early 1920s it proved to be ultimately unsustainable. This failure compelled Heisenberg to renounce a geometrical representation of any quantum behavior in space and time, including that of stationary states in terms of orbits. This renunciation led him to his discovery of QM. There was no longer any algebraic representation of quantum objects and behavior either.

Heisenberg's approach was grounded in a set of fundamental principles, in part stemming from Bohr's 1913 theory. As in the case of concepts, although one could sometimes surmise the meaning of the term "principle" from its use, it is, similarly to the term "concept," rarely defined or explained in physical or even philosophical literature. His title notwithstanding, Heisenberg did not do so in his first book, *The Physical Principles of the Quantum Theory* [100] nor did Dirac, in his famous *Principles of Quantum Mechanics*, published in the same year [47]. Terms like "principle," "postulate," and "axiom," are often used in physics somewhat indiscriminately, and it is difficult to entirely avoid overlapping between the concepts designated by these terms, or those designated as "laws," especially because physical principles often derive from (or give rise to) postulates or laws. It may also be a matter of the functioning of these concepts. Thus, conservation laws are sometimes seen as conservation principles. For present purposes, I shall adopt the concept of principle from Einstein's concept of a "principle theory," which he introduced by way of juxtaposing this concept to that of a "constructive theory." This concept corresponds to the use of principles by Bohr and Heisenberg, and by quantum-information theorists discussed in the next section. According to Einstein, constructive theories aim "to build up a picture of the more complex phenomena out of the materials of a relatively simple formal scheme from which they start out," which, it follows, also make such theories realist. By contrast, principle theories "employ the analytic, not the synthetic, method. The elements which form their basis and starting point are not hypothetically constructed but empirically discovered ones, general characteristics of natural processes, principles that give rise to mathematically formulated criteria which the separate processes or the theoretical representations of them have to satisfy" [101]. I would add the following qualification, which is likely to have been accepted by Einstein: Principles are not empirically discovered but formulated on the basis of empirically established evidence. A principle theory may also be a constructive theory, but it need not be, and Heisenberg's mechanics was not.

Heisenberg's approach and then Bohr's interpretation of QM were grounded in the following three principles (with Bohr's principle or at least concept of complementarity added in 1927), which fit and even embody the "equation" *QUANTUMNESS* → *PROBABILITY* → *ALGEBRA* and the QPA principle:

(1)   the principle of discreteness, the QD principle, according to which all observable quantum phenomena are individual and discrete in relation to each other, which is different from the discreteness of quantum objects;

(2)   the principle of the probabilistic or statistical nature of quantum predictions, the QP/QS principle, which is maintained, in contrast to classical statistical physics, even in considering elemental individual quantum processes, and is accompanied by a special, nonadditive, character of quantum probabilities and rules, such as Born's rule, for deriving them; and

(3)   the correspondence principle, which, as initially understood by Bohr, required that the predictions of quantum theory must coincide with those of classical mechanics in the classical limit, but which

was given by Heisenberg a form of "the mathematical correspondence principle," requiring that the equations and variables of QM convert into those of classical mechanics in the classical limit.

To connect his formalism (defined over **C**) to the probabilities of outcomes of quantum experiments (which probabilities are real numbers), Heisenberg used a version of the Born rule in the special case of the transitions between stationary states, and not, as Born did, as universally applicable in QM. Referring to stationary states requires caution, because *stationary* only means that the electrons remained in their orbits with the same energy, were in the same "energy-state," but would continuously change their position or their "position-state" along each orbit. On the other hand, the electrons would discontinuously, by quantum jumps, change their energy states, or their other states, by moving from one orbit to another. In Heisenberg, there were no longer orbits but only states and discontinuous transitions between states. As noted from the outset, one was no longer thinking, as in classical mechanics, in terms of predictions, even probabilistic predictions, concerning *a moving object*, say, an electron, free or orbiting the nucleus of an atom, but instead in terms of the probabilities of *transitions between the states of an electron*, transitions that were always discrete. This type of thinking emerged in Bohr's 1913 theory in considering an electron's transitions from one energy level to another, but, following Heisenberg, came to define quantum physics in general as a physics of predicting discrete transitions between states [12]. As Heisenberg said in his letter to Kronig (5 June 1925): "What I really like in this scheme is that one can really reduce *all interactions* between atoms and the external world … to transition probabilities" (cited in [102]).

Heisenberg's scheme, thus, extended Bohr's 1913 concept of discrete transitions or quantum jumps, which, unlike the orbital behavior of electrons in stationary states, had no mechanical or geometrical model, to all quantum behavior. The "concept of orbit," as Heisenberg noted later, "had been somewhat doubtful from the beginning," because of "the discrepancy between the calculated orbital frequency of the electrons and the frequency of the emitted radiation," which had to be interpreted as a limitation to the concept of the electronic orbit" [103]. Accepting this discrepancy and, thus, dissociating these two types of frequencies was a revolutionary move on Bohr's part, emphasized by Bohr himself: "How much the above considerations differ from an interpretation based on the ordinary electrodynamics is perhaps most clearly shown by the fact that we have been forced to assume that a system of electrons will absorb radiation of a frequency different from the frequency of vibration of electrons calculated in the ordinary way" [104]. Heisenberg rethought stationary states as just energy states, permitting no mechanical model or geometrical representation. There were, again, only transitions between quantum states (using the term "state" physically, rather than mathematically, as "a state vector").

By speaking of the "*interactions* between atoms and the external world," Heisenberg's statement in his letter to Kronig also suggests that QM, as he saw it, was about (predicting) these interactions, specifically with the measuring instruments involved, a view manifested in Heisenberg's paper and adopted by Bohr. All that one could say about quantum objects and behavior could only concern their effects on measuring instruments, probabilistically or statistically predictable by QM.

The mathematical correspondence principle motivated Heisenberg's decision to retain the equations of classical mechanics, while, necessarily, introducing different variables, both, however, being now equally parts of his algebra. The correspondence with classical theory could still be maintained because new variables could be substituted for conventional classical variables (such as those of position and momentum) in the classical limit, as for large quantum numbers, when the electrons were far away from the nuclei and when classical concepts, such as orbits, could apply, thus also restoring the geometry of classical mechanics. The electrons' behavior itself was still quantum and certain quantum effects, not observed in dealing with classical objects, could be observed, effects predictable only by the algebra of QM. The old quantum theory was defined by the strategy of retaining the variables of classical mechanics while adjusting the equations to achieve better predictions. Heisenberg's reversal of this strategy was, thus, unexpected, and it required a radical change in the role these equations were to play. They no longer represented the motion of electrons, but served as the mathematical means enabling probabilistic or statistical predictions concerning effects of the interaction between electrons and measuring instruments.

Heisenberg's discovery was a remarkable achievement, ranked among the greatest in the history of physics. A detailed discussion of his derivation of QM is beyond my scope [105]. Several key features of his thinking are, however, worth commenting on, following [105], to further illustrate the algebraic nature of his thinking. Heisenberg's new quantum variables were infinite unbounded matrices with complex elements. Their multiplication, which Heisenberg, who was famously unaware of the existence of matrix algebra and reinvented it, had to define, is in general not commutative. Essentially, these variables are operators in Hilbert spaces over **C**. Such mathematical objects had never been used in physics previously, and their noncommutative nature was, initially, questionable and even off-putting for some, including Heisenberg himself and Pauli [106]. In fact, while matrix algebra, in finite and infinite dimensions, was developed in mathematics by then, *unbounded* infinite matrices were not previously studied. As became apparent later, such matrices are necessary to derive the uncertainty relations for continuous variables. There are further details: for example, as unbounded self-adjoint operators, defined on infinite dimensional Hilbert spaces, these matrices do not form an algebra with respect to the composition as a noncommutative product, although some of them satisfy the canonical commutation relation. These details are, however, secondary. Most crucial was that the concept was used in a fundamentally new way. Heisenberg's variables were algebraic entities enabling probabilistic or statistical predictions concerning *quantum phenomena*, observed in measuring instruments, without providing a mathematically idealized representation, geometrical or other, of the spacetime behavior of *quantum objects* responsible for these phenomena.

In this regard, although understandable historically, the term "observables" is misleading and is especially inadequate if one adopts a RWR-type view. As noted, Bohr saw the quantum-mechanical formalism as "symbolic" in the following, essentially algebraic, sense. While the mathematical symbols used in it appear, as variables, in the same equations as those used in classical mechanics, these symbols did not represent physical quantities pertaining to quantum objects themselves and their behavior, in the way such symbols do in classical mechanics. By the same token, the equations of QM, Schrödinger's equation included, no longer functioned as equations of motion. Instead they are part of the probabilistic algebra of QM, enabling us to compile, in Schrödinger's terms, "expectation-catalogs" concerning events observed in quantum phenomena, which, in RWR-type interpretations, gives Schrödinger's equation an algebraic character along with a probabilistic one [63]. Schrödinger's waves were symbolic waves, symbolizing these expectation-catalogs. If Schrödinger's equation may be seen as "deterministic," as it is sometimes, it is only in the sense that it strictly determines such expectation-catalogs, which are, however, catalogs of predictions that are not deterministic even in realist interpretations.

In his 1925 paper, introducing QM, Heisenberg began his derivation with an observation that reflects a radical departure from the classical ideal of continuous mathematical representation of individual physical processes, which would connect discrete quantum events. He says: "in quantum theory it has not been possible to associate the electron with a point in space, *considered as a function of time*, by means of observable quantities. However, even in quantum theory it is possible to ascribe to an electron the emission of radiation" [the effect of which is observed in a measuring instrument] [107]. Technically, a measurement could associate an electron with a point in space, but not by linking this association to a function of time representing the continuous motion of this electron, in the way it is possible in classical mechanics. Matrix mechanics did not offer a treatment of stationary states, when and only when one could in principle speak of the position of an electron in an atom, although while there are stationary energy states, with the same energy-levels, an electron itself is never stationary. Only an instantly repeated measurement can give the same value of its position, which instant repetition is an idealization. Heisenberg described his next task as follows: "In order to characterize this radiation we first need the frequencies which appear as functions of two variables. In quantum theory these functions are in the form" [107]:

$$v(n, n - \alpha) = 1/h \{W(n) - W(n - \alpha)\}$$

and in classical theory in the form

$$v(n, \alpha) = \alpha v(n) = \alpha/h(dW/dn)$$

This difference leads to a difference between classical and quantum theories as regards the combination relations for frequencies, which, in the quantum case, correspond to the Rydberg-Ritz combination rules, reflecting, to return to Heisenberg's locution, "the discrepancy between the calculated orbital frequency of the electrons and the frequency of the emitted radiation." However, "in order to complete the description of radiation [in correspondence, by the correspondence principle, with the classical Fourier representation of motion] it is necessary to have not only frequencies but also the amplitudes" [107]. On the one hand, then, the new, quantum-mechanical equations must formally contain amplitudes, as well as frequencies. On the other hand, these amplitudes could no longer serve their classical physical function (as part of a continuous representation of motion) and are instead related to discrete transitions between stationary states. In Heisenberg's theory and in QM since then, these "amplitudes" are no longer amplitudes of physical motions, which makes the name "amplitude" a *symbolic* term. In commenting on linear superposition in quantum mechanics in his classic book, Dirac emphasized this difference: *"the superposition that occurs in quantum mechanics is of an essentially different nature from any occurring in the classical theory"* [108]. In RWR-type interpretations, this superposition is not even physical: it is only mathematical. In classical physics the mathematics of (wave) superpositions represents physical processes; in QM, at least in the nonrealist view, it does not. Amplitudes are instead linked to the probabilities of transitions between stationary states: they are what became known as probability amplitudes. The corresponding probabilities are derived by a form of Born's rule for this limited case (technically, one needs to use the probability density functions). The standard rule for adding the probabilities of alternative outcomes is changed to adding the corresponding amplitudes and deriving the final probability by squaring the modulus of the sum.

The mathematical structure thus emerging is in effect that of vectors and (in general, noncommuting) Hermitian operators in Hilbert spaces over **C**, which are infinite-dimensional, given that one deals with continuous variables. Heisenberg explains the situation in these terms in [100]. In his original paper, which reflect his thinking more directly, he argues as follows [107]:

> The amplitudes may be treated as complex vectors, each determined by six independent components, and they determine both the polarization and the phase. As the amplitudes are also functions of the two variables *n* and *α*, the corresponding part of the radiation is given by the following expressions:
>
> Quantum-theoretical:
>
> $$\text{Re}\{A(n, n - \alpha)e^{i\omega(n, n - \alpha)t}\}$$
>
> Classical
>
> $$\text{Re}\{A_\alpha(n)e^{i\omega(n)\alpha t}\}$$

The problem—a difficult and, "at first sight," even insurmountable problem—is that "the phase contained in *A* would seem to be devoid of physical significance in quantum theory, since in this theory frequencies are in general not commensurable with their harmonics" [109]. As noted, this incommensurability, which is in an irreconcilable conflict with classical electrodynamics, was one of the most radical features of Bohr's 1913 atomic theory, on which Heisenberg builds here. His strategy is still based on the shift from calculating the probability of finding a moving electron in a given state to calculating the probability of an electron's transition from one state to another, without describing the physical mechanism responsible for this transition. Heisenberg's theory is more in harmony with this approach because there are no longer orbits, where the classical approach would still apply.

Heisenberg says next: "However, we shall see presently that also in quantum theory the phase has a definitive significance which is *analogous* to its significance in classical theory" [109]. "Analogous" could only mean here that, rather than being analogous physically, the way the phase enters mathematically is analogous to the way the classical phase enters mathematically in classical theory, in accordance with the *mathematical* form of the correspondence principle, insofar as quantum- mechanical equations are formally the same as those of classical physics. Heisenberg only considered a toy model of an aharmonic quantum oscillator, and thus needed only a Newtonian equation for it, rather than the Hamiltonian equations required for a full-fledged theory, developed by Born and Jordan [8,110]. As Heisenberg explains, if one considers "a given quantity $x(t)$ [a coordinate as a function of time] in classical theory, this can be regarded as represented by a set of quantities of the form" [109]:

$$A_\alpha(n)e^{i\omega(n)\alpha t},$$

which, depending on whether the motion is periodic or not, can be combined into a sum or integral which represents $x(t)$ [60]:

$$x(n,t) = \sum_{\alpha=-\infty}^{\alpha=+\infty} A_\alpha(n)e^{i\omega(n)\alpha t}$$

or:

$$x(n,t) = \int_{-\infty}^{+\infty} A_\alpha(n)e^{i\omega(n)\alpha t}d\alpha$$

Heisenberg next makes his most decisive and most extraordinary move. He notes that "a similar combination of the corresponding quantum-theoretical quantities seems to be impossible in a unique manner and therefore not meaningful, in view of the equal weight of the variables n and $n - \alpha$. However, he says, "one might readily regard the ensemble of quantities $A(n, n - \alpha)e^{i\omega(n, n - \alpha)t}$ [an infinite square matrix] as a representation of the quantity $x(t)$" [109]. The arrangement of the data into these ensembles, in effect square tables, was a remarkable way to handle the transitions between stationary states, although in retrospect it is also a natural way to do so, but only in retrospect. However, it does not by itself establish an *algebra* of these arrangements, for which one needs to find the rigorous rules for adding and multiplying these elements. Otherwise Heisenberg cannot use these variables in the equations of his new mechanics. To produce a quantum-theoretical version of the classical equation of motion considered, which would apply (no longer as an equation of motion) to these variables, Heisenberg needs to be able to construct the powers of such quantities, beginning with $x(t)^2$, which is actually all that he needs for his equation. The answer in classical theory is obvious and, for the reasons just explained, obviously unworkable in quantum theory, where, Heisenberg proposes, "it seems that the simplest and most natural assumption would be to replace classical [Fourier] equations ... by" [111]:

$$B(n, n - \beta)e^{i\omega(n,n-\beta)t} = \sum_{\alpha=-\infty}^{\alpha=+\infty} A(n, n - \alpha)A(n - \alpha, n - \beta)e^{i\omega(n,n-\beta)t}$$

or:

$$= \int_{-\infty}^{+\infty} A(n, n - \alpha)A(n - \alpha, n - \beta)e^{i\omega(n,n-\beta)t}d\alpha$$

This is the main mathematical postulate, the (matrix) multiplication postulate, of Heisenberg's theory, "an almost necessary consequence of the frequency combination rules" [111].

Although it is commutative in the case of x^2, this multiplication is in general noncommutative, expressly for position and momentum variables, and Heisenberg, without quite realizing, used this noncommutativity in solving his equation, as Dirac was the first to notice. Heisenberg spoke of his new algebra of matrices as the "new kinematics." This was not the best choice of term because his new variables no longer described or were even related to motion as the term kinematic would suggest, one of many, historically understandable, but potentially confusing terms. Planck's constant, h, which is a dimensional, dynamic entity, has played no role thus far. Technically, in Einstein's view, the theory wasn't even a mechanics: it did not offer a representation of individual quantum processes, but only predicted, probabilistically or statistically, what is observed in measuring instruments. To make these predictions, one will need Planck's constant, h.

That in general his new variables did not commute, $PQ - QP \neq 0$, was, again, an especially novel feature of Heisenberg's theory, confirming its essentially algebraic nature. This feature, which was, again, an automatic consequence of his choice of variables, proved to be momentous physically. Most famously, it came to represent the uncertainty relations constraining certain simultaneous measurements, such as those of the momentum (P) and the coordinate (Q), associated with a given quantum object in the mathematical formalism of quantum mechanics and (correlatively) the complementary character of such measurements. Given, however, the nature of the situation to which Heisenberg's new mechanics responded, the noncommutative character of quantum variables should not be surprising. Schwinger instructively commented on the subject in his unpublished lecture, cited at length in [112]. He notes the most commonly stated physical feature corresponding to this character, namely, that if one measures two physical properties in one order, and then in the other, the outcome would in general be different. But he goes further in explaining why this is the case and its implications for the formalism [112]:

> If we once recognize that the act of measurement introduces in the [microscopic] object of measurement changes which are not arbitrarily small, and which cannot be precisely controlled ... then every time we make a measurement, we introduced a new physical situation and we can no longer be sure that the new physical situation corresponds to the same physical properties which we had obtained by an earlier measurement. In other words, if you measure two physical properties in one order, and then the other, which classically would absolutely make no difference, these in the microscopic realm are simply two different experiments ...

> So, therefore, the mathematical scheme can certainly not be the assignment, the association, or the representation of physical properties by numbers because numbers do not have this property of depending upon the order in which the measurements are carried out. ... We must instead look for a new mathematical scheme in which the order of performance of physical operations is represented by an order of performance of mathematical operations.

This is not how Heisenberg discovered QM, in particular given that the noncommutativity of some among the operators representing quantum observables was not his starting point but a consequence of the multiplication rule for his matrices. The type of thinking described by Schwinger is more in accord with quantum-informational approaches to deriving quantum theory, primarily QTFD, from the (formalized) structure of quantum measurements, or as Schwinger revealingly put it the "measurement algebra." [113,114]). The difficulty here is that any such scheme requires a mathematics that does not appear to be naturally connected to this measurement algebra, for one thing, because of the use of complex, rather than real, variables, and rules, such as Born's rule, by means of which this scheme would be related to the probabilities or statistics of quantum predictions, which are real numbers. There is no homomorphic (let alone isomorphic) mapping from measurement algebra to the algebra of QM, "the new mathematical scheme." One needs additional pieces of structure to arrive at this scheme. In Heisenberg, these additional pieces we partly borrowed from classical physics, formally defining his equations, and partly invented by Heisenberg in finding the variables needed.

As explained earlier and as Schwinger stresses in his lecture, no identical assignment of the single quantity is ever possible, or in any event ever guaranteed, in two "identically" prepared experiments in the way it can be in classical physics [112]. This is because quantum experiments cannot be controlled so as to identically prepare quantum objects but only so as to identically prepare the measuring instruments involved, because this behavior can be considered classical. The quantum strata of measuring instruments, through which they interact with quantum objects, do not affect these preparations but only the outcomes of measurements. This interaction is uncontrollable. This fact is central to Bohr's argument, which invokes this "finite and incontrollable interaction" at key junctures of his reply to EPR's paper [34]. Hence, as noted earlier, the outcomes of repeated identically prepared experiments, including those involving sequences of measurements, cannot be controlled even ideally (as in classical physics), and these outcomes will, in general, be different. This circumstance makes statistical considerations unavoidable, as reflected, among other things, in the statistical character of the uncertainty relations, inherent in Heisenberg's formula, $\Delta q \Delta p \cong h$. The noncommutative nature of the corresponding variables responds to this character, along with the uncertainty relations themselves.

The quantum-mechanical situation that emerged with Heisenberg's discovery of quantum mechanics and then Bohr's interpretation of it was (sometime in the late 1930s) recast by Bohr in terms of his concept of "phenomenon," defined by what is observed in measuring instruments under the impact of quantum objects, in contradistinction to quantum objects themselves, which could not be observed or represented, or in the present view, even conceived of. According to Bohr [115]:

> I advocated the application of the word phenomenon exclusively to refer to the *observations* obtained under specified circumstances, including an account of the whole experimental arrangement. In such terminology, the observational problem is free of any special intricacy since, in actual experiments, all observations are expressed by unambiguous statements referring, for instance, to the registration of the point at which an electron arrives at a photographic plate. Moreover, speaking in such a way is just suited to emphasize that the appropriate physical interpretation of the symbolic quantum-mechanical formalism amounts only to predictions, of determinate or statistical character, pertaining to individual phenomena appearing under conditions defined by classical physical concepts [describing the observable parts of measuring instruments].

Phenomena are discrete in relation to each other, and, in Bohr's scheme, one cannot assume that there are continuous processes that connect them, especially classically causally, even in dealing with elemental individual processes and events. Part of Bohr's concept of phenomenon and the main reason for its introduction was that this concept "*in principle* exclude[s]" any representation or analysis, even if not a possible conception, of quantum objects and their behavior, at least, by means of QM [35]. The concept is, thus, correlative to the RWR-type view, reached by Bohr at this stage of his thinking, at least the weak RWR view. Physical quantities obtained in quantum measurements and defining the physical behavior of certain (classically described) parts of measuring instruments are *effects* of the interactions between quantum objects and these instruments, and do not pertain to quantum objects themselves. It is often forgotten by those who comment of Bohr's insistence on the role of classical concepts in quantum theory that Bohr clearly realized and relied on the fact that the measuring instruments used in quantum experiments also have quantum parts through which they interact with quantum objects. Otherwise quantum measurements and the effects defining phenomena would not be possible. These effects are manifested by classical states of these parts of measuring instruments, to which these quantum interactions are "irreversibly amplified" [41]. The language of effects (in the absence of classical causes), found throughout Bohr's writings on quantum mechanics, becomes especially prominent in his later articles, presenting his ultimate interpretation (e.g., [116]).

These effects are no longer assumed to correspond to any properties of quantum objects, even to single such properties, rather than only certain joint properties, in accordance with the uncertainty relations. An attribution of *even of a single property*, such as that of "position," "moment in time," "momentum," or "energy," or even invariant properties, such as the rest mass and charge of a particle, which are defined by the fact that they are the same in all measurements, to any quantum object is *never possible—before, during, or after measurement*. One could only rigorously specify measurable quantities physically pertaining to measuring instruments. Even when we do not want to know the momentum or energy of a given quantum object and thus need not worry about the uncertainty relations, neither the exact *position* of this object itself nor the actual time at which this "position" is established is ever available and, hence, in any way verifiable. Any possible information concerning quantum objects as independent entities is lost in "the finite [quantum] and uncontrollable interaction" between them and measuring instruments [34]. However, this interaction can leave a mark in measuring instruments, a mark, a bit of information, that can be treated as part of a permanent, objective record, which can be discussed, communicated, and so forth. The uncertainty relations, too, now apply to the corresponding (classical) variables of suitably prepared measuring instruments, impacted by quantum objects. We can either prepare our instruments so as to measure or predict a change of momentum of certain parts of those instruments or so as to locate the spot that registers an impact by a quantum object, but never do both in the same experiment. The uncertainty relations are correlative to the complementary nature of these arrangements.

Wheeler spoke of "law without law" in quantum theory [117]. One might see this concept, via the combination of the QPA and the RWR principles, as the algebraic probabilistic law of QM, without any law that would be assumed to govern the independent behavior of quantum objects. It is not surprising either that Wheeler eventually linked this "law without law" to quantum information theory, which he helped to usher, along with Feynman, his one-time student. Quantum objects, in their interactions with measuring instruments, create specifically organized collections of information (composed of classical bits) and make possible certain calculations, by using mathematical models, but we cannot know and possibly cannot conceive how quantum processes do this. The ultimate (quantum) constitution of matter is, according to Wheeler, "it from bit," "it" inferred from "bit" [118]. In the present view, this "it," while real, is beyond thought, and as such, cannot ultimately be called "it," any more than anything else. Wheeler's visionary manifesto was inspired by Bohr, whom Wheeler invoked on the same page that announced "it from bit:" "The overarching principle of 20th-century physics, the quantum—and the principle of complementarity that is the central idea of the quantum—leaves us no escape, Niels Bohr tells us, from 'a radical revision of our attitude [towards the problem of] physical reality'" [118] (I correct Wheeler's slight misquotation of Bohr).

Bohr's argument for the necessity of this revision originated in Heisenberg's algebraic method, of which Einstein, by contrast, remained ever skeptical, not the least because the concept of reality that was the product of this revision remained unpalatable to Einstein, or in his words, "while logically possible without contradiction, it [was] so contrary to [his] scientific instinct that [he] could not forego a search for a more complete conception" [119]. From Bohr's perspective, QM is only incomplete when compared with the kind of realist knowledge possible in classical physics or relativity, which may be called the Einstein completeness. Otherwise, it is as complete as possible, as things stand now, which may be called the Bohr completeness. The question is whether nature would allow us to do better. While Einstein thought that it should, Bohr thought that it *might not*, which is not the same as it never will. As we haven't heard nature's last word on this matter, that is to say, nature's next word (the only last word that nature gives us), the debate concerning this question continues with undiminished intensity.

5. From the Algebra of Circuits to the Algebra of Categories in Quantum Information Theory

Although Heisenberg's creativity and inventiveness were remarkable and although it would be difficult to challenge him on the outcome, his derivation of QM may not have been as rigorous as one could ideally wish. While borrowing the *form* of equations from classical mechanics by the mathematical correspondence principle was a logical deduction concerning part of the mathematical structure of QM, Heisenberg virtually "guessed" the variables he needed. The mathematical expression of the principles in question was only partially worked out and sometimes more intuited than properly developed, which was in part remedied in the later work of Born, Jordan, and Heisenberg himself, but only in part. Even the derivation offered, following this more rigorous treatment, by Heisenberg in his Chicago lectures [100] might still be seen as falling short of a rigorous derivation from first principles, because it relied on intuitive moves of the type found in Heisenberg's original derivation, especially as concerns his matrix variables, still essentially a guess and, arguably, the main difficulty for any rigorous (re)construction, especially for continuous variables. One could accordingly envision a more rigorous derivation. Most of the recent work in this direction has been in quantum information theory in dealing with discrete variables and finite-dimensional Hilbert spaces (QTFD). Some of these efforts, however, have affinities with that of Heisenberg, which, as I argue, exhibits a spirit of quantum-informational thinking. I shall now comment on two such cases, by D'Ariano and coworkers and by Hardy.

D'Ariano, Chiribella, and Perinotti's (DACP's) program, developed over the last decade and presented comprehensively in their book [82], belongs to a particular trend in quantum information theory, and as most of the work there, it deals with discrete variables and the corresponding, finite-dimensional, Hilbert spaces (e.g., [120–122]). This is in part because, as DACP note (a view shared by others in this field), "the study of finite-dimensional systems allows one to decouple the conceptual difficulties in our understanding of quantum theory from the technical difficulties of infinite-dimensional systems" [123]. A rigorous (or at least more rigorous than that of founding figures) derivation of QM, let alone QFT, from fundamental principles remains an open and difficult task, to which I return below.

DACP's project is motivated by "a need for a deeper understanding of quantum theory [QTFD] in terms of fundamental principles," and by the aim of deriving QTFD from such principles, which, the authors contend, has never been quite achieved by their predecessors. As indicated earlier, the fluctuations of terms such as principles, axioms, postulates, commonly used in these reconstructive projects may be confusing and obscure this common aim. DACP use the term "axioms" as well. On the other hand, while one can surmise their understanding of the term "principle" from their use of it, they do not define the concept of "principle" either. As earlier, I adopt the concept of principle defined above, via Einstein. This concept is, I would argue, in accord with DACP's use of principles, and their derivation of QTFD, or that of Hardy, may be seen as that of a principle theory in Einstein's sense.

I put aside the question to what degree this (or Hardy's) derivation amounts to a *fully* rigorous derivation, which, or in the first place, the question of what could be considered as a fully rigorous derivation, would require a separate analysis. One might even question the necessity of a "*fully* rigorous definition." After all, Heisenberg, at least as his scheme was developed by Born and Jordan, and then differently by Dirac, did establish a correct theory, and then Dirac similarly invented QED, and various parts of QFT were created similarly. Heisenberg in effect posed this question in commenting on his own derivation of QM: "It should be distinctly understood, however, this [deduction of the fundamental equation of quantum mechanics] cannot be a deduction in the mathematical sense of the word, since the equations to be obtained form themselves the *postulates* of the theory. Although made highly plausible by the following considerations [of the type that led him to his discovery of QM], their ultimate justification lies in the agreement of their predictions with the experiment" [124]. While a derivation of QM, or QTFD, might be made more rigorous than that offered by Heisenberg even there, it is doubtful that any such derivation, *from (physical) first principles*, could ever be as rigorous as "a deduction in the mathematical sense of the word." As Hardy suggested, even by his title, in [120], it may be more a matter what are "reasonable" initial axioms or postulates, although such a "reasonableness" is not a simple or unconditional matter either. It goes without saying that these qualifications in no way diminish the significance of DACP's or Hardy's work, or that of others pursuing this line research. Besides, as I shall explain, more than merely re-deriving QTFD, or QM and QFT is at stake in these programs.

The main new feature of DACP's approach is adding to the view of QTFD as an extension of probability theory (a view found in the works of their predecessors and applicable to QM as well) "the crucial ingredient of *connectivity* among events" by using the operational framework of "circuits" and giving it an algebra [125]. The framework of circuits has been similarly used by others, specifically by Hardy. This addition allows DACP "to derive key results of quantum information theory and general features of quantum theory [QTFD]" without first assuming Hilbert spaces. Unlike von Neumann, to whom DACP refer for a contrast, Heisenberg, as we have seen, did not begin with a formalism either, although he used the equation of classical mechanics by the correspondence principle. He *arrived* at his formalism from fundamental principles, even if, again, not fully rigorously *derived* it from these principles.

Among the principles adopted by DACP, the purification principle plays a unique role as an essentially quantum principle, because conforming to it distinguishes QTFD from classical probabilistic information theories. According to them: "The purification of mixed states is specifically quantum" [126] (it may be a question whether it *uniquely* quantum, on which I shall comment presently). It is the single principle necessary to do so, which may not be surprising given the history of quantum theory and attempts at its axiomatic derivations. Hardy's pioneering derivation also needed only one axiom, the continuity axiom, to do so [120]. On the other hand, that Hardy's continuity axiom is different from DACP's purification postulate suggests that there may not be a single system of postulates or principles from which QTFD could be derived, even though all such systems should capture something that pertains uniquely to quantum phenomena, and thus to quantum objects, even if one assumes that they are beyond representation or conception. Their effects are representable and enable one to distinguish classical and quantum phenomena, and infer from them the existence of quantum objects.

Whether one can do so definitively remains a question, noted from the outset in connection with Planck's constant, *h*. While this question and literature concerning it are beyond my scope, I would like to comment on Spekkens's recent work, which is especially relevant here because it proceeds along the lines of quantum information theory. Spekkens introduced several toy models or theories, "epirestricted theories" (so-called because of epistemic restrictions on the classical theory, assumed as a starting point), that reproduce many quantum phenomena and features of QM or QTFD, such as the presence of *h*, the uncertainty relations, noncommuting operator observables, entanglement, or the purification of mixed states [127–129]. Many but not all! These models expressly fail to reproduce some among the crucial features of quantum theory, specifically some of those dealing with correlations and entanglement, such as violations of Bell inequalities and the existence of the Kochen-Specker theorem, which is, however, predictable given the nature of Spekkens's models, as Spekkens explained [130]. This is crucial because these theorems reflect the essential features of quantum phenomena (independently of any theory, which, however, must, if correct, satisfy them), and are also crucial to the question of realism and locality, all of which is noted by Spekkens [130]. Accordingly, whether actual quantum phenomena found in nature can be captured by models of this type remains an open question. According to Spekkens himself, this is unlikely [131]:

> The investigation of epirestricted theories, therefore, need not—and indeed *should not*-be considered as the first step in a research program that seeks to find a ψ-epistemic ontological model [a realist model that assumes quantum states in the formalism to be epistemic, states of knowledge] of the full quantum theory. Even though such a model could always circumvent any no-go theorems by violating their assumptions, it would be just as unsatisfying as a ψ-ontic model [a realist model that assumes quantum states to be ontic, states of reality] insofar as it would need to be explicitly nonlocal and contextual. Rather, the investigation of epistricted theories is best considered as a first step in a larger research program wherein the framework of ontological models—in particular the use of classical probability theory for describing an agent's incomplete knowledge—is ultimately rejected, but where one holds fast to the notion that a quantum state is epistemic.

Although common in quantum information theory, viewing quantum states (state vectors in the formalism) as states of knowledge could be misleading, unless one strictly refers to this knowledge as probabilistic or statistical. (Even then quantum states are only part of the corresponding predictive machinery, as one needs Born's or related rule, such as Lüders' postulate, to have these probabilities or statistics.) As noted earlier, from the present, RWR-type perspective, in which quantum states in this mathematical sense never refer to any actual, determined knowledge (which is only obtained in measurements), our knowledge, actual (obtained in measurement) or probabilistic is incomplete only when compared with the kind of realist knowledge possible in classical physics or relativity—the Einstein completeness. Otherwise, it is as complete as possible, as things stand now—the Bohr completeness.

I am not sure whether, his skepticism concerning ontological models of quantum theory notwithstanding, Spekkens would be willing to go that far, especially to the strong RWR view. It appears, however, at least to the present reader, that his epirestricted theories and his argument suggest that the difference between the classical and the quantum may ultimate be irreducible, even though finding and rigorously grounding reconstructive programs remains difficult. According to Spekkens, "it may be that there are particularly elegant axiomatic schemes that are not currently in our reach and the road to program involves temporarily setting one's sight a bit lower," moving to partial reconstructive models of the kind he proposes [132]. That may be, especially in the case of continuous variables, to which Spekkens's epirestricted theories apply. DACP's or Hardy's approach may suggest otherwise, even while dealing only with QTFD. There is also a question whether such elegance should be the main criterion here, especially given that one's goals may ultimately be beyond QTFD or QM? While Dirac would have thought so, neither Heisenberg (in his derivation) nor Bohr was much worried about elegance. It is difficult to be certain which trajectories will lead us there, especially if this "there" is beyond QTFD, as it ultimately must. There may not be one such trajectory, just as in the discovery of QM, there were two—Heisenberg's and Schrödinger's. I leave this complex set of subjects on the following note, concerning the role of h.

It is true that there are classical phenomena, in considering which h must be taken into account, even trivially true, although Spekkens's model considered in [129] is nontrivial because it involves other quantum-like features. On the other hand, most classical phenomena do not involve h at all, while *all* quantum phenomena known thus far require taking h into account, just as all (special) relativistic account require taking c into account. This does, I think, tell us something about the nature of the quantum or our technological interactions with that which we call the quantum, even if does not tell us the whole story, assuming that such a story can ever be told. If one adopts an RWR-type view, there is no story to be told about how quantum phenomena come about, and h, too, may only appear in our interaction with nature and not be a property of nature itself. That, however, does not mean that more cannot be said, perhaps even definitively, about what distinguishes quantum from classical phenomena. I now return to DACP's derivation of QTFD and the purification principle, which they see as "specifically quantum".

In nontechnical terms, the purification principle states that "every random preparation of a system can be achieved by a pure preparation of the system with an environment, in a way that is essentially unique" [133]. The principle originates in Schrödinger's insight in his response, in several papers, to the EPR paper and his concept of entanglement [72]. According to DACP [134]:

> The purification principle stipulates that, whenever you are ignorant about the state of a system A, you can always claim that your ignorance comes from the fact that A is part of a large [composite] system AB, of which you have full knowledge. When you do this, the pure state that you have to assign to the composite system AB is determined by the state of A in an essentially unique way.

> The purification of mixed states is a peculiar feature—surely, not one that we experience in our everyday life. How can you claim that you know A *and* B if you don't have A alone? This counterintuitive feature has been noted in the early days of quantum theory, when Erwin Schrödinger famously wrote: "Another way of expressing the peculiar situation is: *the best possible knowledge of a whole does not necessarily include the best possible knowledge of all its parts.*" And, in the same paper: "I would not call that *one* but rather *the* characteristic trait of quantum mechanics, the one that enforced its entire departure from classical lines of thought . . . " [135]

> The purification of mixed states is specifically quantum. But why should we assume it as a fundamental principle of Nature? At first, it looks like a weird feature—and it must look so,

because quantum theory itself is weird and if you squeeze it inside a principle, it is likely that the principle looks weird too. However, on second thought one realizes that purification is a fundamental requirement: essentially, it is the link between physics and information theory. Information theory would not make sense without the notions of probability and mixed state, for the whole point about information is that there are things that we do not know in advance. But in the world of classical physics of Newton and Laplace, every event is determined and there is no space for information at the fundamental level. In principle, it does not make sense to toss a coin or to play a game of chance, for the outcome is already determined and, with sufficient technology and computational power, can always be predicted. In contrast, purification tells us that "ignorance is physical." Every mixed state can be generated in a single shot by a reliable procedure, which consists of putting two systems in a pure state and discarding one of them. As a result of this procedure, the remaining system will be a physical token of our ignorance. This discussion suggests that, only if purification holds, information can aspire to a fundamental role in physics.

Technically, the purification of mixed states is the fundamental principle arising in *our interactions with nature* by means of our experimental technology, rather than of nature itself, except insofar as we and our technologies are also nature. But then, in the present view, there is no other principles of nature than those defined by us in our interactions with it. DACP's formulation of the purification principle and their derivation of QTFD allows for an RWR-type interpretation, and I shall interpret it in this way, without claiming that this necessarily corresponds to DACP's own view.

Thinking in terms of "circuits" is close to Bohr's thinking concerning the role of measuring instruments in the constitution of quantum phenomena, as distinguished from quantum objects, which give rise to quantum phenomena by interacting with measuring instruments but which are never observable. Circuits and their arrangements, too, embody those of measuring instruments capable of detecting quantum events, and thus enabling the probabilistic predictions of future events. Their arrangements and operations, defining their "measurement algebra," are enabled by rules that should ideally be derived from certain sufficiently natural assumptions. They are described classically, and thus embody the structure of quantum information as a particular form of organization of classical information, which can be used, as by DACP (or Hardy), to *derive* the mathematical formalism of QTFD.

While indispensable for the authors' derivation of QTFD, the purification principle is not sufficient to do so. They need five additional postulates (termed "axioms"): *causality* (essentially locality), *local discriminability*, *perfect distinguishability*, *ideal compression*, and *atomicity of composition* [133]. These postulates define a large class of classical probabilistic informational theories, while the purification postulate, giving rise to the purification principle, distinguishes QTFD. The appearance of these additional postulates or principles is not surprising. Heisenberg's grounding principles, the quantum discreteness (QD) principle and the quantum probability or statistics (QP/QS) principle, were not sufficient for him to derive QM either. To do so, he needed the correspondence principle, which gave him half of the mathematical architecture of quantum theory. The other half was supplied by his matrix variables.

There are instructive parallels between DACP's and Heisenberg's approaches. Both the QD and QP/QS principles are present in both cases. As they say in their earlier article: "The operational-probabilistic framework combines the operational language of circuits with the toolbox of probability theory: on the one hand experiments are described by circuits resulting from the connection of physical devices, on the other hand each device in the circuit can have classical outcomes and the theory provides the probability distribution of outcomes when the devices are connected to form closed circuits (that is, circuits that start with a preparation and end with a measurement)" [136]. This is similar to Heisenberg's thinking in his paper introducing QM, as the classical outcomes are discrete in both cases as well. The concept of "circuit" is not found in Heisenberg and is, again, closer to Bohr's view of the role of measuring instruments and his concept of phenomenon, defined by this role. As I explained, however, the idea that in quantum theory we only deal with transition probabilities between the outcomes of the interactions between quantum objects and measuring instruments was introduced by Heisenberg, as part of his approach to QM, and was then adopted by Bohr. Heisenberg discovered that Bohr's frequencies rules are satisfied by, in general, non-commuting matrix variables with complex coefficients, from which one derives, by means of a Born-type rule, probabilities or statistics for transitions between stationary states, manifested in spectra observed in measuring devices. Thus, Heisenberg's derivation depended on measuring instruments as devices with classically describable observable parts, which are akin to "operational circuits," in his case dealing with continuous variables.

DACP aim to arrive at the mathematical structure of QTFD in a more first-principle-like way, for example, independently of classical physics, which, because of the correspondence principle, was central to Heisenberg (classical physics, to begin with, does not have discrete variables, so there is no correspondence principle). The rules governing the structure of operational devices, circuits, should, they argue, allow them to do so, because these rules are *empirical*. They are, however, *not completely empirical*, because circuits *are given* a mathematical structure, in effect algebra, by human agents, even though this algebra may be partially defined by the organization, required by experiments, of the experimental arrangements in which circuits appear. The mathematized structure or algebra of circuits become a necessary condition for establishing the mathematical structure or algebra of QTFD, but it is not sufficient to do so. As noted above in considering Schwinger's argument, these two structures are not isomorphic or even homomorphic, and they do not appear to be in DACP's derivation. This means that additional pieces of structure, provided by additional postulates or axioms, are required for one thing, to get to the (Hilbert-space or other) formalism over **C**. Then, one needs a Born-type rule for the probabilities of predictions. DACP's derivation requires enormous technical work. It is next to impossible to do it justice here. I shall instead close the article by considering the algebra of circuits in considering Hardy's work.

Hardy has at a different set of main assumptions necessary to derive QTFD than those of DACP, but the main strategy is the same: establishing the architecture, algebra, of circuits that, with additional axioms, would allow one to derive the mathematical formalism of QTFD. According to Hardy [137]:

Circuits have:

- A setting, s(H), given by specifying the setting on each operation.
- An outcome set, o(H), given by specifying the outcome set at each operation (equals o(A) × o(B) × o(C) × o(D) × o(E) in this case). We say the fragment "happened" if the outcome is in the outcome set.
- A wiring, w(E), given by specifying which input/output pairs are wired together.

With this algebraic definition in hand, I shall comment on some of Hardy's fundamental assumptions discussed by him in a different paper. Hardy says [138]:

We will make two assumptions to set up the framework in this paper ...

Assumption 1. *The probability, Prob (A), for any circuit, A (this has no open inputs or outputs), is well conditioned. It is therefore determined by the operations and the wiring of the circuit alone and is independent of settings and outcomes elsewhere.*

This is a physical postulate, essentially that of locality, combined with probability or statistics, along the lines of the QP/QS principle. The task now becomes how to derive a QTFD that could correctly predict these probabilities. One needs another assumption:

Assumption 2. *Operations are fully decomposable ...* We assume that any operation $A_{a1b2...c3}^{d4e5...f6}$ can be written, ... in a symbolic notation, $A_{a1b2...c3}^{d4e5...f6} \equiv^{d4e5...f6} A_{a1b2...c3} X_{a1}^{a1} X_{b2}^{b2} \cdots X_{c3\ d4}^{c3} X_{e5}^{d4} X_{...f6}^{e5} X^{f6}$. In words we will say that any operation is equivalent to a linear combination of operations each of which consists of an effect for each input and a preparation for each output ... We allow the possibility that the entries in $_{d4e5...f6} A_{a1b2...c3}$

are negative (and this will, indeed, be the case in quantum theory). Hence, in general, this cannot be thought of as physical mixing . . . (emphasis added).

I only need the italicized sentence for my conceptual point, insofar as it means that one can construct a suitable algebra. I cite the passage at a greater length to illustrate a manifested algebraic view of circuits and operations in Hardy's scheme. Hardy then says [138]:

Assumption 2 introduces a subtly different attitude than the usual one concerning how we think about what an operation is. Usually we think of operations as effecting a transformation on systems as they pass through. Here we think of an operation as corresponding to a bunch of separate effects and preparations. We need not think of systems as things that preserve their identity as they pass through—we do not use the same labels for wires coming out as going in. This is certainly a more natural attitude when there can be different numbers of input and output systems and when they can be of different types. Both classical and quantum transformations satisfy this assumption. In spite of the different attitude just mentioned, we can implement arbitrary transformations, such as unitary transformations in quantum theory, by taking an appropriate sum over such effect and preparation operations.

This rethinking of the concept of operation is important, especially if one adopts an RWR-type view. An "operation" is now defined in terms of observable "effects" of the interactions between quantum objects and measuring instruments, and not in terms of what happens, even in the course of these interactions (let alone apart from them), to the quantum objects or systems, *considered as independent systems*. It is useful that we can treat classical systems in this way as well. In the classical case, however, we can, equivalently, use a more conventional concept of operation mentioned here, which is not the case in quantum theory. After a technical discussion of "duotensors," which I put aside, Hardy suggests a principle [139]:

Physics to mathematics correspondence principle. For any physical theory, there [exist] a small number of simple hybrid statements that enable us to translate from the physical description to the corresponding mathematical calculation such that the mathematical calculation (in appropriate notation) looks the same as the physical description (in appropriate notation).

Such a principle might be useful in obtaining new physical theories (such as a theory of quantum gravity). Related ideas to this have been considered by category theorists. A category of physical processes can be defined corresponding to the physical description. A category corresponding to the mathematical calculation can also be given. The mapping from the first category to the second is given by a functor (this takes us from one category to another).

The language of correspondence should not mislead one into relating this principle to Bohr's correspondence principle, even in Heisenberg's *mathematical* form. Bohr's correspondence principle deals with the correspondence between different physical theories (such as classical mechanics and quantum mechanics), insofar as their predictions would coincide in the regions when both theories could be used. By contrast, apart from the fact that, as explained, there is no correspondence principle in QFDT, Hardy's "hybrid" construction implies that category of physical processes could be functorially "translated" into a proper formalism of QTFD, which can then, through mathematical calculations enabled by this formalism, be related to what is observed. Hardy's principle may be better called "physics to mathematics functoriality principle." Hardy's suggestion, inviting but somewhat speculative and not really worked out, to begin with, would require a separate discussion, which cannot be undertaken here. It is worthwhile, however, to offer a few brief comments, via the role of category theory in the algebraization of physics, without definitively claiming that these comments are fully in accord with Hardy's view of the situation.

Category theory originated in algebraic topology and then was extensively used, especially thanks to Grothendieck's work, in algebraic geometry, in order to study certain algebraic invariants, such as cohomology or homotopy groups, associated with topological spaces [140]. It was then extended, via topos theory, introduced by Grothendieck as well, to mathematical logic, a subject I shall put aside here [141]. Roughly, category theory considers multiplicities (categories) of mathematical objects conforming to a given concept, such as the category of topological spaces or geometrical spaces (say, Riemannian manifolds), and the morphisms, also called arrows ($X \to Y$), which are the mappings between these objects that preserve this structure. Studying morphisms allows one to learn about the individual objects involved more than we would by considering them individually. Thus, in geometry one does not have to start with a Euclidean space. Instead the latter is just one specifiable object of a large categorical multiplicity, such as the category of Riemannian manifolds, an object marked by a particularly simple way we can measure the distance between points. Categories themselves may be viewed as such objects, and in this case one speaks of "functors" rather than "morphisms."

Now, one can more easily think of properly defining mathematically, the second category in Hardy's suggestion, say, as that of Hilbert spaces [142] or some more directly categorical equivalent algebra by means of which the mathematical calculations in question would be performed. On the other hand, the first category, that is, the structure of its objects and the morphisms between them, and, thus, the nature of the functor in question between these two categories is a more complex matter. Hardy's formulations above "for any physical theory, there [exist] a small number of simple hybrid statements that enable us to translate from the physical description to the corresponding mathematical calculation such that the mathematical calculation (in appropriate notation) *looks the same* as the physical description (in appropriate notation)" (emphasis added) might require qualification as concerns the meaning of the expression "looks the same" and the relationships between "calculation" and "description," because this formulation allows for different interpretations. The same may be said about his characterization of the functor in question as "*virtually* direct" (emphasis added) and then the statement "a category of physical processes can be defined corresponding to the physical description."

I shall sketch the reasons for these qualifications, beginning with a possible meaning of "*physical processes*" in "a category of physical processes can be defined corresponding to the physical description." "Physical processes" may refer either to quantum processes, or to circuits, which appears more likely because "the physical description" in Hardy's scheme (or that of DACP) is given only at the level of circuits. This view would also be suggested by the category theorists who work on using categories in quantum theory, specifically QTFD, such as Abramsky, Coecke, and others, to whom Hardy appears to refer here. Most of this work is primarily concerned with recasting the Hilbert-space language of the QTFD formalism in a categorical framework by replacing Hilbert spaces, belonging to the category of the finite-dimensional Hilbert spaces over **C** (e.g., [142]), with objects of monoidal categories and the morphism between Hilbert spaces with morphism between these objects, rather than with deriving the formalism from the first (physical) principles. There are some moves in this direction, say, by starting with a suitable simple monoidal category and then consider which additional pieces of structure one needs to arrive at QTFD (e.g., [143]). This work further testifies to the dominant role of algebra in quantum theory, even though it uses a lot of diagrammatic operations, somewhat akin to Feynman diagrams in QFT. As Coecke and Kissinger's title, "picturing quantum processes," indicates, they appear to hold a realist view in assuming that their diagrammatic picturalism represents the actual quantum behavior, which assumption, I argue here, has complexities, in part because of the role of complex quantities, complexities not addressed by Coecke and Kissinger [143]. In RWR-type views, quantum behavior is beyond representation or even conception, and thus pieces of this behavior (say, between measurements) cannot be assumed to form a category. Hardy's position on this point is not entirely clear. If quantum processes themselves are given some representation, the corresponding category may need to incorporate this representation in one way or another, by combing it with the effects of quantum processes manifested in circuits and their arrangements.

If one adopts an RWR-type view (either weak or strong), one only deals with the physical description of circuits, similarly to dealing only with the description of measuring instruments, according to Bohr. Circuits and their organizations, their algebra, again, embody the arrangements of measuring instruments capable of detecting quantum events and enabling the probabilistic or statistical predictions of future events, in other words, a structure that may be mathematizable and as such *categorially* translatable into the mathematical formalism of QTFD (which enables the necessary mathematical calculations). As discussed earlier in considering both Schwinger's argument and DACP's derivation of QTFD, one need not and, in the RWR-type view, should not expect a representational correspondence between the structures or algebra of circuits and the mathematical structures or the algebra of the formalism of QTFD. For one thing, one needs, again, a formalism, Hilbert-space one, C*-algebra, categorical, or other, over **C**. As classical, the data manifested in circuits is over **R** (in fact, all measurements are rational numbers, while probabilities need not be). Then, one needs a Born-type rule to get to probabilities. It follows, then, as, again, considered earlier, that one needs additional pieces of structure to those defined by circuits.

One could contemplate two approaches. The one, closer to QTFD category theorists, is to use these additional pieces of structure (algebraic relations, axioms, etc.) to build a category of algebraic mathematical objects, which need not be Hilbert spaces, but would, again, have to be over **C**, and rules, which enable proper probabilistic or statistical predictions of quantum phenomena, again, observed in circuits [143]. Alternatively, closer to what Hardy appears to suggest and to Bohr's way of thinking, one could attempt to establish a category of circuits, each defined by an algebraic structure of units and operations, and morphisms between them, and then, again, by using some reasonable axioms, a category of objects defining the formalism, such as Hilbert spaces or some monoidal categories, and morphism between them (along, again, with a Born-like rule for probabilities), and then connect these two categories by means of a well-defined functor. The advantage of the second approach is that it preserves the role of the algebraic structure of circuits, which are observable. The difficulty is, again, that there is no natural categorical structure for the algebra of circuits, morphisms, etc., all of which need to be defined, which, however, is also true in the first approach. To do so requires additional pieces of structure, and thus additional more or less reasonable principles or postulates, rather than intuitive guesses, assuming, again, that the latter could be entirely avoided. In the second approach (which I prefer), it is the question of a functorial relationship between categories, which, again, still needs to be established, rather than morphism between objects, such as the algebras of circuits and the algebras of the formalism of QTFD, or by implication, QM or QFT, a more complex project. There are morphisms in each category, and there functors between categories, and the view just considered is about the latter. From this perspective, Hardy's "physics to mathematics correspondence principle" would, if rigorously established, be the "physics to mathematics *functoriality* principle": it would be realized in terms of the functorial relationship between a category based on circuits (the objects and morphisms of which need, again, to be defined) and a category of the algebraic objects defining the formalism, such as that of Hilbert spaces over **C**, finite-dimensional ones in QTFD, or some other category of objects and morphisms.

One of the most essential aspects of category theory, even its *raison d'être*, at least in its use in algebraic topology and algebraic geometry (mathematical logic is, again, a different matter), consists in establishing the relationships between multiplicities, "categories," of objects of *different* types, for example, geometrical or topological objects and algebraic objects, or between algebraic objects of different types (such as Lie groups and Lie algebras), rather than, apart from trivial cases, directly mapping the objects of the first category on those of the second. Indeed, the concept of category was introduced in the field of algebraic topology and then extended to algebraic geometry to help study certain algebraic invariants, such as groups, associated with topological spaces. In contradistinction to geometry, defined, as a mathematical discipline, by the concept of measurement (geo-*metry*), topology, as a mathematical discipline, is defined by associating an algebraic structure or a set of structures, most especially groups, such as homotopy or cohomology groups (which came to play an important role in QFT) to a topological space. The structure of a topological space is defined by its continuities and discontinuities, and not by its geometry, even if it had a geometry, and not all topological spaces do. Insofar as one deforms a given figure continuously (i.e., insofar as one does not separate points previously connected and, conversely, does not connect points previously separated), the resulting figure or space is treated as mathematically equivalent. Thus, no matter how much you expand or continuously (without separating connected points or joining separated points) deform the two-dimensional surface of a sphere the resulting spaces are topologically equivalent (homeomorphic), while some of these objects are no longer spherical geometrically speaking. Such spaces are, however, topologically distinct from those of topologically deformed two-dimensional surfaces of tori because spheres and tori cannot be converted into each other without disjoining their connected points or joining the separated ones: the holes in tori make this impossible. By contrast, the spaces of spheres (or any spaces) of different dimensions are not homeomorphic. In algebraic topology, mathematical objects of each type are arranged in categories and are related to one another by morphisms, while categories are related by functors. The category of topological spaces (or a subcategory, such as that of Riemannian manifolds) and their morphisms becomes related to a category of algebraic objects, such as groups and their morphisms. The relationships between these two categories, topological and algebraic, allow one to extract an enormous amount of information concerning topological spaces and, conversely, groups (through the structure of topological spaces to which these groups are associated). A good categorical approach to, and a possible derivation of, quantum theory would similarly establish the functorial relationships between two, now algebraic, categories, one based in circuits and the other defining the formalism.

An important and difficult question, which need not depend on a categorical formalization, but may be helped by it, is that of the relationship between the structure of the circuits, defined by the corresponding experimental arrangements, and the infinite-dimensional mathematical architecture of QM, or the same relationship in QFT. Consider the double-slit experiment, say, in the interference pattern setup. It is a circuit, which embodies preparations, measurements, and predictions, all manifested in the emergence of the interference pattern. I would not presume to be able to mathematize it. But it is a circuit nevertheless, a complex one, albeit child's play in comparison to the circuitry found in high-energy quantum physics, such as that of the Large Hadron Collider (LHC), which led to the detection of the Higgs boson.

Such questions will need be addressed if one is to extend the programs to derivations of QTFD to QM or to QED and QFT, or beyond. Indeed, it is hardly sufficient to merely derive already established theories. The ultimate value of these programs lies in what they can do for the future of fundamental physics. Thus, Hardy aims to rethink general relativity in operational terms, analogous to those of QTFD, and then to reach, in principle, quantum gravity, bypassing QFT or even QM, which would then be merely special cases of the ultimate theory [16,83] D'Ariano and co-workers, by contrast, appear first to move from QTFD to QFT. In their more recent work, they aim to develop a new approach to QFT, which, as based on the concept of quantum cellular automata, is different from the operational framework discussed thus far, but shares with it certain key informational features and, most especially, the aim of developing QFT from fundamental (first) principles. D'Ariano and Perinotti's derivation of Dirac's equation is a step in this direction, for now in the absence of an external field, essential to the proper QED [144,145]. Unlike Dirac's own or other previous derivations of the equation, their derivation only uses, along with other principles (homogeneity, isotropy, and unitarity), the principle of locality, rather than special relativity. The approach may, they hope, offer new possibilities for fundamental physics on Planck's scale, suggesting a potential extension of quantum information theory as far as one can envision it now.

Whether the QPA principle (which is in part experimental, empirical) and the RWR principle (which is interpretive) will remain viable or will be defeated, fulfilling the hope of Einstein and those who followed him, is an open question. On the other hand, it appears likely, as no currently known attempts to move beyond QFT (such as string and M-brane theory, or loop quantum gravity) would indicated otherwise, that we will see the emergence of new algebraic and spatial-algebraic structures. We very much need them for quantum gravity, for example. Because algebraic topology and algebraic geometry are likely to play a role in quantum gravity, the age of algebra—the age of Fermat, Descartes, and Leibniz—is likely to continue for the foreseeable future in mathematics and physics alike.

Funding: This research received no external funding.

Acknowledgments: I would like to thank Mauro D'Ariano, Laurent Freidel, Christopher A. Fuchs, Lucien Hardy, Gregg Jaeger, Andrei Khrennikov, and Paolo Perinotti for valuable discussions concerning the subjects addressed in this article. The suggestions of the anonymous reviewers were invaluable in helping to clarify and sharpen the article's argument.

Conflicts of Interest: The author declares no conflict of interest.

References

1. Bohr, N. *The Philosophical Writings of Niels Bohr*; Ox Bow Press: Woodbridge, CT, USA, 1987; Volume 2, p. 53.
2. Bohr, N. *The Philosophical Writings of Niels Bohr*; Ox Bow Press: Woodbridge, CT, USA, 1987; Volume 2, p. 33.
3. Heisenberg, W. Quantum-theoretical re-interpretation of kinematical and mechanical relations (1925). In *Sources of Quantum Mechanics*; Van der Waerden, B.L., Ed.; Dover: New York, NY, USA, 1968; pp. 261–277.
4. Hacking, I. *The Emergence of Probability: A Philosophical Study of Early Ideas about Probability, Induction and Statistical Inference*, 2nd ed.; Cambridge University Press: Cambridge, UK, 2006.
5. Hájek, A. *Interpretation of Probability, Stanford Encyclopedia of Philosophy*. Zalta, E.N., Ed.; 2012. Available online: http://plato.stanford.edu/archives/win2012/entries/probability-interpret/ (accessed on 26 August 2018).
6. Einstein, A. Physics and reality. *J. Frankl. Inst.* **1936**, *221*, 378. [CrossRef]
7. Plotnitsky, A. *The Principles of Quantum Theory, from Planck's Quanta to the Higgs Boson: The Nature of Quantum Reality and the Spirit of Copenhagen*; Springer: New York, NY, USA, 2016; pp. 84–98.
8. Born, M.; Jordan, P. Zur Quantenmechanik. *Z. Phys.* **1925**, *34*, 858–888. [CrossRef]
9. Bohr, N. *The Philosophical Writings of Niels Bohr*; Ox Bow Press: Woodbridge, CT, USA, 1987; Volume 1, p. 48.
10. Kragh, H. *Niels Bohr and the Quantum Atom: The Bohr Model of Atomic Structure 1913–1925*; Oxford University Press: Oxford, UK, 2012.
11. Plotnitsky, A. *The Principles of Quantum Theory, from Planck's Quanta to the Higgs Boson: The Nature of Quantum Reality and the Spirit of Copenhagen*; Springer: New York, NY, USA, 2016; pp. 54–67.
12. Freidel, L. On the discovery of quantum mechanics by Heisenberg, Born, and Jordan. Unpublished work, 2016.
13. Born, M. *The Einstein-Born Letters*; Walker: New York, NY, USA, 2005; p. 78.
14. Plotnitsky, A. Quantum Atomicity and Quantum Information: Bohr, Heisenberg, and Quantum Mechanics as an Information Theory. In *Quantum Theory: Reexamination of Foundations 2*; Khrennikov, A., Ed.; Växjö University Press: Växjö, Sweden, 2002; pp. 309–343.
15. Plotnitsky, A. *The Principles of Quantum Theory, from Planck's Quanta to the Higgs Boson: The Nature of Quantum Reality and the Spirit of Copenhagen*; Springer: New York, NY, USA, 2016; pp. 72–73.
16. Hardy, L. Towards quantum gravity: A framework for probabilistic theories with non-fixed causal structure. *J. Phys. A* **2007**, *40*, 3081–3099. [CrossRef]
17. Heisenberg, W. *The Physical Principles of the Quantum Theory*; Courier Corporation: Dover, UK; New York, NY, USA, 1930.
18. Plotnitsky, A. *Niels Bohr and Complementarity: An Introduction*; Springer: New York, NY, USA, 2012.
19. Wilczek, F. In search of symmetry lost. *Nature* **2005**, *432*, 239–247. [CrossRef] [PubMed]
20. Borel, E. La logique et l'intuition en mathématique. *Revue de Métaphysique et de Morale* **1907**, *15*, 273–283.
21. Gray, J. *Plato's Ghost: The Modernist Transformation of Mathematics*; Princeton University Press: Princeton, NJ, USA, 2008; p. 202.
22. Bohr, N. *The Philosophical Writings of Niels Bohr*; Ox Bow Press: Woodbridge, CT, USA, 1987; Volume 2, p. 57.
23. Einstein, A. *Autobiographical Notes*; Open Court: La Salle, IL, USA, 1949; p. 47.
24. Van Dongen, J. *Einstein's Unification*; Cambridge University Press: Cambridge, UK, 2010; pp. 89–95.
25. Riemann, B. On the Hypotheses That Lie at the Foundations of Geometry. In *Beyond Geometry: Classic Papers from Riemann to Einstein*; Pesic, P., Ed.; Dover: Mineola, NY, USA, 1854; pp. 23–40.
26. Plotnitsky, A. Comprehending the Connection of Things: Bernhard Riemann and the Architecture of Mathematical Concepts. In *From Riemann to Differential Geometry and Relativity*; Ji, L., Yamada, S., Papadopoulos, A., Eds.; Springer: Berlin, Germany, 2017; pp. 329–363.
27. Deleuze, G.; Guattari, F. What Is Philosophy? Columbia University Press: New York, NY, USA, 1994.

28. Heisenberg, W. *The Physical Principles of the Quantum Theory*; Courier Corporation: Dover, UK; New York, NY, USA, 1930; pp. 10, 13.

29. Bohr, N. Causality and complementarity. In *The Philosophical Writings of Niels Bohr, Volume 4: Causality and Complementarity, Supplementary Papers*; Faye, J., Folse, H.J., Eds.; Ox Bow Press: Woodbridge, CT, USA, 1999; pp. 83–91.

30. Plotnitsky, A. *Niels Bohr and Complementarity: An Introduction*; Springer: New York, NY, USA, 2012; pp. 173–179.

31. Plotnitsky, A. *The Principles of Quantum Theory, from Planck's Quanta to the Higgs Boson: The Nature of Quantum Reality and the Spirit of Copenhagen*; Springer: New York, NY, USA, 2016; pp. 107–120.

32. Bohr, N. *The Philosophical Writings of Niels Bohr*; Ox Bow Press: Woodbridge, CT, USA, 1987; Volume 1, p. 53.

33. Bohr, N. *The Philosophical Writings of Niels Bohr*; Ox Bow Press: Woodbridge, CT, USA, 1987; Volume 2, p. 73.

34. Bohr, N. Can Quantum-Mechanical Description of Physical Reality Be Considered Complete? *Phys. Rev.* **1935**, *48*, 696–703. [CrossRef]

35. Bohr, N. *The Philosophical Writings of Niels Bohr*; Ox Bow Press: Woodbridge, CT, USA, 1987; Volume 2, pp. 61–62.

36. Heisenberg, W. *Physics and Philosophy: The Revolution in Modern Science*; Harper & Row: New York, NY, USA, 1962.

37. Plotnitsky, A. *The Principles of Quantum Theory, from Planck's Quanta to the Higgs Boson: The Nature of Quantum Reality and the Spirit of Copenhagen*; Springer: New York, NY, USA, 2016; pp. 4–23.

38. Plotnitsky, A. The Visualizable, the Representable, and the Inconceivable: Realist and Non-Realist Mathematical Models in Physics and Beyond. *Philos. Trans. R. Soc. A* **2016**, *374*, 20150101. [CrossRef] [PubMed]

39. Frigg, R.; Hartmann, S. Models in Science. *The Stanford Encyclopedia of Philosophy*; Zalta, E.N., Ed.; 2012. Available online: http://plato.stanford.edu/archives/fall2012/entries/models-science/ (accessed on 26 August 2018).

40. Frigg, R. *Theories and Models*; Acumen: Slough, UK, 2014.

41. Bohr, N. *The Philosophical Writings of Niels Bohr*; Ox Bow Press: Woodbridge, CT, USA, 1987; Volume 2, p. 51.

42. Bohr, N. Can Quantum-Mechanical Description of Physical Reality Be Considered Complete? *Phys. Rev.* **1935**, *48*, 697. [CrossRef]

43. Plotnitsky, A. *The Principles of Quantum Theory, from Planck's Quanta to the Higgs Boson: The Nature of Quantum Reality and the Spirit of Copenhagen*; Springer: New York, NY, USA, 2016.

44. Plotnitsky, A.; Khrennikov, A. Reality without Realism: On the Ontological and Epistemological Architecture of Quantum Mechanics. *Found. Phys.* **2015**, *25*, 1269–1300. [CrossRef]

45. Bohr, N. *The Philosophical Writings of Niels Bohr*; Ox Bow Press: Woodbridge, CT, USA, 1987; Volume 1, pp. 54–57.

46. Bohr, N. *The Philosophical Writings of Niels Bohr*; Ox Bow Press: Woodbridge, CT, USA, 1987; Volume 1, pp. 76–77.

47. Dirac, P.A.M. *The Principles of Quantum Mechanics*; Oxford University Press: Oxford, UK, 1930.

48. Von Neumann, J. *Mathematical Foundations of Quantum Mechanics*; Princeton University Press: Princeton, NJ, USA, 1932.

49. Plotnitsky, A. *The Principles of Quantum Theory, from Planck's Quanta to the Higgs Boson: The Nature of Quantum Reality and the Spirit of Copenhagen*; Springer: New York, NY, USA, 2016; pp. 66–68, 124–131, 159.

50. Kant, I. *Critique of Pure Reason*; Cambridge University Press: Cambridge, UK, 1997; p. 115.

51. Cabello, A. Interpretations of quantum theory: A map of madness. In *What Is Quantum Information?* Lombardi, O., Fortin, S., Holik, F., López, C., Eds.; Cambridge University Press: Cambridge, UK, 2017; pp. 138–144.

52. Fuchs, C.A.; Mermin, N.D.; Schack, R. An introduction to QBism with an application to the locality of quantum mechanics. *Am. J. Phys.* **2014**, *82*, 749. [CrossRef]

53. Werner, R.F. Comment on 'What Bell did'. *J. Phys. A* **2014**, *47*, 424011. [CrossRef]

54. Heisenberg, W. *Physics and Philosophy: The Revolution in Modern Science*; Harper & Row: New York, NY, USA, 1962; pp. 47, 145.

55. Heisenberg, W. *Physics and Philosophy: The Revolution in Modern Science*; Harper & Row: New York, NY, USA, 1962; pp. 178–179.

56. Heisenberg, W. *The Physical Principles of the Quantum Theory*; Courier Corporation: Dover, UK; New York, NY, USA, 1930; p. 11.

57. Bohr, N. *The Philosophical Writings of Niels Bohr*; Ox Bow Press: Woodbridge, CT, USA, 1987; Volume 1, pp. 51, 98–100, 108.

58. Bohr, N. *The Philosophical Writings of Niels Bohr*; Ox Bow Press: Woodbridge, CT, USA, 1987; Volume 2, p. 59.

59. Wigner, E.P. On unitary representations of the inhomogeneous Lorentz group. *Ann. Math.* **1939**, *40*, 149–204. [CrossRef]

60. Heisenberg, W. *Physics and Philosophy: The Revolution in Modern Science*; Harper & Row: New York, NY, USA, 1962; pp. 145, 167–186.

61. Bohr, N. *The Philosophical Writings of Niels Bohr*; Ox Bow Press: Woodbridge, CT, USA, 1987; Volume 2, p. 64.

62. Plotnitsky, A. *The Principles of Quantum Theory, from Planck's Quanta to the Higgs Boson: The Nature of Quantum Reality and the Spirit of Copenhagen*; Springer: New York, NY, USA, 2016; pp. 207–246.

63. Schrödinger, E. The present situation in quantum mechanics (1935). In *Quantum Theory and Measurement*; Wheeler, J.A., Zurek, W.H., Eds.; Princeton University Press: Princeton, NJ, USA, 1983; pp. 152–167.

64. Bohr, N. *The Philosophical Writings of Niels Bohr*; Ox Bow Press: Woodbridge, CT, USA, 1987; Volume 2, p. 34.

65. Wittgenstein, L. *Tractatus Logico-Philosophicus*; Routledge: London, UK, 1924; p. 175.

66. Aaronson, S. *Quantum Computing Since Democritus*; Cambridge University Press: Cambridge, UK, 2013.

67. Khrennikov, A. Quantum probabilities and violation of CHSH-inequality from classical random signals and threshold type detection scheme. *Prog. Theor. Phys.* **2012**, *128*, 31–58. [CrossRef]

68. Khrennikov, A. *Interpretations of Probability*; de Gruyter: Berlin, Germany, 2009.

69. De Finetti, B. *Philosophical Lectures on Probability*; Springer: Berlin, Germany, 2008.

70. Jaynes, E.T. *Probability Theory: The Logic of Science*; Cambridge University Press: Cambridge, UK, 2003.

71. Plotnitsky, A. *The Principles of Quantum Theory, from Planck's Quanta to the Higgs Boson: The Nature of Quantum Reality and the Spirit of Copenhagen*; Springer: New York, NY, USA, 2016; pp. 180–184.

72. Einstein, A.; Podolsky, B.; Rosen, N. Can Quantum-Mechanical Description of Physical Reality be Considered Complete? In *Quantum Theory and Measurement*; Wheeler, J.A., Zurek, W.H., Eds.; Princeton University Press: Princeton, NJ, USA, 1935; pp. 138–141.

73. Plotnitsky, A. *The Principles of Quantum Theory, from Planck's Quanta to the Higgs Boson: The Nature of Quantum Reality and the Spirit of Copenhagen*; Springer: New York, NY, USA, 2016; pp. 136–154.

74. Aspect, A.; Dalibard, J.; Roger, G. Experimental test of Bell's inequalities using time varying analyzers. *Phys. Rev. Lett.* **1982**, *49*, 1804–1807. [CrossRef]

75. Bell, J.S. *Speakable and Unspeakable in Quantum Mechanics*; Cambridge University Press: Cambridge, UK, 2004.

76. Cushing, J.T.; McMullin, E. *Philosophical Consequences of Quantum Theory: Reflections on Bell's Theorem*; Notre Dame University Press: Notre Dame, IN, USA, 1989.

77. Ellis, J.; Amati, D. *Quantum Reflections*; Cambridge University Press: Cambridge, UK, 2000.

78. Brunner, N.; Gühne, O.; Huber, M. Special issue on 50 years of Bell's theorem. *J. Phys. A* **2014**, *42*, 424024.

79. Bohr, N. Can Quantum-Mechanical Description of Physical Reality Be Considered Complete? *Phys. Rev.* **1935**, *48*, 699. [CrossRef]

80. Plotnitsky, A. *The Principles of Quantum Theory, from Planck's Quanta to the Higgs Boson: The Nature of Quantum Reality and the Spirit of Copenhagen*; Springer: New York, NY, USA, 2016; pp. 203–206.

81. Brukner, C. Quantum Causality. *Nat. Phys.* **2014**, *10*, 259–263. [CrossRef]

82. D'Ariano, G.M.; Chiribella, G.; Perinotti, P. *Quantum Theory from First Principles: An Informational Approach*; Cambridge University Press: Cambridge, UK, 2017.

83. Hardy, L. A formalism-local framework for general probabilistic theories, including quantum theory. *arXiv* **2010**, arXiv:1005.5164.

84. Bohr, N. *The Philosophical Writings of Niels Bohr*; Ox Bow Press: Woodbridge, CT, USA, 1987; Volume 2, p. 41.

85. Weyl, H. *The Continuum: A Critical Examination of the Foundation of Analysis*; Dover: Mineola, NY, USA, 1928; p. 108.

86. Darigold, O. *From c-Numbers to q-Numbers: The Classical Analogy in the History of Quantum Theory*; University of California Press: Berkeley, CA, USA, 1993; p. 307.

87. Connes, A. *Noncommutative Geometry*; Academic Press: San Diego, CA, USA, 1994.
88. Plotnitsky, A. *Epistemology and Probability: Bohr, Heisenberg, Schrödinger and the Nature of Quantum-Theoretical Thinking*; Springer: New York, NY, USA, 2009; pp. 111–112.
89. Suppes, P. Is Visual Space Euclidean. *Synthese* **1977**, *35*, 397–421. [CrossRef]
90. Foley, J.; Ribeiro-Filho, N.; Da Silva, J. Visual Perception of extend and the geometry of visual space. *Vis. Res.* **2004**, *44*, 147–156. [CrossRef] [PubMed]
91. Erkelens, C. The Perceptual Structure of Visual Space. *i-Perception* **2015**, *6*. [CrossRef] [PubMed]
92. Schweber, S.S. *QED and the Men Who Made It: Dyson, Feynman, Schwinger, and Tomonaga*; Princeton University Press: Princeton, NJ, USA, 1994; pp. 465–466.
93. Bohr, N. *The Philosophical Writings of Niels Bohr*; Ox Bow Press: Woodbridge, CT, USA, 1987; Volume 2, p. 72.
94. Hilbert, D. *Foundations of Geometry*; Open Court: La Salle, IL, USA, 1999.
95. Silverman, J.; Tate, J. *Rational Points on Elliptic Curves*; Springer: Heidelberg, Germany; New York, NY, USA, 2015; p. 277.
96. Newton, I. *The Principia: Mathematical Principles of Natural Philosophy*; University of California Press: Berkeley, CA, USA, 1999.
97. Riemann, B. On the Hypotheses That Lie at the Foundations of Geometry. In *Beyond Geometry: Classic Papers from Riemann to Einstein*; Pesic, P., Ed.; Dover: Mineola, NY, USA, 1854; pp. 23–40.
98. Mehra, J.; Rechenberg, H. *The Historical Development of Quantum Theory*; Springer: Berlin, Germany, 2001; Volume 3, p. 69.
99. Dirac, P.A.M. The fundamental equations of quantum mechanics. In *Sources of Quantum Mechanics*; van der Warden, B.L., Ed.; Dover: New York, NY, USA, 1925; pp. 307–320.
100. Heisenberg, W. *The Physical Principles of the Quantum Theory*; Dover: New York, NY, USA, 1930.
101. Einstein, A. What is the Theory of Relativity? In *Ideas and Opinions*; Einstein, A., Ed.; Bonanza Books: New York, NY, USA, 1954; p. 228.
102. Mehra, J.; Rechenberg, H. *The Historical Development of Quantum Theory*; Springer: Berlin, Germany, 2001; Volume 2, p. 242.
103. Heisenberg, W. *Physics and Philosophy: The Revolution in Modern Science*; Harper & Row: New York, NY, USA, 1962; p. 41.
104. Bohr, N. On the constitution of atoms and Molecules (Part 1). *Philos. Mag.* **1913**, *26*, 1–25. [CrossRef]
105. Plotnitsky, A. *The Principles of Quantum Theory, from Planck's Quanta to the Higgs Boson: The Nature of Quantum Reality and the Spirit of Copenhagen*; Springer: New York, NY, USA, 2016; pp. 68–83.
106. Plotnitsky, A. *Epistemology and Probability: Bohr, Heisenberg, Schrödinger and the Nature of Quantum-Theoretical Thinking*; Springer: New York, NY, USA, 2009; pp. 90, 111, 116.
107. Heisenberg, W. Quantum-theoretical re-interpretation of kinematical and mechanical relations (1925). In *Sources of Quantum Mechanics*; Van der Waerden, B.L., Ed.; Dover: New York, NY, USA, 1968; p. 3.
108. Dirac, P.A.M. *The Principles of Quantum Mechanics*, 4th ed.; Oxford University Press: Oxford, UK, 1958; p. 14.
109. Heisenberg, W. Quantum-theoretical re-interpretation of kinematical and mechanical relations (1925). In *Sources of Quantum Mechanics*; Van der Waerden, B.L., Ed.; Dover: New York, NY, USA, 1968; p. 264.
110. Born, M.; Heisenberg, W.; Jordan, P. On quantum mechanics. In *Sources of Quantum Mechanics*; Van der Waerden, B.L., Ed.; Dover: New York, NY, USA, 1926; pp. 321–385.
111. Heisenberg, W. Quantum-theoretical re-interpretation of kinematical and mechanical relations (1925). In *Sources of Quantum Mechanics*; Van der Waerden, B.L., Ed.; Dover: New York, NY, USA, 1968; p. 265.
112. Schweber, S.S. *QED and the Men Who Made It: Dyson, Feynman, Schwinger, and Tomonaga*; Princeton University Press: Princeton, NJ, USA, 1994; pp. 360–361.
113. Schwinger, J. *Quantum Mechanics: Symbolism of Atomic Measurement*; Springer: New York, NY, USA, 2001.
114. Jaeger, G. Grounding the randomness of quantum measurement. *Philos. Trans. R. Soc. A* **2016**. [CrossRef] [PubMed]
115. Bohr, N. *The Philosophical Writings of Niels Bohr*; Ox Bow Press: Woodbridge, CT, USA, 1987; Volume 2, p. 63.
116. Bohr, N. *The Philosophical Writings of Niels Bohr*; Ox Bow Press: Woodbridge, CT, USA, 1987; Volume 2, pp. 46–47, 56, 62.
117. Wheeler, J.A. Law without law. In *Quantum Theory and Measurement*; Wheeler, J.A., Zurek, W.H., Eds.; Princeton University Press: Princeton, NJ, USA, 1983; pp. 182–216.

118. Wheeler, J.A. Information, physics, quantum: The search for links. In *Complexity, Entropy, and the Physics of Information*; Zurek, W.H., Ed.; Addison-Wesley: Redwood City, CA, USA, 1990; pp. 309–336.

119. Einstein, A. Physics and reality. *J. Frankl. Inst.* **1936**, *221*, 375. [CrossRef]

120. Hardy, L. Quantum mechanics from five reasonable axioms. *arXiv* **2001**, arXiv:quant-ph/0101012v4.

121. Fuchs, C.A. Quantum mechanics as quantum information, mostly. *J. Mod. Opt.* **2003**, *50*, 987–1023. [CrossRef]

122. Chiribella, G.; Spekkens, R. *Quantum Theory: Informational Foundations and Foils*; Springer: New York, NY, USA, 2016.

123. D'Ariano, G.M.; Chiribella, G.; Perinotti, P. *Quantum Theory from First Principles: An Informational Approach*; Cambridge University Press: Cambridge, UK, 2017; p. 4.

124. Heisenberg, W. *The Physical Principles of the Quantum Theory*; Dove: New York, NY, USA, 1930; p. 108.

125. D'Ariano, G.M.; Chiribella, G.; Perinotti, P. *Quantum Theory from First Principles: An Informational Approach*; Cambridge University Press: Cambridge, UK, 2017; pp. 4–5.

126. D'Ariano, G.M.; Chiribella, G.; Perinotti, P. *Quantum Theory from First Principles: An Informational Approach*; Cambridge University Press: Cambridge, UK, 2017; p. 169.

127. Spekkens, R.W. Evidence for the epistemic view of quantum states: A toy theory. *Phys. Rev. A* **2007**, *75*, 032110. [CrossRef]

128. Spekkens, R.W. Quasi-quantization: Classical statistical theories with an epistemic restriction. In *Quantum Theory: Informational Foundations and Foils*; Chiribella, J., Spekkens, R.W., Eds.; Springer: New York, NY, USA, 2016; pp. 83–136.

129. Bartlett, S.D.; Rudolph, T.; Spekkens, R.W. Reconstruction of Gaussian quantum mechanics from Liouville mechanics with an epistemic restriction. *Phys. Rev. A* **2012**, *86*, 012103. [CrossRef]

130. Spekkens, R.W. Quasi-quantization: Classical statistical theories with an epistemic restriction. In *Quantum Theory: Informational Foundations and Foils*; Chiribella, J., Spekkens, R.W., Eds.; Springer: New York, NY, USA, 2016; p. 91.

131. Spekkens, R.W. Quasi-quantization: Classical statistical theories with an epistemic restriction. In *Quantum Theory: Informational Foundations and Foils*; Chiribella, J., Spekkens, R.W., Eds.; Springer: New York, NY, USA, 2016; p. 95.

132. Spekkens, R.W. Quasi-quantization: Classical statistical theories with an epistemic restriction. In *Quantum Theory: Informational Foundations and Foils*; Chiribella, J., Spekkens, R.W., Eds.; Springer: New York, NY, USA, 2016; p. 96.

133. D'Ariano, G.M.; Chiribella, G.; Perinotti, P. *Quantum Theory from First Principles: An Informational Approach*; Cambridge University Press: Cambridge, UK, 2017; p. 6.

134. D'Ariano, G.M.; Chiribella, G.; Perinotti, P. *Quantum Theory from First Principles: An Informational Approach*; Cambridge University Press: Cambridge, UK, 2017; pp. 168–169.

135. Schrödinger, E. Discussion of probability relations between separated systems. *Proc. Camb. Philos. Soc.* **1953**, *31*, 555–563. [CrossRef]

136. Chiribella, G.; D'Ariano, G.M.; Perinotti, P. Informational derivation of quantum theory. *Phys. Rev. A* **2011**, *84*, 012311-1–012311-39. [CrossRef]

137. Hardy, L. Foliable operational structures for general probabilistic theory. In *Deep beauty: Understanding the Quantum World through Mathematical Innovation*; Halvorson, H., Ed.; Cambridge University Press: Cambridge, UK, 2011; pp. 409–442.

138. Hardy, L. A formalism-local framework for general probabilistic theories, including quantum theory. *arXiv* **2010**, arXiv:1005.5164; pp. 19–20.

139. Hardy, L. A formalism-local framework for general probabilistic theories, including quantum theory. *arXiv* **2010**, arXiv:1005.5164; p. 39.

140. MacLane, S. *Categories for the Working Mathematician*; Springer: Berlin, Germany, 2013.

141. MacLane, S.; Moerdijk, I. *Sheaves in Geometry and Logic: A First Introduction to Topos Theory*; Springer: Berlin, Germany, 1994.

142. Hasegawa, M.; Hofmann, M.; Plotkin, G. Finite dimensional vector spaces are complete for traced symmetric monoidal categories. *Lect. Notes Comput. Sci.* **2008**, *4800*, 367–385.

143. Coecke, R.; Kissinger, A. *Picturing Quantum Processes: A First Course in Quantum Theory and Diagrammatic Reasoning*; Cambridge University Press: Cambridge, UK, 2017.

144. D'Ariano, G.M.; Perinotti, P. Derivation of the Dirac equation from principles of information processing. *Phys. Rev. A* **2014**, *90*, 062106. [CrossRef]

145. D'Ariano, G.M. Physics without Physics. *Int. J. Theor. Phys.* **2017**, *56*, 97–128. [CrossRef]

MDPI

Article

$SU(2)$ Decomposition for the Quantum Information Dynamics in 2d-Partite Two-Level Quantum Systems

Francisco Delgado †

Escuela de Ingeniería y Ciencias, Tecnológico de Monterrey, Atizapán 52926, Mexico; fdelgado@itesm.mx;
Tel.: +52-55-5864-5670
† Current address: Departamento de Física y Matemáticas, Tecnológico de Monterrey,
  Campus Estado de México, Atizapán, Estado de México, Mexico.

Received: 1 June 2018; Accepted: 2 August 2018; Published: 17 August 2018

✓ check for updates

Abstract: The gate array version of quantum computation uses logical gates adopting convenient forms for computational algorithms based on the algorithms classical computation. Two-level quantum systems are the basic elements connecting the binary nature of classical computation with the settlement of quantum processing. Despite this, their design depends on specific quantum systems and the physical interactions involved, thus complicating the dynamics analysis. Predictable and controllable manipulation should be addressed in order to control the quantum states in terms of the physical control parameters. Resources are restricted to limitations imposed by the physical settlement. This work presents a formalism to decompose the quantum information dynamics in $SU(2^{2d})$ for 2d-partite two-level systems into 2^{2d-1} $SU(2)$ quantum subsystems. It generates an easier and more direct physical implementation of quantum processing developments for qubits. Easy and traditional operations proposed by quantum computation are recovered for larger and more complex systems. Alternating the parameters of local and non-local interactions, the procedure states a universal exchange semantics on the basis of generalized Bell states. Although the main procedure could still be settled on other interaction architectures by the proper selection of the basis as natural grammar, the procedure can be understood as a momentary splitting of the 2d information channels into 2^{2d-1} pairs of 2 level quantum information subsystems. Additionally, it is a settlement of the quantum information manipulation that is free of the restrictions imposed by the underlying physical system. Thus, the motivation of decomposition is to set control procedures easily in order to generate large entangled states and to design specialized dedicated quantum gates. They are potential applications that properly bypass the general induced superposition generated by physical dynamics.

Keywords: quantum information; quantum dynamics; entanglement

1. Introduction

Quantum information is generating new applications and tentative future technologies such as quantum computation [1–3] and quantum cryptography, based on disruptive phenomena in its main trends: quantum key distribution [4,5], quantum secret sharing [6], and quantum secure direct communication [7,8]. All these trends highlight the importance of entangled states—a basic aspect involved in the current work in order to achieve quantum information processing tasks. In this arena, the understanding of quantum information dynamics and the control of quantum systems is a compulsory development to manage the quantum resources involved. Applications require a tight control of resources and interactions—especially those related with coherence and entanglement. They are fundamental in most applications. Quantum control has developed the fine management of physical variables to prepare, maintain, and transform quantum states in order to exploit them for concrete purposes. The outstanding high-tech commercial appliances D-Wave and IBM-Q use qubits in the form of two-level systems, either with superconducting circuits or ions as well as several approaches to their interconnecting architecture.

For multipartite systems, research in control is oriented to achieve different goals in quantum applications. Most of them are numerical approaches rather than analytical due to the inherited complexity in the quantum information dynamics when the number of parts grows. For a single system with a two-level spectrum, the control problem has been extensively studied in terms of exact optimal control for energy or time cost [9,10]. Recently, research of the anisotropic Heisenberg–Ising model for bipartite systems in $SU(4)$ [11] has shown how this model exhibits $SU(2)$ block decomposition when it is written in the non-local basis of Bell states instead of the traditional computational basis. This means that $\mathcal{H}^2$ becomes a direct sum of two subspaces, each one generated by a pair of Bell states, while U underlies in the direct product $U(1) \times SU(2)^2$. Thus, control can be reduced to $SU(2)$ control problems, each one in each block. Then, exact solutions for some control procedures can be found [12,13]. There, controlled blocks can be configured by the direction of external driven interactions introduced. That scheme allows controlled transformations between Bell states on demand and therefore on a general state. Thus, the procedure sets a method of control to manipulate quantum information on magnetic systems, where the computational grammar is based on Bell states instead of the traditional computational basis. It allows an easier programmed transformation among any pair of elements in that basis. This result provided the inspiration to reproduce similar decomposition schemes for larger systems in terms of simpler problems based on quasi-isolated two-level subsystems, developing easier and universal (not necessarily optimal) controlled manipulation procedures for quantum information. Technology to set up the possible architecture of these generic systems is currently being achieved through trapped-ion qubits [14] and superconducting qubits [15].

Thus, the generalization of $SU(2)$ block decomposition is a convenient formalism to express dynamics, revealing certain quantum information states algebraically free of the complexity introduced by the entangling operations (doing few convenient the use of the computational basis). Nevertheless, they still conserve their entangled properties. This reveals how the probability exchange happens together with the structure of entanglement behind the randomness introduced by the complexity of large quantum information systems. Still, as for their predecessor, those bases maintain a certain degree of universality, including several alternative local and non-local interactions. As for their $SU(4)$ predecessor, when they are combined, it states a series of punctual operations that can be set: (a) fine control based on well-known $SU(2)$ control procedures; (b) the construction of universal gates for the entire process based on two-channel like operations; and (c) the design of more complex dedicated multi-channel gates by factorization.

The general aim of this paper is to show that such decomposition and reduction is achievable for large qubit systems, not only for those in [11,13]. The second section states the general Hamiltonian to be analyzed. The third section shows how the $SU(2)$ decomposition procedure can be generalized on general n-partite two-level systems (not only for the driven Heisenberg–Ising interactions), reducing them to 2^{n-1} selectable transformations between pairs of specific quantum energy states. Then, these transformations can be based on known control schemes for $SU(2)$ systems such as those in [9,10]. The selection of these 2^{n-1} pairs of states can be based on the convenience of the quantum process being considered and the resources involved. Thus, the basis on which the decomposition can be established works as a computational grammar for the quantum procedures being attained. These bases are not completely arbitrary, and thus the fourth section shows how a kind of generalized Bell basis is able to generate the $SU(2)$ decomposition for an even number of parts, $n = 2d$. The fifth section is devoted to analyzing the restrictions on the Hamiltonian to get the $SU(2)$ decomposition, the inherited states, and the block properties. This analysis includes a classification of interactions able to generate the $SU(2)$ decomposition. Because the presented procedure can reproduce complex quantum gates, generate large entangled states, and introduce control procedures in $SU(2^{2d})$ if the grammar is based on the proposed basis, the sixth section analyzes potential applications in these trends. The final section concludes, summarizing the findings and settling the related future work to be developed. Because several aspects in the work may be complex for the reader, an appendix with some critical concepts has been included to clarify the contents.

2. Generalized Hamiltonian

The problem can be established for a general Hamiltonian for n coupled two-level systems on $U(2^n)$ forming a closed system. It can be written as a general combination of tensor products of Pauli matrices for each subsystem (for a more detailed discussion of this Hamiltonian, please see Appendix A.1):

$$\tilde{H} = \sum_{\{i_k\}} h_{\{i_k\}} \bigotimes_{k=1}^{n} \sigma_{i_k} = \sum_{\mathcal{I}=0}^{4^n-1} h_{\mathcal{I}_4^n} \bigotimes_{k=1}^{n} \sigma_{\mathcal{I}_{4,k}^n}, \tag{1}$$

where $\{i_k\} = \{i_1, i_2, \ldots, i_n\}$, $i_k = 0,1,2,3$, and $h_{\{i_k\}}$ is a general set of time-dependent real functions. Sometimes, as in the second expression in (1), $\{i_k\}$ will be represented as the number $\mathcal{I} \in \{0, 1, \ldots, 4^n - 1\}$, as it is expressed in base-4 with n digits, $\mathcal{I}_4^n$. Then, $\mathcal{I}_{4,k}^n = i_k$ for $k = 1, 2, \ldots, n$. σ_{i_k} for $i_k = 0,1,2,3$ are the traditional Pauli matrices in the computational basis $|0\rangle, |1\rangle \in \mathcal{H}^2$ for each part k. Note that due to the $SU(2)$ algebra of Pauli matrices, this Hamiltonian comprises all analytical Hamiltonians based on two-level systems with n parts. The Hamiltonian obeys the Schrödinger equation for its associated evolution operator $\tilde{U}$:

$$\tilde{H}\tilde{U} = i\hbar \frac{\partial \tilde{U}}{\partial t}. \tag{2}$$

Although $h_{\{0,\ldots,0\}}$ is not necessarily zero, if $\{\tilde{E}_j \,|\, j = 1, \ldots, 2^n\}$ are the eigenvalues of $\tilde{H}$ and $\mathcal{E} \equiv \sum_{j=1}^{2^n} \tilde{E}_j = 2^n h_{\{0,\ldots,0\}}$, then defining

$$H \equiv \tilde{H} - h_{\{0,\ldots,0\}} \bigotimes_{k=1}^{n} \sigma_0, \quad U \equiv \tilde{U} e^{\frac{i}{\hbar} h_{\{0,\ldots,0\}} t}, \tag{3}$$

these operators become the equivalent traceless Hamiltonian and its corresponding evolution operator with eigenvalues $E_j = \tilde{E}_j - h_{\{0,\ldots,0\}}$, both fulfilling (2) as well. H and $\tilde{H}$ have the same set of eigenvectors $\{|b_j\rangle \in \mathcal{H}^{2^n} | j = 1, \ldots, 2^n\}$. Thus, the Hamiltonian can be written alternatively as $H = \sum_{j=1}^{2^n} E_j |b_j\rangle \langle b_j|$. Thus, in the following, the Hamiltonian can be assumed traceless without loss of generality. Note that while $\tilde{U} \in U(2^n)$, then $U \in SU(2^n)$. In the following, only H and U symbols will be used as equivalent to $\tilde{H}$ and $\tilde{U}$. H can be split in two parts—the local H_l and the non-local H_{nl} interactions:

$$H_l = \sum_{k=1}^{n} \sum_{i=1}^{3} h_{(i4^{k-1})_4^n} \bigotimes_{s=1}^{n} \sigma_{(i4^{k-1})_{4,s}^n} \rightarrow \tilde{H} = \tilde{H}_{nl} + H_l, \tag{4}$$

where $(i4^{k-1})_4^n$ is the number $i4^{k-1}$ represented in base-4 with n digits and $(i4^{k-1})_{4,s}^n$ is its sth term in that basis.

3. $SU(2)$ Decomposition Generalities

In order to support the understanding of some aspects in the further discussion, Appendix A.2 contains a brief of group theory that is relevant for this work as well as some critical points in the decomposition procedure being presented here. Delgado [11] found that the $SU(2)$ decomposition procedure can be induced by considering a set of 2^n orthogonal states: $\{|\alpha_i\rangle\}$ and 2^{n-1} pairs

$\{j(i), k(i)\}, i = 1, 2, ..., 2^{n-1}$, with $k(i) = j(i) + 1 \in \{2, 4, ..., 2^n\}$ related with the eigenvalues through a mixing matrix, in such way that they fulfill:

$$2\left|b_{2i-1}\right\rangle = A_i\left|\alpha_{j(i)}\right\rangle + B_i\left|\alpha_{k(i)}\right\rangle \qquad \rightarrow \qquad \left|\alpha_{j(i)}\right\rangle = A_i^*\left|b_{2i-1}\right\rangle - B_i e^{-i\phi}\left|b_{2i}\right\rangle,$$

$$\left|b_{2i}\right\rangle = -B_i^* e^{i\phi}\left|\alpha_{j(i)}\right\rangle + A_i^* e^{i\phi}\left|\alpha_{k(i)}\right\rangle \qquad \rightarrow \qquad \left|\alpha_{k(i)}\right\rangle = B_i^*\left|b_{2i-1}\right\rangle + A_i e^{-i\phi}\left|b_{2i}\right\rangle, \tag{5}$$

with $|A_i|^2 + |B_i|^2 = 1$, where last relations are clearly true because of orthogonality (note that energies E_j become ordered as the states are paired). States $\{|\alpha_j\rangle\}$ are then defined by the selection of A_i, B_i. Each pair sets one of the orthogonal subspaces:

$$\mathcal{H}_i^2 = \mathrm{span}(\{|b_{2i-1}\rangle, |b_{2i}\rangle\}) = \mathrm{span}(\{\left|\alpha_{j(i)}\right\rangle, \left|\alpha_{k(i)}\right\rangle\}) \rightarrow \mathcal{H}^{2^n} = \bigoplus_{i=1}^{2^{n-1}} \mathcal{H}_i^2. \tag{6}$$

There are many possibilities for this selection, but not necessarily all practical bases fit in this construction. In particular, separability or entanglement properties are not necessarily assured for $\{|\alpha_i\rangle\}$ as in [11]. Clearly, because these states are assumed unitary, then $A_i = \left\langle b_{2i}|\alpha_{k(i)}\right\rangle e^{i\phi} = \left\langle b_{2i-1}|\alpha_{j(i)}\right\rangle^*$, $B_i = \left\langle b_{2i-1}|\alpha_{k(i)}\right\rangle^* = -\left\langle b_{2i}|\alpha_{j(i)}\right\rangle e^{i\phi}$. By applying H on (5) and considering that $|b_i\rangle$ has the eigenvalue E_i, it is possible to arrive at the following expressions:

$$H\left|\alpha_{j(i)}\right\rangle = (|A_i|^2 E_{2i-1} + |B_i|^2 E_{2i})\left|\alpha_{j(i)}\right\rangle + A_i^* B_i(E_{2i-1} - E_{2i})\left|\alpha_{k(i)}\right\rangle,$$

$$H\left|\alpha_{k(i)}\right\rangle = A_i B_i^*(E_{2i-1} - E_{2i})\left|\alpha_{j(i)}\right\rangle + (|A_i|^2 E_{2i} + |B_i|^2 E_{2i-1})\left|\alpha_{k(i)}\right\rangle, \tag{7}$$

giving the Hamiltonian components in this basis:

$$\left\langle \alpha_{j(i)}|H|\alpha_{j(i)}\right\rangle = |A_i|^2 E_{2i-1} + |B_i|^2 E_{2i},$$

$$\left\langle \alpha_{k(i)}|H|\alpha_{k(i)}\right\rangle = |A_i|^2 E_{2i} + |B_i|^2 E_{2i-1}, \tag{8}$$

$$\left\langle \alpha_{j(i)}|H|\alpha_{k(i)}\right\rangle = A_i B_i^*(E_{2i-1} - E_{2i}),$$

which can be alternatively obtained from (5). Note that the phase ϕ is non-physical. This basis transformation changes the diagonal structure for the basis $\{|b_i\rangle\}$ into a 2×2 diagonal block structure for the basis $\{|\alpha_j\rangle\}$. For simplicity, we define the following quantities:

$$A_i = r_{A_i} e^{i\gamma_{A_i}}, B_i = r_{B_i} e^{i\gamma_{B_i}},$$

$$\Delta_i^{\pm} = \frac{1}{2\hbar}(E_{2i} \pm E_{2i-1}), \tag{9}$$

$$\Gamma_i = \gamma_{A_i} - \gamma_{B_i}.$$

Then, each 2×2 block in H (labeled as $\mathbb{S}_{Hi}$) can be written in matrix form as (see Appendix A.2):

$$\begin{aligned} \mathbb{S}_{Hi} &= \begin{pmatrix} \Delta_i^+ - (r_{A_i}^2 - r_{B_i}^2)\Delta_i^- & -2r_{A_i} r_{B_i} \Delta_i^- e^{i\Gamma_i} \\ -2r_{A_i} r_{B_i} \Delta_i^- e^{-i\Gamma_i} & \Delta_i^+ + (r_{A_i}^2 - r_{B_i}^2)\Delta_i^- \end{pmatrix} \\ &= \Delta_i^+ \mathbb{I}_i - 2r_{A_i} r_{B_i} \Delta_i^- \cos\Gamma_i \mathbb{X}_i + 2r_{A_i} r_{B_i} \Delta_i^- \sin\Gamma_i \mathbb{Y}_i - (r_{A_i}^2 - r_{B_i}^2)\Delta_i^- \mathbb{Z}_i \\ &\equiv \Delta_i^+ \mathbb{I}_i + \mathbb{S}_{Hi}^0, \end{aligned} \tag{10}$$

where although $r_{A_i}^2 + r_{B_i}^2 = 1$, we use both terms r_{A_i} and r_{B_i} for the symmetry. In addition, $\mathbb{I}_i, \mathbb{X}_i, \mathbb{Y}_i$, and $\mathbb{Z}_i$ are respectively the 2×2 unitary matrix and the Pauli matrices settled as basis for the block $\mathbb{S}_{Hi}$.

Thus, H can be written as a sum of operators acting on the different subspaces $\mathcal{H}_i^2$ or as the following direct sum structure of 2^{n-1} 2×2 block-diagonal matrices:

$$
H = \overset{2^{n-1}}{\underset{i=1}{\bigoplus}} \mathbb{S}_{Hi} = \left(
\begin{array}{c|c|c|c}
\mathbb{S}_{H1} & 0 & \cdots & 0 \\
\hline
0 & \mathbb{S}_{H2} & \cdots & 0 \\
\hline
\vdots & \vdots & \ddots & \vdots \\
\hline
0 & 0 & \cdots & \mathbb{S}_{H2^{n-1}}
\end{array}
\right), \tag{11}
$$

with $\mathbf{0}$, the 2×2 zero matrix. Because this structure is preserved under matrix products, it is inherited by the evolution matrix U. In particular, if the Hamiltonian (1) is not time-dependent, then $U = \sum_{j=1}^{2^n} e^{-\frac{i}{\hbar}E_j t} |b_j\rangle \langle b_j|$. Thus, when the basis is changed (see Appendix A.2):

$$
\begin{aligned}
\mathbb{S}_{Ui} &= e^{-\frac{i}{\hbar}E_{2i-1}t} |b_{2i-1}\rangle \langle b_{2i-1}| + e^{-\frac{i}{\hbar}E_{2i}t} |b_{2i}\rangle \langle b_{2i}| \\
&= e^{-i\Delta_i^+ t} \Big((e^{i\Delta_i^- t} - 2ir_{B_i}{}^2 \sin\Delta_i^- t) |\alpha_j(i)\rangle \langle \alpha_j(i)| + \\
&\quad 2ir_{A_i}r_{B_i} \sin\Delta_i^- t (e^{i\Gamma_i} |\alpha_j(i)\rangle \langle \alpha_k(i)| + e^{-i\Gamma_i} |\alpha_k(i)\rangle \langle \alpha_j(i)|) + \\
&\quad (e^{-i\Delta_i^- t} + 2ir_{B_i}{}^2 \sin\Delta_i^- t) |\alpha_k(i)\rangle \langle \alpha_k(i)| \Big).
\end{aligned} \tag{12}
$$

Similarly, in matrix form or in terms of $\mathbb{I}_i$, $\mathbb{X}_i$, $\mathbb{Y}_i$, and $\mathbb{Z}_i$:

$$
\begin{aligned}
\mathbb{S}_{Ui} &= e^{-i\Delta_i^+ t} \begin{pmatrix} \cos\Delta_i^- t + i(r_{A_i}{}^2 - r_{B_i}{}^2)\sin\Delta_i^- t & 2ir_{A_i}r_{B_i}e^{i\Gamma_i}\sin\Delta_i^- t \\ 2ir_{A_i}r_{B_i}e^{-i\Gamma_i}\sin\Delta_i^- t & \cos\Delta_i^- t - i(r_{A_i}{}^2 - r_{B_i}{}^2)\sin\Delta_i^- t \end{pmatrix} \\
&= e^{-i\Delta_i^+ t} \Big(\cos\Delta_i^- t\,\mathbb{I}_i + 2ir_{A_i}r_{B_i}\cos\Gamma_i \sin\Delta_i^- t\,\mathbb{X}_i - \\
&\quad 2ir_{A_i}r_{B_i}\sin\Gamma_i \sin\Delta_i^- t\,\mathbb{Y}_i + i(r_{A_i}{}^2 - r_{B_i}{}^2)\sin\Delta_i^- t\,\mathbb{Z}_i \Big) \equiv e^{-i\Delta_i^+ t}\mathbb{S}_{Ui}^0.
\end{aligned} \tag{13}
$$

Note that the election of Γ_i lets us simplify the last expression to contain only one operator between $\mathbb{X}_i$ and $\mathbb{Y}_i$ (as in [11,12]). This property is useful to set the optimal control in [9] in each block. Then, similar to H:

$$
U = \overset{2^{n-1}}{\underset{i=1}{\bigoplus}} \mathbb{S}_{Ui} = \left(
\begin{array}{c|c|c|c}
\mathbb{S}_{U1} & 0 & \cdots & 0 \\
\hline
0 & \mathbb{S}_{U2} & \cdots & 0 \\
\hline
\vdots & \vdots & \ddots & \vdots \\
\hline
0 & 0 & \cdots & \mathbb{S}_{U2^{n-1}}
\end{array}
\right), \tag{14}
$$

where in general for the time-dependent Hamiltonian:

$$
\mathbb{S}_{Ui} = \tau\{e^{-\frac{i}{\hbar}\int_0^t \mathbb{S}_{Hi}dt'}\} = e^{-i\Delta_i^+ t}\tau\{e^{-\frac{i}{\hbar}\int_0^t \mathbb{S}_{Hi}^0 dt'}\} \equiv e^{-i\Delta_i^+ t}\mathbb{S}_{Ui}^0, \tag{15}
$$

where τ is the time-ordered integral. This implies that U is an element in the direct product $U(1)^{2^{n-1}-1} \times SU(2)^{2^{n-1}} \subset SU(2^n)$ (because any factor phase $e^{-i\Delta_i^+ t}$ depends on the remaining phase factors through $\mathcal{E}$, see Appendix A.2). In the following, we will informally call this factorization the $SU(2)$ decomposition (in reality, each block has the form $U(1) \times SU(2)$) due to the block structure. Consequently, the Hilbert space $\mathcal{H}^n$ becomes the direct sum of 2^{n-1} subspaces generated by each pair $\{|\alpha_{j(i)}\rangle, |\alpha_{k(i)}\rangle\}$, $i = 1, 2, \ldots, 2^{n-1}$. In each subspace, dynamics mixes the probabilities, but probabilities among subspaces remain unmixed if there is no rearrangement in the pairing between $\{|b_i\rangle\}$ and $\{|\alpha_j\rangle\}$ (clearly, in this rearrangement, the basis $\{|\alpha_j\rangle\}$ could change).

Thus, if $|\psi_0\rangle = \sum_{i=1}^{2^{n-1}} |\psi_{0i}\rangle$ is the initial state with $|\psi_{0i}\rangle = \psi_{0i,j(i)} \left|\alpha_{j(i)}\right\rangle + \psi_{0i,k(i)} \left|\alpha_{k(i)}\right\rangle$, then each component is evolved in the subspace $i = 1, 2, ..., 2^{n-1}$, fulfilling $\| |\psi_{ti}\rangle \| = \|\mathbb{S}U_i^0 |\psi_{0i}\rangle \| = \| |\psi_{0i}\rangle \|$.

Finally, note that the $SU(2)$ decomposition is not the only one available, although it is the most valuable for the binary inheritance from the classical computation. In fact, other decompositions involving bigger subgroups are possible, whether using bigger systems than two-level ones and/or simply involving more than two eigenvectors in (5). Inclusively, a mixed-sized block structure can be realized.

4. GBS: A Non-Local Basis Fitting in $\{|\alpha_j\rangle\}$

Non-local bases are used as a theoretical resource to explicitly show how evolution [16] and measurement [17] can generate entangled states. In [11], it was shown that the Heisenberg–Ising model including driven magnetic fields in a fixed direction allows the generation of the block structure in the traditional Bell basis. Thus, the Bell basis for two-level bipartite systems has been shown to fit in the $U(1) \times SU(2)^2$ decomposition of $SU(4)$. Despite the added complexity to manage non-local states, recent work has moved towards the control of entangled states [18]. This basis works as a universal basis for the Heisenberg–Ising interaction, including an external magnetic field in any specific direction on a couple of qubits [11–13]. This model includes other interaction models, such as XXX [19], XY [20], and XXZ [21]. In the current development, the most obvious guess is the generalized Bell states (GBS) basis for $n = 2d$ presented in [22] as tensor products of Bell states. In the next sections, some further useful formulas are obtained to show how the GBS basis fits in the $SU(2)$ decomposition for larger systems than bipartite ones.

4.1. GBS Basis and Hamiltonian Components

For $n = 2d$, the GBS basis [22] forms an orthogonal basis of partial entangled states for $2d$ particles. A more extended treatment for this basis is given in Appendix A.3 in order to ease further understanding in the current context, particularly related with the underlying single Bell states in their construction together with their index notation—a key aspect in the remaining development. Each element in this basis can be written briefly as:

$$
\begin{aligned}
\left|\Psi_{\mathcal{I}_4^d}\right\rangle &= \bigotimes_{s=1}^{d} \frac{1}{\sqrt{2}} \sum_{\epsilon_s,\delta_s=0}^{1} (\tilde{\sigma}_{i_s})_{\epsilon_s,\delta_s} |\epsilon_s \delta_s\rangle \\
&= \frac{1}{\sqrt{2^d}} \sum_{\{\epsilon_j\},\{\delta_k\}} (\tilde{\sigma}_{i_1} \otimes \dots \otimes \tilde{\sigma}_{i_d})_{\epsilon_1 \dots \epsilon_d \delta_1 \dots \delta_d} |\epsilon_1 \dots \epsilon_d\rangle \otimes |\delta_1 \dots \delta_d\rangle \\
&= \frac{1}{\sqrt{2^d}} \sum_{\mathcal{E},\mathcal{D}=0}^{2^d-1} (\tilde{\sigma}_{i_1} \otimes \dots \otimes \tilde{\sigma}_{i_d})_{\mathcal{E}_2^d,\mathcal{D}_2^d} \left|\mathcal{E}_2^d\right\rangle \otimes \left|\mathcal{D}_2^d\right\rangle,
\end{aligned}
\tag{16}
$$

where $\{\epsilon_j\} = \{\epsilon_1, \dots, \epsilon_d\}$, $\{\delta_k\} = \{\delta_1, \dots, \delta_d\}$; $\epsilon_j, \delta_k = 0, 1$. At this point, $\tilde{\sigma}_i$ can be considered as proportionally unitary to the traditional Pauli matrices [22]. In addition, $\mathcal{I}_4^d$ is a brief expression of $\{i_1, i_2, \dots, i_d\}$ as the digits set of $\mathcal{I} \in \{0, 1, \dots, 4^d - 1\}$ when it is written in base-4 with d digits. Similarly, $\mathcal{E}_2^d, \mathcal{D}_2^d$ are numbers written in base-2 with d digits ($\mathcal{E}, \mathcal{D} \in \{0, 1, \dots, 2^d - 1\}$) representing $\{\epsilon_1, \dots, \epsilon_d\}, \{\delta_1, \dots, \delta_d\}$, respectively (note that digits are used inverted, as they commonly appear in $\mathcal{E}_2^d$ or $\mathcal{I}_4^d$ expressions). In the following, for simplicity, we use $\mathcal{I}_b^d$ and $\mathcal{I}$ interchangeably because the base b can normally be inferred from the context. Each element in this basis is not maximally entangled. Instead, they have maximally entangled bipartite subsystems (see Appendix A.3), which are separable from the remaining system. Separable pairs contain the parts $[s, s+d]$, $s = 1, 2, ..., d$ (in the following, square brackets will be used to point out a subsystem of parts in the whole system).

In order for $\{\left|\Psi_{\mathcal{I}_4^d}\right\rangle\}$ ($\mathcal{I} \in \{0, 1, \dots, 4^d - 1\}$) to reach the kind of sets $\{|\alpha_j\rangle\}$ stated in the previous section where H and U achieve the $SU(2)$ block structure, H should fulfill some restrictions. We are interested in setting these in the current subsections. Combining expressions (1) and (16), we can express the components of H in the GBS basis. First, we note [23]:

$$\langle \Psi_{\mathcal{I}_4^d} | \sigma_{j_1} \otimes \ldots \otimes \sigma_{j_{2d}} | \Psi_{\mathcal{K}_4^d} \rangle = \prod_{s=1}^{d} \frac{1}{\sqrt{2}} \sum_{\epsilon_s, \delta_s = 0}^{1} (\tilde{\sigma}_{i_s}^*)_{\epsilon_s, \delta_s} \frac{1}{\sqrt{2}} \sum_{\gamma_s, \phi_s = 0}^{1} (\tilde{\sigma}_{k_s})_{\gamma_s, \phi_s} \langle \epsilon_s | \sigma_{j_s} | \gamma_s \rangle \langle \delta_s | \sigma_{j_{s+d}} | \phi_s \rangle$$

$$= \frac{1}{2^d} \sum_{\substack{\mathcal{E}, \mathcal{D} \\ \mathcal{F}, \mathcal{G}}} (\tilde{\sigma}_{i_1}^* \otimes \ldots \otimes \tilde{\sigma}_{i_d}^*)_{\mathcal{E}_2^d, \mathcal{D}_2^d} (\sigma_{j_1} \otimes \ldots \otimes \sigma_{j_{2d}})_{\mathcal{E}_2^d \mathcal{D}_2^d, \mathcal{F}_2^d \mathcal{G}_2^d} (\tilde{\sigma}_{k_1} \otimes \ldots \otimes \tilde{\sigma}_{k_d})_{\mathcal{F}_2^d, \mathcal{G}_2^d} \qquad (17)$$

$$= \frac{1}{2^d} \prod_{s=1}^{d} \text{Tr}(\tilde{\sigma}_{i_s}^* \sigma_{j_{d+s}} \tilde{\sigma}_{k_s}^T \sigma_{j_s}^T),$$

where combined subscripts as $\mathcal{E}_2^d \mathcal{D}_2^d$ represent the set of subscripts obtained by merging $\{\epsilon_1 \ldots \epsilon_d\}$ and $\{\delta_1 \ldots \delta_d\}$. Therefore, the final and notable expression for the Hamiltonian components becomes [23]:

$$\langle \Psi_{\mathcal{I}_4^d} | H | \Psi_{\mathcal{K}_4^d} \rangle = \frac{1}{2^d} \sum_{\mathcal{J}=0}^{4^{2d}-1} h_{\mathcal{J}_4^{2d}} \prod_{s=1}^{d} \text{Tr}(\tilde{\sigma}_{i_s}^* \sigma_{j_{d+s}} \tilde{\sigma}_{k_s}^T \sigma_{j_s}^T), \qquad (18)$$

where $\mathcal{J} \in \{0, 1, \ldots, 4^{2d} - 1\}$ (here, $\mathcal{J} = 0$ can be removed in spite of the discussion in the first section). In the last expressions, the product $\tilde{\sigma}_{i_s}^* \sigma_{j_{d+s}} \tilde{\sigma}_{k_s}^T \sigma_{j_s}^T$ has some properties inherited from Pauli matrices. Because $\sigma_1, \sigma_2, \sigma_3$ are traceless and $\sigma_i^T = \pm \sigma_i$ (negative sign only if $i = 2$), then $\text{Tr}(\tilde{\sigma}_{i_s}^* \sigma_{j_{d+s}} \tilde{\sigma}_{k_s}^T \sigma_{j_s}^T)$ is non-zero only if i_s, j_{d+s}, k_s, j_s are: (a) completely different between them; or (b) equal by pairs.

A remark is convenient at this point. In some works (e.g., [22]), GBS are preferred to be defined using $\tilde{\sigma}_i = \sigma_i$ for $i = 0, 1, 3$ and $\tilde{\sigma}_2 = i\sigma_2$ because it allows real coefficients when they are expressed in the computational basis $|0\rangle, |1\rangle$ (alternative definitions introduce specific phase factors in $\tilde{\sigma}_i$). We adopt the last definition in the following, which does not produce changes in the previous discussion. Note that $\tilde{\sigma}_i^* = \sigma_i^T = \sigma_i$. The last expression should be fitted to (11), in particular with the non-diagonal block entries. In the following sections, we will show that the GBS basis naturally generates the $SU(2)$ decomposition if the Hamiltonian fulfills certain restrictions. The use of the GBS basis allows the management of this analysis because it is based on Pauli matrices.

4.2. Case $d = 1$

For $d = 1$ there are three possibilities to arrange the pairs in the corresponding GBS basis (reduced in this case to the traditional Bell states: $\{|\beta_{00}\rangle, |\beta_{01}\rangle, |\beta_{10}\rangle, |\beta_{11}\rangle\}$). A direct but large analysis shows that by fitting (18) to (11), the Hamiltonian should be reduced to the forms shown in Table 1 (assuming always $h_{0_4^{2d}} = 0$ and $H_0 = \sum_{j=1}^{3} h_{jj} \sigma_j \otimes \sigma_j$). The first column shows the pairs arrangement to construct the blocks. These results generalize those found in [11,12] for the anisotropic Heisenberg–Ising model reached if the crossed interaction terms such as $h_{ij} \sigma_i \otimes \sigma_j$ with $i, j = 1, 2, 3; i \neq j$ are not present. These terms are similar to the Dzyaloshinskii–Moriya model [24,25], opening additional possibilities for control in the pair exchange. Case $d = 1$ is special in the current context because for $d > 1$ crossed terms can be present only for a unique pair in order to keep the $SU(2)$ decomposition.

Table 1. Basis pairs and Hamiltonian required to get the $SU(2)$ block decomposition for case $d = 1$.

Basis Arrangement	Hamiltonian
$\{\{|\beta_{00}\rangle, |\beta_{01}\rangle\}, \{|\beta_{11}\rangle, |\beta_{10}\rangle\}\}$	$H = H_0 + h_{01} \sigma_0 \otimes \sigma_1 + h_{10} \sigma_1 \otimes \sigma_0 + h_{23} \sigma_2 \otimes \sigma_3 + h_{32} \sigma_3 \otimes \sigma_2$
$\{\{|\beta_{00}\rangle, |\beta_{11}\rangle\}, \{|\beta_{01}\rangle, |\beta_{10}\rangle\}\}$	$H = H_0 + h_{02} \sigma_0 \otimes \sigma_2 + h_{20} \sigma_2 \otimes \sigma_0 + h_{13} \sigma_1 \otimes \sigma_3 + h_{31} \sigma_3 \otimes \sigma_1$
$\{\{|\beta_{00}\rangle, |\beta_{10}\rangle\}, \{|\beta_{01}\rangle, |\beta_{11}\rangle\}\}$	$H = H_0 + h_{03} \sigma_0 \otimes \sigma_3 + h_{30} \sigma_3 \otimes \sigma_0 + h_{12} \sigma_1 \otimes \sigma_2 + h_{21} \sigma_2 \otimes \sigma_1$

Although the eigenvalues $\{E_j\}$ do not follow a specific order, expressions in (18) can be arranged in several orders as functions of the pairs selected $\{|\alpha_{j(i)}\rangle, |\alpha_{k(i)}\rangle\}$, being related with the decomposition process. In general, there are $\frac{(2^{2d})!}{(2^{2d}-1)!2^{2^{2d}-1}}$ combinations for these pairs, which grow very quickly with d (3 for $d = 1$; 2,027,025 for $d = 2$, etc.), making the cases for $d > 1$ unmanageable in an analogous direct analysis.

4.3. Case $d > 1$

The exponential growth of the problem with d makes an exhaustive analysis for $d > 1$ based on a large algebraic equation system impossible, as in the previous case. The previous case and the results in [11,12] suggest some possible Hamiltonians for more complex cases. Thus, some of the following forms (see Appendix A.1) could allow the $SU(2)$ decomposition for the basis (16):

$$H_0 = \sum_{j=1}^{3} H_0^{(j)}, \qquad H_0^{(j)} = h_{(j\frac{4^{2d}-1}{3})_4^{2d}}\sigma_j^{\otimes 2d}, \tag{19}$$

$$H_{\mathrm{nl}_i} = \sum_{k'>k=1}^{2d} H_{\mathrm{nl}_i}^{(k,k')}, \qquad H_{\mathrm{nl}_i}^{(k,k')} = h_{(i(4^{k-1}+4^{k'-1}))_4^{2d}}\bigotimes_{s=1}^{2d}\sigma_{(i(4^{k-1}+4^{k'-1}))_{4,s}^{2d}}, \tag{20}$$

$$H_{\mathrm{cnl}_i} = \sum_{k'>k=1}^{2d} H_{\mathrm{cnl}_i}^{(k,k')}, \qquad H_{\mathrm{cnl}_i}^{(k,k')} = \sum_{p=0}^{1} h_{(j_p4^{k-1}+k_p4^{k'-1})_4^{2d}}\bigotimes_{s=1}^{2d}\sigma_{(j_p4^{k-1}+k_p4^{k'-1})_{4,s}^{2d}}, \tag{21}$$

$$H_{1_i} = \sum_{k=1}^{2d} H_{1_i}^{(k)}, \qquad H_{1_i}^{(k)} = h_{(i4^{k-1})_4^{2d}}\bigotimes_{s=1}^{2d}\sigma_{(i4^{k-1})_{4,s}^{2d}}, \tag{22}$$

where $(i4^{k-1})_4^{2d}$ is the base-4 representation with $2d$ digits of $i4^{k-1}$, a number with only one i in position k and zero in the other; $(i4^{k-1})_{4,s}^{2d}$ is its element s; and $(j\frac{4^{2d}-1}{3})_4^{2d}$ is the base-4 representation with $2d$ digits of $j\frac{4^{2d}-1}{3}$, a number with j in each digit position (by using the geometric partial sums properties). Note that $i \in \{1,2,3\}$ is fixed in all expressions. Physically, H_0 represents a full simultaneous interaction between all particles (as in the bipartite Heisenberg–Ising anisotropic interaction). Although this kind of interaction is non-physical for $d > 1$, it is included here for reference. H_{nl_i} represents the interaction between the component i of the spin for pairs of particles as in the Heisenberg–Ising model. H_{cnl_i} is the crossed non-local interactions by pairs in the direction i (as those for $d = 1$ in Table 1), a label used to characterize these interactions (as in the Dzyaloshinskii–Moriya model). Note that i, j_p, k_p is a permutation of $1, 2, 3$ with parity $p = 0, 1$, even and odd, respectively. Finally, H_{1_i} is the component i of the local interactions with $h_{(i4^{k-1})_4^{2d}}$ as strengths (e.g., magnetic fields in the direction i for magnetic systems). These cases generalize the bipartite models presented in [11,12] and those found for $d = 1$.

Some observations are useful at this point: (a) $\tilde{\sigma}_i = \alpha_i\sigma_i$, $\alpha_i \in \{1,i\}$; (b) $\sigma_i^T = \beta_i\sigma_i$, $\beta_i \in \{-1,1\}$; (c) $\sigma_i\sigma_j = \gamma_{i,j}\sigma_j\sigma_i$, $\gamma_{i,j} \in \{-1,1\}$. Thus, $2c_{j_s,j_{d+s}}^{i_s,k_s} \equiv \mathrm{Tr}(\tilde{\sigma}_{i_s}\sigma_{j_{d+s}}\tilde{\sigma}_{k_s}^T\sigma_{j_s}^T) = \alpha_{i_s}\alpha_{k_s}\beta_{j_s}\beta_{k_s}\gamma_{k_sj_s}\gamma_{k_si_s}\mathrm{Tr}(\sigma_{i_s}\sigma_{k_s}\sigma_{j_{d+s}}\sigma_{j_s}) \in \{0,\pm2,\pm2i\}$. We do not provide extensive formulas for the coefficients $\alpha_i, \beta_i, \gamma_{i,j}, c_{j_s,j_{d+s}}^{i_s,k_s}$, but they are trivially constructed departing from the Pauli matrices properties.

At this point, a convenient definition is introduced for the following cases. We will say that two particles or parts, i, j, are *correspondents* if $j = i + d$, with $i, j - d \in \{1, 2, ..., d\}$. This means simply that one is in the same position of the first group of the Hamiltonian subscripts $1, 2, ..., d$ as the other is in the second group $d + 1, d + 2, ..., 2d$. Then, the analysis of $\langle \Psi_{\mathcal{I}_4^d}|H_0|\Psi_{\mathcal{K}_4^d}\rangle$, $\langle \Psi_{\mathcal{I}_4^d}|H_{1_i}|\Psi_{\mathcal{K}_4^d}\rangle$, $\langle \Psi_{\mathcal{I}_4^d}|H_{\mathrm{cnl}_i}|\Psi_{\mathcal{K}_4^d}\rangle$ and $\langle \Psi_{\mathcal{I}_4^d}|H_{\mathrm{nl}_i}|\Psi_{\mathcal{K}_4^d}\rangle$ is conducted with the following results.

4.3.1. Analysis of $\langle \Psi_{\mathcal{I}_4^d}|H_0|\Psi_{\mathcal{K}_4^d}\rangle$

Because $\mathcal{J} = j\frac{4^{2d}-1}{3}$ in (18), then $j_{d+s} = j_s = j$ $\forall s = 1, 2, ..., d$, implying $c_{j_s,j_{d+s}}^{i_s,k_s} \neq 0$ only if $i_s = k_s$ $\forall s = 1, 2, ..., d$. Thus, H_0 is diagonal in the GBS basis representation and each entry will contain the same three terms $h_{(j\frac{4^{2d}-1}{3})\frac{2d}{4}}$ for $j = 1, 2, 3$, but each with diverse signs. Despite the similitude of H_0 with the bipartite case ($d = 1$), for multipartite cases this interaction is non-physical, but it allows the main idea to be introduced and understood in the remaining analysis.

4.3.2. Analysis of $\left\langle \Psi_{\mathcal{I}_4^d} | H_{1_i} | \Psi_{\mathcal{K}_4^d} \right\rangle$

The treatment for the remaining cases is compressed in the explanation of the current case. By first considering only an isolated term $H_{1_i}^{(k)}$ (in this case $\mathcal{J} = i4^{k-1}$ for some $i \in \{1, 2, 3\}$ and $k = 1, 2, \ldots, 2d$ in (18)), then $\mathcal{J}$ in the base-4 representation contains only one i (in the position k) while other digits are zero. Thus, there are only two meaningful possibilities for each correspondent part: (1) $j_{d+s} = j_s = 0$ in most cases, so $i_s = k_s$ is the only case with $c_{j_s, j_{d+s}}^{i_s, k_s} \neq 0$; or (2) one and only one position $s = k$ or $d + s = k$ in $\mathcal{J}_4^d$ has $j_k = i$, either for j_s or j_{d+s}, while the other is zero. This last case implies only two possibilities for i_s, k_s: Case (A) one of i_s, k_s is i and other is zero (both possibilities are possible); or Case (B) i, i_s, k_s are different among them and from zero, thus they are a permutation i, i', i'' of $1, 2, 3$. In this case, there are two possibilities, $i_s = i', k_s = i''$ or $i_s = i'', k_s = i'$.

Case A is depicted in Figure 1 for indexes $\mathcal{I}, \mathcal{K}$ being considered in $\left\langle \Psi_{\mathcal{I}_4^d} | H_{1_i} | \Psi_{\mathcal{K}_4^d} \right\rangle$. There is a pair of entries whose labels for rows and columns have 0 or i in the position $s = k$: $((i_1, i_2, \ldots, i, \ldots, i_d), (i_1, i_2, \ldots, 0, \ldots, i_d))$ and $((i_1, i_2, \ldots, 0, \ldots, i_d), (i_1, i_2, \ldots, i, \ldots, i_d))$. This will be named the $0 \leftrightarrow i$ association (or index exchange) rule.

Case A

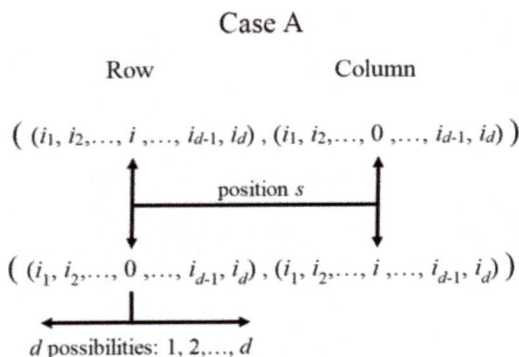

Figure 1. First case for a pair of entries in which $\left\langle \Psi_{\mathcal{I}_4^d} | H_{1_i} | \Psi_{\mathcal{K}_4^d} \right\rangle$ is non-zero. In them, for a fixed position $s = k$ in the row and the column labels appears i or 0, while the other corresponding positions in the row and in the column have the same values.

Case B is depicted in Figure 2. Here, there is a pair of entries whose labels for rows and columns have i' or i'' in the position $s = k$: $((i_1, i_2, \ldots, i', \ldots, i_d), (i_1, i_2, \ldots, i'', \ldots, i_d))$ and $((i_1, i_2, \ldots, i'', \ldots, i_d), (i_1, i_2, \ldots, i', \ldots, i_d))$, i, i', i'' being a permutation of $1, 2, 3$. This will be named the $i' \leftrightarrow i''$ association (or index exchange) rule.

Clearly, in each case (A or B), for each pair of correspondent interaction terms with i and k fixed ($k \leq d$ and $k + d$ positions), there are only two pairs on non-zero entries in rows $(i_1, i_2, \ldots, i, \ldots, i_d)$, $(i_1, i_2, \ldots, 0, \ldots, i_d)$ for case A and in rows $(i_1, i_2, \ldots, i', \ldots, i_d)$, $((i_1, i_2, \ldots, i'', \ldots, i_d))$ for case B (with the corresponding column labels exchanged in both cases). Together with the diagonal entries generated by other adequate Hamiltonians (e.g., H_0 or H_{nl_i} as it will be seen), they will form 2×2 blocks. In fact, each non-zero entry for H_{1_i} will have only two $h_{\mathcal{J}}$ terms corresponding with $h_{0,0\ldots0,i,0,\ldots,0}$ with i in positions s or $d + s$ (meaning local interaction with each element of the pair of correspondent parts in position k). Noting that labels in the position $s = k$ in $\mathcal{I}$ (row) and $\mathcal{K}$ (column) for the non-zero entries are $0, i; i, 0; i', i''$; or i'', i', they cover all possibilities $i_k = 0, 1, 2, 3$. Thus, for a fixed column and defined i, k values in $\left\langle \Psi_{\mathcal{I}_4^d} | H_{1_i} | \Psi_{\mathcal{K}_4^d} \right\rangle$, there is exactly one non-zero row. Still, if two correspondent k elements are considered (local interactions on each element of a correspondent pair), they still generate only one non-zero row (each one with the two terms explained before).

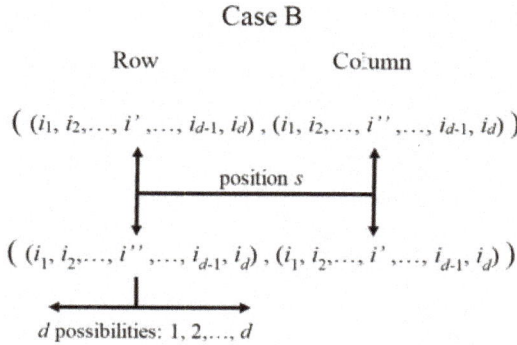

Case B

Row   Column

$$\Big(\, (i_1, i_2, \ldots, i'\,, \ldots, i_{d-1}, i_d) \, , \, (i_1, i_2, \ldots, i''\,, \ldots, i_{d-1}, i_d) \, \Big)$$

position s

$$\Big(\, (i_1, i_2, \ldots, i''\,, \ldots, i_{d-1}, i_d) \, , \, (i_1, i_2, \ldots, i'\,, \ldots, i_{d-1}, i_d) \, \Big)$$

d possibilities: $1, 2, \ldots, d$

Figure 2. Second case for a pair of entries in which $\left\langle \Psi_{\mathcal{I}_4^d} | H_{1_i} | \Psi_{\mathcal{K}_4^d} \right\rangle$ is non-zero. In them, for a fixed position $s = k$ in the row and the column labels appears i' or i'' alternatively (i, i', i'' being a permutation of $1, 2, 3$), while other corresponding positions in the row and in the column have the same values.

Although there are $2d$ possibilities to select the position $s = k$ in (22), they do not count as separate blocks because they appear in other entries in (18). Instead, each term in non-correspondent terms will appear in a different non-zero row, giving d non-zero rows as total. For each i direction of the interaction being included, additional non-zero rows will appear. This implies that $3d$ rows could appear when all parts have local interactions in the three spatial directions at time, destroying in this case the 2×2 block structure. Thus, maintaining local interactions in only one direction and on only one correspondent pair of elements, together, cases A and B form $\frac{1}{2} 4^d$ 2×2 blocks as was required in the previous section. In any case, each non-zero entry will have the same 2 terms $h_{(i4^{k-1})_4^{2d}}$ with different signs depending on $c_{j_s, j_{d+s}}^{i_s, k_s}$ involved in each factor of $H_{1_i}^{(k)}$. Clearly, blocks can be rearranged to adequately order the GBS basis elements getting the form (11). A brief analysis shows that there are no more diagonal-off elements in addition to last cases being generated by local terms. Additional diagonal-off elements come from the non-local terms, such as those in Table 1.

4.3.3. Analysis of $\left\langle \Psi_{\mathcal{I}_4^d} | H_{\mathrm{nl}_i} | \Psi_{\mathcal{K}_4^d} \right\rangle$

With the correspondent parts definition and the analysis for H_{1_i}, we can identify two cases for the different terms $H_{\mathrm{nl}_i}^{(k,k')}$: (a) non-local interactions between correspondent parts; and (b) non-local interactions between non-correspondent parts. The discussion is similar to the previous subsection. **Correspondent terms** $H_{\mathrm{nl}_i}^{(k,k+d)}$. This term in the Hamiltonian H_{nl_i} contains $\sigma_0 \otimes \ldots \otimes \sigma_i \otimes \ldots \otimes \sigma_i \otimes \ldots \otimes$ σ_0 with σ_i in positions k and $k + d$, and σ_0 in any other. When this term is allocated in $\left\langle \Psi_{\mathcal{I}_4^d} | H_{\mathrm{nl}_i} | \Psi_{\mathcal{K}_4^d} \right\rangle$ in agreement with (18), it does not cancel if each factor in the product become different from zero, implying $i_s = k_s$ $\forall s = 1, 2, \ldots, d$. Thus, this term gives non-zero entries only in the diagonal elements. Thus, each non-zero entry of H_{nl_i} will have d different terms in each diagonal element (one for each pair of interacting correspondent particles). Those terms will appear with different signs in each diagonal element in spite of $c_{j_s, j_{d+s}}^{i_s, k_s}$. At this point, note that results for $H_{\mathrm{nl}_i}^{(k,k+d)}$ and $H_{1_i}^{(k)}$ were expected due to the results in [11,26] and the separability of the GBS basis in their constitutive entangled pairs. **Non-correspondent terms** $H_{\mathrm{nl}_i}^{(k,k' \neq k+d)}$. These terms have a different behavior. Each term contains $\sigma_0 \otimes \ldots \otimes \sigma_i \otimes \ldots \otimes \sigma_i \otimes \ldots \otimes \sigma_0$, with σ_i in positions k and k', and σ_0 in any other. It defines two pairs of correspondent parts involving σ_i: $[k, k+d, k', k'+d]$ if $k < k' \leq d$ or $k, k+d, k'-d, k'$ if $k \leq d < k' \leq 2d$. Then, each factor in (18) related with those two pairs ($s = k, k'$ or $s = k, k'-d$) will now include $\text{Tr}(\sigma_{i_s}\sigma_i\sigma_{k_s})$ (until unitary factors), which is non-zero only if: (a) i_s or k_s are one of the pairs 0 and i or i and 0; (b) i, i_s, k_s are a permutation i, i', i'' of $1, 2, 3$ (having two cases depending on the parity). The last situation is similar to the local terms in the previous subsection, but in two parts simultaneously. The remaining factors for $s \neq k, k'$ or $s \neq k, k'-d$ will require $i_s = k_s$ in order to become non-zero. The latter scenario gives 16 possibilities for each term $h_{(i(4^{k-1}+4^{k'-1}))_{2^d}}$, which will appear in diagonal-off positions obtained departing from the diagonal position $(i_1, ..., i_d; i_1, ..., i_d)$ in $\left\langle \Psi_{\mathcal{I}_4^d} | H_{nl_i} | \Psi_{\mathcal{K}_4^d} \right\rangle$, by changing each index in the pair (i_k, i'_k) in the row, following the rules depicted in cases A and B. Thus, for each column and with i, k, k' fixed, only one row becomes non-zero, in agreement with the previous rule. Each entry of this kind involves four terms, including the four combinations of each pair of non-correspondent parts selected from the set $[k, k', k+d, k'+d]$. Instead, when all values i and k, k' are considered, a total of $3 \cdot \frac{1}{2}d(d-1)$ non-zero rows appear in each column (clearly, by considering all these terms, $SU(2)$ decomposition is not achieved).

4.3.4. Analysis of $\left\langle \Psi_{\mathcal{I}_4^d} | H_{cnl_i} | \Psi_{\mathcal{K}_4^d} \right\rangle$

Correspondent terms $H_{cnl_i}^{(k,k+d)}$. For each term $H_{cnl_i}^{(k,k+d)}$, the behavior is similar as for $H_{l_i}^{(k')}$. Because only one correspondent pair has $j_p = j_s \neq 0 \neq j_{d+s} = k_p$ in (18), then i_s, k_s for $s = k'$ should be $0, i$ or j_p, k_p. For $s \neq k'$, $i_s = k_s$. As before, it means that each term is diagonal-off by combining the values of index k' in $\mathcal{I}$ and $\mathcal{K}$ as before: $0, i; i, 0; j_p, k_p;$ and k_p, j_p. For a fixed column and i, k, it will give four possibilities and two $SU(2)$ blocks. Each entry will have two terms corresponding to the different parities p. Note that only one i and k' can be considered to achieve the $SU(2)$ decomposition. Otherwise, for each column, $3d$ rows different from zero could appear, breaking the $SU(2)$ decomposition as for the local interaction case.

Non-correspondent terms $H_{cnl_i}^{(k,k' \neq k+d)}$. As for $H_{nl_i}^{(k,k' \neq k+d)}$, in this case the only non-zero terms have $i_s = k_s$ for $s \neq k, k', k-d, k'-d$. Meanwhile, for the two remaining cases $s \in \{k, k', k-d, k'-d\} \cap \{1, 2, ..., d\}$, each i_s, k_s should be selected from the set $0, j_p; j_p, 0; i, k_p; k_p, i$ or $0, k_p; k_p, 0; i, j_p; j_p, i$. In a specific column and fixing i, it will give 16 possibilities and 8 blocks in $SU(2)$, as for the $H_{nl_i}^{(k,k' \neq k+d)}$ case. Note that parity p should be fixed in this case because each one gives a different decomposition. Each entry will contain four terms for each parity p combining the four possible interaction terms. Again, if all options for i and k, k', p are considered, then $3 \cdot d(d-1)$ non-zero rows will appear for each column, breaking the $SU(2)$ decomposition. These terms are not commonly introduced in models such as Heisenberg–Ising and those related. Instead, for magnetic systems they are the first-order approximation in the spin–orbit coupling, introducing antisymmetric exchange as in the Dzyaloshinskii–Moriya model: $H_{DM} = \vec{D} \cdot (\vec{\sigma_1} \times \vec{\sigma_2})$. There, $\vec{D}$ is the Dzyaloshinskii–Moriya vector defining the orientation of coupling. Here, as only one term can be included in order to preserve the $SU(2)$ reduction property, this coupling should be strictly oriented.

4.4. Explicit Analytical Formulas for Hamiltonian Components

After the last analysis, it is clear that other candidates to generate $SU(2)$ decomposition are possible, but they involve more than two parts at a time (as in the case of H_0), which are non-physical for common point-like interactions. Nevertheless, these terms could appear for the quantum mechanical extended objects in which (1) is a mere expansion of the interactions. Therefore, we will restrict our remaining discussion to local or pairwise interactions. In this section, analytical formulas for $\left\langle \Psi_{\mathcal{I}_4^d} | H_{l_i} | \Psi_{\mathcal{K}_4^d} \right\rangle$, $\left\langle \Psi_{\mathcal{I}_4^d} | H_{nl_i} | \Psi_{\mathcal{K}_4^d} \right\rangle$, and $\left\langle \Psi_{\mathcal{I}_4^d} | H_{cnl_i} | \Psi_{\mathcal{K}_4^d} \right\rangle$ are provided to summarize the previous findings and because of their utility for optimal computer simulation purposes for larger systems. In order to simplify the expressions, we introduce the definition of the following generalized Kronecker delta:

$$\delta^S_{\mathcal{IK}} \equiv \prod_{\substack{s=1 \\ s \notin S}}^{d} \delta_{i_s k_s},$$ (23)

where S is a set of scripts of the excluded parts in the product. Thus, for H_{1_i}:

$$\left\langle \Psi_{\mathcal{I}^d_4} | H_{1_i} | \Psi_{\mathcal{K}^d_4} \right\rangle = \sum_{k'=1}^{d} \delta^{\{k'\}}_{\mathcal{IK}} \mathcal{H}^{k'}_{1_i \; \mathcal{I}^d_4, \mathcal{K}^{d'}_4}$$

$$\text{with}: \quad \mathcal{H}^{k'}_{1_i \; \mathcal{I}^d_4, \mathcal{K}^d_4} = \sum_{t'=0}^{1} h_{(i 4^{k'} + dt' - 1) 2^d_4} \mathcal{F}^{i\delta_{0,t'}, i\delta_{1,t'}}_{i,k'},$$ (24)

by noting $c^{i_s, i_s}_{0,0} = 1$. In $\mathcal{H}^{k'}_{1_i \; \mathcal{I}^d_4, \mathcal{K}^{d'}_4}$ $[k', k' + d]$ is the correspondent pair where each local interaction is being applied. There, the exchange factor generating the diagonal-off entries in the $SU(2)$ blocks is:

$$\mathcal{F}^{j,k}_{i,k'} = \delta_{i_{k'} 0} \delta_{k_{k'} i} c^{0,i}_{j,k} + \delta_{i_{k'} i} \delta_{k_{k'} 0} c^{i,0}_{j,k} + \sum_{i',i''=1}^{3} \epsilon^2_{i i' i''} \delta_{i_{k'} i'} \delta_{k_{k'} i''} c^{i',i''}_{j,k}.$$ (25)

For H_{nl_i}:

$$\left\langle \Psi_{\mathcal{I}^d_4} | H_{\mathrm{nl}_i} | \Psi_{\mathcal{K}^d_4} \right\rangle = \sum_{k'=1}^{d} \delta^{\{k'\}}_{\mathcal{IK}} \mathcal{H}^{c,k'}_{\mathrm{nl}_i \; \mathcal{I}^d_4, \mathcal{K}^d_4} + \sum_{k''>k'=1}^{d} \delta^{\{k',k''\}}_{\mathcal{IK}} \mathcal{H}^{nc,k'k''}_{\mathrm{nl}_i \; \mathcal{I}^d_4, \mathcal{K}^{d'}_4}$$

$$\text{with}: \quad \mathcal{H}^{c,k'}_{\mathrm{nl}_i \; \mathcal{I}^d_4, \mathcal{K}^d_4} = h_{(i(4^{k'} - 1 + 4^{k'+d} - 1)) 2^d_4} \delta_{i_{k'} k_{k'}} c^{i_{k'}, i_{k'}}_{i,i},$$ (26)

$$\mathcal{H}^{nc,k'k''}_{\mathrm{nl}_i \; \mathcal{I}^d_4, \mathcal{K}^d_4} = \sum_{t',t''=0}^{1} h_{(i(4^{k'} + dt' - 1 + 4^{k''} + dt'' - 1)) 2^d_4} \mathcal{F}^{i\delta_{0,t'}, i\delta_{1,t'}}_{i,k'} \mathcal{F}^{i\delta_{0,t''}, i\delta_{1,t''}}_{i,k''}.$$

Each term belongs to correspondent and non-correspondent interactions, respectively. In $\mathcal{H}^{c,k'}_{\mathrm{nl}_i \; \mathcal{I}^d_4, \mathcal{K}^d_4}$ and $\mathcal{H}^{nc,k'k''}_{\mathrm{nl}_i \; \mathcal{I}^d_4, \mathcal{K}^{d'}_4}$ $[k', k'']$ are the parts with non-local interactions between them. Similarly, for $\mathcal{H}_{\mathrm{cnl}_i}$:

$$\left\langle \Psi_{\mathcal{I}^d_4} | H_{\mathrm{cnl}_i} | \Psi_{\mathcal{K}^d_4} \right\rangle = \sum_{k'=1}^{d} \delta^{\{k'\}}_{\mathcal{IK}} \mathcal{H}^{c,k'}_{\mathrm{cnl}_i \mathcal{I}^d_4, \mathcal{K}^d_4} + \sum_{p=0}^{1} \sum_{k''>k'=1}^{d} \delta^{\{k',k''\}}_{\mathcal{IK}} \mathcal{H}^{nc,k'k''p}_{\mathrm{cnl}_i \; \mathcal{I}^d_4, \mathcal{K}^{d'}_4}$$

$$\text{with}: \quad \mathcal{H}^{c,k'}_{\mathrm{cnl}_i \mathcal{I}^d_4, \mathcal{K}^d_4} = \sum_{p=0}^{1} h_{(j_p 4^{k'} - 1 + k_p 4^{k'+d} - 1) 2^d_4} \mathcal{F}^{j_p, k_p}_{i,k'},$$ (27)

$$\mathcal{H}^{nc,k'k''p}_{\mathrm{cnl}_i \; \mathcal{I}^d_4, \mathcal{K}^d_4} = \sum_{t',t''=0}^{1} h_{(j_p 4^{k'} + dt' - 1 + k_p 4^{k''} + dt'' - 1) 2^d_4} \mathcal{F}^{j_p \delta_{0,t'}, j_p \delta_{1,t'}}_{j_p, k'} \mathcal{F}^{k_p \delta_{0,t''}, k_p \delta_{1,t''}}_{k_p, k''}.$$

Again, $\mathcal{H}^{c,k'}_{\mathrm{cnl}_i}$ and $\mathcal{H}^{nc,k'k''p}_{\mathrm{cnl}_i}$ are the correspondent and non-correspondent interactions in the Hamiltonian, $[k', k'']$ being the parts where there are non-local interactions. This explicitly shows the existence of four (for $\mathcal{H}^{k'}_{1_i}$ and $\mathcal{H}^{c,k'}_{\mathrm{cnl}_i}$) and sixteen (for $\mathcal{H}^{nc,k'k''}_{\mathrm{nl}_i}$ and $\mathcal{H}^{nc,k'k''p}_{\mathrm{cnl}_i}$) diagonal-off entries, respectively, in agreement with cases A and B depicted by Figures 1 and 2 (if only single specific values of i, k', k'' are considered instead of the whole sum), generating $2 \times 4^{d-1} = \frac{1}{2} \times 4^d$ and $8 \times 4^{d-2} = \frac{1}{2} \times 4^d$ blocks, respectively. Then, the $SU(2)$ decomposition could be achieved only by: (a) including any desired non-local terms $\mathcal{H}^{c,k'}_{\mathrm{nl}_i}$ (to generate the diagonal elements); and (b) including only one type of interaction among $\mathcal{H}^{k'}_{1_i}, \mathcal{H}^{nc,k'k''}_{\mathrm{nl}_i}, \mathcal{H}^{c,k'}_{\mathrm{cnl}_i}$ or $\mathcal{H}^{nc,k'k''p}_{\mathrm{cnl}_i}$ for concrete values for i, k', k'', and p.

An important property used later for $\mathcal{F}_{i,k'}^{j_s,j_{d+s}}$ is that only one term in (25) remains with the election of $i_{k'}$ and $k_{k'}$. Because each $c_{j,k}^{j_s,j_{d+s}}$ is real or imaginary, and more concretely as a brief analysis shows, if it is not zero, then it becomes imaginary only if j_s or j_{d+s} is equal to 2, this property is transferred to $\mathcal{F}_{i,k'}^{j_s,j_{d+s}}$.

5. Specific Interactions Generating $SU(2)$ Decomposition

In this section, we summarize and organize the global findings to reach the $SU(2)$ block structure on the GBS basis. Finally, we conclude that there are three great types of interactions that are able to generate the block structure depicted in Section 3.

5.1. General Depiction of Interactions Having $SU(2)$ Decomposition for the GBS Basis

Based on the previous discussion, there are three groups of interactions that are able to generate the $SU(2)$ decomposition on the GBS basis. The first one (Type I) involves all kinds of non-local and non-crossed interactions between any two correspondent parts in any direction. These terms generate the diagonal terms depicted previously in the Hamiltonian. Together, only two local interactions in only one specific direction and on only one pair of correspondent parts, k_l, should be included to generate the diagonal-off entries. Thus, this group of interactions generates the $SU(2)$ blocks. Note that local interaction terms could be intended as external driven fields as in [11,26]. The second interaction (Type II) is obtained by substituting the previous local interactions with non-local interactions among only those non-correspondent elements included in two pairs of correspondent parts. This means that if $k, k', k+d, k'+d$ with $k < k' \leq d$ are these elements in the two correspondent parts, then only the interactions between the following non-correspondent elements are allowed: $[k, k']$, $[k, k'+d]$, $[k', k+d]$, and $[k+d, k'+d]$. This group of four interactions generates the diagonal-off terms to conform the $SU(2)$ blocks. Nevertheless, the Type II interaction should normally be understood as a non-driven process of control. Note that Type II interaction could be classified into two other subclasses: (a) Type IIa for non-crossed interactions $\mathcal{H}_{nl_i}^{nc,k'k''}$; and (b) Type IIb for crossed interactions $\mathcal{H}_{cnl_i}^{nc,k'k''p}$. Finally, the third interaction (Type III) involves both the non-local and non-crossed interactions, with the inclusion of crossed interactions between one specific correspondent pair.

In order to clarify the structure of those notable interaction architectures as special cases of Hamiltonian (1), we make some remarks as follows. Figure 3 summarizes the three types of interactions depicted above by listing the $2d$ qubits involved and then relating them with arrows in agreement with their mutual interactions. Then:

A: Curved arrows point out those qubits related through entangling operations in any case.

B: All curved arrows in the bottom refer to Heisenberg–Ising-like (non-crossed) interactions involving the three possible spatial directions together. Those interaction relations set the correspondent pairs.

C: For the curved arrows in the top, two kinds of entangling operations can be considered according to the text: Heisenberg–Ising-like (non-crossed) interactions or Dzyaloshinskii–Moriya-like (crossed) interactions. Only one characteristic spatial direction is allowed.

D: Type II interactions can be split into Type IIa and Type IIb if interactions in the top are non-crossed or crossed (between parts of two different correspondent pairs), respectively. Type IIb interactions in the top admits only one possible parity from the two possible.

E: Type III interactions admit only crossed interactions in the top between parts of one specific correspondent pair, but the two possible parities together are allowed.

F: For Figure 3a, the right arrows correspond to external local interactions such as those generated by magnetic fields on spin-based qubits. Due to their locality, they are referred to as driven interactions, although it actually depends on the available control of the interactions.

Figure 4a shows a pictorial representation of each interaction, where the pairing is graphically represented. Therein, yellow rays with blue contour are non-crossed interactions in the three spatial directions [B]. Yellow rays with red contour represent one interaction from non-crossed or crossed entangling interactions in only one spatial direction [D]. Blue rays with red contour indicate non-crossed interaction in three spatial directions together with a crossed interaction in only one direction [E]. Yellow triangles indicate local interactions on the respective qubits in only one correspondent pair [F].

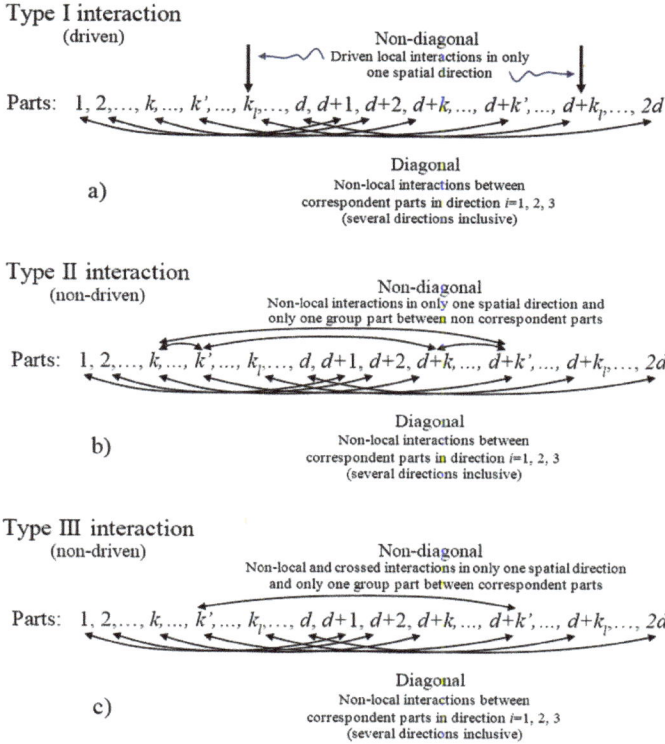

Figure 3. Three types of physical interactions able to generate the block decomposition. Non-local and non-crossed interactions among any correspondent parts combined with: (a) local interactions on only two correspondent parts $(k_l, k_l + d)$; (b) any two non-correspondent parts in only two specific pairs of correspondent parts of only one subtype, non-crossed or crossed; and (c) crossed interactions between a specific pair of correspondent parts.

In particular, note that this description is in agreement with the results in Table 1 for $d = 1$, although it is a special case because diagonal-off entries for Type I, II, and III coincide in the same diagonal-off entries, so both interactions could be combined at the same time, preserving the $SU(2)$ decomposition. This case has a richer structure for control in terms of the number of free parameters involved with respect to the number of parts to be controlled. Note that while Types I and III are only able to modify the inner entanglement of the correspondent pairs, Type II interaction (Type IIa and IIb) allows the modification of the global entanglement between different correspondent pairs, thus letting it spread on the entire system by switching the pairs involving interactions generating diagonal-off entries.

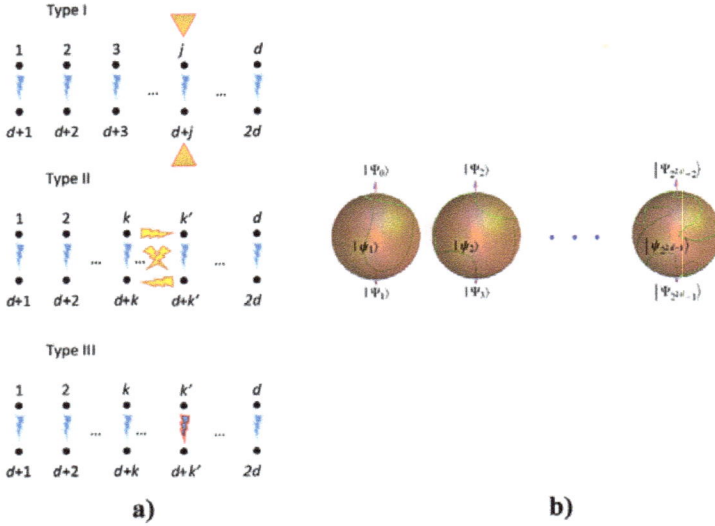

Figure 4. Representation of qubit interactions able to generate $SU(2)$ decomposition: (a) Type I, II, and III interactions among $2d$ qubits (Type III assumes the inclusion of crossed interactions in the pair k'); and (b) Distributed evolution on 2^{2d-1} Bloch spheres, each one for the states $|\psi_j\rangle$.

5.2. General Structure of SU(2) Blocks

A complementary analysis of $SU(2)$ blocks obtained for the last interactions is given in this subsection. Their form is particularly useful as a connection with optimal control schemes, such as those presented in [9]. In any case (Type I, II, or III), each block $\mathbb{S}_{H\mathcal{I},\mathcal{I}'}$ (with $\mathcal{I}, \mathcal{I}'$ the rows in which is situated) has the form:

$$
\mathbb{S}_{H\mathcal{I},\mathcal{I}'} = \begin{pmatrix} h_{11} & h_{12} \\ h_{12}^* & h_{22} \end{pmatrix}
$$

$$
= \frac{h_{11} + h_{22}}{2} \mathbb{I}_{\mathcal{I},\mathcal{I}'} + \mathrm{Re}(h_{12}) \mathbb{X}_{\mathcal{I},\mathcal{I}'} - \mathrm{Im}(h_{12}) \mathbb{Y}_{\mathcal{I},\mathcal{I}'} + \frac{h_{11} - h_{22}}{2} \mathbb{Z}_{\mathcal{I},\mathcal{I}'},
$$

(28)

where $\{\mathbb{I}_{\mathcal{I},\mathcal{I}'}, \mathbb{X}_{\mathcal{I},\mathcal{I}'}, \mathbb{Y}_{\mathcal{I},\mathcal{I}'}, \mathbb{Z}_{\mathcal{I},\mathcal{I}'}\}$ is the Pauli basis for the $SU(2)$ block. If the Hamiltonian coefficients involved in the block are time-independent, then the corresponding $\mathbb{S}_{U\mathcal{I},\mathcal{I}'}$ block in the evolution matrix becomes:

$$
\mathbb{S}_{U\mathcal{I},\mathcal{I}'} = e^{i\mathbb{S}_{H\mathcal{I},\mathcal{I}'}\frac{t}{\hbar}} = e^{i\frac{h_{11}+h_{22}}{2\hbar}t} e^{i\omega \mathbf{n} \cdot \mathbf{s}_{\mathcal{I},\mathcal{I}'}t} = e^{i\frac{h_{11}+h_{22}}{2\hbar}t}(\cos\omega t + i\sin\omega t \mathbf{n} \cdot \mathbf{s}_{\mathcal{I},\mathcal{I}'})
$$

$$
= e^{i\frac{h_{11}+h_{22}}{2\hbar}t} \begin{pmatrix} \cos\omega t + i\frac{h_{11}-h_{22}}{2\hbar\omega}\sin\omega t & i\frac{h_{12}}{\hbar\omega}\sin\omega t \\ i\frac{h_{12}^*}{\hbar\omega}\sin\omega t & \cos\omega t - i\frac{h_{11}-h_{22}}{2\hbar\omega}\sin\omega t \end{pmatrix},
$$

$$
\text{with :} \mathbf{n} = \frac{1}{\hbar\omega}(\mathrm{Re}(h_{12}), -\mathrm{Im}(h_{12}), \frac{h_{11}-h_{22}}{2}),
$$

(29)

$$
\mathbf{s}_{\mathcal{I},\mathcal{I}'} = (\mathbb{X}_{\mathcal{I},\mathcal{I}'}, \mathbb{Y}_{\mathcal{I},\mathcal{I}'}, \mathbb{Z}_{\mathcal{I},\mathcal{I}'}),
$$

$$
\hbar\omega = \sqrt{|h_{12}|^2 + \frac{1}{4}|h_{11} - h_{22}|^2},
$$

clearly belonging to $U(1) \times SU(2)$ (see Appendix A.2). As stated previously, $\mathcal{F}_{j,k'}^{j_s,j_{d+s}}$ is imaginary only if j_s or j_{d+s} is 2. Thus, only one component from n_1 or n_2 is different from zero because non-diagonal entries of block in (24), (26), and (27) are always real or imaginary. This reduces the optimal control to the second case reported by [9]. An additional analysis shows that $h_{11} \pm h_{22} \neq 0$ in general (without imposing restrictions on the non-local strengths $h_\mathcal{J}$). This aspect will be relevant later.

5.3. Structure of $SU(2)$ Blocks for Each Interaction

Several classical interactions fitting in the current procedure were analyzed. All them generate blocks (not necessarily $SU(2)$ blocks) when they are expressed in the GBS basis, denoting a kind of universality for this basis due to its ability to gather similar interactions through simplified representations. For the sake of the search for $SU(2)$ decomposition, we discuss finally closed forms for the specific Hamiltonians able to achieve the $SU(2)$ decomposition. These formulas are quite useful for computer simulation purposes.

5.3.1. Blocks in Type I Interaction

This interaction includes non-crossed spin interactions between correspondent particles in all spatial directions and external local interactions on the pair $[k', k' + d]$ of correspondent particles in direction j. From (24)–(26), it can be written as:

$$H_I = H_D + H_{ND_I}^{(j,k')},$$

$$\text{with :} H_D \equiv \sum_{i'=1}^{3} \sum_{k=1}^{d} h_{(i'(4^{k-1}+4^{k+d-1}))_4^{2d}} \bigotimes_{s=1}^{2d} \sigma_{(i'(4^{k-1}+4^{k+d-1}))_{4,s}^{2d}}, \tag{30}$$

$$H_{ND_I}^{(j,k')} = \sum_{t'=0}^{1} h_{(j4^{k'}+dt'-1)_4^{2d}} \bigotimes_{s=1}^{2d} \sigma_{(j4^{k'}+dt'-1 \frac{2d}{4,s})}$$

generating $SU(2)$ blocks with the diagonal terms from non-local interactions between correspondent parts and the non-diagonal terms from local interactions. Departing from (24)–(26), we obtain for the Hamiltonian components:

$$\left\langle \Psi_{\mathcal{I}_4^d} | H_I | \Psi_{\mathcal{K}_4^d} \right\rangle = \delta_{\mathcal{IK}} \sum_{i'=1}^{3} \sum_{k''=1}^{d} \left((-1)^{\delta_{i',2}+(1-\delta_{i',i_{k''}})(1-\delta_{0,i_{k''}})} h_{(i'(4^{k''-1}+4^{k''+d-1}))_4^{2d}} \right) +$$

$$\sum_{t'=0}^{1} h_{(j4^{k'}+dt'-1)_4^{2d}} \delta_{\mathcal{IK}}^{\{k'\}} \mathcal{F}_{j,k'}^{j\delta_{0,t'},j\delta_{1,t'}} \equiv H_{D\mathcal{IK}} + H_{ND_I \mathcal{IK}}^{(j,k')}. \tag{31}$$

The last formula is obtained noting that $c_{i,i}^{i_{k''},i_{k''}} = (-1)^{\delta_{i,2}+(1-\delta_{i,i_{k''}})(1-\delta_{0,i_{k''}})}$. The first term of the last expressions denotes the diagonal terms of interaction. This formula shows that the pair of entries in the diagonal of each $SU(2)$ block are generally different. Because the block is formed by switching an index $i_{k''}$ in the row labels (or two as in the following cases) in agreement with the association rules $0 \leftrightarrow j$ or $i \leftrightarrow k$ (j is the direction associated to the interaction and i, j, k a permutation of $1, 2, 3$), then for $i' \neq j$ the terms in $H_{D\mathcal{IK}}$ have a sign change. This implies that in general $h_{11} \neq h_{22}$ in (28), generating non-diagonal $\mathbb{S}_{H\mathcal{I},\mathcal{I}'}$-blocks. The second term contains the four diagonal-off elements generating two blocks with two terms each. Note that Hamiltonian terms ($h_\mathcal{I}$) are real together with $c_{i,i}^{i_{k''},i_{k''}}$, so diagonal terms are real, as expected. Diagonal-off terms will be real or imaginary depending on $\mathcal{F}_{j,k'}^{j,0}, \mathcal{F}_{j,k'}^{0,j}$. In any case, concretely, they are imaginary only if $j = 2$.

Note that this interaction (when it is applied to a combination of correspondent pairs with bipartite entangled states) generates only non-local operations on each correspondent pair, such as those presented in [11,13]. Still switching the direction j and the correspondent pair k' on which the local interaction is applied, this kind of Hamiltonian cannot generate extended entanglement between correspondent pairs more than that included in the initial state. This means that if the initial state is separable by correspondent pairs, it will remain separable at this level (but should be able to entangle or untangle the parts of each pair). Conversely, it cannot disentangle each correspondent pair from the remaining state in more complex cases. We dedicate a later section to analyzing these topics.

5.3.2. Blocks in Type II Interaction

Type IIa: In this case, the interaction is completely non-local between correspondent pairs to generate the diagonal entries, and in only one direction between non-correspondent parts in two correspondent pairs to generate the diagonal-off entries. The Hamiltonian becomes:

$$
\begin{aligned}
H_{IIa} &= H_D + H_{ND_{IIa}}^{(j,k'k'')}, \\
\text{with} : H_{ND_{IIa}}^{(j,k'k'')} &= \sum_{t',t''=0}^{1} h_{(j(4^{k'+dt'-1}+4^{k''+dt''-1}))_4^{2d}} \bigotimes_{s=1}^{2d} \sigma_{(j(4^{k'+dt'-1}+4^{k''+dt''-1}))_{4,s}^{2d}},
\end{aligned}
\tag{32}
$$

with a non-local and non-crossed interaction in the direction j for the group of non-correspondent terms defined by $k' < k'' \leq d$. The Hamiltonian entries are similar to those in (24)–(26), but with the last restriction for the non-correspondent terms of interaction. Due to discussion in the previous subsection, diagonal-off entries in the Hamiltonian are now always real. The components become:

$$
\begin{aligned}
\left\langle \Psi_{\mathcal{I}_4^d} | H_{IIa} | \Psi_{\mathcal{K}_4^d} \right\rangle &= H_{D\mathcal{I}\mathcal{K}} + H_{ND_{IIa}}^{(j,k'k'')}{}_{\mathcal{I}\mathcal{K}}, \\
H_{ND_{IIa}}^{(j,k'k'')}{}_{\mathcal{I}\mathcal{K}} &\equiv \sum_{t',t''=0}^{1} h_{(j(4^{k'+dt'-1}+4^{k''+dt''-1}))_4^{2d}} \delta_{\mathcal{I}\mathcal{K}}^{\{k',k''\}} \mathcal{F}_{j,k'}^{j\delta_{0,t'},j\delta_{1,t'}} \mathcal{F}_{j,k''}^{j\delta_{0,t''},j\delta_{1,t''}}.
\end{aligned}
\tag{33}
$$

Type IIb: For this interaction, the non-diagonal part generated by the non-local interaction between non-correspondent parts is supplied by a non-local and crossed interaction among non-correspondent parts of two correspondent pairs:

$$
\begin{aligned}
H_{IIb} &= H_D + H_{ND_{IIb}}^{(i,k'k''p)}, \\
\text{with} : H_{ND_{IIb}}^{(i,k'k''p)} &\equiv \sum_{t',t''=0}^{1} h_{(j_p 4^{k'+dt'-1}+k_p 4^{k''+dt''-1})_4^{2d}} \bigotimes_{s=1}^{2d} \sigma_{(j_p 4^{k'+dt'-1}+k_p 4^{k''+dt''-1})_{4,s}^{2d}}.
\end{aligned}
\tag{34}
$$

As before, i, j_p, k_p is a permutation of $1, 2, 3$ with parity $p = 0, 1$ (even and odd, respectively). Thus, the components become:

$$
\begin{aligned}
\left\langle \Psi_{\mathcal{I}_4^d} | H_{IIb} | \Psi_{\mathcal{K}_4^d} \right\rangle &= H_{D\mathcal{I}\mathcal{K}} + H_{ND_{IIb}}^{(i,k'k''p)}{}_{\mathcal{I}\mathcal{K}}, \\
H_{ND_{IIb}}^{(i,k'k''p)}{}_{\mathcal{I}\mathcal{K}} &= \sum_{t',t''=0}^{1} h_{(j_p 4^{k'+dt'-1}+k_p 4^{k''+dt''-1})_4^{2d}} \delta_{\mathcal{I}\mathcal{K}}^{\{k',k''\}} \mathcal{F}_{j_p,k'}^{j_p\delta_{0,t'},j_p\delta_{1,t'}} \mathcal{F}_{k_p,k''}^{k_p\delta_{0,t''},k_p\delta_{1,t''}}.
\end{aligned}
\tag{35}
$$

The non-diagonal entries are now imaginary, except for $i = 2$.

5.3.3. Blocks in Type III Interaction

Finally, for Type III interaction, the non-diagonal part is generated by the non-local and crossed interaction between a pair of correspondent parts k':

$$
\begin{aligned}
H_{III} &= H_D + H_{ND_{III}}^{(i,k')}, \\
\text{with} : H_{ND_{III}}^{(i,k')} &= \sum_{p=0}^{1} h_{(j_p 4^{k'-1}+k_p 4^{k'+d-1})_4^{2d}} \bigotimes_{s=1}^{2d} \sigma_{(j_p 4^{k'-1}+k_p 4^{k'+d-1})_{4,s}^{2d}},
\end{aligned}
\tag{36}
$$

with the Hamiltonian components:

$$\left\langle \Psi_{\mathcal{I}_4^d} \middle| H_{III} \middle| \Psi_{\mathcal{K}_4^d} \right\rangle = H_{D\mathcal{IK}} + H_{ND_{III}\mathcal{IK}}^{(i,k')},$$

$$H_{ND_{III}\mathcal{IK}}^{(i,k')} = \sum_{p=0}^{1} h_{(j_p 4^{k'-1} + k_p 4^{k'+d-1})_4^{2d}} \delta_{\mathcal{IK}}^{\{k'\}} \mathcal{F}_{i,k'}^{j_p,k_p},$$

(37)

where non-diagonal entries are imaginary only if $i = 2$.

Figure 4b shows a distributed evolution on 2^{2d-1} Bloch spheres for the states $\left| \psi_j \right\rangle = \alpha_{2j-2} \left| \Psi_{2j-2} \right\rangle + \alpha_{2j-1} \left| \Psi_{2j-1} \right\rangle$, which are part of the global state $\left| \psi \right\rangle = \sum_{j=1}^{2^{2d-1}} \left| \psi_j \right\rangle$, where each $\left| \Psi_k \right\rangle$ is an element of the GBS basis. Each state $\left| \psi_j \right\rangle$ evolves as a different curve on each Bloch sphere depending on parameters $h_{\mathcal{J}}$.

Finally, we should note that each of the previous interactions involves labels to be completely identified, namely: $H_I^{(j,k')}, H_{IIa}^{(j,k',k'')}, H_{IIb}^{(j,k',k'',p)}$, and $H_{III}^{(j,k',k'')}$. These labels will be omitted by simplicity unless their specification becomes needed. In any case, closed expressions (31), (33), (35), and (37) are computationally efficient to generate matrix representations of Hamiltonians $H_I, H_{IIa,b}, H_{III}$, and for their respective U, inclusively in the time-dependent case, although a numerical approach to construct could also be necessary.

5.4. Available Parameters and Structure of Entries

The number of free parameters (coefficients $h_{\mathcal{I}}$ of Hamiltonian) and their availability are important to set control procedures. In this section, we count the entries and terms for each Hamiltonian, summarizing the previous findings. If $D \leq 3$ is the number of spatial dimensions involved in each interaction, then the accounting of free parameters generating the $SU(2)$ decomposition, together with the maximum number of entries by column able to generate it (breaking the $SU(2)$ decomposition) is reported in Table 2. Note that the number of entries by column for all Hamiltonians (labeled with i, in some sense the direction of the interaction) can be increased by a factor D if all directions are considered at time. In the table, each Hamiltonian analyzed is reported, arriving at the main Hamiltonians $H_I, H_{IIa,b}$, and H_{III}. Accounting shows few free parameters at time (compared with the exponential growth of the matrix with the system size d) to set a whole control (over all blocks) in one period of constant driven parameters, suggesting the use of time-dependent or at least constant-piecewise parameters to increase the control.

Table 2. Rows generated and free parameters in each interaction considered in the text.

Hamiltonian	Entries Type	Entries by Column/Row	Parameters by Entry
H_0	Diagonal	1	$D \leq 3$
H_{1_i}	Non-diagonal	d	2
$H_{\mathrm{nl}_i}^c$	Diagonal	1	d
$H_{\mathrm{nl}_i}^{nc}$	Non-diagonal	$\frac{1}{2}d(d-1)$	4
$H_{\mathrm{cnl}_i}^c$	Non-diagonal	d	2
$H_{\mathrm{cnl}_i}^{nc}$	Non-diagonal	$d(d-1)$	4
H_I	2×2 block	2	$2 + Dd \leq 2 + 3d$
$H_{IIa,b}$	2×2 block	2	$4 + Dd \leq 4 + 3d$
H_{III}	2×2 block	2	$2 + Dd \leq 2 + 3d$

5.4.1. Structure of Diagonal Entries Belonging to a Specific Block

Other aspects should be discussed. The first is related to terms in diagonal entries generated by non-local interactions $\mathcal{H}_{\mathrm{nl}_j}^{c,s}$ among correspondent parts. Note that blocks are generated by interactions other than those, which are prescribed as a difference in one ($\mathcal{H}_{\mathrm{nl}_i}^{c,k'}$ or $\mathcal{H}_{\mathrm{cnl}_i}^{c,k'}$) or two ($\mathcal{H}_{\mathrm{nl}_i}^{nc,k'k''}$ or $\mathcal{H}_{\mathrm{cnl}_i}^{nc,k'k''p}$) terms in the scripts labels, in agreement with the rules depicted in Figures 1 and 2. This implies that there will be two or eight blocks, each one relating rows (and columns) differing in only one or two terms of their scripts, respectively. Note the diagonal entries for $\mathcal{H}_{nl_j}^{c,s}$ in (18) for the GBS defined as in [22]: $\text{Tr}(\tilde{\sigma}_{i_s}^* \sigma_j \tilde{\sigma}_{i_s}^T \sigma_j^T) = 2(-1)^{\delta_{j,2} + (1-\delta_{j,i_s})(1-\delta_{0,i_s})}$. Then, for each three strengths for a fixed correspondent pair, there will be only four sign combinations (none is the negative of another) depending on: (a) the direction of the interaction involved (on the correspondent pair s) is $j = 2$ or $j \neq 2$; and (b) i_s for the sth script is in the set $\{0, j\}$ or in $\{i, k\}$ (with i, j, k a permutation of 1, 2, 3). There, the factors corresponding to other correspondent pairs will be equal to one. Then, for the $3d$ terms included in all diagonal entries there will be 4^d combinations for the whole terms—precisely the number of rows. This implies that all diagonal entries are different (but not independent because there are only $3d$ parameters). For two rows differing in only one or two terms in their scripts, only the three or six terms corresponding with the strengths of $\mathcal{H}_{nl_j}^{c,s}$ for such correspondent pairs (associated with those terms in the scripts) will change their signs in the diagonal terms in their block. Consequently, for such 4^{d-1} or 4^{d-2} groups of blocks having the same scripts exchanged and generated by the whole combinations in the other $d-1$ or $d-2$ terms in their scripts, they will have the same $h_{11} - h_{22}$ parameters, respectively. Thus, it will be only two or eight different $h_{11} - h_{22}$ parameters for the entire H. Meanwhile, $h_{11} + h_{22}$ parameters could be different.

5.4.2. Structure of Diagonal-Off Entries Belonging to a Specific Block

The second aspect is related to the explicit calculation of $c_{j_s, j_{d+s}}^{i_s, k_s}$ for the basic cases of interest in the diagonal-off entries. (a) For H_I and $H_{IIa,b}$: $j_s = j, j_{d+s} = 0$, or $j_s = 0, j_{d+s} = j$ (j being the direction label involved in the local and non-local interactions between non-correspondent parts); and (b) for H_{III}: $j_s = j_p, j_{d+s} = k_p$. Table 3 explicitly shows these values. Note the parallelism between their two halves (vertically and horizontally).

Table 3. Values of $c_{j_s, j_{d+s}}^{i_s, k_s}$ for all exchange scripts in $H_I, H_{IIa,b}, H_{III}$. i, j, k is an even permutation of 1, 2, 3.

(j_s, j_{s+d})	(i_s, k_s)	$c_{j_s, j_{d+s}}^{i_s, k_s}$	(i_s, k_s)	$c_{j_s, j_{d+s}}^{i_s, k_s}$	(i_s, k_s)	$c_{j_s, j_{d+s}}^{i_s, k_s}$	(i_s, k_s)	$c_{j_s, j_{d+s}}^{i_s, k_s}$
$(0, 2)$	$(0, 2)$	$-i$	$(2, 0)$	i	$(1, 3)$	i	$(3, 1)$	$-i$
$(2, 0)$	$(0, 2)$	i	$(2, 0)$	$-i$	$(1, 3)$	i	$(3, 1)$	$-i$
$(0, j \neq 2)$	$(0, j)$	1	$(j, 0)$	1	(i, k)	$-(-1)^{\delta_{2k}}$	(k, i)	$-(-1)^{\delta_{2k}}$
$(j \neq 2, 0)$	$(0, j)$	1	$(j, 0)$	1	(i, k)	$(-1)^{\delta_{2k}}$	(k, i)	$(-1)^{\delta_{2k}}$
$2 \in (j, k)$	(j, k)	$-i$	(k, j)	i	$(0, i)$	$-i(-1)^{\delta_{2k}}$	$(i, 0)$	$i(-1)^{\delta_{2k}}$
$2 \in (k, j)$	(j, k)	i	(k, j)	$-i$	$(0, i)$	$-i(-1)^{\delta_{2k}}$	$(i, 0)$	$i(-1)^{\delta_{2k}}$
$(1, 3)$	$(1, 3)$	1	$(3, 1)$	1	$(0, 2)$	-1	$(2, 0)$	-1
$(3, 1)$	$(1, 3)$	1	$(3, 1)$	1	$(0, 2)$	1	$(2, 0)$	1

These cases generate the diagonal-off entries in each block in agreement with the exchange rules depicted previously for the sth scripts of such entries' rows: $(i_s, k_s) \in \{(0, j), (j, 0); (i, k), (k, i)\}$, with i, j, k a permutation from 1, 2, 3 and j the associated direction for the corresponding interaction being used from H_I and $H_{II_{a,b}}$; $(i_s, k_s) \in \{(j_p, k_p), (k_p, j_p); (0, i), (i, 0)\}$, with i, j_p, k_p a permutation of parity p from 1, 2, 3 and j_p, k_p are the associated directions for the interaction H_{III}.

First, we should note that the signs for each term in the diagonal-off entries do not depend on the entries' scripts in positions other than the parts in which the interaction is being applied, k', k'' in the expressions of the previous section (30), (32), (34), and (36). This is because $\text{Tr}(\tilde{\sigma}_{i_s}^* \sigma_{j_{d+s}} \tilde{\sigma}_{k_s}^T \sigma_{j_s}^T) = \text{Tr}(\tilde{\sigma}_{i_s}^* \sigma_0 \tilde{\sigma}_{k_s}^T \sigma_0^T) = 2$. Instead, signs only depend on the type of exchange indexes shown in Table 3. It has already been noted that $c_{j_s, j_{d+s}}^{i_s, k_s}$ is imaginary only if $j_s = 2$ or $j_{d+s} = 2$. This property is then transferred to the corresponding $\mathcal{F}_{j,s}^{j_s, j_{d+s}}$, and then transformed to h_{12} as a function of the number of those factors in (31), (33), (35), and (37). Thus, by exchanging i_s, k_s (block transposing), only the cases with $h_{12} \in \mathbb{I}$ will change their sign.

The final fact is related with the different signs appearing in the terms of diagonal-off entries. This will be important to analyze the number of independent blocks in the entire evolution matrix. For H_I and H_{III}, the two different terms are obtained by the exchange of j_s, j_{d+s}. Thus, for H_I, only the $c^{i_s,k_s}_{j_s,j_{d+s}}$ with $(j_s, j_{s+d}) = (0,2),(2,0)$ and $(i_s, k_s) \in \{(0,2),(2,0)\}$ or $(j_s, j_{s+d}) = (0, j \neq 2),(j \neq 2, 0)$ and $(i_s, k_s) \in \{(i,k),(k,i)\}$ will change their sign (in the first four rows of Table 3). For H_{III}, if $(j_s, j_{s+d}) = (1,3),(3,1)$ and $(i_s, k_s) \in \{(0,2),(2,0)\}$ or $(j_s, j_{s+d}) = (0,2),(2,0)$ and $(i_s, k_s) \in \{(i,k),(k,i)\}$, then $c^{i_s,k_s}_{j_s,j_{d+s}}$ will change their sign (in the last four rows of Table 3). For $H_{IIa,b}$, two terms in the scripts are involved, so different aspects contribute: the location of interacting parts, the type of exchange, and their order in the scripts.

Last properties exhibits the way in which each term in h_{12} will change its sign. The three aspects mentioned in the previous paragraph allow us to understand the diagonal-off structure of H_I, $H_{IIa,b}$, and H_{III} (considering that their diagonal components follow the properties discussed above). In the following subsections, we analyze this structure for each interaction, particularly discussing the independence of blocks in terms of the free parameters, making a distinction between the effective parameters (those appearing in the final expression of (28)) and the physical parameters (those appearing as coefficients h_I in the Hamiltonian). They are not the same because many physical parameters appear clustered in the same way in (28), because the entries of $\mathbb{S}_U$ depend only on the parameters $h_{11} \pm h_{22}, h_{12}$. As a result, by grouping finally in the $U(1) \times SU(2)$ blocks, there will be only two or eight different blocks $\mathbb{S}_U$ in U.

5.4.3. Block Entries of H_I

The diagonal-off entries have exactly the two terms $h_{\cdot j4^{s}+dt-1)\frac{2d}{4}}$ for $t' = 0, 1$, and there are only two combinations: adding or subtracting terms. As was stated previously, they are imaginary only if local interactions are in the direction $j = 2$. In this case, we separate the factor $\pm i$ for $j = 2$ cases in the diagonal-off entries, and the remaining coefficients in the opposite corners in each block are equal as expected from (28). Then, there is generally one term with the same sign through all diagonal-off entries (when $k = 2$, or otherwise when $j_s = 2$ in the first four rows in the Table 3), leaving only two possibilities for the remaining term. Thus, in each H_I matrix there are blocks with only two different diagonal-off entries, depending only on the index exchange type in the local interaction position and not on the remaining indexes. Thus, for a fixed set of indexes for the positions unrelated to the part on which the interaction is applied, a pair of blocks exists, one each for the exchanges $(0, j), (j, 0)$ and $(i, k), (k, i)$, with different relative signs in their diagonal-off terms. For the corresponding diagonal entries, in (29), only the difference $h_{11} - h_{22}$ is relevant. As previously stated by analyzing equation (31), it is also possible realize that in each diagonal entry there are only two terms from the $3d$ terms changing their sign with respect to other rows. Block scripts differ in only one index, those corresponding with $i' \neq j$ (the local interaction direction) and $k = k'$ (the correspondent pair on which the local interaction is being applied), leaving only two terms and two different combinations for $h_{11} - h_{22}$. This implies that there are only two different blocks for (29) through all U, each one operating with different exchange rules, $(0, j), (j, 0)$ or $(i, k), (k, i)$. Each one is the same (until unitary factors, which can be different) for all entries with different indexes in positions other than k'. This fact can be attributed, depending on the number of disposable parameters (five, including the time and excluding the parameters in the unitary factor of each block), to the independence between the two types of blocks in the evolution matrix (29).

5.4.4. Block Entries of H_{IIa}

For the non-diagonal entries, because the exchange factor $\mathcal{F}^{j_s, j_{d+s}}_{j_s}$ appears two times for each j, all of them are real, so the opposite corners of each block are always equal. Each entry has four terms with alternating signs, in agreement with the Table 3, as a function of the rows' subscripts. Signs only depend on both indexes exchanged: either they are the same type $(0, j), (j, 0)$ or $(i, k), (k, i)$, or otherwise opposite with an exchange of each type. This will give only four sign combinations (a calculation not developed explicitly here), except for $j = 2$, where the appearance of two factors i will change the overall factor, giving eight combinations, one half of them with opposite overall sign to the remaining. For the diagonal entries, based on the ideas in the previous case for H_I, there will be four terms changing their relative signs with respect to other associated diagonal entries in the same block, but now differing in two part indexes (due to the related non-local interaction). As before, one term has a fixed sign, so there are only eight combinations for the three remaining terms from the 16 possible. This means eight different combinations for $h_{11} - h_{22}$ in (29), due to the values $k = k', k''$ for the non-correspondent parts with non-local interactions in H_D for this case. Thus, similar to H_I, in this case there will be eight different blocks in U for (29): one for each one of the eight different combinations of the exchange rules. Each type applies in the same way for all entries with different indexes in positions other than k', k''. There are nine free parameters, including the time and excluding the parameters in the unitary factor for the block, so independence among the eight types of blocks can be more elusive. Despite all this, located operations not involving all GBS basis states appear as achievable.

5.4.5. Block Entries of H_{IIb}

Although the exchange factors $\mathcal{F}_{j,s}^{j_s,j_{d+s}}$ are crossed and j takes two different values in the subscripts, the discussion regards certain similitude to that for H_{IIa}. For the diagonal-off entries, in agreement with Table 3, it implies that only if $j = 2$ is not included in the crossed interaction ($i = 2$ in (35)) will they become real. Each entry will have four terms with alternating signs, in agreement with the outcomes of products of exchange factors in Table 3 as a function of the rows' subscript involved. Here, there will be eight combinations (four and four with opposite overall signs), except for $j = 2$ with only four combinations. For the diagonal entries, $h_{11} - h_{22}$ in (29), the situation is identical to H_{IIa}. Then, there will be eight different block types in U for each combination of exchange rules on the indexes k', k''. Again, nine free parameters for the $SU(2)$ blocks are available.

5.4.6. Block Entries of H_{III}

This is a special case exception of the previous remark where j_s, j_{d+s} is not of the forms $0, j$ or $j, 0$. Nevertheless, $\mathcal{F}_{i,s}^{j_p,k_p}$ becomes in the same way on of $c_{j_p,k_p}^{0,i}$, $c_{j_p,k_p}^{i,0}$, $c_{j_p,k_p}^{j_p,k_p}$, or $c_{j_p,k_p}^{k_p,j_p}$. However, several aspects are identical to the H_I case. A brief analysis shows that entries become real only for $i = 2$ (see the last four rows of Table 3). Each diagonal-off entry has two terms with alternating signs as functions of entry labels. For the diagonal entries, again only two types of terms change their sign in H_D from (31) for the rows forming the $SU(2)$ blocks with the exchange rules. This gives only two types of $h_{11} - h_{22}$ in (29), again generating only two different blocks in the whole U—each one for a kind of exchange rule involved here, containing five free parameters.

To resume the findings, Figure 5 shows the relations exhibited in the exchange indexes for each interaction. This figure depicts each of the exchange index relations of GBS basis states under the interaction. Thus, Figure 5a,d, depicts the two groups of exchange states for H_I and H_{III} generated by the two different blocks through the whole $SU(2^{2d})$ evolution matrix, both independent up to five parameters and with h_{12} in (29). Figure 5b,c depict the double exchange indexes induced by the eight blocks generated by H_{IIa} and H_{IIb}. These eight blocks are independent up to nine parameters. All representations in Figure 5 are for a single GBS basis state, but clearly one specific block is operating on any of them simultaneously. Note finally that for all cases there are a complementary number of free physical parameters in $h_{11} + h_{22}$: $3d + 2 - 4 = 3d - 2$ for H_I and H_{III} and $3d + 4 - 8 = 3d - 4$ for H_{IIa} and H_{IIb} (time t is not accounted because it was considered in the $SU(2)$ fitting). Then, there is a linearly growing space to fit the blocks into a programmed operation in terms of the physical parameters, although there is an exponential growth of those blocks.

Type I

$$H_I = H_D + H_{ND_I}^{(j,k')}$$

$|\Psi_{\alpha_1 \ldots \alpha_{k'} \ldots \alpha_d}\rangle$

$\mathbb{S}^0_{U_{\mathfrak{z},\mathfrak{z}'}}$ $h_{12} \in \mathbb{I}$ if $j = 2$

0   i

   5 parameters

j k

$|\Psi_{\alpha_1 \ldots \alpha'_{k'} \ldots \alpha_d}\rangle$

a)

Type IIa

$$H_{IIa} = H_D + H_{ND_{IIa}}^{(j,k'k'')}$$

$|\Psi_{\alpha_1 \ldots \alpha_{k'} \ldots \alpha_{k''} \ldots \alpha_d}\rangle$

00 0i i0 ii $\mathbb{S}^0_{U_{\mathfrak{z},\mathfrak{z}'}}$ $h_{12} \in \mathbb{R}$

jj jk kj kk 9 parameters

$|\Psi_{\alpha_1 \ldots \alpha'_{z'} \ldots \alpha'_{k''} \ldots \alpha_d}\rangle$

b)

Type IIb

$$H_{IIb} = H_D + H_{ND_{IIb}}^{(i,k'k''p)}$$

$|\Psi_{\alpha_1 \ldots \alpha_{k'} \ldots \alpha_{k''} \ldots \alpha_d}\rangle$

00 0i i0 ii $\mathbb{S}^0_{U_{\mathfrak{z},\mathfrak{z}'}}$ $h_{12} \in \mathbb{R}$ if $i = 2$

jk jj kk kj 9 parameters

$|\Psi_{\alpha_1 \ldots \alpha'_{k'} \ldots \alpha'_{k''} \ldots \alpha_d}\rangle$

c)

Type III

$$H_I = H_D + H_{ND_{III}}^{(i,k')}$$

$|\Psi_{\alpha_1 \ldots \alpha_{k'} \ldots \alpha_d}\rangle$

0   j

   $\mathbb{S}^0_{U_{\mathfrak{z},\mathfrak{z}'}}$ $h_{12} \in \mathbb{R}$ if $i = 2$

i k 5 parameters

$|\Psi_{\alpha_1 \ldots \alpha'_{k'} \ldots \alpha_d}\rangle$

d)

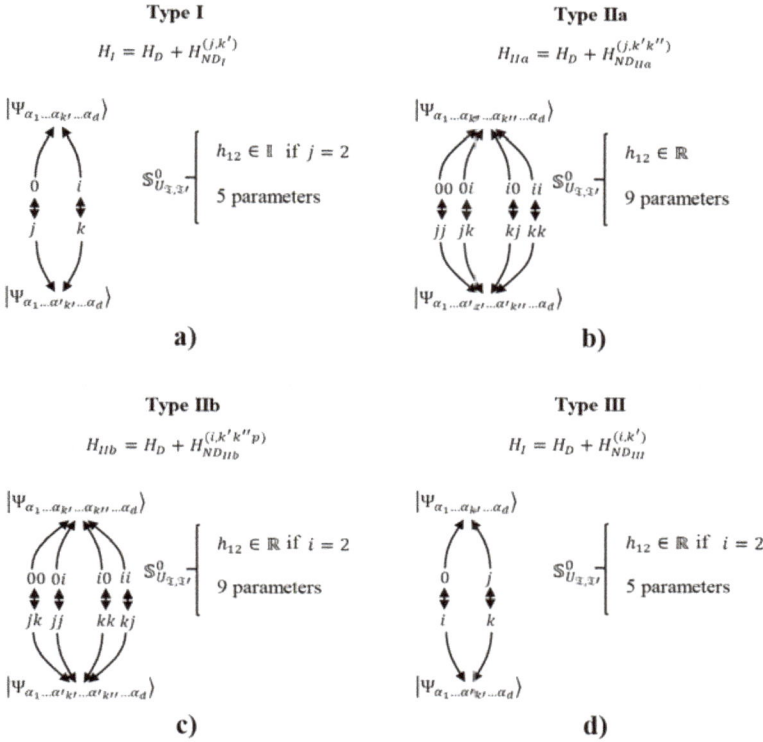

Figure 5. Exchange index relations involved for each interaction and highlighted properties for their correspondent $\mathbb{S}^0_{U_{I,I'}}$: (a) H_I; (b) H_{IIa}; (c) H_{IIb}; and (d) H_{III}. Exchange relations in (**b,d**) are doubled by considering the vertical switching in one of the indexes for each pair shown.

6. Connectedness, Superposition, Entanglement and Separability

To understand how dynamics is addressed under the interactions $H_I, H_{IIa,b}, H_{III}$ (used independently or combined), some complementary analysis is convenient. In order to prepare the reader, some illustrative examples are included in Appendix A.4 for $d = 1$ and $d = 2$, depicting some notable properties of dynamics in such cases by including several kinds of entangling operations.

6.1. Exchange Connectedness under Interactions

Under the $SU(2)$ decomposition, pairs of states in GBS basis become related, showing a probability exchange between them. As it was seen, each one of the $H_I, H_{IIa,b}, H_{III}$ interactions has rules for this exchange. In any case, it should be clear this exchange is achievable between any pair by combining all types of interactions obtained by switching the value of: (a) interaction direction and correspondent pair j, k' in (31) for H_I; (b) interaction direction and correspondent pairs j, k', k'' in (33) for H_{IIa}; (c) interaction directions, correspondent pairs, and parity i, k', k'', p in (35) for H_{IIb}; and (d) interaction direction and correspondent pair i, k' in (37) for H_{III}. Several types of interactions can be combined in a sequence. The combination of interactions is not precise for the basis element connectedness, but it is necessary to increase the entanglement, and thus to connect two arbitrary quantum states. In those terms, there are only two types of states: (1) those exchanging one script (H_I and H_{III}), and (2) those exchanging two scripts (H_{IIa} and H_{IIb}) in the GBS basis elements under the rules depicted in Figure 5 (although the rules and connections are different). All basis states become connected under one or several interactions applied consecutively, depending on the number of necessary exchanges in their scripts. Figure 6 shows a graph with these relations for the cases $d = 1, 2, 3$. Green edges indicate one script exchange and red lines indicate two script exchanges. The connection can only be achieved with a single interaction in the first two cases, due to the low entanglement level. Figure 6a corresponds to the figure presented in [11] for Bell states in $SU(4)$ systems.

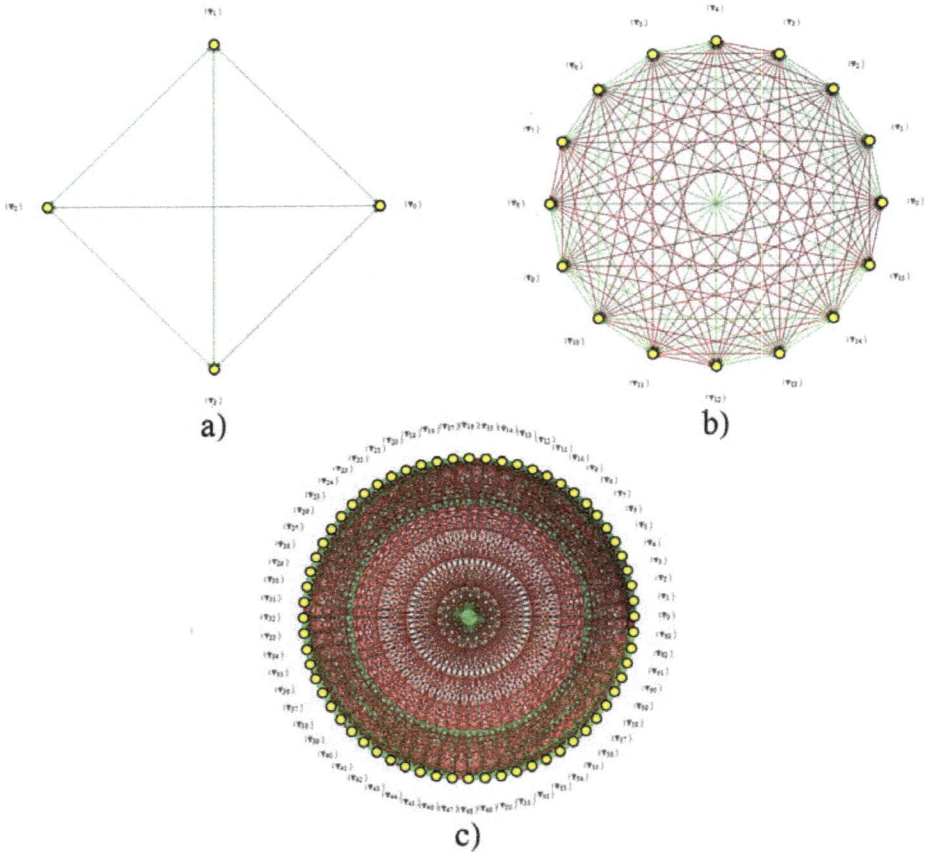

Figure 6. Connectedness graphs between states under $SU(2)$ decomposition for one (green) and two (red) exchange scripts for all generalized Bell state (GBS) basis states: (a) $d = 1$; (b) $d = 2$; and (c) $d = 3$.

Connectedness in a finite number of steps by applying some or all cases in each type of interaction (piecewise with constant parameters or with time-dependent parameters in each case) warrants the full probability exchange between the occupancy level of each state in terms of the discussion included in Section 3. Nevertheless, not all interactions are able to reach an arbitrary evolution. As is obvious, H_I and H_{III} are not able to generate extended entanglement out of the correspondent pair on which they operate (this assumes no rearrangements are made in the correspondent pairs and their elements). We discuss this aspect in the next subsection.

6.2. Notable Quantum Processing Operations Achievable under $SU(2)$ Decomposition

Departing from $\mathbb{S}_U$, then by fixing $\omega t = \frac{2n+1}{2}\pi$, $\frac{h_{11}-h_{22}}{2\hbar\omega} = \epsilon$, $\frac{h_{12}}{\hbar\omega} = i^c\delta$, $\frac{h_{11}+h_{22}}{2\hbar\omega} = 2(m-\frac{1}{2})$; $n, m \in \mathbb{Z}$, $c \in \{0,1\}, \delta \in \mathbb{R}$, where $\epsilon^2 + \delta^2 = 1$ in (29). Note that the parameter c depends on each kind of interaction in the terms discussed in the previous section. Then, we get the $\mathbb{S}_{U_{\mathcal{I},\mathcal{I}'}}$ block [27]:

$$\mathbb{H}_m^c(\delta,\epsilon)_{\mathcal{I},\mathcal{I}'} \equiv (-1)^m \begin{pmatrix} \epsilon & i^c\delta \\ (-i)^c\delta & -\epsilon \end{pmatrix}, \tag{38}$$

operating on the GBS basis. Note that this form cannot always be achieved independently in all blocks in terms of the free parameters and the possible restriction $h_{0,0,...,0} = 0$ (here $\det(\mathbb{H}_m^c(\delta,\epsilon)_{\mathcal{I},\mathcal{I}'}) = -1$, although it is not decisive in the following development). Nevertheless, we need only achieve it in some blocks in the immediate discussion. We are using the time-independent case, but other more practical cases with time-dependent Hamiltonian coefficients can be implemented. The last form is highly versatile. If $s_\epsilon|\epsilon| = \delta = \frac{1}{\sqrt{2}}$ ($s_\epsilon = \text{sign}(\epsilon)$, referred in the notation as $-,+$), we get a Hadamard-like gate $H_{\mathcal{I},\mathcal{I}'}^{m,c,\text{sign}(\epsilon)} \equiv \mathbb{S}_{U_{\mathcal{I},\mathcal{I}'}}$ (in particular, if $c = 0$, but this condition can be relaxed). When $\delta = 1$, we get an exchange-like gate [12,26] for the pair in the $SU(2)$ block, $E_{\mathcal{I},\mathcal{I}'}^{m,c} \equiv \mathbb{S}_{U_{\mathcal{I},\mathcal{I}'}}$. Note that this case is a limit case for the time-independent case (29) when $h_{12} \gg h_{11} - h_{22}$. Otherwise, it can be achieved in two steps of time-independent piecewise Hamiltonians (as in [12]) or as a continuous time-dependent Hamiltonian. These gates are:

$$H_{\mathcal{I},\mathcal{I}'}^{m,c,s_\epsilon} = \frac{(-1)^m}{\sqrt{2}} \begin{pmatrix} s_\epsilon & i^c \\ (-i)^c & -s_\epsilon \end{pmatrix}, \quad E_{\mathcal{I},\mathcal{I}'}^{m,c} = (-1)^m \begin{pmatrix} 0 & i^c \\ (-i)^c & 0 \end{pmatrix}. \tag{39}$$

Note additionally that when $\frac{h_{11}+h_{22}}{2\hbar\omega} = (\frac{\alpha}{m} - 1)\pi$, $\omega t = m\pi$; $n, m \in \mathbb{Z}$, we get the quasi-identity gate $\mathbb{S}_{U_{\mathcal{I},\mathcal{I}'}} = e^{i\alpha\pi}\mathbb{I}_{\mathcal{I},\mathcal{I}'} \equiv I_{\mathcal{I},\mathcal{I}'}^\alpha$. The combination of these blocks (allowed because the block independence previously discussed) allows important quantum processing operations to be set.

6.3. $SU(2)$ Decomposition in the Context of $n-$Qubit Controlled Gates

Transformation between quantum states can generally be achieved by means of linear and anti-linear operators. Anti-linear operators are particularly useful to depict time-reversal operations or the action of some Einstein-Podolsky-Rosen channels. If these kinds of operations are being considered in the processing, an extension of the Hamiltonian (1) should be considered by the inclusion of anti-linear operations [28]. In this work, we have restricted our development to linear operators, as was settled in Sections 2 and 3.

Below of such context, it should be advised that $SU(2)$ decomposition is compatible with the most quantum information developments in the literature. Nevertheless, many of those works do not consider that such proposed processing forms are rarely compatible with the dynamics of physical systems if the computational basis continues to be used (the natural basis based on physical properties of local systems such as spin and polarization). The nature of entangling operations naturally induces both superposition and entanglement, thus generating a complex dynamics evolution in such basis compared with the structured gates proposed in the quantum information developments (whose authors were clearly not always concerned with the underlying physics). $SU(2)$ decomposition (mainly the part developed in the Sections 2 and 3) naturally proposes a better basis to set the quantum processing grammar for certain interaction architectures (e.g., those developed in Sections 4 and 5). The induced 2×2 block structure allows such processing structures to be set more easily, mainly based on binary processing.

In the context of quantum computation, the most common trend is the settlement of universal gates in the sense of a quantum Turing machine. A set of universal quantum gates for two-qubit processing was established by [29] as a set of local gates together with the *CNot* gate. Despite universality, this trend is not optimal because for a given processing, it is not clear how to express it in terms of those elements in the universal set. In an alternative trend, [30] has settled a more optimal gate decomposition by factorization in terms of $P-$unitary matrices. In the last two trends, $SU(2)$ decomposition for $SU(4)$ ($d = 1$, meaning two-qubit processing) has shown how to adapt those results for the physics of Heisenberg–Ising interactions including driven magnetic fields: (a) in [31], a set of alternative universal gates has been proposed on the grammar of Bell states; and (b) in [26], an optimal set of six gates ($P-$unitary matrices) is proposed using the forms of $SU(2)$ decomposition on a Bell states basis to reproduce any other gate for two-qubit processing. In the current context, those outcomes are automatically applicable to Type I and III interactions. Type II interactions are excluded because they require at least $d = 2$. In any case, the contribution of the $SU(2)$ reduction is in the proposal of Bell basis as a grammar instead of the computational one so that the physical evolution fulfills the forms required by the processing gates.

Although two-qubit processing is still universal, more powerful processing is possible by attaining more than two qubits at a time. In this approach, [32,33] have stated universal processing gates in terms of local rotations and $n-$qubits controlled gates. In the computational basis, rotations are obtained by local interactions by turning off the entangling operations, but controlled gates can be physically difficult to reproduce. In the $SU(2)$ reduction scheme, the form of rotations in those works ($R_y(\alpha)$ and $R_z(\alpha)$) are achieved by the forms (29) as follows. First, $R_y(\alpha)$ is mainly achieved by settling $h_{11} = h_{22}$ and $h_{12} \in \mathbb{I}$. $R_z(\alpha)$ is obtained by fixing $h_{12} = 0$. Notably, those rotations are not necessarily physical neither local, they could operate among entangled states. Instead, they can be determined as rotations on the informational states being used (elements of GBS basis). Other basic forms are also easily obtained, for example, $\mathrm{Ph}(\delta)$ is obtained by settling $\cos \omega t = \pm 1$. For the controlled gates $\Lambda_n(U)$ proposed in [32], authors in [33] turn to a long factorization in terms of rotations and controlled gates $\Lambda_1(U)$ (which can also be obtained departing from the *CNot* gate and rotations). In any case, if a computational basis is used, the reproduction of the *CNot* gate can still bring certain difficulties in many quantum systems [34]. In the context of $SU(2)$ reduction, *CNot* gate and inclusively $\Lambda_1(U)$ are directly obtained if the Bell basis is used as grammar:

$$\Lambda_1(U) = \left(\begin{array}{c|c} \mathbb{S}_{U_1} \to \mathbb{I} & \mathbf{0} \\ \hline \mathbf{0} & \mathbb{S}_{U_2} \to U \end{array} \right), \tag{40}$$

where U is a general matrix in $SU(2)$ as in (29). Because of the independence of blocks stated in Section 5, the achievement of $\Lambda_1(U)$ is warranted. Then, the construction of $\Lambda_n(U)$ follows immediately as proposed in [32,33], but considering those forms working on the grammar basis of the Bell states or on the GBS basis in general. Clearly, in the $SU(2)$ decomposition scheme, other controlled gates are achievable by the alternative selection of the elements on which interaction is being applied. If more optimal factorization methods are possible for $d > 1$ (where blocks are repeated by groups), based on the set of matrices U as in (14) by including all the possible forms generated by Type I, IIa, IIb, and III interactions, it is still an open question.

6.4. Generating Superposition and Entanglement

In the following, we will use an arrow to depict a certain group of quantum operations. On the top of the arrow, we set the type of interaction being used. On the bottom, we set the subspace on which they apply or the generic form of each operation, together with their prescriptions. For instance, if an operation for $d = 4$ (8 qubits and 256 elements in the GBS basis) generated by the Type IIa interaction is applied in the associated direction y and on the pairs 1 and 4 ($j = 2, k' = 2, k'' = 4$ in (32)) with prescriptions for a Hadamard gate mixing the basis states $|\Psi_0\rangle = |\Psi_{0,0,0,0}\rangle$ and $|\Psi_{130}\rangle = |\Psi_{2,0,0,2}\rangle$ (i.e.,

$H_{0,130}^{0,0,+}$) and an exchange gate between the basis states $|\Psi_1\rangle = |\Psi_{1,0,0,0}\rangle$ and $|\Psi_{131}\rangle = |\Psi_{3,0,0,2}\rangle$ (i.e., $E_{0,131}^{0,0}$), we will write:

$$\xrightarrow[H_{0,130}^{0,0,+}\oplus E_{0,131}^{0,0}]{H_{IIa}^{(2,2,5)}}. \tag{41}$$

Although other operations can be defined between the remaining basis states, if they are not specified, it is because some operations are repeated for other certain groups of scripts (e.g., for $|\Psi_{20}\rangle = |\Psi_{0,1,1,0}\rangle$ and $|\Psi_{150}\rangle = |\Psi_{2,1,1,2}\rangle$, $H^{0,0,+}$ is also being applied) or because the concrete operation being developed does not require such specification (e.g., there is no specification for the operation between $|\Psi_{67}\rangle = |\Psi_{3,0,0,1}\rangle$ and $|\Psi_{193}\rangle = |\Psi_{1,0,0,3}\rangle$). In some cases, complex families of subsequent operations are required, and then one family is specified by a group of indexes defining it.

6.4.1. Generating 2−Separable Superposition

By using the general block operations $\mathbb{S}_{U_{\mathcal{I},\mathcal{I}'}} \in U(1) \times SU(2)$ (29), it is possible to arrive at a state exhibiting complete superposition through all of the basis elements. Thus, for example, departing from the simple state $|\Psi_0\rangle^{2d} = |\Psi_0\rangle_1|\Psi_0\rangle_2...|\Psi_0\rangle_d$ (easily obtained from $|00...0\rangle$), a couple of local operations $H_I^{(i,k)}$ on each correspondent pair k are sufficient to generate a state containing representatives from each basis element:

$$
|\Psi_0\rangle^{2d} \xrightarrow[\substack{\oplus_{s,s'}\mathbb{S}_{U_{s,s'}}\\s'-s=4^{k-1},\\s_{4,k}^d=0}]{\substack{H_I^{(1,k)}\\k=1,2,...,d}} \bigotimes_{k=1}^d \sum_{i=0}^1 \alpha_{i,0}^k |\Psi_i\rangle_k ,
$$

$$
\xrightarrow[\substack{\oplus_{s,s'}\mathbb{S}_{U_{s,s'}}\oplus_{s'',s'''}\mathbb{S}_{U_{s'',s'''}}\\s'-s=3\cdot4^{k-1},s'''-s''=4^{k-1}\\s_{4,k}^d=0,s_{4,k}''=1}]{\substack{H_I^{(3,k)}\\k=1,2,...,d}} \bigotimes_{k=1}^d \sum_{i=0}^1 \sum_{j(i)\in\{i,3-i\}} \alpha_{i,0}^k\beta_{j(i),i}^k \left|\Psi_{j(i)}\right\rangle_k \equiv \sum_{\mathcal{I}=0}^{4^d-1} \gamma_{\mathcal{I}}|\Psi_{\mathcal{I}}\rangle , \tag{42}
$$

$$
\text{with}: \gamma_{\mathcal{I}} = \prod_{k=1}^d \alpha_{j-1(\mathcal{I}_{4,k}^d),0}^k \beta_{\mathcal{I}_{4,k}^d j^{-1}(\mathcal{I}_{4,k}^d)'}^k
$$

where $j^{-1}(i)$ is the inverse of $j(i)$ and directions $i = 1,3$ were used as examples. In addition, $\alpha_{i,j}^k$ are the components of $\mathbb{S}_{U_{s,s'}}$ in the first operations, and $\beta_{j(i),i}^k$ are the components of $\mathbb{S}_{U_{s,s'}}$, $\mathbb{S}_{U_{s'',s'''}}$ for the second operations with $i = 0,1$, respectively. Figure 7 depicts each step of the process, using the local operations (alternatively, crossed interactions in H_{III} could be considered).

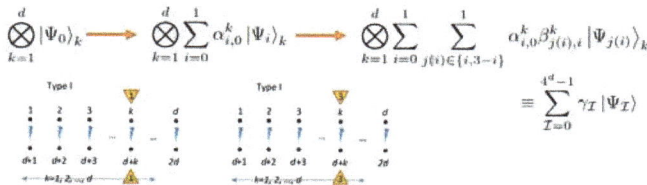

Figure 7. Processes to build 2−separable states with complete superposition.

The last process is a particular case of more general operations by considering $O_J^{(i,\{s\})} = \mathbb{S}_{U_{\mathcal{I},\mathcal{I}'}}$ to mix the states through the momentary associated blocks changing the indexes $\{s\}$ with some interaction $H_J, J \in \{I, IIa, IIb, III\}$ in the associated direction i. We coin the term $k-$local operation when $\mathbb{S}_{U_{\mathcal{I},\mathcal{I}'}}$ generates entanglement at the most in k parts. In our basic interactions scheme, there are only $2-$local and $4-$local operations, as was discussed previously. Thus, following the previously-introduced notation, we set a family of procedures to develop superposition including the previous procedure. Departing from the $|\Psi_0\rangle^{2d}$, it is possible to apply several alternate $2-$local operations to generate superposition involving all GBS basis states. By defining a sequence of paired directions for the H_I evolution involving all pairs $s = 1, 2, ..., d$ (this process can alternatively be achieved by H_{III}): $\{\{i_s, k_s(i_s)\}|\{1, 2, 3\} \ni i_s \neq k_s(i_s) \in \{1, 2, 3\} \setminus \{i_s\}; s = 1, 2, ..., d\}$. Additionally, $j_s(i_s) \in \{1, 2, 3\}, i_s \neq j_s(i_s) \neq k_s(i_s)$. Then, following the evolution process:

$$|\Psi_0\rangle^{2d} \xrightarrow[O_I^{(i_1,\{1\})}]{H_I^{(i_1,1)}} \sum_{t \in \{0,i_1\}} \alpha_{t,0}^1 |\Psi_{t,0,...,0}\rangle \xrightarrow[O_I^{(k_1(i_1),\{1\})}]{H_I^{(k_1(i_1),1)}} \sum_{\epsilon_1=0}^{3} \alpha_{p_1(\epsilon_1),0}^1 \beta_{\epsilon_1,p_1(\epsilon_1)}^1 |\Psi_{\epsilon_1,0,...,0}\rangle,$$

$$\xrightarrow[O_I^{(i_2,\{2\})}]{H_I^{(i_2,2)}} \cdots \xrightarrow[O_I^{(k_d(i_d),\{d\})}]{H_I^{(k_d(i_d),d)}} \sum_{\epsilon_1,...,\epsilon_d=0}^{3} \left(\prod_{s=1}^{d} \alpha_{p_s(\epsilon_s),0}^s \beta_{\epsilon_s,p_s(\epsilon_s)}^s \right) |\Psi_{\epsilon_1,\epsilon_2,...,\epsilon_d}\rangle \equiv |\Psi_{f_{sep}}\rangle, \tag{43}$$

where $p_s(\epsilon_s)$ are the inverses of the association rules for the one index exchanges depicted in Figure 5a (or 5d for H_{III}): $p_s(i_s) = i_s = p_s(j_s(i_s)), p_s(0) = 0 = p_s(k_s(i_s))$. Additionally, $|\alpha_{0,0}^s|^2 + |\alpha_{i_s,0}^s|^2 = 1$, $|\beta_{0,0}^s|^2 + |\beta_{k_s(i_s),0}^s|^2 = 1, |\beta_{i_s,i_s}^s|^2 + |\beta_{j_s(i_s),i_s}^s|^2 = 1$. $\text{Tr}^S(\rho_{IJ})$ represents the partial trace with respect to the entire system except the $s \in S$ parts. As expected, $|\Psi_{f_{sep}}\rangle$ is $2-$separable:

$$\text{Tr}^{\{k'\}} \left(|\Psi_{f_{sep}}\rangle \langle \Psi_{f_{sep}}| \right) = \left(\sum_{\epsilon_{k'}=0}^{3} \alpha_{p_{k'}(\epsilon_{k'}),0}^{k'} \beta_{\epsilon_{k'},p_{k'}(\epsilon_{k'})}^{k'} |\Psi_{\epsilon_{k'}}\rangle \right) \left(\sum_{\epsilon_{k'}=0}^{3} \alpha_{p_{k'}(\epsilon_{k'}),0}^{k'} \beta_{\epsilon_{k'},p_{k'}(\epsilon_{k'})}^{k'} |\Psi_{\epsilon_{k'}}\rangle \right)^{\dagger} \tag{44}$$

due to the limited nature of operations involved, which cannot be able to generate more extended entanglement. In addition, superposition can be limited to the $SU(2)$ blocks coverage through the number of parameters introduced, $\alpha_{p_s(\epsilon_s),0}^s, \beta_{\epsilon_s,p_s(\epsilon_s)}^s$, and their physical scope. As shown in [11], a richer superposition coverage on $SU(2^{2d})$ can be achieved with additional $2-$local operations on each part, introducing extra parameters and probability mixing. As in [11], **n** in (29) is limited to take the two forms $(n_x, 0, n_z)$ or $(0, n_y, n_z)$ (for the time-independent case), but by combining both forms we arrive at two general forms with arbitrary $\mathbf{n} = (n_x, n_y, n_z)$ (this also fulfills the time-dependent case with adequate $h_{ij}(t)$).

Although this procedure can include a general full $2-$separable state together with entangled segments between correspondent pairs, it cannot exhibit states with more extended entanglement, requiring more extended entangling operations such as H_{IIa} and H_{IIb}. The quest is to obtain general states departing from a simple resource, which is still an open challenge—particularly for the possible entanglement degree there (a more ambitious challenge is the transformation between two general states [35], but it can always be reduced in two steps of this kind). We discuss this issue in the remaining subsection, and we develop some procedures to generate some maximal entangled states of arbitrary size.

6.4.2. Entanglement Dynamics under Interactions

Now, we analyze the entanglement generation under the interactions being considered. We employ the partial trace criterion [19] for pure states by considering a single $SU(2)$ combination of two GBS basis states $|\phi_{IJ}\rangle = \alpha_I |\Psi_I\rangle + \alpha_J |\Psi_J\rangle$. In addition, the explicit form for coefficients will be written as $\alpha_I = \cos \theta/2, \alpha_J = e^{i\phi} \sin \theta/2$. Then, we construct their associated density matrix $\rho_{IJ} = |\phi_{IJ}\rangle \langle \phi_{IJ}|$ to conveniently take partial traces in order to analyze the entanglement of specific subsystems in this quantum state under concrete interactions. Because the rules in the exchange scripts (in the GBS basis states to form the $SU(2)$ blocks) are basically the same for the three interactions $H_I, H_{IIa,b}, H_{III}$, the analysis is reduced to only two cases. The first is for a pair of GBS basis elements

$|\Psi_{\mathcal{I}}\rangle$, $|\Psi_{\mathcal{J}}\rangle$ differing in only one subscript between $\mathcal{I}$ and $\mathcal{J}$: $i_s = j_s \forall s \in \{1,\dots,d\}, s \neq k'$ (in H_I, H_{III} interactions). Thus, in this case (omitting the base b and the size d for simplicity in the scripts):

$$\left|\phi^1_{\mathcal{I}\mathcal{J}}\right\rangle = \frac{1}{\sqrt{2^d}} \sum_{\mathcal{E},\mathcal{D}=0}^{2^d-1} \left(\bigotimes_{k' \neq s=1}^{d} \tilde{\sigma}_{i_s} \otimes (\alpha_{\mathcal{I}} \tilde{\sigma}_{i_{k'}} + \alpha_{\mathcal{J}} \tilde{\sigma}_{j_{k'}}) \right)_{\mathcal{E},\mathcal{D}} |\mathcal{E}\rangle \otimes |\mathcal{D}\rangle. \tag{45}$$

The second case is for a pair of elements in the GBS basis differing in two subscripts of $\mathcal{I}$ and $\mathcal{J}$: $i_s = j_s \forall s = 1,\dots,d, s \neq k',k''$ (in $H_{IIa,b}$ interactions):

$$\left|\phi^2_{\mathcal{I}\mathcal{J}}\right\rangle = \frac{1}{\sqrt{2^d}} \sum_{\mathcal{E},\mathcal{D}=0}^{2^d-1} \left(\bigotimes_{k',k'' \neq s=1}^{d} \tilde{\sigma}_{i_s} \otimes (\alpha_{\mathcal{I}} \tilde{\sigma}_{i_{k'}} \otimes \tilde{\sigma}_{i_{k''}} + \alpha_{\mathcal{J}} \tilde{\sigma}_{j_{k'}} \otimes \tilde{\sigma}_{j_{k''}}) \right)_{\mathcal{E},\mathcal{D}} |\mathcal{E}\rangle \otimes |\mathcal{D}\rangle. \tag{46}$$

Then, we analyze the entanglement of several subsystems in each case by taking the partial trace with respect to its complement. Calculations are direct. At the end, the association rules $0 \leftrightarrow i$ and $j \leftrightarrow k$ should be applied to explicitly denote the viable relations between $\mathcal{I}$ and $\mathcal{J}$, and to reduce some traces on parts k', k'' in (45) and (46). In Table 4 we report the generalized bipartite concurrence for pure states [36]:

$$\mathcal{C}^2(\mathrm{Tr}^S(\rho_{\mathcal{I}\mathcal{J}})) = 2(1 - \mathrm{Tr}^S(\rho^2_{\mathcal{I}\mathcal{J}})), \tag{47}$$

where, j is assumed as the direction label of the interaction involved. If $m = \min(m_1, m_2)$, where m_1, m_2 are the Hilbert space dimensions of each subsystem, then this measure changes smoothly from 0 for separable states to $2(m-1)/m$ for maximally entangled states. Note that we take $\tilde{\sigma}_i \equiv e^{i\phi_i}\sigma_i$, although it is only relevant for σ_2. With this distinction, we introduce $\phi' = \phi + \phi_{i_k'} - \phi_{j_k'}$.

Table 4. Bipartite concurrence $\mathcal{C}^2(\mathrm{Tr}^S(\rho_{\mathcal{I}\mathcal{J}}))$ for several subsystems in the $SU(2)$ mixing of some pairs of GBS basis states.

	Case	S	$\mathcal{C}^2(\mathrm{Tr}^S(\rho_{\mathcal{I}\mathcal{J}}))$
(a)	$\left|\phi^1_{\mathcal{I}\mathcal{J}}\right\rangle$	$[s \notin \{k', k'+d\}]$	1
(b)	$\left|\phi^1_{\mathcal{I}\mathcal{J}}\right\rangle$	$[s \in \{k', k'+d\}]$	$1 - \sin^2\theta(\cos\phi'\delta_{0,i_{k'}\cdot j_{k'}} + (-1)^{\epsilon_{i_{k'}j_{k'}}}(1-\delta_{0,i_{k'}\cdot j_{k'}})\sin\phi')^2$
(c)	$\left|\phi^1_{\mathcal{I}\mathcal{J}}\right\rangle$	$[k', k'+d]$	0
(d)	$\left|\phi^2_{\mathcal{I}\mathcal{J}}\right\rangle$	$[k', k'+d]$	$\sin^2\theta$
(e)	$\left|\phi^2_{\mathcal{I}\mathcal{J}}\right\rangle$	$[k', k'']$	$\frac{3}{2} - \frac{1}{2}\sin^2\theta(\cos^2\phi'\delta_{i_{k'}j_{k'}}\delta_{i_{k''}j_{k''}} + \sin^2\phi'(1-\delta_{i_{k'}j_{k'}}\delta_{i_{k''}j_{k''}}))$

Table 4 includes some obvious results for "local" interactions on single parts (H_I) or on correspondent pairs (H_{III}): (a) any part is maximally entangled with respect to the remaining system (through its correspondent pair) if there are currently no active local or non-local crossed interactions in H_I and H_{III}, respectively, so $\mathcal{C}^2(\mathrm{Tr}^S(\rho_{\mathcal{I}\mathcal{J}})) = 1$; (b) nevertheless, if these local or non-local crossed interactions act on the correspondent pair, each part of it can become separable or partially entangled to the remaining system; and (c) any correspondent pair (as a subsystem) is separable from the remaining system in any GBS basis state, so $\mathcal{C}^2(\mathrm{Tr}^S(\rho_{\mathcal{I}\mathcal{J}})) = 0$. Note that in the cases (b) and (c) that the subsystem comprises two parts $[k', k'+d]$ being compared with the remaining system, so the Hilbert space dimension is four ($m = 4$). Similarly, the most important results here: (d) shows how interactions between non-correspondent parts (crossed or non-crossed) affect the original separability of each correspondent pair with respect to the remaining system, letting it become entangled with the remaining system. Finally, (e) exhibits the change of entanglement between non-correspondent parts. They are clearly originally entangled with their respective pair outside of the subsystem, but that entanglement becomes reduced ($\mathcal{C}^2(\mathrm{Tr}^S(\rho_{\mathcal{I}\mathcal{J}})) \leq 3/2$) due to the non-local interactions.

6.4.3. Generating Larger Maximal Entangled Systems

The generation of extended entanglement can be shown with a couple of introductory examples [27]. If $\left|\beta_{ij}\right\rangle = \left|\Psi_{2i+i\oplus j}\right\rangle$ are the GBS basis elements for $d = 1$ corresponding to the Bell states [22], then considering the $\left|GHZ\right\rangle$ and $\left|W\right\rangle$ states of size $2d$ expressed in the GBS basis:

$$\left|GHZ\right\rangle^{2d} = \frac{1}{\sqrt{2}}\sum_{i=0}^{1}\bigotimes_{j=1}^{d}\left|i,i\right\rangle_{j} = \frac{1}{2^{\frac{d+1}{2}}}\sum_{i=0}^{1}\bigotimes_{j=1}^{d}\left(\left|\Psi_0\right\rangle_j + (-1)^i\left|\Psi_3\right\rangle_j\right), \tag{48}$$

$$\left|W\right\rangle^{2d} = \frac{1}{\sqrt{2d}}\sum_{i=1}^{2d}\bigotimes_{j=1}^{d}\left|\delta_{i,2j-1},\delta_{i,2j}\right\rangle_{j} = \frac{d^{-\frac{1}{2}}}{2^{\frac{d-1}{2}}}\sum_{i=1}^{d}\bigotimes_{\substack{j=1\\j\neq i}}^{d}\left(\left|\Psi_0\right\rangle_j + \left|\Psi_3\right\rangle_j\right)\otimes\left|\Psi_1\right\rangle_i, \tag{49}$$

where j sums over correspondent pairs. Note that we are alternating the notation in the kets by convenience: $\left|\Psi_k\right\rangle_j$ is the Bell state $\left|\Psi_k\right\rangle$ on the jth correspondent pair, while $\left|\Psi_{i_1,i_2,\ldots,i_d}\right\rangle = \left|\Psi_\mathcal{I}\right\rangle$ is the $\mathcal{I} = 4^{d-1}i_d + \ldots + 4i_2 + i_1$ element in the GBS basis. For $d = 2$, they are simply:

$$\left|GHZ\right\rangle^4 = \frac{1}{\sqrt{2}}\left(\left|\Psi_{0,0}\right\rangle + \left|\Psi_{3,3}\right\rangle\right) = \frac{1}{\sqrt{2}}\sum_{\mathcal{I}\in\{0,15\}}\left|\Psi_\mathcal{I}\right\rangle, \tag{50}$$

$$\left|W\right\rangle^4 = \frac{1}{2}\left(\left|\Psi_{1,0}\right\rangle + \left|\Psi_{0,1}\right\rangle + \left|\Psi_{3,1}\right\rangle + \left|\Psi_{1,3}\right\rangle\right) = \frac{1}{2}\sum_{\mathcal{I}\in\{1,4,7,13\}}\left|\Psi_\mathcal{I}\right\rangle. \tag{51}$$

Then, we can depart from the basic state $\left|0000\right\rangle = \frac{1}{2}\left(\left|\Psi_0\right\rangle_1 + \left|\Psi_3\right\rangle_1\right)\otimes\left(\left|\Psi_0\right\rangle_2 + \left|\Psi_3\right\rangle_2\right)$ for $d = 2$. We arrive at the $\left|GHZ\right\rangle$ by applying the following operations (as before, the interaction Hamiltonian is indicated in the upper position, while the operation is written below):

$$
\begin{aligned}
\left|0000\right\rangle &\xrightarrow[H_{0,3}^{0,0,+}\oplus H_{12,15}^{0,0,+}]{H_I^{(3,1)}} \frac{1}{\sqrt{2}}\left|\Psi_0\right\rangle_1\otimes\left(\left|\Psi_0\right\rangle_2 + \left|\Psi_3\right\rangle_2\right),\\[4pt]
&\xrightarrow[H_{0,12}^{0,0,+}]{H_I^{(3,2)}} \left|\Psi_0\right\rangle_1\otimes\left|\Psi_0\right\rangle_2 = \left|\Psi_{0,0}\right\rangle,\\[4pt]
&\xrightarrow[H_{0,15}^{0,0,+}]{H_{IIa}^{(3,1,2)}} \frac{1}{\sqrt{2}}\left(\left|\Psi_{0,0}\right\rangle + \left|\Psi_{3,3}\right\rangle\right) = \frac{1}{\sqrt{2}}\left(\left|\Psi_0\right\rangle + \left|\Psi_{15}\right\rangle\right) = \left|GHZ\right\rangle^4.
\end{aligned}
\tag{52}
$$

The first operation requires action on two sets of GBS basis states. They are of the same form, so they are easily achieved in terms of prescriptions for $H_{\mathcal{I},\mathcal{I}'}^{m,c,s_\epsilon}$. Note that no more specifications are needed in complementary blocks. They are free because their effect will work on states that are not included. Similarly, for example:

$$
\begin{aligned}
\left|GHZ\right\rangle^4 &\xrightarrow[I_{0,10}^0\oplus E_{5,15}^{0,0}]{H_{IIa}^{(2,1,2)}} \frac{1}{\sqrt{2}}\left(\left|\Psi_{0,0}\right\rangle + \left|\Psi_{1,1}\right\rangle\right),\\[4pt]
&\xrightarrow[E_{0,4}^{0,0}\oplus E_{1,5}^{0,0}]{H_I^{(1,2)}} \frac{1}{\sqrt{2}}\left(\left|\Psi_0\right\rangle_1\otimes\left|\Psi_1\right\rangle_2 + \left|\Psi_1\right\rangle_1\otimes\left|\Psi_0\right\rangle_2\right),\\[4pt]
&\xrightarrow[H_{4,7}^{0,0,+}\oplus I_{1,2}^{2p}]{H_I^{(3,1)}} \frac{1}{\sqrt{2}}\left(\frac{1}{\sqrt{2}}\left(\left|\Psi_0\right\rangle_1 + \left|\Psi_3\right\rangle_1\right)\otimes\left|\Psi_1\right\rangle_2 + \left|\Psi_1\right\rangle_1\otimes\left|\Psi_0\right\rangle_2\right),\\[4pt]
&\xrightarrow[I_{4,8}^{2q}\oplus I_{7,11}^{2r}\oplus H_{1,13}^{0,0,+}]{H_I^{(3,2)}} \frac{1}{2}\left(\left(\left|\Psi_0\right\rangle_1 + \left|\Psi_3\right\rangle_1\right)\otimes\left|\Psi_1\right\rangle_2 + \left|\Psi_1\right\rangle_1\otimes\left(\left|\Psi_0\right\rangle_2 + \left|\Psi_3\right\rangle_2\right)\right)\\[4pt]
&= \frac{1}{2}\left(\left|\Psi_4\right\rangle + \left|\Psi_7\right\rangle + \left|\Psi_1\right\rangle + \left|\Psi_{13}\right\rangle\right) = \left|W\right\rangle^4,
\end{aligned}
\tag{53}
$$

where $p, q, r \in \mathbb{Z}$. In the last operations, the block independence discussed in the previous section was applied to justify the construction of some simultaneous operations.

6.4.4. Recursive Generation of Larger Maximal Entangled Systems

In the previous subsection, we obtained the larger maximal entangled states $|GHZ\rangle^4$ and $|W\rangle^4$ departing from the more basic states such as $|0000\rangle$. The enlargement of entangled states can be stated in a more impressive way as recursive processes. In each case, these processes are based on the control of the parameters involved and the independence among block types generated in each interaction.

Thus, the process shown in Figure 8 combines some of the operations depicted previously to develop $|GHZ\rangle^{2(d+1)}$ departing from $|GHZ\rangle^{2d}$, stating a procedure to get larger versions of these maximal entangled states. The first step begins by using the state $|\Psi_0\rangle_{d+1} \otimes |GHZ\rangle^{2d}$. Then, a local operation is applied on each pair in the original state $k = 1, 2, ..., d$ to reduce the factors $(|\Psi_0\rangle_k + |\Psi_3\rangle_k)$ and $(|\Psi_0\rangle_j - |\Psi_3\rangle_j)$ in (48) into $|\Psi_0\rangle_j$ and $|\Psi_3\rangle_j$, respectively. Then, we exchange the indexes $30 \leftrightarrow 21$ for the non-correspondent pairs k' and $d + 1$ with a non-local operation. This transformation is followed by a couple of local operations changing the indexes $2 \leftrightarrow 3$ for the pair k' and $1 \leftrightarrow 3$ for the pair $d + 1$ (which adds a factor i). In this last case, we transform the index 0 by itself, but adding the factor i. Finally, we revert for $k = 1, 2, ..., d + 1$ the initial transformation between $|\Psi_0\rangle_k, |\Psi_3\rangle_k$ and $(|\Psi_0\rangle_k \pm |\Psi_3\rangle_k)$, respectively. All additional index transformations are settled as the identity. The state obtained will be $i |GHZ\rangle^{2(d+1)}$. It is notable that only one 4-entangling operation between the added pair with another arbitrary pair from the original $2d$-partite system has become necessary in this case. This reflects the low robustness of the genuine entanglement for these states. Considering the expression for $|GHZ\rangle^{2d}$ in (48), the precise prescriptions are:

$$
|\Psi_0\rangle_{d+1} \otimes |GHZ\rangle^{2d} \xrightarrow[\substack{\oplus_{s,s'} H^{0,0,+}_{s,s'} \\ s'-s=3\cdot4^{k-1}, \\ s,s'\in\{3p\leq N|p\in\mathbb{N}\}}]{\substack{H_I^{(3,k)} \\ k=1,2,...,d}} \frac{1}{\sqrt{2}}\left(|\Psi_0\rangle^{d+1} + |\Psi_0\rangle_{d+1} \otimes |\Psi_N\rangle^d\right),
$$

$$
\xrightarrow[\substack{I^0_{0,u}\oplus E^{0,0}_{N,u'} \\ u=4^{k'}-1+4^d \\ u'=N-4^{k'}-1+4^d}]{H_{IIa}^{(1,k',d+1)}} \frac{1}{\sqrt{2}}\left(|\Psi_0\rangle^{d+1} + |\Psi_{u'}\rangle^{d+1}\right),
$$

$$
\xrightarrow[\substack{I^0_{0,4^{k'}-1}\oplus E^{0,0}_{u',u''} \\ u''=u'+4^{k'}-1}]{H_I^{(1,k')}} \frac{1}{\sqrt{2}}\left(|\Psi_0\rangle^{d+1} + |\Psi_{u''}\rangle^{d+1}\right), \tag{54}
$$

$$
\xrightarrow[\substack{I^{\frac{1}{2}}_{0,2\cdot4^d}\oplus E^{0,1}_{u'',N'}}]{H_I^{(2,d+1)}} \frac{i}{\sqrt{2}}\left(|\Psi_0\rangle^{d+1} + |\Psi_{N'}\rangle^{d+1}\right),
$$

$$
\xrightarrow[\substack{\oplus_{s,s'} H^{0,0,+}_{s,s'} \\ s'-s=3\cdot4^{k-1}, \\ s,s'\in\{3p\leq N'|p\in\mathbb{N}\}}]{\substack{H_I^{(3,k)} \\ k=1,2,...,d+1}} i |GHZ\rangle^{2(d+1)},
$$

where $|\Psi_{\mathcal{I}}\rangle^n = |\Psi_{i_1}\rangle_1 \otimes |\Psi_{i_2}\rangle_2 \otimes ... \otimes |\Psi_{i_n}\rangle_n$. In addition, $N = 4^d - 1$ and $N' = 4^{d+1} - 1$. Note that the first and last operations are actually a set of operations for $k = 1, 2, ..., d$ and $k = 1, 2, ..., d + 1$ through several correspondent pairs. They exploit the Hadamard-like block operations for $H_I^{(3,k)}$ to switch first the $|GHZ\rangle^{2d}$ into versions where only the states $|\Psi_0\rangle$ and $|\Psi_3\rangle$ appear. Thus, operations generated with $H_{IIa}^{(1,k',d+1)}$ between two different correspondent pairs are used as exchange operations entangling the added state $|\Psi_0\rangle_{d+1}$. Then, the additional operations $H_I^{(1,k')}$ and $H_I^{(2,d+1)}$ generate a state expressed only in terms of $|\Psi_0\rangle$ and $|\Psi_3\rangle$, to finally be transformed into $|GHZ\rangle^{2(d+1)}$ with the same kind of initial operations.

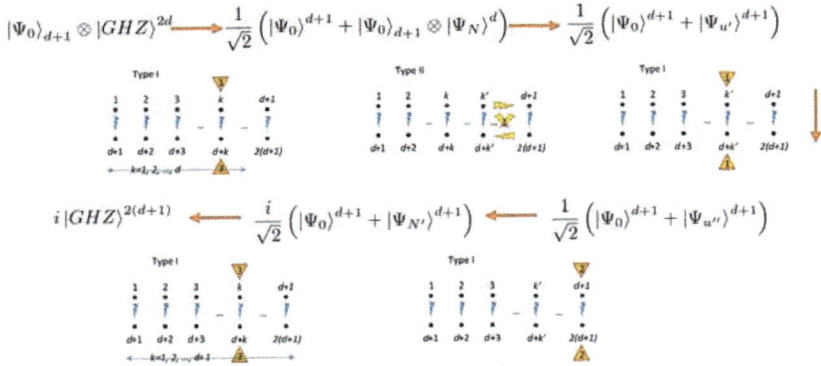

Figure 8. Processes to build recursive enlargement of $|GHZ\rangle$ entangled states.

To obtain the $|W\rangle^{2(d+1)}$ state, we begin with $|\Psi_0\rangle_{d+1} \otimes |W\rangle^{2d}$, then we use the same local transformation to reduce the factors $(|\Psi_0\rangle_k + |\Psi_3\rangle_k)$ in (48) into $|\Psi_0\rangle_k$ for each $k = 1, 2, ..., d$. Then, we apply a sequence of non-local transformations between the pairs $k, d+1$ for $k = 1, 2, ..., d$ to transfer probability between states with indexes $01 \leftrightarrow 10$ there, in such a way as to reach the coefficient $\frac{1}{\sqrt{d+1}}$ in each term. Finally, we revert the initial transformation for $k = 1, 2, ..., d+1$, changing $|\Psi_0\rangle_k$ into $(|\Psi_0\rangle_k + |\Psi_3\rangle_k)$. The final result is $|W\rangle^{2(d+1)}$, as is shown in Figure 9. Note how the entangling operations need to go through the overall pairs. It reflects the robustness of genuine entanglement in these states. By considering the expression for $|W\rangle^{2d}$ in (48), the following process gives the prescriptions to reach $|W\rangle^{2(d+1)}$ from $|W\rangle^{2d}$:

$$|\Psi_0\rangle_{d+1} \otimes |W\rangle^{2d} \xrightarrow[\substack{\oplus_{s,s'} I^0_{s,s'} \oplus_{s'',s'''} H^{0,0,+}_{s'',s'''} \\ s'-s=4^{k-1}, \\ s-4^{k-1},s'-2\cdot4^{k-1}\in\{3p\leq N|p\in\mathbb{N}\} \\ s'''-s''=3\cdot4^{k-1}, \\ s''-4^{i-1},s'''-4^{i-1}\in\{3p\leq N|p\in\mathbb{N}\}, \\ k\neq i\in\{1,2,...,d\}}]{\substack{H^{(3,k)}_i \\ k=1,2,...,d}} \frac{|\Psi_0\rangle_{d+1}}{\sqrt{d}} \otimes \sum_{i=1}^{d} |\Psi_{4i-1}\rangle^d,$$

$$\xrightarrow[\substack{\oplus_{u,u'} I^0_{u,u'} \oplus \mathbb{H}^0_0(\delta_k, \epsilon_k)_{4^{k-1},4^d} \\ u=4^{i-1},u'=u+4^{k-1}+4^d \\ k\neq i\in\{1,2,...,d\}}]{\substack{H^{(1,k,d+1)}_{IIa} \\ k=1,2,...,d}} \frac{1}{\sqrt{d+1}} \sum_{i=1}^{d+1} \bigotimes_{\substack{j=1 \\ j\neq i}}^{d+1} |\Psi_0\rangle_j \otimes |\Psi_1\rangle_i, \qquad (55)$$

$$\xrightarrow[\substack{\oplus_{s,s'} I^0_{s,s'} \oplus_{s'',s'''} H^{0,0,+}_{s'',s'''} \\ s'-s=4^{k-1}, \\ s-4^{k-1},s'-2\cdot4^{k-1}\in\{3p\leq N'|p\in\mathbb{N}\} \\ s'''-s''=3\cdot4^{k-1}, \\ s''-4^{i-1},s'''-4^{i-1}\in\{3p\leq N'|p\in\mathbb{N}\}, \\ k\neq i\in\{1,2,...,d+1\}}]{\substack{H^{(3,k)}_i \\ k=1,2,...,d,d+1}} |W\rangle^{2(d+1)}.$$

As before, Hadamard-like block operations for $H^{(3,k)}_i$ allow the states $|W\rangle^{2d}$ and $|W\rangle^{2(d+1)}$ to be switched, at the beginning and at the end, in terms of $|\Psi_0\rangle$ and $|\Psi_1\rangle$. The remarkable set of operations are obtained with $H_{IIa}^{(1,k,d+1)}$ to entangle the added state $|\Psi_0\rangle_{d+1}$ through the operations (38), which progressively transfer the probability to the state $|\Psi_{4d}\rangle$, completing a state easily transformed into $|W\rangle^{2(d+1)}$ with the final set of operations. Additional exchange in the indexes is settled in the identity. The adequate set of ϵ_k values for each step of operations should fulfill the $d+1$ equations:

$$
\begin{aligned}
g_0 &\equiv 0, \\
\sqrt{d} &= \sqrt{d+1}(\epsilon_j + \delta_j g_{j-1}), \quad j = 1,2,...,d, \\
g_j &= \delta_j - \epsilon_j g_{j-1}, \quad j = 1,2,..,d-1, \\
\sqrt{d} &= \sqrt{d+1}(\delta_d - \epsilon_d g_{d-1}).
\end{aligned}
\tag{56}
$$

These equations can be solved numerically for any d. Figure 10 shows the $-\log_{10}\epsilon_i$ solutions for $d = 1,2,...,60$ by taking $\epsilon_i, \delta_i > 0$. Note that ϵ_i drops rapidly to zero when d and i grow.

$$
|\Psi_0\rangle_{d+1} \otimes |W\rangle^{2d} \longrightarrow \frac{|\Psi_0\rangle_{d+1}}{\sqrt{d}} \otimes \sum_{i=1}^{d}|\Psi_{4^{i-1}}\rangle^d \longrightarrow \frac{1}{\sqrt{d+1}}\sum_{i=1}^{d+1}\bigotimes_{\substack{j=1\\j\neq i}}^{d+1}|\Psi_0\rangle_j \otimes |\Psi_1\rangle_i \longrightarrow |W\rangle^{2(d+1)}
$$

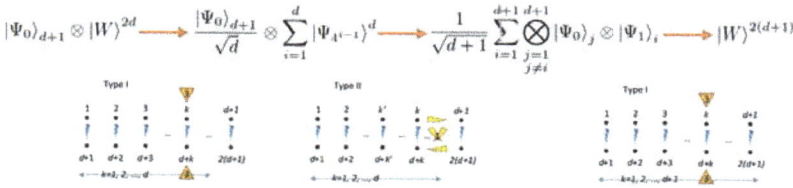

Figure 9. Processes to build recursive enlargement of $|W\rangle$ entangled states.

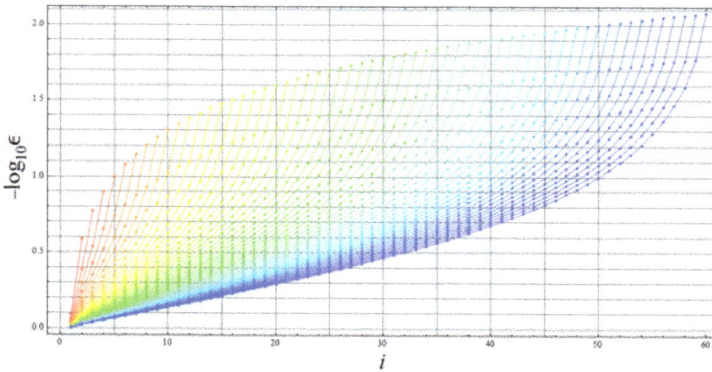

Figure 10. Solutions for ϵ_i in $\mathbb{H}_0^0(\delta_i,\epsilon_i)_{\mathcal{I},\mathcal{I}'}$ involved in the enlargement of $|W\rangle^d$ into $|W\rangle^{d+1}$ for values of $d \in \{2,...,60\}$.

6.5. Multipartite Entanglement and General States

In a previous subsection we described how to generate extended superposition using type I interactions. However, that process does not reach genuine entangled states. The use of type IIa, IIb interactions is mandatory to extend the entanglement as a set of operations involving elements of two pairs. Nevertheless, it is clear that many operations and combinations are necessary and possible.

For instance, by considering the permutation i,j,k from $1,2,3$ and departing from the state $|\Psi_0\rangle^{2d}$, the process to reach an entangled state based on a complete combination from the basis elements for two correspondent pairs is as follows (note that the process is not unique). First, we apply a $2-$local operation on the pair s and direction i followed by another on the pair s' in the direction j. A linear combination from four basis elements is obtained. Then, we apply a $4-$local operation in the direction

k and for pairs s, s', obtaining a state of eight terms. Finally, we again apply a $2-$local operation on pair s in the direction j. At the end, we obtain the desired state of sixteen terms with the pairs s, s', genuinely entangled as it was seen in Table 4.

$$
|\Psi_0\rangle^{2d} \xrightarrow[O_I^{(i,\{s\})}]{H_I^{(i,s)}} \sum_{t \in \{0,i\}} \alpha_{t,0}^s |\Psi_{0,\dots,t,\dots,0,\dots,0}\rangle \xrightarrow[O_I^{(j,\{s'\})}]{H_I^{(j,s')}} \sum_{\substack{t \in \{0,i\} \\ t' \in \{0,j\}}} \alpha_{t,0}^s \alpha_{t',0}^{s'} |\Psi_{0,\dots,t,\dots,t',\dots,0}\rangle ,
$$

$$
\xrightarrow[O_{IIa}^{(k,\{s,s'\})}]{H_{IIa}^{(k,s,s')}} \sum_{\substack{\epsilon,\epsilon' \in C_4 \\ t \in \{0,i\} \\ t' \in \{0,j\}}} \alpha_{t,0}^s \alpha_{t',0}^{s'} \beta_{\epsilon,\epsilon';t,t'}^{s,s'} \delta_{t,p_{s,k}^{t,\epsilon}} \delta_{t',p_{s',k}^{t',\epsilon'}} |\Psi_{0,\dots,\epsilon,\dots,\epsilon',\dots,0}\rangle ,
$$

$$(57)$$

$$
\xrightarrow[O_I^{(j,\{s\})}]{H_I^{(j,s)}} \sum_{\substack{\chi,\epsilon,\epsilon' \in C_4 \\ t \in \{0,i\} \\ t' \in \{0,j\}}} \alpha_{t,0}^s \alpha_{t',0}^{s'} \beta_{\epsilon,\epsilon';t,t'}^{s,s'} \alpha_{\chi,\epsilon}^s \delta_{t,p_{s,k}^{t,\epsilon}} \delta_{t',p_{s',k}^{t',\epsilon'}} \delta_{\epsilon,p_{s,j}^{t,\chi}} |\Psi_{0,\dots,\chi,\dots,\epsilon',\dots,0}\rangle ,
$$

where $C_4 = \{0,1,2,3\}$ and $p_{s,j}^{t,\epsilon}$ is the extension of the inverse exchange rule presented before $p_s(\epsilon)$, but specifying the rule j as a function of the direction of the interaction involved. The script $t \in \{0,i\}$ is a label specifying each possible inverse. This means that if j is the characteristic direction of the interaction, then: $p_{s,j}^{0,i} = k = p_{s,j}^{0,k}$, $p_{s,j}^{0,0} = 0 = p_{s,j}^{0,j}$ and $p_{s,j}^{i,i} = i = p_{s,j}^{i,k}$, $p_{s,j}^{i,0} = j = p_{s,j}^{i,j}$. This single process could be improved using additional interactions to grow the spectrum of coefficients $\alpha_{\beta,\alpha}^s$, $\beta_{\beta,\beta';\alpha,\alpha'}^{s,s'}$ in order to have a wider coverage of $SU(4)$. In addition, it is clear the last process (or another alternative) should be repeated, varying one or two pairs in order to generate more complex entanglement. The question about how to generate a specific state under this procedure or to generate certain kind or level of entanglement is clearly open, mainly due to the poorly understood complexity to measure this property for large states in general.

7. Conclusions

Quantum gate array computation is based on the transformation of quantum states under certain universal operations. These operations are used to manipulate the information settled on quantum systems to simulate or reproduce computer processing, and normally use separable states as primary resources. Quantum systems involved—light or matter—are manipulated around entanglement generation in this kind of processing. Then, commonly involved interactions are non-local, implying that their parts become entangled when they are being manipulated. In the process, several slightly differentiated interactions are applied, each one with a different set of eigenvalues. This does not allow a common grammar to be set through the entire quantum information processing problem.

$SU(2)$ decomposition provides a procedure not only to reduce control in the quantum manipulation states. Together, it provides a common language to address the evolution through several kinds of similar interactions in order to manage a wider processing. Upon the selection of a compatible basis, it allows the recovery of two-state processing despite the inclusion of the necessary entangling interactions. Although we developed the procedure for certain types of well-known interactions (i.e., Heisenberg–Ising and Dzyaloshinskii–Moriya), the process can be extended to other interactions and architectures (the arrangement of qubits under interaction) by the adequate selection of the basis on which dynamics should be expressed conveniently. In addition, it is advised that other configurations based on qudits are possible using alternative group decompositions to $SU(2^{2d})$ and $SU(2)$. Finally, the development only proposes the change of quantum information grammar being used as function of the physical system in the deployment, preserving their applicability for most quantum information proposals in the literature.

Some applications of $SU(2)$ decomposition are foreseen. It can be exploited in the quantum control of larger systems in which control schemes are not as well-developed as those of $SU(2)$ dynamics.

The previously established decomposition allows the establishment of exact control when blocks are reduced to the standard forms I, NOT, H, etc. The success of such strategy for exact control depends on the number of free parameters involved, which can be reached using a sequence of pulses instead of a single one, or otherwise time-dependent controlled parameters in the Hamiltonian, although the block structure is conserved. Similarly, optimal control in terms of energy or time can be achieved when procedures such as those in [9,10] are adapted to each block in the depicted structure. More ambitious ideas about the control of quantum processing such as the use of traveling waves, ion traps, resonant cavities, or superconducting circuits [37–40] could be adapted to the architectures presented here.

Note that the selectivity of pairing in the blocks is related to the arisen non-diagonal elements (i.e., with the interactions generating diagonal-off entries in all cases). This approach to quantum evolution will allow analytical control of the flow of quantum information in different adaptive geometrical arrangements. The use of more feasible external fields (other than stepwise fields) is compulsory, which is completely compatible with the current $SU(2)$ reduction scheme [41].

In a related but not necessarily equivalent direction, selective block decomposition could be useful for unitary factorization in quantum gate design (e.g., that developed for the $SU(4)$ case [26]), particularly for large dedicated gates involving the processing of several qubits. A current challenge in the mathematical arena is solving how to express certain $SU(2^{2d})$ matrices as a finite product in $U(1)^{2^{2d-1}-1} \times SU(2)^{2^{2d-1}}$, such as those developed here.

Finally, other applications in quantum processing could be engineered for multichannel quantum information storage, using certain subspaces to store differentiated information which could be processed simultaneously in other subspaces (e.g., in quantum image processing or quantum machine learning). Additional research should be developed to adapt this procedure to specific gate operations, and the translation of the most common algorithms into equivalent ones based on entangled resources like those shown here.

Funding: This research received no external funding.

Acknowledgments: I gratefully acknowledge the support from Escuela de Ingeniería y Ciencias of Tecnológico de Monterrey in the development of this research work.

Conflicts of Interest: The author declares no conflict of interest.

Appendix A

The Appendix is divided into four parts to develop a more detailed understanding of some critical aspects in the paper. The first is the motivation of the Hamiltonian used here, which is expressed in terms of Pauli operators (or Pauli matrices) together with the identity. Because another central aspect is the terminology around the group theory, the second Appendix briefly explains some terms and developments used in the paper, always centered in the *special unitary groups*, $SU(n)$. Special attention is given to the concept of the *groups product*, which is central in the paper. The third Appendix explains the GBS basis developed by Sych and Leuchs [22]—a set of quantum states with partial entanglement setting a basis that is useful for our development. Because this paper contains sections which may make it difficult to understand the generality of the proposal for larger values of d, the fourth Appendix presents the two more basic examples: $d = 1$, which has already been indirectly presented in the literature [11]; and $d = 2$, which comprises aspects not encountered with $d = 1$, while they are present for the $d > 1$ cases.

Appendix A.1. Generic Hamiltonian Expressed in Terms of Pauli Operators

The Hamiltonian for the interaction between a magnetic object and an external magnetic field is $-\vec{\mu} \cdot \vec{B}$, where $\vec{\mu}$ is the dipole moment of the object. For quantum particles, this dipole moment is precisely the spin, commonly expressed in terms of the Pauli operators $\vec{\sigma} = (\sigma_x, \sigma_y, \sigma_z)$ as $\vec{\mu} = \frac{ge}{2m}\vec{s} = \frac{ge\hbar}{4m}\vec{\sigma}$. Thus, the interaction reads $H = -\vec{\sigma} \cdot \vec{B}$ by absorbing the physical constants $\frac{\hbar}{2}$ in $\vec{B}$.

For quantum magnetic systems, the most common interaction between two level systems is the Heisenberg–Ising interaction. This interaction is a low-order approximation for the far-field interaction between two magnetic dipoles in terms of the spin particles: $\vec{s_1} \cdot \mathbf{J} \cdot \vec{s_2}$ or $\vec{\sigma_1} \cdot \mathbf{J} \cdot \vec{\sigma_2}$, when the spins $\vec{s_i} = \frac{\hbar}{2}\vec{\sigma_i}$ are expressed in terms of Pauli operators by absorbing the factors $\frac{\hbar}{2}$ in $\mathbf{J}$, which is generally a tensor. When $\mathbf{J}$ becomes diagonal, we get the anisotropic Heisenberg–Ising interaction. Moreover, if those diagonal elements become equal, the interaction becomes isotropic.

In the context of this work, another kind of interaction appears—the Dzyaloshinskii–Moriya interaction, which is a contribution to the total magnetic exchange interaction between two neighboring magnetic spins [24,25]: $H = \vec{D} \cdot (\vec{\sigma_1} \times \vec{\sigma_2})$. $\vec{D}$ is a vector expressed in terms of the sources' orientation. Clearly, this Hamiltonian will contain terms such as $D_1\sigma_{1_y} \otimes \sigma_{2_z}, D_2\sigma_{1_z} \otimes \sigma_{2_x}, D_3\sigma_{1_x} \otimes \sigma_{2_y}$, where the script denotes the part and the subscript denotes the component (tensor product symbol $\otimes$ is introduced to remark that the product is between the spins of different quantum objects).

It is clear from the previous examples that the interactions contain terms involving one or two Pauli operators from the different physical parts. Although most terms with two spins appear there because interactions directly occur between a pair of physical objects, the motivation to include all classes of products of Pauli operators in the Hamiltonian (1) is to consider the most extensive types of interaction Notably, in the development of this work, precisely the previous interactions set a special kind of interactions, making the $SU(2)$ decomposition possible Particularly we should note that each term in the Heisenberg–Ising and Dzyaloshinskii–Moriya interactions are only able to generate entangling operations between the pair of objects involved. However, extended entanglement could be generated by including many of those interactions between other pairs, as in the case of Ising chains. A conclusion from this paper is that non-physical terms containing more than two spin factors in the interaction could automatically generate more extended and inclusively genuine entanglement. For instance, for $2d$ qubits, one term containing $2d$ factors in one term of the Hamiltonian $\sigma_{1_z} \otimes \sigma_{2_z} \otimes ...\sigma_{2d_z}$, which can generate genuine entanglement in fewer steps than are necessary in Section 6. Note also that powers or additional factors for each operator are not necessary because of their algebraic properties (any product of them for each part can be reduced to only one operator until unitary factors).

Although these examples are for magnetic systems, these kinds of Hamiltonians are not exclusive to those systems. Thus, for instance, the dipole interaction for a two-level system (an atom or ion restricted to excitation between two energy levels) in a radiation trap, particles in a double-well potential, etc., also have Hamiltonians expressed in terms of the Pauli matrices because they are the basis of $SU(2)$ dynamics, common for all two-level systems (see A.2). Finally, there is a mathematical reason for the form of Hamiltonian (1). For all two-level quantum systems, dynamics are ruled by transformations given by elements of the *unitary group of order* 2, $U(2)$ (see Appendix A.2), as solutions from the Schödinger equation for the evolution operator. In group theory, those elements can be depicted as the exponential of the *generators* of the group, which are precisely the Pauli operators defining an associated Lie algebra: $\exp(i\sum_{k\in 0,1,2,3}\alpha_k\sigma_k)$ (see Appendix A.2). For composed systems, the set of generators (see Appendix A.2) and the basis elements (see Appendix A.3) for their dynamics are precisely the different products between the generators for each part: $\{\otimes_{j=1}^{d}\sigma_{j_k}|k \in 0,1,2,3\}$. Thus, through the Schrödinger equation, we can identify the exponent with the Hamiltonian in (1), thus representing the most general Hamiltonian for the current system composed of $2d$ two-level quantum systems.

Appendix A.2. Group Theory Basics in the Context of the $SU(2)$ Decomposition

In the current Appendix we deliver a minimum of the group theory context to understand some aspects in this work. For a deeper treatment, [42,43] is a modern introductory resource. We begin by remarking on the notion of a *group*. It is a set G of elements $g_i \in G$ together with a defined product operation $\cdot$ fulfilling the properties: (a) *Closure*: $g_1 \cdot g_2 \in G$ for all $g_1, g_2 \in G$ (otherwise with a defined map: $G \times G \to G$); (b) *Associativity*: $g1 \cdot (g_2 \cdot g_3) = (g1 \cdot g_2) \cdot g_3$ for all $g_1, g_2, g_3 \in G$; (c) *Identity element*: there is a unique $e \in G$ such that $g \cdot e = e \cdot g = g$ for all $g \in G$; (d) *Inverse*: for each $g \in G$ there exist $g^{-1} \in G$ such that $g \cdot g^{-1} = g^{-1} \cdot g = e$. If $G' \subset G$ is itself a group, then we say G' is a subgroup of G.

In two-level quantum systems we are interested in states defined in terms of a superposition of two orthonormal states $|\psi_0\rangle, |\psi_1\rangle$: $|\psi\rangle = \alpha_0 |\psi_0\rangle + \alpha_1 |\psi_1\rangle$. Equivalently, we can use the matrix notation:

$$\psi = \begin{pmatrix} \alpha_0 \\ \alpha_1 \end{pmatrix} \tag{A1}$$

to depict such states. Those states have a time evolution $|\psi(t)\rangle$ via the evolution operator $U(t)$ obeying the Schrödinger Equation (2) in terms of the Hamiltonian operator, H. $U(t)$ is an operator acting on the original state to evolve it: $|\psi(t)\rangle = U(t) |\psi\rangle$. It fulfills: (a) the outcome $|\psi(t)\rangle$ belongs to the set depicted by a superposition of $|\psi_0\rangle, |\psi_1\rangle$; and (b) the norm of the new state is preserved: $\langle\psi(t)|\psi(t)\rangle = \langle\psi|U(t)^\dagger U(t)|\psi\rangle = \langle\psi|\psi\rangle$, then $U(t)^\dagger = U(t)^{-1}$. In the context of the Dirac notation, the dual $U(t)^\dagger$ is another operator related with $U(t)$. The evolution operator should clearly fulfill: (i) $U(0) = \mathbb{I}$, the identity operator which leaves $|\psi\rangle$ without change; and (ii) $U(t_2)U(t_1) = U(t_1 + t_2)$. Then, the reader can easily note that the set of operators $U(t)$ for different values of t should form a group with the property $U(t)^\dagger = U(t)^{-1}$. This group is said the *unitary group of order 2*, $U(2)$. Similarly, we can define a $n-$level system, where the evolution operators define the *unitary group of order n*, $U(n)$.

For $U(2)$, because the action of $U(t)$ on $|\psi\rangle$ is again a linear combination of $|\psi_0\rangle, |\psi_1\rangle$, then we know that:

$$U(t) = \sum_{i,j=0}^{1} u_{i,j} |\psi_i\rangle \langle\psi_j| \quad \text{or}: \quad U(t) = \begin{pmatrix} u_{00} & u_{01} \\ u_{10} & u_{11} \end{pmatrix} \tag{A2}$$

(in the following, for simplicity, we will adopt both representations as equivalent). Because of the norm definition for quantum states, we know that:

$$U(t)^\dagger = \sum_{i,j=0}^{1} u_{i,j}^* |\psi_j\rangle \langle\psi_i| \quad \text{or}: \quad U(t)^\dagger = \begin{pmatrix} u_{00}^* & u_{10}^* \\ u_{01}^* & u_{11}^* \end{pmatrix} = \frac{1}{|U(t)|} \begin{pmatrix} u_{11} & -u_{01} \\ -u_{10} & u_{00} \end{pmatrix} = U(t)^{-1}, \text{(A3)}$$

which clearly shows that entries for $U(t)$ should fulfill the restrictions: $u_{11} = u_{00}^* e^{2i\phi}, u_{10} = -u_{01}^* e^{2i\phi}, |u_{00}|^2 + |u_{01}|^2 = 1$ (then, $|U(t)| = e^{2i\phi}$, with $\phi \in \mathbb{R}$ arbitrary):

$$U(t) = \begin{pmatrix} u_{00} & u_{01} \\ -e^{2i\phi}u_{01}^* & e^{2i\phi}u_{00}^* \end{pmatrix} = e^{i\phi} \begin{pmatrix} e^{-i\phi}u_{00} & e^{-i\phi}u_{01} \\ -(e^{-i\phi}u_{01})^* & (e^{-i\phi}u_{00})^* \end{pmatrix} \equiv e^{i\phi}\tilde{U}(t). \tag{A4}$$

We advise in the last structure that both $e^{i\phi}$ and $\tilde{U}(t)$ form groups separately. The set of numbers $e^{i\phi}$ are clearly the $U(1)$ group under the standard multiplication of complex numbers. We skip the demonstration that $\tilde{U}(t)$ with the standard matrix multiplication forms a group, which is trivial for the associativity, identity element, and inverse properties. Demonstration for the closure property is direct. We note that elements in this group fulfill the property $|\tilde{U}(t)| = 1$. This group is said to be the *special unitary group*, $SU(2)$. Normally, in quantum mechanics we select $U(t) \in SU(2)$ because the phase $e^{i\phi}$ is non-physical. For this reason we drop the tilde indistinctly. Because (A4), we say that $U(2)$ is the *direct product* of $U(1)$ and $SU(2)$: $U(2) = U(1) \times SU(2)$ (the reader is advised that the term product is not due to the scalar product underlying in (A4), but instead to a pairing in terms of the Cartesian product of the elements of each group. For a formal definition, consult [42,43]; this concept will be relevant later). $SU(2)$ is clearly a subgroup of $U(2)$.

Another important property of the $SU(2)$ group is that any element of it can be written as a linear combination of the Pauli matrices (this aspect is widely used in the text). This means that they are a basis for matrices in $SU(2)$ (then also for $U(2)$). In fact:

$$\tilde{U}(t) = \Re(e^{-i\phi}u_{00})\sigma_0 + i\Im(e^{-i\phi}u_{01})\sigma_1 + i\Re(e^{-i\phi}u_{01})\sigma_2 + \Im(e^{-i\phi}u_{00})\sigma_3, \qquad (A5)$$

where $\Re$ and $\Im$ are the real and imaginary part functions (in addition, note we are using here indistinctly the notation $\sigma_0 = \mathbb{I}, \sigma_1 = \sigma_x = \mathbb{X}, \sigma_2 = \sigma_y = \mathbb{Y}, \sigma_3 = \sigma_z = \mathbb{Z}$, although in the text they have several meanings as the basis of the different $SU(2)$ elements appearing there). Moreover, in several parts of the text, the following property is used, derived from the algebra fulfilled by Pauli matrices (obtained from the fact: $(\mathbf{n} \cdot \vec{\sigma})^{2s} = 1, s \in \mathbb{Z}$):

$$e^{i\alpha\mathbf{n}\cdot\vec{\sigma}} = \cos\alpha\,\sigma_0 + i\sin\alpha\,\mathbf{n}\cdot\vec{\sigma} = \begin{pmatrix} \cos\alpha + in_3\sin\alpha & i\sin\alpha(n_1 - in_2) \\ i\sin\alpha(n_1 + in_2) & \cos\alpha - in_3\sin\alpha \end{pmatrix}, \qquad (A6)$$

where $\mathbf{n}$ is a unitary vector with real components. From the last expression, it is easy to demonstrate that $|e^{i\alpha\mathbf{n}\cdot\vec{\sigma}}| = 1$, so by comparing with (A4), it is advisable that (A6) is a parametrization for the elements in $SU(2)$. Moreover, from the Baker–Campbell–Hausdorff formula [42,43]:

$$e^{i(\phi\sigma_0 + \alpha\mathbf{n}\cdot\vec{\sigma})} = e^{i\phi}e^{i\alpha\mathbf{n}\cdot\vec{\sigma}}. \qquad (A7)$$

Then, it is said that $\sigma_1, \sigma_2, \sigma_3$ are the *generators* of $SU(2)$, while $\sigma_0, \sigma_1, \sigma_2, \sigma_3$ are the generators of $U(2)$. This fact was mentioned in Appendix A.1 to suggest the generality of Hamiltonian (1), although some steps remain to arrive at the $SU(2^{2d})$ group.

The reader can note that (13) and (29) adjust to those structures, and thus the blocks in the decomposition belong to $U(1) \times SU(2)$. Similar arguments in group theory show that $U(n) = U(1) \times SU(n)$, where $SU(n)$ is the *special unitary group of order n*, the group of unitary matrices $U^{\dagger} = U^{-1}$ with determinant equal to 1, $|U| = 1$. Although there are generators for such groups, the development of the article shows that the combination of $2d$ two-level quantum systems requires evolution matrices in $U(2^{2d})$ (or in $SU(2^{2d})$). As a result of the decomposition, we show that the evolution matrices belonging to such groups form the product group $U(2)^{2^{2d-1}}$ (while the Hilbert space of the quantum states is decomposed into the direct sum of 2^{2d-1} two-level subspaces of dimension 2). Precisely, the elimination of the term $h_{\{00...0\}}$ in the Hamiltonian induces directly in (14) that $U \in SU(2^{2d})$. In addition, the $SU(2)$ decomposition shows in this case that $U \in U(1)^{2^{2d-1}-1} \times SU(2)^{2^{2d-1}}$ (due to the dependence of one $U(1)$-term from the remaining). The reader should consult the formal definition of a *direct product* in [42,43].

Appendix A.3. Generalized Bell States Basis in Context

GBS states (generalized Bell states) as introduced by Sych and Leuchs [22] (16) were expressed in terms of Pauli operators. That original expression is highly convenient for the development of the current work because it allows it to be easily connected with the form of the Hamiltonian (1) in the same terms, allowing the important result to be easily obtained (18). Nevertheless, a more simple expression could be given for the understanding of such states. In fact, it is easy to note that each element in the GBS basis for $2d$ qubits can be written as:

$$\left|\Psi_{\mathcal{I}_4^d}\right\rangle = \bigotimes_{j=1}^{d}\left|\Psi_{\mathcal{I}_{4,j}^d}\right\rangle = |\Psi_{i_1}\rangle \otimes |\Psi_{i_2}\rangle \otimes ... \otimes |\Psi_{i_d}\rangle, \qquad (A8)$$

where $\mathcal{I}_{4,j}^d = i_j$, and $\left|\Psi_{2\gamma+(\gamma\oplus\delta)}\right\rangle = |\beta_{\gamma,\delta}\rangle$ or $|\Psi_i\rangle = \left|\beta_{(\frac{i}{2}\text{mod}2),(i-2(\frac{i}{2}\text{mod}2))\oplus(\frac{i}{2}\text{mod}2)}\right\rangle$, the well-known single Bell states (in the last expressions, $\oplus$ represents the module-2 sum). Thus, each element of the GBS basis is in reality a tensor product of d Bell states identified through their scripts in base-4. These states are 2−separable (meaning the smallest separable subsystems still contains two entangled parts). Thus, when we apply a Hadamard-like operation involving only one script (Type I or III interactions), we are consequently able to convert the involved Bell state into a separable state. When we look at this version of the GBS basis, it is clearer why Type I and III interactions become in entangling or unentagling operations on only one correspondent pair. Both types of operations actually resemble the effect on two-qubits processing with $SU(4)$ operations such as those developed in [11], while the remaining system is not involved. Only the Type II operations provide more extended entangling properties.

Appendix A.4. Illustrative Examples of SU(2) Decomposition

In the following two subsections, we develop examples of the aspect of the evolution operators for the specific cases $d = 1$ and $d = 2$. The latter case is of special importance because it depicts how Type II interactions extend the entanglement, as shown in Section 6.

Appendix A.4.1. Case $d = 1$

This case has been developed in the literature [11,13]. In the current context, only Type I and Type III interactions are possible (because there is only one correspondent pair). The corresponding GBS basis has four elements: $|\Psi_0\rangle, |\Psi_1\rangle, |\Psi_2\rangle, |\Psi_3\rangle$, the Bell states precisely. In the next expressions, we assume a lexicographic order in the components of the basis, so any arrangement of them is supposed (in contrast to how it was considered in (14)). The Hamiltonian H_I contains at the most five terms:

$$H_I = \sum_{m=1}^{3} h_{m,m} \sigma_{1_m} \otimes \sigma_{2_m} + h_{k,0} \sigma_{1_k} + h_{0,k} \sigma_{2_k}, \tag{A9}$$

where k is the direction of the local interaction. Te Hamiltonian H_{III} (with the crossed interaction in the direction k) also contains utmost five terms. If i, j, k is an even permutation of $1, 2, 3$:

$$H_{III} = \sum_{m=1}^{3} h_{m,m} \sigma_{1_m} \otimes \sigma_{2_m} + h_{i,j} \sigma_{1_i} \otimes \sigma_{2_j} + h_{j,i} \sigma_{1_j} \otimes \sigma_{2_i}. \tag{A10}$$

Although it is an special case accepting the combination of the two Hamiltonians (Table 1), we set them separately. Because of the space and complexity, we do not express $U_k(t)$ in terms of the original coefficients in (A9) and (A10). In any case, formulas (31)–(37) are sufficiently efficient to reproduce the entries of each Hamiltonian. Instead, after expressing both Hamiltonians in the GBS basis, they become (as in (28)):

$$H_1 = \begin{pmatrix} h_{11}^1 & h_{12}^1 & 0 & 0 \\ h_{12}^{1*} & h_{22}^1 & 0 & 0 \\ 0 & 0 & h_{11}^2 & h_{12}^2 \\ 0 & 0 & h_{12}^{2*} & h_{22}^2 \end{pmatrix} = \mathbb{S}_{H_{0,1}} \oplus \mathbb{S}_{H_{2,3}} \implies U_1(t) = \mathbb{S}_{U_{0,1}} \oplus \mathbb{S}_{U_{2,3}},$$

$$H_2 = \begin{pmatrix} h_{11}^1 & 0 & h_{12}^1 & 0 \\ 0 & h_{11}^2 & 0 & h_{12}^2 \\ h_{12}^{1*} & 0 & h_{22}^1 & 0 \\ 0 & h_{12}^{2*} & 0 & h_{22}^2 \end{pmatrix} = \mathbb{S}_{H_{0,2}} \oplus \mathbb{S}_{H_{1,3}} \implies U_2(t) = \mathbb{S}_{U_{0,2}} \oplus \mathbb{S}_{U_{1,3}}, \tag{A11}$$

$$H_3 = \begin{pmatrix} h_{11}^1 & 0 & 0 & h_{12}^1 \\ 0 & h_{11}^2 & h_{12}^2 & 0 \\ 0 & h_{12}^{2*} & h_{22}^2 & 0 \\ h_{12}^{1*} & 0 & 0 & h_{22}^1 \end{pmatrix} = \mathbb{S}_{H_{0,3}} \oplus \mathbb{S}_{H_{1,2}} \implies U_3(t) = \mathbb{S}_{U_{0,3}} \oplus \mathbb{S}_{U_{1,2}},$$

where the superscript i in h^i_{mn} points out the consecutive number of blocks and $S_{H_{I,I'}}$ fulfills the syntactic notation followed in (28) and (29). We will exploit this notation in the following section for simplicity, where the matrix notation will become hardly extensive.

Appendix A.4.2. Case $d = 2$

We develop two cases for the case $d = 2$. The first considers the Type I interaction and the second pertains to the Type IIa interaction. The last case involves a different situation not appearing in $d = 1$: the possibility of generating extended entanglement among the four qubits involved in this case.

By considering the four qubits under the Type I interaction with the local interaction terms on the pair $k' = 2$ in the direction j in (30), the Hamiltonian has the form:

$$H_I = \sum_{m=1}^{3} h_{m,0,m,0}\sigma_{1_m} \otimes \sigma_{3_m} + \sum_{m=1}^{3} h_{0,m,0,m}\sigma_{2_m} \otimes \sigma_{4_m} + h_{0,j,0,0}\sigma_{2_j} + h_{0,0,0,j}\sigma_{4_j}. \tag{A12}$$

There are 16 GBS elements in the basis: $|\Psi_{0,0}\rangle, |\Psi_{0,1}\rangle, ..., |\Psi_{0,3}\rangle, |\Psi_{1,0}\rangle, |\Psi_{1,1}\rangle, ..., |\Psi_{3,3}\rangle$. Interaction generates exchanges between the GBS basis elements as follows (forming eight blocks but only two types of them). If i, j, k is a permutation of $1, 2, 3$ with $i < k$, then: (a) $|\Psi_{m,0}\rangle \longleftrightarrow |\Psi_{m,j}\rangle$; (b) $|\Psi_{m,i}\rangle \longleftrightarrow |\Psi_{m,k}\rangle$, with $m = 0, ..., 3$. Due to the extension, we do not write matrix expressions as in the case $d = 1$ (which already settled an illustrative orientation to the reader). Instead, we use the notation of direct sum for block matrices as before. Thus, remembering the scripts are numbers becoming from the base-4 scripts in the GBS basis, $|\Psi_{a,b}\rangle = |\Psi_{\mathcal{I}}\rangle$ with $\mathcal{I} = a + 4b \in 0, 1, ..., 15$, the decomposition for the evolution operator will become for $U_j(t)$:

$$U_j(t) = \left(\bigoplus_{m=0}^{3} S_{U_{m,m+4j}} \right) \oplus \left(\bigoplus_{m=0}^{3} S_{U_{m+4i,m+4k}} \right),$$

then :

$$U_1(t) = S_{U_{0,4}} \oplus S_{U_{1,5}} \oplus S_{U_{2,6}} \oplus S_{U_{3,7}} \oplus S_{U_{8,12}} \oplus S_{U_{9,13}} \oplus S_{U_{10,14}} \oplus S_{U_{11,15}},$$

$$U_2(t) = S_{U_{0,8}} \oplus S_{U_{1,9}} \oplus S_{U_{2,10}} \oplus S_{U_{3,11}} \oplus S_{U_{4,12}} \oplus S_{U_{5,13}} \oplus S_{U_{6,14}} \oplus S_{U_{7,15}},$$

$$U_3(t) = S_{U_{0,12}} \oplus S_{U_{1,13}} \oplus S_{U_{2,14}} \oplus S_{U_{3,15}} \oplus S_{U_{4,8}} \oplus S_{U_{5,9}} \oplus S_{U_{6,10}} \oplus S_{U_{7,11}}. \tag{A13}$$

These exchanges only involve qubits in the same correspondent pair, so they cannot extend the entanglement beyond this pair. Additionally, we remark that the first four blocks have the same form, as do the last four. Here, only two different types of blocks exist.

We develop the case $d = 2$ for a Type IIa interaction involving additional non-local and non-crossed interactions between the pairs $k' = 1$ and $k'' = 2$ (note that the situation will be similar for the cases with $d > 2$). Assuming the interaction in the direction j and in (30), the Hamiltonian becomes:

$$H_{IIa} = \sum_{m=1}^{3} h_{m,0,m,0}\sigma_{1_m} \otimes \sigma_{3_m} + \sum_{m=1}^{3} h_{0,m,0,m}\sigma_{2_m} \otimes \sigma_{4_m} +$$

$$h_{j,j,0,0}\sigma_{1_j} \otimes \sigma_{2_j} + h_{j,0,0,j}\sigma_{1_j} \otimes \sigma_{4_j} + h_{0,j,j,0}\sigma_{2_j} \otimes \sigma_{3_j} + h_{0,0,j,j}\sigma_{3_j} \otimes \sigma_{4_j}. \tag{A14}$$

Then, there exist eight types of exchanges and blocks (i, j, k is a permutation of $1, 2, 3$): (a) $|\Psi_{0,0}\rangle \longleftrightarrow |\Psi_{j,j}\rangle$; (b) $|\Psi_{0,i}\rangle \longleftrightarrow |\Psi_{j,k}\rangle$; (c) $|\Psi_{0,j}\rangle \longleftrightarrow |\Psi_{j,0}\rangle$; (d) $|\Psi_{0,k}\rangle \longleftrightarrow |\Psi_{j,i}\rangle$; (e) $|\Psi_{i,0}\rangle \longleftrightarrow |\Psi_{k,j}\rangle$; (f) $|\Psi_{i,i}\rangle \longleftrightarrow |\Psi_{k,k}\rangle$; (g) $|\Psi_{i,j}\rangle \longleftrightarrow |\Psi_{k,0}\rangle$; (h) $|\Psi_{i,k}\rangle \longleftrightarrow |\Psi_{k,i}\rangle$. In this case, all blocks will become different, but it is not a general situation when d grows. As before $|\Psi_{a,b}\rangle = |\Psi_{\mathcal{I}}\rangle$, with $\mathcal{I} = a + 4b \in 0, 1, ..., 15$. Then, the evolution operator can be written as:

$$U_j(t) = S_{U_{0,j+4j}} \oplus S_{U_{4i,j+4k}} \oplus S_{U_{4j,j}} \oplus S_{U_{4k,j+4i}} \oplus S_{U_{i,k+4j}} \oplus S_{U_{i+4i,k+4k}} \oplus S_{U_{i+4j,k}} \oplus S_{U_{i+4k,k+4i}}, \tag{A15}$$

noting that the exchange involving two scripts implies the generation of entanglement between the two correspondent pairs (i.e., among the four qubits as a whole). In this case, eight blocks are different (but not necessarily independent,—because there are only 11 parameters free, including time t).

These two examples show in detail how the $SU(2)$ decomposition is established. For cases $d > 2$, the situation becomes similar and they are easily understood using the last synthetic notation in terms of direct sums of blocks. It should finally be remarked that formulas (30) and (37) are computationally useful and efficient to connect the original Hamiltonian coefficients with the entries for each block.

References

1. Feynman, R.P. Simulating Physics with Computers. *Int. J. Theor. Phys.* **1982**, *21*, 467–488. [CrossRef]
2. Deutsch, D. Quantum theory, the Church-Turing principle and the universal quantum computer. *Proc. R. Soc. Lond. A* **1985**, *400*, 97–117. [CrossRef]
3. Steane, A. Error Correcting Codes in Quantum Theory. *Phys. Rev. Lett.* **1996**, *77*, 793–797. [CrossRef] [PubMed]
4. Bennett, C.H.; Brassard, G. Quantum cryptography: Public key distribution and coin tossing. In Proceedings of the International Conference on Computers, Systems & Signal Processing, Bangalore, India, 9–12 December 1984; IEEE: New York, NY, USA, 1984; pp. 175–179.
5. Ekert, A. Quantum cryptography based on Bell's theorem. *Phys. Rev. Lett.* **1991**, *67*, 661–663. [CrossRef] [PubMed]
6. Hillery, M.; Vladimír, B.; Berthiaume, A. Quantum secret sharing. *Phys. Rev. A* **1999**, *59*, 1829–1834. [CrossRef]
7. Long, G.; Liu, X. Theoretically efficient high-capacity quantum-key-distribution scheme. *Phys. Rev. A* **2002**, *65*, 032302. [CrossRef]
8. Deng, F.; Long, G.; Liu, X. Two-step quantum direct communication protocol using the Einstein-Podolsky-Rosen pair block. *Phys. Rev. A* **2003**, *68*, 042317. [CrossRef]
9. D'Alessandro, D.; Dahleh, M. Optimal control of two-level quantum systems. *IEEE Trans. Autom. Control* **2001**, *46*, 866–876. [CrossRef]
10. Boscain, U.; Mason, P. Time minimal trajectories for a spin $1/2$ particle in a magnetic field. *J. Math. Phys.* **2006**, *47*, 062101. [CrossRef]
11. Delgado, F. Algebraic and group structure for bipartite anisotropic Ising model on a non-local basis. *Int. J. Quantum Inf.* **2015**, *13*, 1550055. [CrossRef]
12. Delgado, F. Generation of non-local evolution loops and exchange operations for quantum control in three dimensional anisotropic Ising model. *ArXiv* **2016**, arXiv:1410.5515.
13. Delgado, F. Stability of Quantum Loops and Exchange Operations in the Construction of Quantum Computation Gates. *J. Phys.* **2016**, *648*, 012024.
14. McConnell, R.; Bruzewicz, C.; Chiaverini, J.; Sage, J. Characterization and Mitigation of Anomalous Motional Heating in Surface-Electrode Ion Traps. In Proceedings of the 46th Annual Meeting of the APS Division of Atomic, Molecular and Optical Physics, Columbus, OH, USA, 8–12 June 2015; Volume 60.
15. Gambetta, J.M.; Jerry, M.; Steffen, M. Building logical qubits in a superconducting quantum computing system. *npj Quantum Inf.* **2017**, *3*, 1–7. [CrossRef]
16. Fubini, A.; Roscilde, T.; Tognetti, V.; Tusa, M.; Verrucchi, P. Reading entanglement in terms of spin-configuration in quantum magnet. *Eur. Phys. J. D* **2006**, *38*, 563–570. [CrossRef]
17. Pfaff, W.; Taminiau, T.; Robledo, L.; Bernien, H.; Matthew, M.; Twitchen, D.; Hanson, R. Demonstration of entanglement-by-measurement of solid-state qubits. *Nat. Phys.* **2013**, *9*, 29–33. [CrossRef]
18. Magazzu, L.; Jamarillo, J.; Talkner, P.; Hanggi, P. Generation and stabilization of Bell states via repeated projective measurements on a driven ancilla qubit. *ArXiv* **2018**, arXiv:1802.04839v1.
19. Nielsen, M.; Chuang, I. *Quantum Computation and Quantum Information*; Cambridge University Press: Cambridge, UK, 2011.
20. Lieb, E.; Schultz, T.; Mattis, D. Two soluble models of an antiferromagnetic chain. *Ann. Phys.* **1961**, *16*, 407–466. [CrossRef]
21. Baxter, R.J. *Exactly Solved Models in Statistical Mechanics*; Academic Press: New York, NY, USA, 1982.
22. Sych, D.; Leuchs, G. A complete basis of generalized Bell states. *New J. Phys.* **2009**, *11*, 013006. [CrossRef]

23. Delgado, F. Modeling the dynamics of multipartite quantum systems created departing from two-level systems using general local and non-local interactions. *J. Phys.* **2017**, *936*, 012070. [CrossRef]
24. Dzyaloshinskii, I. A thermodynamic theory of weak ferromagnetism of antiferromagnetics. *J. Phys. Chem. Solids* **1958**, *4*, 241–255. [CrossRef]
25. Moriya, T. Anisotropic Superexchange Interaction and Weak Ferromagnetism. *Phys. Rev.* **1960**, *120*, 91–98. [CrossRef]
26. Delgado, F. Two-qubit quantum gates construction via unitary factorization. *Quantum Inf. Comput.* **2017**, *17*, 721–746.
27. Delgado, F. Generalized Bell states map physical systems' quantum evolution into a grammar for quantum information processing. *J. Phys.* **2017**, *936*, 012083. [CrossRef]
28. Uhlmann, A. Anti-(conjugate) linearity. *Sci. China Phys. Mech. Astron.* **2016**, *59*, 630301. [CrossRef]
29. Boykin, P.; Mor, T.; Pulver, M.; Roychowdhury, V.; Vatan, F. On universal and fault tolerant quantum computing. *ArXiv* **1999**, arXiv:9906054.
30. Li, C.; Jones, R.; Yin, X. Decomposition of unitary matrices and quantum gates. *Int. J. Quantum Inf.* **2013**, *11*, 1350015. [CrossRef]
31. Delgado, F. Universal Quantum Gates for Quantum Computation on Magnetic Systems Ruled by Heisenberg-Ising Interactions. *J. Phys.* **2017**, *839*, 012014. [CrossRef]
32. Barenco, A.; Bennett, C.; Cleve, R.; DiVincenzo, D.; Margolus, N.; Shor, P.; Sleator, T.; Smolin, J.; Weinfurter, H. Elementary gates for quantum computation. *Phys. Rev. A* **1995**, *52*, 3457. [CrossRef] [PubMed]
33. Liu, Y.; Long, G.; Sun, Y. Analytic one-bit and CNOT gate constructions of general n-qubit controlled gates. *Int. J. Quantum Inf.* **2008**, *6*, 447–462. [CrossRef]
34. Hou, S.; Wang, L.; Yi, X. Realization of quantum gates by Lyapunov control. *Phys. Lett. A* **2014**, *378*, 699–704. [CrossRef]
35. Gurvits, L. Classical complexity and quantum entanglement. *J. Comput. Syst. Sci.* **2004**, *69*, 448–484. [CrossRef]
36. Uhlmann, A. Fidelity and concurrence of conjugated states. *Phys. Rev. A* **2000**, *62*, 032307. [CrossRef]
37. Serikawa, T.; Shiozawa, Y.; Ogawa, H.; Takanashi, N.; Takeda, S.; Yoshikawa, J.; Furusawa, A. Quantum information processing with a travelling wave of light. In Proceedings of the SPIE-OPTO, San Francisco, CA, USA, 19–23 August 2018.
38. Britton, J.; Sawyer, B.; Keith, A.; Wang, J.; Freericks, J.; Uys, H.; Biercuk, M.; Bollinger, J. Engineered two-dimensional Ising interactions in a trapped-ion quantum simulator with hundreds of spins. *Nature* **2012**, *484*, 489–492. [CrossRef] [PubMed]
39. Bohnet, J.; Sawyer, B.; Britton, J.; Wall, M.; Rey, A.; Foss-Feig, M.; Bollinger, J. Quantum spin dynamics and entanglement generation with hundreds of trapped ions. *Science* **2016**, *352*, 1297–1301. [CrossRef] [PubMed]
40. De Sa Neto, O.; de Oliveira, M. Hybrid Qubit gates in circuit QED: A scheme for quantum bit encoding and information processing. *ArXiv* **2011**, arXiv:1110.1355.
41. Delgado, F.; Rodríguez, S. Modeling quantum information dynamics achieved with time-dependent driven fields in the context of universal quantum processing. *ArXiv* **2018**, arXiv:1805.05477.
42. Hall, B. *Lie Groups, Lie Algebras, and Representations: An Elementary Introduction*; Graduate Texts in Mathematics 222; Springer: Cham, Switzerland, 2015.
43. Hall, B. An Elementary Introduction to Groups and Representations. *ArXiv* **2000**, arXiv:quant-ph/0005032.

logo

MDPI

Article

An Information-Theoretic Perspective on the Quantum Bit Commitment Impossibility Theorem

Marius Nagy [1,2,*] and Naya Nagy [3]

[1] College of Computer Engineering and Science, Prince Mohammad Bin Fahd University, Al Khobar 31952, Saudi Arabia
[2] School of Computing, Queen's University, Kingston, ON K7L 2N8, Canada
[3] College of Computer Science and Information Technology, Imam Abdulrahman Bin Faisal University, Dammam 34212, Saudi Arabia; nmnagy@iau.edu.sa
* Correspondence: mnagy@pmu.edu.sa or marius@cs.queensu.ca; Tel.: +966-13-849-9272

Received: 16 January 2018; Accepted: 12 March 2018; Published: 13 March 2018

Abstract: This paper proposes a different approach to pinpoint the causes for which an unconditionally secure quantum bit commitment protocol cannot be realized, beyond the technical details on which the proof of Mayers' no-go theorem is constructed. We have adopted the tools of quantum entropy analysis to investigate the conditions under which the security properties of quantum bit commitment can be circumvented. Our study has revealed that cheating the binding property requires the quantum system acting as the safe to harbor the same amount of uncertainty with respect to both observers (Alice and Bob) as well as the use of entanglement. Our analysis also suggests that the ability to cheat one of the two fundamental properties of bit commitment by any of the two participants depends on how much information is leaked from one side of the system to the other and how much remains hidden from the other participant.

Keywords: quantum information theory; bit commitment; protocol; entropy; entanglement

---

1. Introduction

Bit commitment refers to a cryptographic protocol that can be described informally as follows. In the first phase, Alice decides on a binary value (0 or 1) that she locks in a safe, keeping the key for herself and then hands over the locked safe to Bob. This is referred to as the commit phase of the protocol. Later on, during the decommit phase, Alice reveals her commitment by presenting Bob with the safe key. Bob can now use the key to unlock the safe and learn the value of the bit that Alice has previously committed to.

Any secure implementation of bit commitment must satisfy two crucial properties:

1. Alice should no longer be able to change the value of her commitment, once the safe is in Bob's hands. This requirement is known as the binding property of bit commitment.
2. Bob should not be able to learn the content of the safe before the decommit phase. This requirement is known as the hiding property of bit commitment.

If Alice and Bob employ quantum means to work out the details of a bit commitment procedure, then the resulting protocol falls into the category of quantum bit commitment (QBC). Unconditionally secure QBC is known to be impossible [1], meaning that no protocol can ever be devised such that both binding and hiding properties are guaranteed, if no restriction is placed on the capabilities of the two participants, Alice and Bob. This result is a direct consequence of the Schmidt decomposition theorem for composite systems [2], but the essence of the impossibility theorem for QBC is not easy to grasp and understand intuitively, beyond the technical details outlined in the papers of Mayers [3] and Lo and Chau [4]. That explains why even after the publication of the "no-go" theorem for QBC, quantum cryptographers were still looking for a protocol that would not fall under its scope. Therefore, in the present investigation, we wish to gain further insight into why hiding and binding are mutually exclusive properties for QBC by adopting the perspective and tools of quantum information theory.

The remainder of this paper is organized as follows. The next section summarizes the most important turns and results that have shaped the history of quantum bit commitment. Section 3 defines a general framework for QBC protocols and arrives at a formulation of the hiding property in terms of the entropy accumulated within the quantum safe. In Section 4, we show that entanglement is a necessary condition for cheating the binding property, using the tools of quantum information theory. A detailed procedure of how Alice is able to cheat the binding property is described in Section 5, with a concrete exemplification for the QBC protocol proposed in the BB84 paper [5]. Section 6 extends the analysis of the two security properties of bit commitment to protocols initiated by Bob. Finally, Section 7 concludes the paper with a summary of our findings.

2. Brief History of Quantum Bit Commitment

The feasibility of quantum bit commitment has huge implications for the field of cryptography. Classical cryptography is able to offer solutions (based on bit commitment) to a wide variety of situations classified as discreet decision problems [6]. All these situations share an important characteristic, namely that discretion is vital to achieving agreements. Examples range from negotiating arms treaties to forming business partnerships or organizing mergers.

Classic cryptographic solutions to these applications do exist, but since they involve public-key systems, they are inevitably based on unproven assumptions about the difficulty of factoring large numbers and other related problems. What was expected from quantum cryptography was a totally secure system, guaranteed by the laws of physics, similar to what was already achieved in the case of quantum key distribution [7]. To this end, Claude Crépeau and Joe Kilian have shown how oblivious transfer (or 1-out-of-2 oblivious transfer) can be used as a building block for solving two-party problems requiring discretion [8]. In turn, to provide totally secure quantum oblivious transfer, one would need a secure form of bit commitment. Consequently, much of the research effort in quantum cryptography in the early 1990s was devoted to finding a protocol for quantum bit commitment that is absolutely and provably secure. That result (known as BCJL after the authors' names) was reported in 1993 [9] and became the foundation for numerous applications in quantum cryptography, pertaining to discreet decision making.

The surprise came in 1995 when Dominic Mayers discovered how Alice could cheat in the BCJL bit commitment protocol by using entanglement [1]. Furthermore, Mayers [3] and, independently, Lo and Chau [4] proved that it would be possible for Alice to cheat in any protocol for quantum bit commitment that guarantees the hiding property. An intuitive explanation is that the description of the quantum safe she hands over to Bob must give nothing away about the committed bit inside. Consequently, regardless of the particular bit commitment scheme employed, the quantum states of the safe containing either 0 or 1 must be very similar (if not identical) since otherwise Bob would be able to discern the difference and gain knowledge about the committed bit prematurely. However, the very fact that the two states are virtually the same gives Alice the possibility to keep her options open and postpone her commitment for later on. Although in their 1996 review paper of quantum cryptography, Brassard and Crépeau [10] argued that for the time being the practical implications of the flaw discovered in the quantum bit commitment protocol are minimal, the weakness definitely affected the entire edifice of quantum cryptography built upon quantum bit commitment.

Perhaps the importance of bit commitment for the general field of cryptography or the intuition that the success of quantum key distribution could be replicated for quantum bit commitment still pushed people to look for a solution. Several protocols were proposed that try to restrict the behavior of the cheater in some way so as to obtain a secure bit commitment scheme [10–12]. It turned out that all these protocols were falling under the scope of Mayers' impossibility result. Building on Mayers' work, Spekkens and Rudolph [13] proved that the two fundamental properties of bit commitment, binding and hiding, are mutually exclusive. The more a protocol is hiding, the less it is binding and vice-versa. This led to a general belief that the principles of quantum mechanics alone cannot be used to create an unconditionally secure bit commitment protocol. Therefore, recent advances on the topic either exploit realistic physical assumptions like the dishonest party being limited by "noisy storage" for quantum information [14] or combine the power of Einstein's relativity with quantum theory [15–17]. Yet another direction explored by researchers in the field is the class of "cheat-sensitive" quantum bit commitment protocols. Since the hope of designing an unconditionally secure QBC protocol had to be abandoned, researchers focused instead on protocols in which the probability of detecting a dishonest participant is merely required to be non-zero. Properties of such cheat-sensitive protocols are explored in [18–23].

Ultimately, secure bit commitment using quantum theory alone remains unattainable. Although we know that entanglement and Schmidt's decomposition theorem are key ingredients in this impossibility result, these are technical details that fail to provide a deep and intuitive understanding on why, for example, unconditionally secure quantum key distribution is possible, but quantum bit commitment is not. In the following, we try to shed some light into what the conditions for successful hiding and binding properties are and to describe the complex relationship between these properties, with the help of quantum information theory and quantum entropy.

3. An Information-Theoretic Formulation of the Hiding Property

In a typical QBC framework, Alice encodes some classical information into a quantum system using one out of two possible encoding bases: B_0 if she decides to commit to 0 or B_1 if the commit value is 1. The quantum system represents the safe, while the classical information that Alice keeps secret from Bob plays the role of the key in the description of bit commitment above. In order to prevent Bob from squeezing any information from the quantum safe through some clever measurement, B_0 and B_1 must be complementary bases, such as the normal computational base and the Hadamard base. In terms of a practical implementation, these represent the rectilinear and diagonal polarizations of a photon.

However, to characterize the hiding property formally, we also need to define more precisely Alice's secret key that she encodes in the quantum safe. In general, this key can be any string over a certain alphabet, so without loss of generality, let us assume that the key is a binary string of a certain length n. Then the hiding property is an expression of Bob's total uncertainty over the quantum state of the safe. From his point of view, the safe must be in a complete mixture of all possible bitstrings that Alice might have encoded, regardless of the basis used. This is expressed formally, using density matrices, as follows:

$$\rho_0^{Bob} = \sum_{k_i \in (0|1)^n} \frac{1}{2^n} |B_0(k_i)\rangle\langle B_0(k_i)| = \frac{I}{2^n}$$
$$= \sum_{k_i \in (0|1)^n} \frac{1}{2^n} |B_1(k_i)\rangle\langle B_1(k_i)| = \rho_1^{Bob} \tag{1}$$

where ρ_0^{Bob} represents Bob's view of the system when Alice commits to 0, and ρ_1^{Bob} represents Bob's view when Alice commits to 1. Since the density matrices corresponding to Alice's two commitments are identical, there is not even a theoretical chance for Bob to distinguish between a commitment to 0 and a commitment to 1, no matter what measurement(s) he may try to perform. This uncertainty on Bob's side can also be quantified using the information theoretic concept of entropy:

$$S(B) = -\text{tr}(\rho_0^{Bob} \log \rho_0^{Bob}) = -\text{tr}(\rho_1^{Bob} \log \rho_1^{Bob}) = -\text{tr}(\frac{I}{2^n} \log \frac{I}{2^n}) = n \tag{2}$$

In other words, the amount of uncertainty in the quantum state of the safe, from Bob's perspective, is equal to n bits and this is the maximum it can be, given that the key used by Alice is of length n. Equation (2) therefore captures the hiding property of bit commitment in the language of quantum information theory.

So far, we have concentrated on Bob's perspective on the quantum system representing the safe, which is just a part of the whole ensemble Alice–Bob. The other part, which remains in Alice's possession, contains information about the key chosen by Alice and the bit commitment. The two parts are not independent, as the key together with the encoding basis completely determine the quantum state of the subsystem given to Bob. This means that the conditional entropy $S(B|A) = 0$. For the reader not familiar with the various forms of entropy, we mention that $S(B|A)$ is the amount of information present in system B that does not come from system A, and is defined as:

$$S(B|A) = S(A, B) - S(A). \tag{3}$$

In a hypothetical situation, where systems A and B are completely independent, $S(B|A) = S(B)$. An "entropy Venn diagram" of the whole system depicting the two components and the relationship between them is given below:

The number in each region of the diagram reflects the quantum entropy or amount of uncertainty characterizing the part of the system represented graphically by that region. The mutual information content of A and B, depicted as the intersection of A and B in the diagram, measures how much information systems A and B have in common and is defined as:

$$S(A : B) = S(A) + S(B) - S(A, B). \tag{4}$$

Information which is common to both systems is counted twice in the summation $S(A) + S(B)$, while information which is not common is counted just once. Therefore, by subtracting the joint information present in both systems (namely, $S(A, B)$) from this summation, we obtain just the common or mutual information of A and B. We note that the whole uncertainty present in subsystem B actually comes from subsystem A or, equivalently, there are n bits of uncertainty shared in the system between the two components (the mutual information $S(A : B) = n$). On the other hand, subsystem A has one more bit of uncertainty compared to subsystem B, which is enough to ensure the hiding property, even if Bob is able to dispel the uncertainty characterizing the quantum state of the safe in his view.

It is crucial here to make the observation that the above diagram is constructed solely from Bob's perspective on the system. Consequently, the fact that $S(A|B) = 1$ cannot be interpreted in the sense that if Bob knows the exact quantum state of the safe, as prepared by Alice, there is still some uncertainty left about the key and/or the bit commitment on Alice's side. Dispelling the n bits of uncertainty describing the quantum state of the safe, simply means that Bob's knowledge of this state advances from a complete mixture of all 2^n possible terms to a single, precise, pure state characterizing the quantum safe.

Such a sharp decrease in the entropy of the quantum safe can be triggered, for example, by Bob learning the outcome of a projective measurement applied on all n qubits composing the quantum safe, where each possible key k_i becomes a projector $P_i = |k_i\rangle\langle k_i|$. However, even after such a measurement is performed, there still remains one full bit of uncertainty about Alice's choice for the bit commitment and the corresponding key used.

To better understand the point, let us exemplify with the trivial case of a 1-qubit safe and the two encoding bases being the normal computational base and the Hadamard base. In this simple scenario, Alice encodes either 0 or 1 in the basis of her choice and sends the resulting qubit to Bob. Alice keeps a record of the basis chosen (representing her commitment) and the bit encoded (representing the key used). From Bob's point of view, the whole system is in a mixed state:

$$\rho^{AB} = \tfrac{1}{4}|00\rangle|0\rangle\langle 00|\langle 0| + \tfrac{1}{4}|01\rangle|1\rangle\langle 01|\langle 1| + \tfrac{1}{4}|10\rangle\tfrac{|0\rangle+|1\rangle}{\sqrt{2}}\langle 10|\tfrac{\langle 0|+\langle 1|}{\sqrt{2}} + \tfrac{1}{4}|11\rangle\tfrac{|0\rangle-|1\rangle}{\sqrt{2}}\langle 11|\tfrac{\langle 0|-\langle 1|}{\sqrt{2}}, \tag{5}$$

where the first qubit denotes the encoding basis, the second one represents the encoded key and the third qubit plays the role of the quantum safe passed over to Bob. Thus, the first two qubits come from subsystem A, while the last qubit makes up subsystem B. The amount of uncertainty present in the system equals two bits, corresponding to the four choices Alice has with respect to the encoding basis and key used:

$$S(\rho^{AB}) = -\text{tr}(\rho^{AB} \log \rho^{AB}) = 2. \tag{6}$$

In order to see how the quantum safe appears to Bob, we can trace out subsystem A from the global state ρ^{AB}:

$$\rho^B = \text{tr}_A(\rho^{AB}) = \frac{1}{2}|0\rangle\langle 0| + \frac{1}{2}|1\rangle\langle 1| = \frac{I}{2}. \tag{7}$$

This state entails one bit of uncertainty, since it is a mixture of both possible terms $|0\rangle$ and $|1\rangle$, each with probability $1/2$. Naturally, this bit of uncertainty comes entirely from subsystem A, as it can easily be checked that the entropy of subsystem A equals the entropy of the entire system:

$$\rho^A = \text{tr}_B(\rho^{AB}) = \frac{1}{4}(|00\rangle\langle 00| + |01\rangle\langle 01| + |10\rangle\langle 10| + |11\rangle\langle 11|) = \frac{I}{4}. \tag{8}$$

Dispelling the bit of uncertainty that Bob sees in the quantum safe amounts to Bob being informed or finding out through a measurement which of the two possible pure states the safe actually finds itself in. Regardless of the answer ($|0\rangle$ or $|1\rangle$), this state can come from both commitments with equal probability, as it can be seen from Equation (5). So the uncertainty on the bit commitment still remains for Bob ($S(A|B) = 1$), thus ensuring the hiding property. On the other hand, if Alice plays by the rules and does prepare a quantum state for the safe based on a specific key and bit commitment, then from her perspective there is not a shred of uncertainty anywhere in the system. All entropies depicted in Figure 1 would be zero in this case.

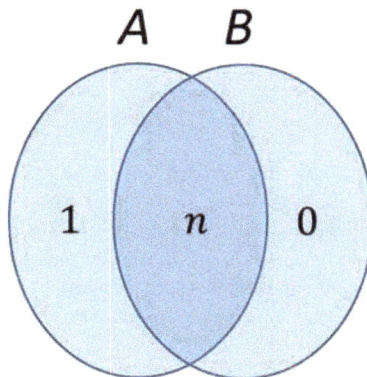

Figure 1. Entropy diagram of the whole system Alice–Bob viewed from Bob's perspective. All uncertainty present in subsystem B (quantum safe) comes from subsystem A (Alice's own quantum register), which has one extra bit of uncertainty.

This is reminiscent of the so-called "relational interpretation" of quantum mechanics (RQM) [24,25], which refutes the idea of an objective or absolute reality (state of a physical system) and proclaims that different observers can give different accounts on the properties of the same physical system. In the case of bit commitment, different accounts arise due to different amounts of information the participants (observers in RQM) have about a particular system, which motivates the use of information theoretic tools in order to get further insight into the problem. It would appear that there is a deep connection between information theory and quantum mechanics which is brought to light in the relational interpretation of the latter.

4. Cheating Requires Entanglement

In the previous section, we have clearly stated the condition that needs to be satisfied in order to ensure the hiding property and preclude any possibility of Bob gaining premature knowledge about Alice's bit commitment. The condition was formulated in terms of both density matrices and using the language of quantum information theory. In this section, we turn our attention to Alice and use the concept of entropy to prove that, if the hiding property is guaranteed, then Alice can cheat and change her commitment in the decommit phase if and only if she is endowed with the ability to generate and manipulate multi-party entangled states.

In a perfectly hiding quantum bit commitment protocol, Alice can cheat in the decommit phase by applying a transformation that will rotate the state of the whole system Alice–Bob from $|\psi_0\rangle$ (global quantum state corresponding to a commitment to 0) to $|\psi_1\rangle$ (quantum state describing the status of the entire system in the case of a commitment to 1). Since, in the decommit phase, she no longer has access to the safe, which was given to Bob (subsystem B), Alice must be able to effect this transformation only by acting on her side (subsystem A). This is only possible if subsystem B looks the same for both possible commitments, not only for Bob, but from Alice's point of view as well:

$$\rho_0^{Alice}(B) = \rho_1^{Alice}(B) = \frac{I}{2^n}. \tag{9}$$

Then, the transformation $|\psi_0\rangle \longrightarrow |\psi_1\rangle$ is guaranteed to exist as a direct consequence of the Schmidt decomposition theorem. Condition (9) is not met if Alice prepares the quantum state of the safe using a specific key k_i:

$$\rho_0^{Alice}(B) = |B_0(k_i)\rangle\langle B_0(k_i)| \neq |B_1(k_i)\rangle\langle B_1(k_i)| = \rho_1^{Alice}(B). \tag{10}$$

Consequently, in such a case, there is no transformation that can rotate $|\psi_0\rangle$ into $|\psi_1\rangle$ just from Alice's side. Therefore, maintaining a full uncertainty on the quantum state of the safe is a necessary condition for Alice to cheat:

$$S(B) = -\text{tr}(\rho^{Alice}(B) \log \rho^{Alice}(B)) = n. \tag{11}$$

We now show that this condition implies that the entire system Alice–Bob must be in an entangled state.

Regardless of how Alice chooses to prepare the state of the safe (subsystem B), she has full knowledge of its state in relation to her own quantum register (subsystem A). In other words, since she is the one preparing both subsystems, Alice has complete knowledge of the state of the ensemble Alice–Bob. This means the whole system is in a pure state according to Alice:

$$S(A, B) = 0. \tag{12}$$

At this point, the Schmidt decomposition theorem can be applied and it follows that both subsystems must have the same eigenvalues. In addition, since quantum entropy is completely determined by the eigenvalues, then it must be the case that the entropy of subsystem A is the same as that of subsystem B:

$$S(A) = S(B) = n. \tag{13}$$

Based on this last equality and Equation (12), the entropy diagram of the system looks like the one depicted in Figure 2. The mutual information $S(A : B) = 2n$, while both conditional entropies $S(A|B) = S(B|A) = -n$. A conditional entropy can be negative if and only if the two subsystems are entangled. Equivalently put, a supercorrelation indicated by negative values of conditional entropy is the unmistakable hallmark of entanglement.

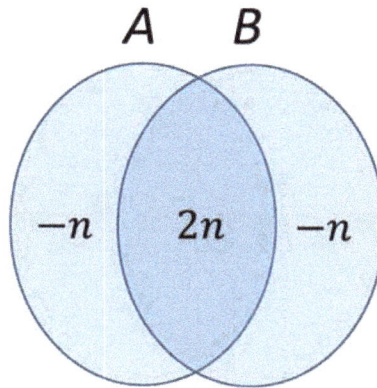

Figure 2. Entropy diagram of the whole system Alice–Bob viewed from Alice's perspective. The negative conditional entropies indicate that the two components of the system must be entangled.

In conclusion, we formulate the observation that the binding property can be expressed in the language of quantum information theory as $S_A(B) = 0$, forcing Alice to commit to a specific key. A value of n for the same entropy guarantees a cheating strategy that completely circumvents the binding requirement of bit commitment. Any intermediate value for $S_A(B)$ (between 0 and n) denotes some degree of entanglement between the quantum safe and Alice's register, which ultimately leads to some probability of cheating from her part and consequently, a partial realization of the binding requirement.

It is interesting to note that, from an information-theoretic perspective, both basic properties of bit commitment are formulated by quantifying the entropy of subsystem B (the quantum safe), albeit from two different points of view:

$$S_B(B) = n \tag{14}$$

ensures the hiding requirement by maximizing the entropy of the safe from Bob's perspective, while

$$S_A(B) = 0 \tag{15}$$

is the condition that prevents Alice from cheating the binding property by minimizing the entropy of the safe from her point of view.

5. Something up Her Sleeve

Having established that entanglement is an essential ingredient in Alice's cheating strategy, let us now detail how she can take advantage of this important quantum resource in order to avoid commitment and keep her options open until the decommit step.

The initial state of the system prepared by Alice has to take into consideration the two requirements: using entanglement and keeping a full uncertainty on the key used to "lock" the quantum safe. Consequently, Alice prepares an entangled superposition in which each term corresponds to one possible key and the encoding corresponds to a commitment to 0:

$$|\psi_0\rangle = \frac{1}{\sqrt{2^n}} \sum_i |k_i\rangle \otimes |B_C(k_i)\rangle. \tag{16}$$

The term to the left of the tensor product describes Alice's own quantum register, while the term on the right characterizes the quantum state of the safe.

This departure from the original protocol in which Alice is supposed to commit to a specific key value is sometimes labeled as the "purified" version of the protocol, due to the fact that the global state of the system is now a pure state. However, this modification which is essential for Alice's cheating strategy is transparent for Bob. He cannot distinguish between the original and purified version just based on the state of the quantum safe, in both cases the reduced density matrix being the same and equal to ρ_0^{Bob} from Equation (1). In a way, this is part of the reason why cheating is always possible in a QBC protocol.

In the decommit phase, if Alice wants to keep her commitment to 0, she just measures her quantum register (in the normal computational basis) and announces to Bob the values obtained as the encoding key. On the other hand, if she wishes to change her commitment to 1, she will have to first apply a transformation on her register that will rotate the overall state of the system to

$$|\psi_1\rangle = \frac{1}{\sqrt{2^n}} \sum_i |k_i\rangle \otimes |B_1(k_i)\rangle \tag{17}$$

and then apply the measurement. Let us illustrate this point by showing what happens in the case of the QBC protocol proposed by Bennett and Brassard in their seminal paper which launched the field of quantum cryptography [5].

The BB84 QBC protocol adheres to the generic framework outlined at the beginning of Section 3 with the particularizations that base B_0 represents rectilinear polarization of a photon and B_1 represents diagonal polarization. For the purpose of a theoretical analysis abstracted away from implementation details, we will use the normal computational basis and the Hadamard basis as B_0 and B_1, respectively.

The cheating strategy described in the original BB84 manuscript involves Alice preparing n Bell states $\frac{1}{\sqrt{2}}|00\rangle + \frac{1}{\sqrt{2}}|11\rangle$, keeping the first qubit from each entangled state as her own quantum register and sending the second ones to Bob as the quantum safe. Formally, the initial state of the ensemble Alice–Bob is therefore:

$$|\psi_0\rangle = (\frac{1}{\sqrt{2}}|0_A 0_B\rangle + \frac{1}{\sqrt{2}}|1_A 1_B\rangle)^{\otimes n}, \tag{18}$$

where the labels A and B are used to identify the two parts of the system. A closer look to state $|\psi_0\rangle$ reveals that it is actually an entangled superposition of all 2^n possible keys with their encodings in the normal computational basis:

$$|\psi_0\rangle = \frac{1}{\sqrt{2^n}} \sum_i |k_i^A\rangle \otimes |k_i^B\rangle, \tag{19}$$

so Alice is actually preparing the initial state according to the purified version of the protocol as explained above.

Since a Bell state always yields perfectly correlated outcomes when the two qubits are measured in the same basis, regardless of what this basis is, Alice can claim commitment to either of the two possible bit values in the decommit step. This is how cheating is explained in the BB84 paper. Yet again, at a closer inspection, measuring a tensor product of Bell states either in the normal computational basis or in the Hadamard basis conforms exactly to the general cheating procedure described above.

If Alice wishes to claim a commitment to 0, she simply measures her register (subsystem A) in the normal computational basis, collapsing the superposition of all possible keys to a specific key k_i. From Equation (19) it is obvious that when Bob measures his subsystem (the safe) also in the normal basis, he will obtain the same outcome k_i due to the entanglement present in state $|\psi_0\rangle$.

According to our previously described cheating procedure, if Alice wants to claim a commitment to 1, she has to first rotate the state of the system from $|\psi_0\rangle$ to

$$|\psi_1\rangle = \frac{1}{\sqrt{2^n}} \sum_i |k_i\rangle \otimes H^{\otimes n}|k_i\rangle, \tag{20}$$

by applying a quantum transformation on her register. After that, she can measure the register in the normal computational basis and whatever outcome k_i she obtains, it will coincide with the outcome obtained by Bob after a measurement in the Hadamard basis on his side. We claim that the transformation that Alice has to operate on her quantum register before measurement is the Hadamard transform. In other words, the Hadamard measurement invoked in the BB84 paper can be seen as a Hadamard transform followed by a measurement in the normal computational basis.

We can now verify formally that the Hadamard transform does the cheating trick by proving the following equality:

$$(H^{\otimes n} \otimes I^{\otimes n})|\psi_0\rangle = |\psi_1\rangle. \tag{21}$$

One way to prove this is by showing that the dot product between the vectors on the left-hand side and the right-hand side of the above equality is 1:

$$(|\psi_1\rangle, (H^{\otimes n} \otimes I^{\otimes n})|\psi_0\rangle) = (\sum_i \frac{1}{\sqrt{2^n}}|k_i\rangle \otimes H^{\otimes r}|k_i\rangle, \sum_j \frac{1}{\sqrt{2^n}} H^{\otimes n}|k_j\rangle \otimes |k_j\rangle)$$

$$= \sum_{ij} \frac{1}{2^n} \langle k_i|H^{\otimes n}|k_j\rangle (H^{\otimes n}|k_i\rangle)^{\dagger}|k_j\rangle \tag{22}$$

$$= \sum_{ij} ((\langle k_i|H^{\otimes n}|k_j\rangle)^2 = \sum_{i=1}^{2^n} \sum_{j=1}^{2^n} \frac{1}{2^{2n}} = 1.$$

Therefore, if Alice wants to change her commitment (from 0 to 1), she just has to apply $H^{\otimes n}$ on her quantum register and measure it in the normal computational basis or equivalently, measure her register in the Hadamard basis. In general, the transformation that will rotate $|\psi_0\rangle$ into $|\psi_1\rangle$ depends on the particulars of the respective protocol, but it will always exist as guaranteed by the Schmidt decomposition theorem.

6. Extensions to Protocols Initiated by Bob

We have seen that, if the entropy of the quantum safe (as it appears to both participants) is maximal and equal to the length of the key used by Alice, then the hiding property is guaranteed, but Alice can cheat the binding property with certainty, taking advantage of her entanglement with the quantum safe. At the first glance, this may appear as a consequence of the fact that Alice is the one initiating the protocol and having full control over how she prepares the quantum state of the system acting as the safe. In other words, she has all the cards in her hand and this may be perceived as an unfair advantage in realizing an unconditionally secure quantum bit commitment protocol.

Therefore, in this section, we extend the discussion to protocols in which the procedure is initiated by Bob, in the hope to deny Alice any opportunity of acting dishonestly. The main idea of the framework we take into consideration here is that Bob should have a choice in preparing the initial state of the quantum safe, prior to Alice encoding her commitment into the safe through the use of a key. Consequently, the commit phase of the protocol should consist of two steps, the first performed by Bob and the second one by Alice:

1. Bob chooses one of m initial states $|\phi_j\rangle$, $j = 1, ..., m$ for the quantum system acting as the safe. He then sends the n qubits composing the quantum system over to Alice through a quantum channel.
2. Upon receiving the quantum safe, Alice chooses a key k_i, $i = 1, ..., p$ and applies a unitary transformation $U_0(k_i)$ or $U_1(k_i)$ on the qubits composing the safe, depending on whether she wants to commit to 0 or to 1. Afterwards, she sends the qubits back to Bob through the same communication channel and waits for the decommit phase.

In an effort to keep our model as general as possible, we allow different values for the size of the safe (n qubits), number of possible initial states (m) and number of different quantum transformations (p) that Alice can apply on the system prepared by Bob for a given commitment. The latter also coincides with the number of keys k_i that Alice can choose from to encode in the quantum safe. Naturally, some (or all) of these three variables can actually be related to each other, depending on the particular characteristics of a specific protocol.

Since the choice of j is intended as a protection against cheating attempts by Alice, states $|\phi_j\rangle$, $j = 1, ..., m$ should be chosen such that they cannot be reliably distinguished. In one possible instance of the protocol, Bob could prepare the initial state of the safe by encoding one of the 2^n bitstrings of length n in one of two complementary bases. This means that Bob is choosing the initial state from a set with 2×2^n elements, thus making m equal to 2^{n+1}.

Similarly, the number of keys p Alice is encoding in the quantum safe may also equal 2^n. Each key may again represent a length n bitstring such that

$$U_0(k_i) = U_0(b_{n-1}) \otimes U_0(b_{n-2}) \otimes \cdots \otimes U_0(b_0), \tag{23}$$

where $U_0(b_i)$ is either the identity transformation, if $b_i = 0$, or a certain single-qubit gate G, if $b_i = 1$. Assuming that Alice's strategy is to avoid commitment and encode all possible keys in quantum parallel, the entropies of the three subsystems (from the point of view of Alice) are as follows:

$$S(Bob) = \log m = n + 1, \tag{24}$$

$$S(Alice) = \log p = n, \tag{25}$$

$$S(Safe) = n. \tag{26}$$

There are $n + 1$ bits of uncertainty on Bob's side, corresponding to the 2^{n+1} choices he has in preparing the initial state of the safe. Similarly, Alice's quantum register (entangled with the quantum safe) is characterized by n bits of uncertainty, reflecting the 2^n possible keys Alice can encode in the quantum safe. Finally, the quantum safe itself appears to be in a fully mixed state following all possible initial state preparations $|\phi_j\rangle$, $j = 1, ..., 2^{n+1}$ and remains in this state after Alice encodes all possible keys k_i, $i = 1, ..., 2^n$ in quantum parallel. Consequently, the entropy of the quantum safe is maximal and equal to the number of qubits n composing the safe.

The entropy of the global system Alice–Bob-Safe is also $n + 1$, since the state of the entire system (from Alice's point of view) consists of a different entanglement for each possible initial state $|\phi_j\rangle$:

$$\rho_0^{global} = \sum_{j=1}^{2^{n+1}} \frac{1}{2^{n+1}} \left(\sum_{i=1}^{2^n} \frac{1}{\sqrt{2^n}} |k_i\rangle U_0(k_i) |\phi_j\rangle \right) \left(\sum_{i=1}^{2^n} \frac{1}{\sqrt{2^n}} \langle\phi_j| U_0^\dagger(k_i) \langle k_i| \right). \tag{27}$$

For a fixed initial state prepared by Bob, the ensemble Alice-Safe is described by an entangled state (pure state) with zero entropy. Consequently, the Schmidt decomposition theorem dictates that both subsystems have equal entropy:

$$S(Safe|Bob) = S(Alice|Bob) = \frac{n}{2}. \tag{28}$$

The full entropy diagram of the whole system and its components, as it appears from Alice's perspective, is depicted in Figure 3.

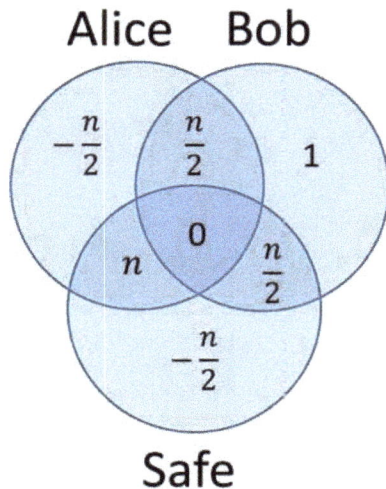

Figure 3. Entropy diagram of the whole system Alice–Bob–Safe viewed from Alice's perspective. The n bits of uncertainty originally encapsulated in the safe by Bob are split between Alice's quantum register and the safe through entanglement.

When Bob sends to Alice the n qubits representing the quantum safe, the entropy of Bob's subsystem ($(n+1)$ bits) can be decomposed into n bits of uncertainty characterizing the quantum safe handed over to Alice and one bit left for Bob's own register keeping track of the actual initial state prepared. Subsequently, through the entanglement generated by Alice, the maximal entropy of the quantum safe is spread uniformly among the two entangled parties: Alice's subsystem (quantum register keeping track of the actual key encoded) and the safe. Consequently, the mutual information between Bob and Alice, respectively, Bob and the safe is

$$S(Alice : Bob) = S(Safe : Bob) = \frac{n}{2}, \tag{29}$$

while the conditional entropy $S(Bob|Alice, Safe)$ remains 1. We also notice in Figure 3 the negative conditional entropies:

$$S(Alice|Bob, Safe) = S(Safe|Alice, Bob) = -\frac{n}{2}, \tag{30}$$

revealing the entanglement between Alice's own quantum register and the quantum safe, entanglement which ultimately allows Alice to steer the state of the system towards a commitment to 1 in the decommit phase, if desired. However, in order for Alice to be able to cheat the binding property, the hiding property must be enforced. In terms of the framework considered in this section, the protocol is hiding if, no matter what its initial state was, the safe looks identical to Bob at the end of the commit phase for both possible commit values:

$$\rho_0^{Bob} = \sum_{i=1}^{2^n} \frac{1}{2^n} U_0(k_i) |\phi_j\rangle \langle \phi_j| U_0^\dagger(k_i) =$$

$$= \sum_{i=1}^{2^n} \frac{1}{2^n} U_1(k_i) |\phi_j\rangle \langle \phi_j| U_1^\dagger(k_i) = \rho_1^{Bob}, \ \forall j = 1, ..., 2^{n+1}. \tag{31}$$

In this new scenario in which Bob is the one initiating the protocol, the hiding property is not easy to achieve, since it has to hold for every possible initial state $|\phi_j\rangle$. However, one straightforward way to ensure it, is to choose as the single-qubit gate G a Hermitian operator (like Hadamard, for example) and then set

$$U_1(k_i) = G^{\otimes n}U_0(k_i), \forall i = 1, ..., 2^n. \tag{32}$$

In this way, when Alice wants to change her commitment to 1 just before the decommit phase, she just needs to flip (apply the negation operator X on) all qubits in her quantum register.

For example, consider a three-qubit safe initially in state $H|0\rangle \otimes H|1\rangle \otimes H|0\rangle$. Moreover, assume that the key Alice encodes in the state of this quantum safe is 110 by applying the Hadamard gate on the first two qubits of the safe. The state of the quantum safe will consequently change to $|0\rangle \otimes |1\rangle \otimes H|0\rangle$. Now, this state is the result of following the procedure corresponding to a commitment to 0. However, we notice that the same state of the quantum safe could have resulted as a consequence of committing to 1, if the encoded key is not 110, but its opposite 001. Furthermore, the same technique of complementing the content of her own quantum register allows Alice to change her commitment regardless of the initial state $|\phi_j\rangle$ and/or the encoded key k_i.

Once again, we can formulate the observation that the entropy of the quantum safe appears maximized (equal to n) to both observers. This guarantees the hiding property, but leaves the door wide open for Alice to cheat the binding property of bit commitment, despite the fact that in our new setting Alice is forced to work with a quantum safe whose initial state is not known exactly.

In the setting considered above, Alice is still able to elude the binding requirement at the cost of choosing an operator G that is Hermitian. This constraint is a counterpart measure to the advantage that Bob now has through the choice of the initial state for the quantum safe. Certainly, there are other protocol frameworks that can be designed to annihilate Bob's advantage of trying to prepare an initial quantum state that would allow him to distinguish between the two possible commitments. We describe in the following an alternative protocol that adopts a different approach in enforcing the hiding property for Bob. Subsequently, we analyze the cheating strategies available to Alice and compare the new scenario with the framework we have just investigated, especially from the point of view of entropies and other information-theoretic measures characterizing the various subsystems involved.

In an effort to keep our QBC protocol as simple as possible, our next scenario considers a quantum safe consisting of only one qubit. Bob starts off the protocol by preparing four qubits, one in each of the states $|0\rangle$, $|1\rangle$, $H|0\rangle$ and $H|1\rangle$. He sends a random permutation of these four qubits to Alice, who then randomly selects one as the quantum safe. For a commitment to 0, she sends the unaltered qubit back to Bob. A commitment to 1 requires the application of the Hadamard gate before handing Bob the single-qubit safe back.

As in the previous case, we will focus on the purification of the protocol in which Alice acts as if committing to 0, but selects all four qubits received from Bob in quantum parallel. This is done through an entanglement between Alice's own quantum register (who plays the role of a pointer to the actual qubit selected) and the quantum safe. In order to distinguish among the four possibilities, the quantum register in Alice's possession must span two qubits. It is easy to check that this protocol is hiding, since, for Bob, the quantum safe appears to be in a fully mixed state, regardless of what particular permutation he is initially sending over to Alice:

$$\begin{aligned}\rho_0^{Safe} &= \frac{1}{4}|0\rangle\langle0| + \frac{1}{4}|1\rangle\langle1| + \frac{1}{4}H|0\rangle\langle0|H + \frac{1}{4}H|1\rangle\langle1|H = \frac{I}{2} \\ &= \frac{1}{4}H|0\rangle\langle0|H + \frac{1}{4}H|1\rangle\langle1|H + \frac{1}{4}|0\rangle\langle0| + \frac{1}{4}|1\rangle\langle1| = \rho_1^{Safe}.\end{aligned} \tag{33}$$

However, despite the fact that the hiding requirement is realized, Alice can no longer cheat the binding requirement, since in this setting, she needs a different transformation T to rotate $|\psi_0\rangle$

(global quantum state corresponding to a commitment to 0) to $|\psi_1\rangle$ (quantum state describing the status of the entire system in the case of a commitment to 1), for each possible permutation prepared by Bob. For instance, if the four-qubit sequence received from Bob is $|0\rangle \otimes |1\rangle \otimes H|0\rangle \otimes H|1\rangle$, then the transformation Alice has to apply on her quantum register is:

$$T_1 = \begin{bmatrix} 0 & 0 & 1 & 0 \\ 0 & 0 & 0 & 1 \\ 1 & 0 & 0 & 0 \\ 0 & 1 & 0 & 0 \end{bmatrix}. \tag{34}$$

However, if Bob prepares and sends the sequence $|0\rangle \otimes |1\rangle \otimes H|1\rangle \otimes H|0\rangle$, then it is the following transformation that does the job of rotating $|\psi_0\rangle$ to $|\psi_1\rangle$ just from Alice's subsystem:

$$T_2 = \begin{bmatrix} 0 & 0 & 0 & 1 \\ 0 & 0 & 1 & 0 \\ 0 & 1 & 0 & 0 \\ 1 & 0 & 0 & 0 \end{bmatrix}. \tag{35}$$

Therefore, unlike in the previous instance of QBC, here, there is no unique transformation that can help Alice elude the binding requirement, even though in both protocols she makes use of entanglement in order to apply all her options in quantum parallel. Let us now take a look at the entropy diagram for our current protocol, depicted in Figure 4, and see how this change is reflected in the entropies of the three subsystems involved.

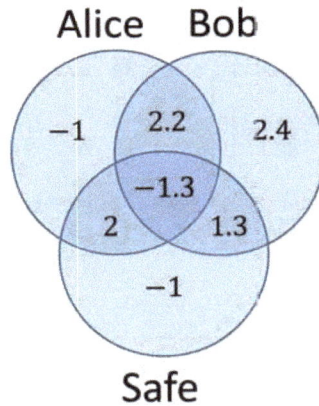

Figure 4. Entropy diagram at the end of the commit phase for an alternative Bob-initiated QBC (quantum bit commitment) protocol, viewed from Alice's perspective. The amount of uncertainty left in Bob's subsystem and not shared with the other subsystems prevents Alice from cheating.

The degree of uncertainty in Bob's subsystem is $\log 24$ (approximately 4.6), corresponding to the 24 degrees of freedom (number of permutations) Bob has in preparing the initial sequence given to Alice. The same value of the entropy also characterizes the entire system, because Alice sees the global state of the system as a mixed state having a different entanglement for each of the 24 possible permutations of the four qubits prepared by Bob.

Assuming that Alice knows exactly the quantum state of each qubit received, the ensemble Alice-Safe is characterized by the exact entangled state corresponding to that particular permutation. This means that the conditional entropy $S(Alice, Safe|Bob) = 0$ and following Schmidt's decomposition theorem, each side would entail an uncertainty of one bit. As always, the negative conditional entropies $S(Alice|Bob, Safe) = S(Safe|Alice, Bob) = -1$ witness the use of entanglement by Alice.

Where the entropy diagrams for the two protocols analyzed in this section differ from one another, is the amount of information leaked from Bob towards the other subsystems. In the first case, from the $n + 1$ bits of uncertainty characterizing Bob's subsystem, almost all of it (n bits) gets evenly distributed among Alice and the Safe: $S(Alice : Bob) = S(Safe : Bob) = \frac{n}{2}$. Only one bit of information remains unshared with Alice and the Safe: $S(Bob|Alice, Safe) = 1$. In the second protocol, on the other hand, out of the 4.6 bits of information contained in Bob's subsystem, 2.4 (that is, more than 50%) remains in Bob's hands only, and does not leak towards the other two subsystems. This is due to the fact that the size of the quantum safe (one qubit) is much smaller than the size of the vector space spanned by Bob's subsystem (five qubits, in order to accommodate 24 possible initial states). To make a comparison again, there is no mutual information between the safe and Bob's subsystem in the second scenario, while in the first the two subsystems share $\frac{n}{2}$ bits of information.

The conditional entropy $S(Bob|Alice, Safe)$ measures the amount of uncertainty left in Bob's subsystem once we acquire all the information in the Alice-Safe ensemble. The magnitude of this entropy seems to be the key in preventing Alice from cheating the binding requirement. Without the knowledge of the particular permutation prepared by Bob (most of which remains in his hands), Alice does not have enough information to rotate the state of the global system towards a commitment to 1.

Does this mean that the particular protocol we have just investigated satisfies both crucial security requirements of bit commitment: it is at the same time binding and hiding? The answer is no, as this would also constitute a counterexample to Mayers' impossibility theorem on realizing an unconditionally secure quantum bit commitment protocol [1]. The catch is that Bob can also enlist the help of entanglement to his own advantage. He too could not commit to a particular initial state $|\phi_j\rangle$ and instead prepare the quantum safe in a superposition of all possible initial states, each entangled with a corresponding label (or pointer) in his own ancilla qubits. This ensemble Bob-Safe would then be in a pure state:

$$|\psi_{init}^{Bob-Safe}\rangle = \sum_{j=1}^{24} \frac{1}{\sqrt{24}} P_j(|0\rangle \otimes |1\rangle \otimes H|0\rangle \otimes H|1\rangle) \otimes |l_j\rangle, \tag{36}$$

where each P_j represents one of the 24 possible permutation operators acting on four qubits. After receiving these four qubits composing the safe from Bob, Alice selects all of them in superposition, effectively creating a three-party entanglement among the three subsystems:

$$|\psi_0^{Alice-Safe-Bob}\rangle = \frac{1}{4\sqrt{6}} \sum_{j=1}^{24} \sum_{i=1}^{4} |k_i\rangle \otimes U_0(k_i) P_j(|0\rangle \otimes |1\rangle \otimes H|0\rangle \otimes H|1\rangle) \otimes |l_j\rangle. \tag{37}$$

In the equation above, $U_0(k_i)$, for $i = 0, 1, 2, 3$ is to be interpreted as the selection operator acting on the four-qubit sequence received from Bob whose effect is to select the qubit with index i. It is not difficult to verify that by attaching a label l_j to each possible permutation P_j, the hiding requirement is no longer satisfied, as the reduced density matrix obtained from $|\psi_0^{Alice-Safe-Bob}\rangle\langle\psi_0^{Alice-Safe-Bob}|$ by tracing out the two qubits from Alice's own quantum register is different from the similar reduced density matrix obtained from $|\psi_1^{Alice-Safe-Bob}\rangle\langle\psi_1^{Alice-Safe-Bob}|$. Consequently, Bob would be able to distinguish (with a certain probability $p > 1/2$) between a commitment to zero and a commitment to one, which ultimately translates into a non-zero probability of detecting a dishonest Alice.

7. Conclusions

The effort that many quantum cryptographers put into developing an unconditionally secure QBC protocol is motivated by (at least) two important factors: the intuition that the success of quantum key distribution can be replicated for bit commitment as well, and perhaps more importantly, the key importance of QBC as a building block in constructing more complex cryptographic applications.

These two factors may, at least in part, explain why people kept trying to devise new ways of achieving a protocol with the desired security properties, even after this was proved to be impossible. Another factor may be the relative difficulty in grasping the essence of this impossibility result, especially since other quantum realizations of cryptographic protocols proved superior to their classical counterparts.

In this paper, we adopted a novel approach in trying to get new insight into the problem. With the help of the tools provided by quantum information theory, we have investigated the conditions under which the two fundamental properties of bit commitment, the hiding property and the binding property cannot be reconciled. Our investigation has revealed that both properties can be expressed in terms of the entropy of the quantum system playing the role of the safe into which Alice is supposed to lock her commitment.

The hiding property dictates that the quantum safe must appear to Bob as harboring a certain level of uncertainty. The entropy of the quantum safe is usually maximized (especially when Alice is the one preparing the initial state of the safe) in order to hide the commit value behind the fully mixed state of the quantum system acting as the safe. When dealing with a dishonest Alice, the quantum safe actually achieves maximum entropy with respect to both observers, as Alice needs to keep all her options open. We have also proved that cheating requires enlisting the help of entanglement, a necessary condition without which any cheating strategy is impossible, whether we are talking about breaking the binding requirement by Alice or the hiding requirement by Bob.

Naturally, the ability to circumvent one or the other of the security requirements of bit commitment ultimately depends on the particular details and structure of the protocol at hand. However, our investigation, performed with the specific tools of quantum information theory, seems to point to two important factors influencing the ability to adopt a successful cheating strategy. On one hand, it is easier for any of the two participants to the protocol to mount an effective cheating strategy, if they are the ones initiating the protocol and preparing the initial state of the safe. This gives them the opportunity to avoid committing to a particular initial state and prepare a superposition of all possible initial states, each entangled with a pointer (or label) in their own ancilla qubits. On the other hand, devising a successful cheating scheme or successfully annihilate a cheating attempt by the adversary, also depends on how much information they manage to keep hidden inside their own subsystem and not share it (through the quantum safe) with the other participant.

Author Contributions: Marius Nagy and Naya Nagy analyzed the security properties of quantum bit commitment protocols using entropic measures. Marius Nagy wrote the paper.

Conflicts of Interest: The authors declare no conflict of interest.

References

1. Mayers, D. Unconditionally secure quantum bit commitment is impossible. In Proceedings of the Fourth Workshop on Physics and Computation—PhysComp '96, Boston, MA, USA, 22–24 November 1996.
2. Nielsen, M.A.; Chuang, I.L. *Quantum Computation and Quantum Information*; Cambridge University Press: Cambridge, UK, 2000.
3. Mayers, D. Unconditionally secure quantum bit commitment is impossible. *Phys. Rev. Lett.* **1997**, *78*, 3414–3417.
4. Lo, H.K.; Chau, H. Is quantum bit commitment really possible? *Phys. Rev. Lett.* **1997**, *78*, 3410–3413.
5. Bennett, C.H.; Brassard, G. Quantum cryptography: Public key distribution and coin tossing. In Proceedings of the IEEE International Conference on Computers, Systems and Signal Processing, Bangalore, India, 9–12 December 1984; pp. 175–179.
6. Brown, J. *The Quest for the Quantum Computer*; Simon & Schuster: New York, NY, USA, 2001.
7. Biham, E.; Boyer, M.; Boykin, P.O.; Mor, T.; Roychowdhury, V. A proof of the security of quantum key distribution. *J. Cryptol.* **2006**, *19*, 381–439.
8. Crépeau, C.; Kilian, J. Achieving oblivious transfer using weakened security assumptions. In Proceedings of the 29th Annual IEEE Symposium on Foundations of Computer Science, White Plains, NY, USA, 22–26 October 1988; pp. 42–52.

9. Brassard, G.; Crépeau, C.; Jozsa, R.; Langlois, D. A quantum bit commitment scheme provably unbreakable by both parties. In Proceedings of the 34th Annual IEEE Symposium on Foundations of Computer Science, Palo Alto, CA, USA, 3–5 November 1993; pp. 362–371.

10. Brassard, G.; Crépeau, C. 25 years of quantum cryptography. *SIGACT News* **1996**, *27*, 13–24.

11. Crépeau, C. What is going on with quantum bit commitment? In Proceedings of the Pragocrypt '96: 1st International Conference on the Theory and Applications of Cryptology, Prague, Czech Republic, 30 September–3 October 1996.

12. Kent, A. Permanently secure quantum bit commitment protocol from a temporary computation bound. *arXiv* **1997**, arXiv:quant-ph/9712002.

13. Spekkens, R.W.; Rudolph, T. Degrees of concealment and bindingness in quantum bit commitment protocols. *Phys. Rev. A* **2001**, *65*, 012310.

14. Ng, N.H.Y.; Joshi, S.K.; Ming, C.C.; Kurtsiefer, C.; Wehner, S. Experimental implementation of bit commitment in the noisy-storage model. *Nat. Commun.* **2012**, *3*, 1326.

15. Kent, A. Unconditionally secure bit commitment with flying qudits. *New J. Phys.* **2011**, *13*, 113015–113029.

16. Lunghi, T.; Kaniewski, J.; Bussières, F.; Houlmann, R.; Tomamichel, M.; Kent, A.; Gisin, N.; Wehner, S.; Zbinden, H. Experimental bit commitment based on quantum communication and special relativity. *Phys. Rev. Lett.* **2013**, *111*, 180504.

17. Adlam, E.; Kent, A. Device-independent relativistic quantum bit commitment. *Phys. Rev. A* **2015**, *92*, 022315.

18. Hardy, L.; Kent, A. Cheat sensitive quantum bit commitment. *Phys. Rev. Lett.* **2004**, *92*, 157901.

19. Buhrman, H.; Christandl, M.; Hayden, P.; Lo, H.K.; Wehner, S. Possibility, impossibility, and cheat sensitivity of quantum-bit string commitment. *Phys. Rev. A* **2008**, *78*, 022316.

20. Shimizu, K.; Fukasaka, H.; Tamaki, K.; Imoto, N. Cheat-sensitive commitment of a classical bit coded in a block of $m \times n$ round-trip qubits. *Phys. Rev. A* **2011**, *84*, 022308.

21. He, G.P. Comment on "Cheat-sensitive commitment of a classical bit coded in a block of $m \times n$ round-trip qubits". *Phys. Rev. A* **2014**, *89*, 056301.

22. Li, Y.B.; Wen, Q.Y.; Li, Z.C.; Qin, S.J.; Yang, Y.T. Cheat sensitive quantum bit commitment via pre- and post-selected quantum states. *Quantum Inf. Process.* **2014**, *13*, 141–149.

23. Li, Y.B.; Xu, S.W.; Huang, W.; Wan, Z.J. Quantum bit commitment with cheat sensitive binding and approximate sealing. *J. Phys. A Math. Theor.* **2015**, *48*, 135302.

24. Rovelli, C. Relational quantum mechanics. *Int. J. Theor. Phys.* **1996**, *35*, 1637–1678.

25. Grinbaum, A. The Significance of Information in Quantum Theory. Ph.D. Thesis, Ecole Polytechnique, Paris, France, 2004.

entropy

MDPI

Review

Developments in Quantum Probability and the Copenhagen Approach

Gregg Jaeger [1,2]

[1] Quantum Communication and Measurement Laboratory, Department of Electrical and Computer Engineering, Boston University, Boston, MA 02215, USA; jaeger@bu.edu
[2] Division of Natural Science and Mathematics, Boston University, Boston, MA 02215, USA

Received: 2 May 2018; Accepted: 28 May 2018; Published: 31 May 2018

Abstract: In the Copenhagen approach to quantum mechanics as characterized by Heisenberg, probabilities relate to the statistics of measurement outcomes on ensembles of systems and to individual measurement events via the actualization of quantum potentiality. Here, brief summaries are given of a series of key results of different sorts that have been obtained since the final elements of the Copenhagen interpretation were offered and it was explicitly named so by Heisenberg—in particular, results from the investigation of the behavior of quantum probability since that time, the mid-1950s. This review shows that these developments have increased the value to physics of notions characterizing the approach which were previously either less precise or mainly symbolic in character, including complementarity, indeterminism, and unsharpness.

Keywords: quantum probability; potentiality; complementarity; uncertainty relations; Copenhagen interpretation; indefiniteness; indeterminism; causation; randomness

1. Introduction

The orthodox approach to quantum theory emerged primarily from interactions in Copenhagen and elsewhere from the work of Niels Bohr, Werner Heisenberg, and Wolfgang Pauli, depending also on contributions of Max Born, and was largely set out by 1927, cf. [1–4]. After various criticisms of the initial form, Bohr focused more strongly on complementarity in the 1930s, and Heisenberg—in a strong response of 1955 in which the basis of the approach can be considered to have been essentially finalized—added the new element of actualization of potentiality to its approach to quantum probability [5]. Since then, much attention has been paid to newly emerging alternative treatments of quantum physics, such as Bohmian mechanics and collapse-free (e.g., many-worlds) mechanics, that use quantum probability and the quantum formalism. Indeed, with a few exceptions (e.g., [2,6–8] and others mentioned below) relatively little attention has been paid to the implications of new research for the more orthodox, Copenhagen approach. Here, a non-exhaustive but wide-ranging series of theoretical results, several directly related to experiment, obtained since the work of Heisenberg related to quantum probability, is discussed, which, by articulating more precisely and better clarifying the application of its basic notions, including complementarity, uncertainty, and indeterminacy, which were before either less precise or even merely symbolic, further demonstrates the value of this approach.

Beyond the basic notion that the probability of any observed future physical event in a system can be found via the quantum state using the Born rule given the results of a complete set of measurements on it, the character of quantum probabilities on this Copenhagen approach is discussed in the next section; experimentally verified results demonstrating new quantum complementarities are considered in Section 3; theoretical developments involving unsharpness and quantum measurement are considered in Section 4; novel explications of indeterminism and randomness are considered in Section 5.

2. Probability in the Copenhagen Approach

The Copenhagen approach to quantum theory generally gives primacy to measured phenomena—with any results regarding the measured system being given in relation to the entire experimental situation in which each arises and with the records of measuring devices being classical describable—without being positivist and remaining essentially realist. Although Bohr's notion of complementarity was the greatest influence early in the development of the approach—as noted, for example, by Jan Faye [9] and Arkady Plotnitsky [10,11]—it was later supplanted in a development toward more precise and mathematically advanced treatments of the effect, as exhibited in Section 3 (also cf. [12])—and even explicitly extended to situations involving entanglement as, arguably, thought by Bohr to be the case, as Don Howard has argued [13]. The Copenhagen approach explicitly named as such is that circumscribed by Werner Heisenberg [5] in which measurement corresponds to the actualization of a potential physical situation where a single value appears from among a set of possible values that were not certain (cf. [14,15] for analyses of the fundamentals of this version, specifically considered here). The essential mathematical formalism of non-relativistic QM emerged with work of Paul Dirac and John von Neumann, with Hilbert-space as the space of individual system states (cf. [1–3] and Chapter. 7 in [16]), forming the context for its later extension, discussed in detail below, cf. [2,14,17].

A succinct overview of the role of probability in the Copenhagen interpretation was given by Heisenberg, who gave the interpretation its name: "...the probability function does not in itself represent a course of events in the course of time. It represents a tendency for events and our knowledge of events. The probability function can be connected with reality only if one essential condition is fulfilled: if a new measurement is made to determine a certain property of the system. Only then does the probability function allow us to calculate the probable result of the new measurement" (pp. 46–47, [18] ; cf. [15]).

2.1. Quantum States and Probability

An (O, S, p) formulation of general physical theory serves as a basic formal framework for the non-relativistic theory of QM [19,20]: To each physical system, one can associate with the set of all associated sharp observables (Hermitian self-adjoint operators) O and the set of its states S, a function $p: O \times S \times \mathcal{B}(\mathbb{R}) \to [0,1]$, where $\mathcal{B}(\mathbb{R})$ is the set of all Borel subsets of $\mathbb{R}$, cf. [21]—of values appearing in measurements, cf. [2]. Restricting ourselves specifically to the Hilbert-space formulation of quantum mechanics, each statistical operator ρ is decomposable into a non-trivial weighted sum of quantum pure states represented by normalized vectors $|\psi\rangle_i \in \mathcal{H}$ (cf. [5,18]) as $\rho_i = |\psi_i\rangle\langle\psi_i|$ that have no further state decomposition; each statistical operator ρ also induces an expectation functional $A \mapsto \mathrm{tr}(\rho A)$ on $\mathcal{L}(\mathcal{H})$, the space of linear operators on the Hilbert space $\mathcal{H}$.

The probability p in the Copenhagen approach involves an explicit distinction between objective and subjective aspects of physical states describable in this formalism and compares with that in classical mechanics as follows. In classical mechanics, when needed at all, probability is used only in situations where a detailed knowledge of the system is lacking, i.e., for statistical mechanics; in the quantum context, the subjective aspect of probability also appears in such situations, which involve state mixtures (*Gemische*, cf. [22] p. 9, [23] p. X) that are representable as statistical operators ρ but not as pure ones. However, probability in QM on this approach is found also in the *individual states* (*Zustände*, cf. [23]) as the objective aspect, representable as vectors $|\psi\rangle$ in $\mathcal{H}$ [23,24]. This objective contribution to the quantum probability of a measurement outcome is provided specifically by the state $|\psi\rangle = \sum_i |\psi_i\rangle$ via its complex amplitudes $\{c_i\}$, now known as *probability amplitudes*, as their squared magnitudes $p_i = |c_i|^2$.

The quantities $\{p_i\}$ are understood as the probabilities of the measured system to be found to possess actual respective values of its physical properties according to the rule of Born that is elemental to the Copenhagen approach [25,26]. The measured value of a property is considered definite (actual), as opposed to indefinite (potential) [5], as discussed in great detail in [14,15]; a dynamical property of a quantum system S becomes actual with probability p upon precise measurement wherein the measuring apparatus, A, must be in contact with the greater physical environment (the "the rest of the world") and be classically describable or macroscopic (cf. [27] for a discussion of this notion and its use in the Copenhagen interpretation and elsewhere more recently). In general, some member from a set of *possible* values must occur in measurement, but the specific *actual* value measured appears randomly [28], as discussed in Sections 4 and 5 here.

2.2. Quantum Indeterminacy

Another important element in the initial success of the Copenhagen approach is that it articulates well the behavior of joint probabilities appearing in the "uncertainty relations," a manifestation of complementarity. Indeed, in Born's view, "the factor that contributed [most]... to the speedy acceptance of [the Copenhagen] interpretation of the ψ-function was a paper by Heisenberg [29] that contained his celebrated uncertainty relations" [30]. The result of von Neumann that (sharp) observable quantities are simultaneously measurable if and only if they commute with one another and if and only if they are functions of a single observable later came to play a strong role in this understanding of such relations. It has also been shown by Pekka Lahti [31] that Heisenberg's joint indeterminacy hypothesis—the "uncertainty principle" [29], later generalized by H. P. Robertson, [32], providing corresponding "uncertainty" relations discussed in Section 2.4 below—together with an axiomatic formulation of complementarity, when considered within the (O, S, p) framework formalized by George Mackey [19] and M. J. Maczyński [20], rigorously imply the existence of pairs of observables that cannot be jointly sharply measured. The known set of such relations has recently expanded, as shown in the next section.

The indeterminacy principle contrasts with the determinacy principle, that the magnitude of each continuous quantity is determined by a real number, as is typically assumed in classical mechanics (cf. Michael Dummett's discussion of this principle in [33]). Note that the indeterminacy hypothesis is a statement about the associated indeterminacy of incompatible observables not algebraically commuting with each other (see Section 4 below), rather than the epistemic uncertainty regarding an independent quantity *per se*. It is, therefore, a statement to the effect that the associated properties, jointly considered, are *objectively indefinite*, as emphasized by Abner Shimony [34]; (cf. [35] Section 2.1). It is a consequence of this indefiniteness that their measured values are also not precisely predictable, that is, random in the sense explained in Section 5 below.

2.3. Quantum Potentia and Probability

In the Copenhagen approach, according to Heisenberg, the objective probability p, given by the Born rule, relates to "statements about possibilities or better tendencies ('potentia'...)" of the system itself later to have certain actual values of measured properties [18] (p. 53). The subjective content of these probabilities is "negligible" in the pure case, i.e., where $\mathrm{tr}\rho^2 = 1$, exactly when it is a projector, which suffices for the maximal specification of the system's actual properties; in this case, the elements of the set of quantum probabilities $\{p_i(O)\}$ for the outcomes $\{o_i\}$ in a measurement of the observable (i.e., Hermitian operator) O are

$$p_i(O) = \mathrm{tr}(P_i(O)\rho) = \langle\psi|P_i(O)|\psi\rangle = \langle\psi|\phi_i\rangle\langle\phi_i|\psi\rangle \tag{1}$$

which are the squared magnitudes of the corresponding complex-valued state-vector amplitudes $\{c_i = \langle\phi_i|\psi\rangle\}$, that is, of the components of $|\psi\rangle$ in the eigenbasis $\{|\phi_i\rangle\}$ for O—the exclusivity of this form being demonstrated rigorously by Andrew Gleason [21]; see Section 4.1 below. (Here, we consider for simplicity the case of discrete properties; analogous relations hold in the continuous case.) In particular, the complex probability amplitude c_k corresponds to the *potentia* for actualization of the specific property value o_k upon the measurement of O, with $p_k(O) = |c_k|^2$, something discussed further in Section 4.3 below—cf. [14]. (Very recently, the notion of quantum potentia in a sense of *res potentia* related to that of Heisenberg has been used by Ruth Kastner et al. to offer a novel analysis of quantum measurement, in combination with *res extensa*, purely physical substance, as "implicative constituents of every quantum measurement event" [36].)

Recall that the novel characteristic of the quantum probability first to be discovered (n.b. a mathematically precise treatment of its novelty in general terms has been given by Luigi Accardi in [37]), which motivates the continual reconsideration of its relationship to traditional probability, involves ostensibly disjoint events: An empirically measurable difference between the quantum mechanical probability and the classical mechanical probability of a disjunction of such events is that the associated probability is not, in general, additive in the quantum case. This is exhibited in the appearance of a particle such as an electron at a spatial location x in the basic double-slit experiment (cf. [38]) by passing first through a slit (Slit 1) and/or the other slit (Slit 2), which occurs with a probability $p_{12}(x)$ that is related to the probability densities of reaching x in the alternatives of either first passing through Slit 1, $p_1(x)$, or of first passing through Slit 2, $p_2(x)$:

$$p_{12}(x) \not\propto p_1(x) + p_2(x). \tag{2}$$

This quantum probability is the magnitude squared of the sum $c_{12}(x) = c_1(x) + c_2(x)$ of the complex amplitudes $\{c_i(x)\}$ of those alternatives, rather than a simple sum of the probabilities of the two alternative situations consistent with the future event, here detection at x.

Thus, the quantum probabilities do not arise by direct calculation from, for example, prior probabilities of particle detection as in classical mechanics as a sum such as $p_{12}(x) \propto p_1(x) + p_2(x)$ in the situation of this double-slit experiment. (Another difference of quantum probability from classical probability is found in *joint* probabilities is discussed in Section 3 below.)

2.4. Quantum Interference and Dispersion

The difference between the two sorts of probability, classical and quantum, is reflected clearly in the corresponding probability density distribution in the detection plane in the double-slit experiment: There is an additional modulated "interference term" arising because the $c_i(x)$ are complex-valued, which precludes these probabilities from being given a straightforward Kolmogorovian representation under a single probability measure (cf. [37] and [39] p. 125, [37] for detail regarding this), so that

$$p_{12}(x) \propto p_1(x) + p_2(x) + \sqrt{p_1(x)p_2(x)} \cos\left(\theta_2(x) - \theta_1(x)\right). \tag{3}$$

In this situation, there is a range of possible values for the detected position, as well as of the momentum of the system approaching it, that is, a certain dispersion of values due to its indeterminacy.

More generally, any observable given as an Hermitian operator A will have a dispersion $\mathrm{Disp}_\rho A = \langle (A - \langle A \rangle \mathbb{I})^2 \rangle = \langle A^2 \rangle - \langle A \rangle^2$, for any system in state ρ; the indeterminacy relation of Heisenberg for momentum and position relevant in this experiment was generalized by H. P. Robertson so as to apply two any two observables A and B [29,32]:

$$\langle (\Delta A)^2 \rangle \langle (\Delta B)^2 \rangle \geq \frac{1}{4} |\langle [A, B] \rangle|^2 \tag{4}$$

where the "uncertainty" of A for state ρ is $\Delta A \equiv \sqrt{\mathrm{Disp}_\rho A}$. These relations are connected with single-particle interferometric complementarities, i.e., between visibility and particle path, as shown in the 1980s and 1990s; see the following section, Jaeger et al. [40] and references therein for more on the relation to interferometry, and a recent analysis of Paul Busch and Christopher Shilladay [41] for a detailed discussion of the various forms of complementarity.

Significant new developments regarding quantum probability that allow for the clearer explication of these central aspects of the Copenhagen approach are discussed in the next two sections.

3. Complementarity and Entanglement

The extraordinary behavior of quantum probabilities regarding compound systems due to entanglement was brought to the forefront relatively early in the history of quantum mechanics (QM) by Erwin Schrödinger [42,43] and remained the subject ongoing entirely theoretical discussions until after John Bell produced his now-famous inequality, which was subsequently rendered experimentally testable in a reformulation by John Clauser, Michael Horne, Abner Shimony, and Richard Holt (CHSH) [44] and shown to be violated in an interferometric setting in the early 1980s by Alain Aspect et al. [45].

An entanglement-related manifestation of complementarity involving joint probabilities was noted later in the 1990s: When the two-particle interference visibility is unity, the one-particle visibility is zero, and conversely as noted in the work of Marlan Scully, Berge Englert, and Herbert Walther [6] and of Shimony, Horne, and Anton Zeilinger [46]. In the mid-1990s, it was shown that the interferometric phenomena involved in the violation of CHSH inequalities obey a precise *trade-off relation*, later experimentally verified in the 2000s [40,47,48], over a full range of different experimental arrangements. This is also related to the fact that entanglement can be understood as an instance of the uncertainty of quantum properties, cf. [41]. In particular, it was found by Jaeger and Shimony that there is a general quantum interferometric complementarity relation between single-system interference visibility, v_1, and compound-system interference visibility, v_{12}, for pairs of two-level systems, further illustrating the surprising nature of quantum correlations exhibited in two-particle interference due to the presence of entanglement [47], as first verified at the Boston University in 2001 [48].

This novel exhibition of complementarity can be understood concretely in terms of the washing out of photon self-interference due to indeterminacy in the initial direction of individual particles in a *doubled* discrete (Mach–Zehnder, MZ) two-"slit" arrangement with a source of particle pairs at center simultaneously feeding *two* Mach–Zehnder interferometers, symmetrically oriented, with one particle moving in one interferometer involving a generalized beamsplitter (transducer) at left and similarly for the other particle moving in a second such interferometer at right (see [47] for figures). Consider two particles A and B in this arrangement, with A taken to be that in beams 0 and/or 1, and similarly for particle B (but indicated by primes below). Let each particle pair of the ensemble involved be produced by the centrally located source in a possibly entangled two-particle pure state $|\Theta\rangle = \gamma_1|0\rangle_A|0'\rangle_B + \gamma_2|0\rangle_A|1'\rangle_B + \gamma_3|1\rangle_A|0'\rangle_B + \gamma_4|1\rangle_A|1'\rangle_B$, with $\gamma_i \in \mathbb{C}$ such that $|\gamma_1|^2 + |\gamma_2|^2 + |\gamma_3|^2 + |\gamma_4|^2 = 1$, $|0\rangle_A$ and $|1\rangle_A$ being basis vectors in the Hilbert space $\mathcal{H}_A$ of the first particle corresponding to the propagation of A to the left in beams 0 and 1 and $|0'\rangle_B$ and $|1'\rangle_B$ being similar vectors in space $\mathcal{H}_B$ of B moving to the right.

Let beams 0 and 1 be brought together at a transducer, T_A (inducing a general unitary transformation in the state space, not only a phase-shift+reflection/transmission as in a simple MZ apparatus), feeding two output beams, an upper U and lower L beam in the MZ interferometer at left, and let similar beams in the other interferometer to the right be brought together into another transducer, T_B (inducing a similar unitary transformation in the other particle's state space), that feeds two corresponding output beams U' and L'; the joint, local operation of this pair of transducers is described by the general pair of local unitary operations induced by them separately: $T = T_A \otimes T_B$. As these transducers T_A and T_B are varied, the probabilities $P(UU')$ of coincidence detection in beams U and U', and $P(UL')$, $P(UL')$, and $P(LL')$, as well as single-detection probabilities $P(U), P(L), P(U')$, and $P(L')$—corresponding to particle coincidence detection and single detection rates, respectively, in the output beams of the pair of interferometer—are modulated.

The corresponding visibilities of interference were found to obey trade-off relations, quantifying their complementaries. As T is varied over the full range of parameters for the two general local unitary transformations involved, continuously altering the apparatus, the one-particle interferometric fringe visibility V_i ($i = A, B$) is found from the maximum and minimum probabilities of detection:

$$V_i = \frac{[P(Y)]_{\max} - [P(Y)]_{\min}}{[P(Y)]_{\max} + [P(Y)]_{\min}} , \tag{5}$$

where $Y = U, L$. V_{12}, the two-particle interferometric visibility in the sense of variations of detection probability as the T is changed, is similarly calculable from the probabilities $P(YY')$ of occupation of the joint-paths YY', generalizing the case of the single paths Y. For example,

$$V_{12} = \frac{[\bar{P}(UU')]_{\max} - [\bar{P}(UU')]_{\min}}{[\bar{P}(UU')]_{\max} + [\bar{P}(UU')]_{\min}} \tag{6}$$

where $\bar{P}(UU') = P(UU') - P(U)P(U') + \frac{1}{4}$ represents *nonaccidental* coincidence probabilities; likewise for the three other possible path pairs YY' [47].

When the two systems, A and B, are entangled, one has the non-factoring joint probability

$$P(UU') \neq P(U)P(U'), \tag{7}$$

as do the other joint probabilities $P(UL')$, $P(LU')$, and $P(LL')$; the extraordinarily highly correlated behavior of particles A and B arises due to entanglement, and one finds that a strong complementarity trade-off relation, taking the form of an equality [47], holds for all $|\Theta\rangle$, namely,

$$V_{12}^2 + V_A^2 = 1 \,; V_{12}^2 + V_B^2 = 1 \,; \tag{8}$$

this was subsequently experimentally confirmed by Bahaa Saleh and associates at Boston University [48,49]. This explicitly demonstrates precise quantum complementarity involving entangled systems of the sort violating the Bell and CHSH inequalities.

4. Quantum Measurement

In the Copenhagen approach to quantum mechanics, "the behaviour of the measuring apparatus must be capable of being registered as something actual. . . if the measuring apparatus is to be used as a measuring instrument at all. . . the connection with the external world is. . . necessary" [5] (p. 27); any fully quantum A and S alone would become entangled, something Don Howard has argued Bohr was already noting in 1927 [13], and neither S nor A+S can be considered closed *during measurement*. Indeed, in a fully quantum formal treatment of such a process, the apparatus would itself have to be measured in order to provide an outcome, and so on. In this approach, a fully formal quantum mechanical treatment of the measurement process utilizing a closed system description or without the use of classical descriptions for at least some elements of the measurement process is considered impossible—quantum mechanics is literally incapable of being used to *account for* all details of the process (n.b.: the adjective sometimes used in relation to this is often translated as *uncontrollable* in English, but is better translated as *unaccountable-for*.) Moreover, the objective, probabilistic aspect of quantum state evolution invoked upon actualization is considered irreducible to any amount of ignorance that might be removed using the theory alone, as mentioned in Section 2.

In the Copenhagen approach, any system S, as well as the joint system of S and any other quantum piece of apparatus A thought of as distinct from A, is an open system while being measured, because it must be coupled to the larger world to provide an actual measurement record. A coupling of quantum system S to *classically describable* devices—Heisenberg's *recording system plus the rest of the world*, which physically intervene when using an apparatus in such a way that one among a set of differing outcomes can appear on the resulting record—is held to characterize measurement, and this coupling gives rise to a probabilistic, indeterministic state change (corresponding to the actualization of the system property) that relates to the actual value in a prescribed way (described by the EE link, see Section 4.2, below).

Actualization (discussed in Section 4.3) *begins with the coupling* of the system under measurement to the classically describable observational apparatus in the world, which is physically designed (based on rigorous testing for reliability) to elicit an outcome, and *ends upon decoupling*, leaving a classically describable physical record of the outcome.

If one does consider, *contra* the views of Bohr and Heisenberg, a fully quantum mechanical closed system treatment of measurement, the behavior of the joint system A+S, or a larger chain of interactions systems (see below), this might be expected to suffice for the description of measurement. Such a treatment is often given as follows; cf. e.g., [17], [30] (Section 2.4). One prepares system S via a series of physical interactions, such as filtering, in some well-defined quantum state $|\eta\rangle$, after which it is arranged to interact with a measurement apparatus, A. This apparatus, after being similarly arranged to be in a fiducial initial state $|\chi_0\rangle$, would be required to enter a final state corresponding to the value of a pointer property Z, which must be correlated with the value of the measured property (non-degenerate observable) E of the system. We may consider, for simplicity, a discrete measured property

$$E = \sum_i e_i |\psi_i\rangle\langle\psi_i|$$

where $\{|\psi_i\rangle\}$ is a countable orthonormal basis for the system Hilbert space $\mathcal{H}$ corresponding to its eigenvalues $\{e_i\}$. Another typical requirement for such a measurement is that a "calibration condition" be satisfied, namely, that if a measured property is real, then its value must be exhibited properly, unambiguously, and with certainty: if system S is an eigenstate $|\psi_k\rangle$ of E, then the state of apparatus A after the interaction of the two is an eigenstate of Z (with an eigenbasis $\{|\phi_i\rangle\}$ associated with pointer readings z_i), which serves to indicate the specific value of E present, the free-Hamiltonian function contribution to the evolution of the system being considered negligible relative to that of the measurement interaction. For quantum observables, the calibration condition generally takes the form of a probability reproducibility condition, namely, that **a** probability measure exists for a property be transcribed onto that of the corresponding apparatus pointer property. Finally, registration of the measured property by the measurement apparatus is taken to include the physical reading out of the registered value.

If one formally considers an entire chain of interacting objects connecting the system S up as far as physically conceivable, to the brain of an experimenter, for example X, Y,..., in the environment in addition to the original measurement system S and the experimenter's apparatus A—such as focusing elements, counter or counters, various cables, a computer, output display, etc.—a good measurement would involve all these becoming correlated in their properties for the measurement outcome to be physically indicated. Under the Schrödinger state evolution, which is unitary, upon completion of the measurement interactions, one would then find

$$|\Psi\rangle = \sum c_i |s_i\rangle |a_i\rangle |x_i\rangle |y_i\rangle \cdots , \qquad (9)$$

with $\{s_i\}$, $\{a_i\}$, $\{x_i\}$, $\{y_i\}$ etc. as the Hilbert space eigenbases for S, A, X, Y,..., respectively. The result of considering all physical systems involved entirely quantum mechanically is simply a regress backward from the prepared state, which presents and indefinite value of the quantity to be observed.

Heisenberg had already engaged this difficulty early on (in 1935 [51], cf. [52]) in a Copenhagenist spirit by insisting on a bipartite division of a set of different systems involved, only one of which is to be considered in any one analysis among all those possible, one for each way of making a bipartite division of the above chain, and considering only *one side of the division* quantum mechanically, as described below, cf. [51]. Again, for him, consideration of the entire measurement chain—or even simply the system and portion of apparatus in direct contact with it—as a full accounting of the measurement process as described within the state-vector formalism, as done in the above, without truncation, is an inappropriate use of the quantum formalism, the proper role of which is *to make predictions of measurement outcomes*; the only plausible use of the quantum formalism for the purposes of symbolizing a measurement process requires the introduction of a cut or split (*Schnitt*) between what is considered the measured portion of this chain of systems on the side including the entity S and the remainder, considered then to be a single, classically describable measuring system.

Notably, a change in the location of the cut makes no difference to the statistics obtained for the purposes of prediction. This formal description is strictly speaking only a symbolic description of the elements of the world involved. Heisenberg also imposed an important condition restricting this location: "The claim ... that it is indifferent at which location the cut between the parts of the system to be treated quantum mechanically and the classical measuring devices should be drawn, should thus be made more precise in the sense that this cut may indeed be shifted arbitrarily far in the direction of the observer in the region that is otherwise described according to the laws of classical physics; but that this cut cannot be shifted arbitrarily in the direction of the atomic system. Rather, there are physical systems—and all atomic systems belong among these—that the classical concepts are unsuitable to describe, and whose behaviour can therefore be expressed correctly only in the language of wavefunctions" [51], cf.[53].

In this way, the chain of statistical correlations appearing in the formal representation of the state $|\Psi\rangle$ is considered to be cut in two—into a system S' and the remainder A'—somewhere along this chain of interacting systems, with subsystems to one side of the cut collectively considered the system to be measured: S' subsumes S together with all other subsystems left of the cut, and A'(=A + W) is the collective of those systems right of the cut, that is, the "apparatus plus the rest of the world" W. Thus, A', is removed from the quantum-physical description used to make predictions relating to outcomes, with the cut always being made *somewhere* within S–X–Y–···–A–W, where a classical description is possible for the entire portion including the recording system. The actualization of potentia requires an interaction of S' (the size of S or larger) with the classically describable measuring apparatus, itself in interaction with the rest of the world. Formally, the change of state of the measured system involved is sometimes said to be "projected" to the appropriate component of the initially prepared system state, that is to be attributed by the eigenvalue-eigenstate link (discussed below in Section 4.2). Such a projection involves a change of state differing from the unitary evolution predicted when using the Schrödinger law of motion alone, for any non-trivial measurement.

It should be noted, however, that the Copenhagen approach to measurement can be criticized for not offering, indeed, for *denying the possibility of* a complete, closed system description of the measurement process or, for that matter, even precisely specifying the conditions under which measurement will occur, for example, due to the unclear boundary between the classical and the quantum realms, by its reliance on the requirements of the use of a macroscopic apparatus and the production of classically describable records of reliable measuring instruments not precisely characterizable by quantum mechanics. For this reason, Heisenberg's appeal to actualization (discussed in Section 4.3) can be considered an incomplete quantum mechanical treatment. Moreover, descriptions of the sort given by Heisenberg in the above have been criticized for conflating measurement with state preparation, as Henry Margenau did already in 1936 [54].

Let us turn now to Heisenberg's indeterminacy relations. In the presentation of the indeterminacy relations in Heisenberg's 1929 Chicago lectures, published as *The Physical Principles of the Quantum Theory* [29], in which he spoke only of the *Der Kopenhagener Geist der Quantentheorie* rather than a full *interpretation*, the indeterminacy relations were, strictly speaking, only symbolic in nature and in the process of being generalized (beginning with work by H. P. Robertson [32] discussed in Section 2 above). These relations were placed on a firm mathematical grounding soon thereafter in the Hilbert-space formalism; cf. Section 2.4. Recently, notably since the final explication of the Copenhagen approach by Heisenberg in the 1950s [5], these relations have been analyzed, extended, and clarifed via the notion of *unsharpness* in a way that captures the notion of quantum indeterminacy more efficaciously.

4.1. Unsharpness

The maximally specified state of a quantum system relative to an observable O in the Copenhagen approach can be given as a projector $\rho_{\text{pure}} = |\psi\rangle\langle\psi|$ appearing in the spectral decomposition of an observable O, as discussed in Section 2.1; cf. [2,55]. Nonetheless, in addition to measurements corresponding to such operators, *unsharp measurements* have also been formalized as the class of quantum operations that are described by (normalized) *positive-operator-valued measures* (POVMs) developed by Günther Ludwig, Karl Kraus, Busch, and Lahti [17,56,57].

Given a nonempty set S and a σ-algebra Σ of its subsets X_m, a POVM E is a collection of operators $\{E(X_m)\}$ satisfying the conditions: (i) *Positivity*—$E(X_m) \geq E(\emptyset)$, for all $X_m \in \Sigma$; (ii) *Additivity*—for all countable sequences of disjoint sets X_m in Σ, $E(\cup_m X_m) = \sum_m E(X_m)$; (iii) *Completeness*—$E(S) = \mathbb{I}$. If the *value space* (S, Σ) of a POVM E is a subspace of the real Borel space $(\mathbb{R}, \mathcal{B}(\mathbb{R}))$, then E provides a unique Hermitian operator on $\mathcal{H}$, namely $\int_{\mathbb{R}} \text{Id } dE$, where Id is the identity map. The positive operators $E(X_m)$ in the range of a POVM are referred to as *effects* $\mathcal{E}(\mathcal{H}) = \{A \in \mathcal{L}(\mathcal{H}) : \mathbb{O} \leq A \leq \mathbb{I}\}$, the expectation values of which provide the quantum probabilities.

Given an effect A, one can define properties in general by the following set of conditions. (i) There exists a property $A^{\perp}$; (ii) there exist states ρ and ρ' such that both $\text{tr}(A\rho) > \frac{1}{2}$ and $\text{tr}(A\rho') > \frac{1}{2}$; (iii) if A is regular, for any effect B below A and $A^{\perp}$, $2B \leq A + A^{\perp} = \mathbb{I}$, where a *regular effect* is an effect with spectrum both above and below $\frac{1}{2}$. Thus, the set of properties is $\mathcal{E}_p(\mathcal{H}) = \{A \in \mathcal{E}(\mathcal{H}) | A \nleq \frac{1}{2}\mathbb{I}, A \ngeq \frac{1}{2}\mathbb{I}\} \cup \{\mathbb{O}, \mathbb{I}\}$; the set of *unsharp properties* is $\mathcal{E}_u(\mathcal{H}) \equiv \mathcal{E}(\mathcal{H})_p / \mathcal{L}(\mathcal{H})$. A POVM is an *unsharp observable* if there exists an unsharp property in its range [2]. The POVMs are the natural correspondents in the operator space of quantum mechanics of standard probability measures and thereby make precise the notion of indeterminacy in the Hilbert-space setting.

The probability of a given outcome m upon a (generalized) measurement on a system in a pure state $P(|\psi\rangle)$ is given by

$$p(m) = \langle\psi|E(X_m)|\psi\rangle = \text{tr}((|\psi\rangle\langle\psi|)E(X_m)) ; \tag{10}$$

cf. Equation (1), which holds for the case of sharp measurement. The effects form a convex subset of the space of linear operators on $\mathcal{L}(\mathcal{H})$ on the system Hilbert space, the extremal elements of this subset being the projectors $\{P_i\}$. A collection of effects is said to be *coexistent* if the union of their ranges is contained within the range of a POVM. Any two quantum observables E_1 and E_2 are representable as sharp measures on $(\mathbb{R}, \mathcal{B}(\mathbb{R}))$ exactly when $[E_1, E_2] = \mathbb{O}$, following from the results of von Neumann for Hermitian operators [23] mentioned in Section 2 above; coexistent observables are thus those that can be *measured simultaneously in a common measurement arrangement*, and when two observables are coexistent, there exists an observable, the statistics of which contain those of both observables, the *joint observable*. Typically, the two observables are recoverable as marginals of a joint distribution on the product of the corresponding two outcome spaces.

The introduction of unsharpness allows one to circumvent the requirement of commutativity of jointly measurable properties, which captures only the extremes of complementarity, by including the unsharp properties, and enables a continuous range of complementarity to be captured in the Hilbert space formalism. For POVMs, commutativity remains sufficient but is not necessary for the coexistence of effects (cf. [2]). It has been shown that "smeared versions" of two noncommuting observables can still have a joint observable. For example, the operators $F = \{\frac{1}{2}(I \pm f\sigma_z)\}$ and $G = \{\frac{1}{2}(I \pm g\sigma_z)\}$ have a joint observable precisely when $f^2 + g^2 \leq 1$. Therefore, as a requirement for this pair to be jointly observable, the magnitudes $|f|, |g|$ (their degrees of unsharpness) must be complementary, in accordance with this trade-off relation, as demonstrated by Busch and Shilladay [41].

The introduction of POV measures and unsharpness have thus helped make indeterminacy and mathematical complementarity more precise by exploiting the Hilbert-space setting; cf. [41] for further detail on the intertwined connection and contrast between those two notions and more detail on their role in understanding joint measurability of properties in experimental situations, such as the single-particle Mach-Zehnder interferometry, not discussed here.

4.2. Linking Actual and Possible Values

In the Copenhagen approach to quantum mechanics, the state of a measured system is related to the actual values obtained in measurement by what has come to be known as the eigenvalue-eigenstate

(EE) link. The essence of the EE link was first introduced by Heisenberg and then used by others including Dirac in 1930 and after; cf. the careful explanation of this by Marian Gilton [58]. In the 1930 version of Dirac's *Quantum mechanics*, one finds the following:

"If a state ψ_r and an observable α are such that, when an observation is made of the observable with the system in this state the result is certain to be the number a, we assume this information can be expressed by the equation

$$\alpha\psi_r = a\psi_r \tag{11}$$

Conversely, when an equation of this type is given, we assume it has the physical meaning that a measurement of the observable α with the system in state ψ_r will certainly give for result the number a or that the observable α has the value a for the state ψ_r, to use a classical way of speaking, which is permissible in this case" [24] (p. 30).

In 1958, in the 4th edition of his classic textbook *Quantum Mechanics*, which appeared after Heisenberg's article "The Development of the Interpretation of Quantum Mechanics," one finds the EE link connected explicitly with probability.

"The expression that an observable 'has a particular value' for a particular state is permissible in quantum mechanics in the special case when a measurement of the observable is certain to lead to the particular value, so that the state is in an eigenstate of the observable ... In the general case we cannot speak of an observable having a value for a particular state, but we can speak of its having an average value for the state. We can go further and speak of the probability of its having any specified value for the state, meaning the probability of this specified value being obtained when one makes a measurement of the observable." [59] (p. 253)

4.3. Causation, Possibility, and Actuality

Like other sorts of probability, quantum probability can be viewed as the "graded possibility" of the occurrence of events, as first suggested by Leibniz [60,61]. Moreover, in the Copenhagen approach to quantum mechanics, unlike others, possibility plays a fundamental role in relating theory to experiment. This was explicitly indicated by Heisenberg in his invocation of an aspect of Aristotle's theory of causation, wherein possibility appears prominently in relation to all phenomena: "... in modern physics the concept of possibility, that played such a decisive role in Aristotle's philosophy, has moved again into a central place" [62] (p. 298).

As pointed out in Section 2 above, the isolated quantum system, described by a state-vector $|\psi\rangle \in \mathcal{H}$, "no longer contains features connected with the observer's knowledge... it is also completely abstract ... and the representation becomes a part of the description of Nature only by being linked to the question of how real or possible experiments will result" [5] (p. 26). The objective aspect of probability is that of quantifying the likelihood $|c_i|^2$ of the appearance of each value among any set of possible measurement outcomes as the *actual* result in the actualization of the potential physical state which occurs upon measurement, which according to Heisenberg is independent of any subjectivity: "the transition from the 'possible' to the 'actual' takes place as soon as the interaction between the object and the measuring device, and thereby with the rest of the world, has come into play; it is not connected with the act of registration of the result in the mind of the observer" [18] (pp. 54–55).

Heisenberg explained the objective character of this registration process as follows. "Of course, the introduction of the observer must not be misunderstood to imply that some kind of subjective features are to be brought into the description of Nature. The observer has rather only the function of registering decisions, i.e., processes in space and time, and it does not matter whether the observer is an apparatus or a human being; but the registration, i.e., the transition from the possible to the actual, is absolutely necessary here, and cannot be omitted from the interpretation of the quantum theory.

It must also be pointed out that in this respect the Copenhagen interpretation of quantum theory is in no way positivistic. For whereas positivism is based on sensual perceptions of the observer as elements of reality, the Copenhagen interpretation regards things and processes which are describable in terms of classical concepts, i.e., the actual, as the foundation of any physical interpretation." [5] (p. 22). Thus, Heisenberg neither requires nor refers to the mind, the brain, or human knowledge for the *actualization* of potential values, which appear in successful measurements as actual values of an observable inferrable from a resulting classically describable record in accordance with the EE-link; only the interaction of the measured system with the greater world in a way so as to produce such a classical record is required. It is such a record from which the mind could later acquire knowledge if the recorded, classically describable measurement outcome is later attended to.

Measurement of a quantum system precludes the state change that would otherwise occur were it to remain isolated, that is, the time-evolution according to the Schrödinger law of motion. It is in this way that, in the Copenhagen approach to QM, the possible becomes the actual and can be recorded, according to Heisenberg—a way not necessarily captured by the evolution dictated by the law of motion; it is often said that it is at this point that *causation* fails in QM—that it becomes acausal. However, this non-deterministic change of state-vector evolution arises precisely with the intervention of the measuring apparatus and the rest of the world participating in the production of a classical record of the outcome as its *cause*.

Hence, the Copenhagen approach to non-relativistic quantum mechanics presented by Heisenberg is not one in which there is a genuine *lack* of causation; instead, there is a form of *Aristotelian causation* that is not captured by the fundamental law of motion, which only governs closed systems not being measured—in particular, it is the form Aristotle calls *chance* causation, as I have argued in [14,15]: In the actualization of potentiality, there is a single chance occurrence that lies within the set of possible occurrences for the system in question according to the characteristics of its Hilbert-space description, with a measured value capable of being recorded on a system that is also classically describable but unpredictable [5]. It is in this way that the Copenhagen approach remains, in a specific sense, causative. This approach to quantum probability, where it fundamentally involves possibility, has recently been connected with logic in relation to the possible experience to be gained through measurement, as shown in the next subsection.

4.4. Logic and Indeterminacy

In the Copenhagen approach, the quantum state is taken to characterize as completely as possible the system to which it is attributed. The changes in the world occurring in any measurement are changes of values of quantities that were theretofore potential, that is, only possibly possessed, in accordance with the EE link connecting possible and actual observed values of observables. Empirically, the probabilities of these various outcomes to occur accord with the likelihoods of obtaining the set of possible outcomes for measurements in the future correspond formally to the Hermitian operator O, in measurements on collections of identically prepared systems described by the state $|\psi\rangle$ with the resulting measurement record being classically describable when necessary for their readout. These probabilities have been connected to logic; cf. [1] (Chapter 8). Indeed, the results of the early work of von Neumann and Birkhoff associates propositions to quantum Hilbert (sub)spaces [63] and the field of quantum logic that arose from that pioneering work; cf. [64] for a general summary of later developments.

4.4.1. Completeness of the Quantum State Description

Whether the quantum state was indeed complete, despite the apparent incompatibility of this with the attribution of precise values to *all* observable quantities, remained unclear until the work of Simon Kochen and Ernst Specker in 1967 [65]. Their theorem now known as the Kochen-Specker theorem precludes a consistent truth-valuation from being given to the propositions identified by von Neumann and Birkhoff, which is what is required to satisfy the corresponding value-definiteness thesis, namely, (i) that each and every physical magnitude have a definite value at all times (see Alan Stairs' valuable discussion [66]) and (ii) that measurements reveal those preexisting precise values; cf. [50] (Section 2.3) for a discussion of related issues. The value-definiteness condition can be stated more formally: Each proposition regarding the system, of the form "$O \in \Delta$," where O describes the physical magnitude and Δ is a Borel subset of the real numbers, is given a definite value 0 or 1.

The work of Kochen and Specker followed a very general theorem of Andrew Gleason [21]: All probability measures that can be defined on the lattice of quantum propositions from the quantum statistical operators, that is all quantum probabilities, are of the form $p(P_i) = \text{tr}(\rho P_i)$, for some statistical operator ρ on Hilbert space $\mathcal{H}$, for all $\mathcal{H}$ of dimension greater than two. This result demonstrates that every probability measure over the set of quantum state projectors is one from a quantum state ρ on the $\mathcal{H}$ attributed to the system in question; the trace measure $\text{tr}(\rho P_i)$ assigns to each projector P_i the dimension of its range, which can then be normalized by the dimension of the (finite-d) $\mathcal{H}$. It can, therefore, be obtained by taking ρ to be the maximally mixed state on $\mathcal{H}$. This shows that the only natural generalization of Kolmogorov probability functions of the sort used in quantum mechanics is exactly that of the Hilbert-space formulation of quantum mechanics. The values corresponding to orthogonal projectors thus obey a Born-type rule for the assignment of probabilities.

States for which definite truth-values could be attributed to all observables are the so-called dispersion-free states, states for which projectors take expectation values of only either 0 or 1 under the above mapping. Following a presentation first given by Bell, we can relate this to Gleason's work [67]. The condition $\sum_i \langle P(|\phi_i\rangle)\rangle = 1$ implies that both 0 and 1 occur because (1) there are no other possible values for satisfying the condition and (2) neither alone suffices. However, then, there must be arbitrarily close pairs $|\psi\rangle, |\phi\rangle$ having different expectation values, 0 and 1 respectively; however, such pairs cannot be arbitrarily close, by the above lemma. Hence, there can be *no* dispersion-free states providing the statistics of quantum statistics. Accordingly, no variables parameterizing dispersion-free probability measures can exist for systems having $\mathcal{H}$ [67]. The set of quantum states is therefore complete, because it provides the probability measures definable on the quantum propositional lattice.

Consider the complete set of Hermitian self-adjoint operators for the set of quantum states describing a system attributed a Hilbert space with dimension $d > 2$, constraining their algebra to reflect the values assigned, and take the assignment of the values of real numbers to the quantum operators to reflect corresponding properties of the system. In this setting, the Kochen–Specker theorem demonstrates the impossibility of such an assignment for a finite sublattice of quantum propositions [65]. Take a *value function*, v_ψ connecting an observable O to a value of a physical magnitude O when a system is in a state ψ, that is, $v_\psi : O \to \mathbb{R}$. Define $F(O)$ to be the value associated with $F(O)$ for all functions F with a one-to-one, onto mapping from values of O to O. Taking $v_\psi(F(O)) = F(v_\psi(O))$ has the consequence that v_ψ is additive and multiplicative on commuting operators with the consequence that $v_\psi(\mathbb{I})$ for all states ψ as long as there is at least one magnitude O for which $v_\psi(O) = 0$ (cf. [68], pp. 191–192). Another consequence of this is that $v_\psi(P_i)$ must be either 0 or 1 for all propositions P_i, which have corresponding projectors P_i. Thus, if one considers a resolution of the identity into a set of projectors $\{P_i\}$, that is, this set is such that $\sum_i P_i = \mathbb{I}$, in an interpretation of quantum properties where one and only one of the corresponding magnitudes P_i can take the value 1, no such function exists except for an overly restricted class of properties.

The results of Gleason as well as Kochen and Specker thus support the basic Copenhagen assumption that the quantum state is complete. Note also that the Kochen–Specker result can be extended to general von Neumann algebras, as shown by Andreas Doring et al. in [69], with implications quantum and generalized probabilistic models, as noted by Federico Holik et al. in [70].

4.4.2. Logical Quantum Indeterminacy Relations

More recent work of Itamar Pitowsky has shown that this connection can be placed in the context of indeterminacy to provide a new class of trade-off relations exhibiting complementarity in logical context. In his investigation, Pitowsky began by noting that George Boole, in developing his conception of probability, identified necessary and sufficient conditions for a set of rational numbers $p_1, p_2, ..., p_n$ to represent properly the probabilities, considered (relative) frequencies, of the occurrence of a set of n logically connected events $E_1, ... E_n$ [71] $p_i = \text{prob}(E_i)$ $i = 1, 2, ..., n$ to express what Boole called "conditions of possible experience" [72]. These conditions are either linear inequalities or equalities in $p_1, p_2, ..., p_n$; if the events under consideration are entirely independent, then the fractions corresponding to probabilities might be constrained only by the conditions $p_i \geq 0, p_i \leq 1$, but the expression for sets within possible experience must take the simple form

$$a + \sum_i^N a_i p_i \geq 0 \qquad (12)$$

where a, a_i are constants involving the logical relations constraining them [72].

This set of conditions on probabilities lie within n-dimensional polytopes in the case of probabilities of correlation, the convex hull of a finite number of points in $\mathbb{R}^d$, that is, the set of all convex combinations of its points [73]. Any violation of these conditions is manifested geometrically by the location of points (corresponding to probabilities) outside of the polytope dictated by them. The conditions on possible experience can then be methodically constructed from the logical relations among sets of possible events. Take, for example, a pair of events E_1, E_2 having relative frequencies p_1, p_2, again taking p_{12} to denote the frequency of the *joint event* $E_1 \cap E_2$. Being probabilities, these numbers have the relations: $p_1 \geq p_{12}$, $p_2 \geq p_{12}$, $p_{12} \geq 0$. The frequency of the disjoint event $E_1 \cup E_2$ is then $p_1 + p_2 - p_{12}$ with

$$p_1 + p_2 - p_{12} \leq 1 . \qquad (13)$$

One then has a corresponding three-dimensional space of vectors (p_1, p_2, p_{12}) that can be viewed as a convex polytope with vertices $(0,0,0)$, $(1,0,0)$, $(0,1,0)$, and $(1,1,1)$. Pitowksy considered a set of measurements known to have as outcomes 0 and 1, such as the measurements on a squared value S_i^2 of the component of spin along orthogonal spatial directions for a spin-1 system.

In this case, the basic operators S_i do not commute and so cannot be precisely measured simultaneously, while their squares do, cf. Pitowsky's [74]. Their sum $S^2 = 2I$, where I is the identity, so that in a simultaneous measurement of these spin-squared operators, one and only one of these observables will have the value 0, while the others take value 1. This illustration of the general situation corresponds to measurements with a triple of possible outcomes. Let the events that appear in more than one measurement be written as E_i, and let those that appear in only one triple be F_i. Suppose the noncontextuality of probability—the requirement that probability assignments do not depend on the outcomes of measurements of other observables that might be measured at the same time, cf. [75,76], and assign the same probability to each event above in all cases. Given that the probabilities in each triple of possible outcomes must also sum to one, one then finds

$$
\begin{align}
p(E_1) + p(E_2) + p(F_2) &= 1 \qquad (14) \\
p(E_1) + p(E_3) + p(F_3) &= 1 \qquad (15) \\
p(E_2) + p(E_4) + p(E_6) &= 1 \qquad (16) \\
p(E_3) + p(E_5) + p(E_7) &= 1 \qquad (17) \\
p(E_6) + p(E_7) + p(F_1) &= 1 \qquad (18) \\
p(E_4) + p(E_8) + p(F_4) &= 1 \qquad (19) \\
p(E_5) + p(E_8) + p(F_5) &= 1. \qquad (20)
\end{align}
$$

These requirements on probability then imply trade-off inequalities expressing complementarity, for example,

$$p(E_1) + p(E_8) \leq 3/2 . \qquad (21)$$

One of the two outcomes E_1 and E_8—which cannot arise as alternative outcomes of the same measurement—becomes more certain, that is, there is an increased probability of occurrence. The other outcome becomes less certain, with a decreased probability of occurrence. This is a logic-based indeterminacy relation quantitatively expressing that the likelihoods of positive results in alternative measurement arrangements are complementary quantities [74].

Here again, a novel sort of indeterminacy relation was obtained and given in the clear mathematical form of a trade-off relation—this in addition to the others introduced and developed after the advent of the Copenhagen approach. This new sort of relation arises from the consideration of a collection of sets of alternative events in Boolean logic within single measurements, with the two events involved resulting from a collection of such measurements as outcomes of different measurements, as discussed in [77]. Like those discussed in the previous section, this result furthers the significance the Copenhagen approach's notions of complementarity and indeterminacy by revealing their appearance beyond its original mathematical locus, further demonstrating its fundamental significance.

5. Indeterminism and Randomness

Classical physics has most often been thought to be deterministic in the following sense, introduced by Laplace in the following statement. "We ought... to regard the present state of the universe as the effect of its anterior state and as the cause of the one which is to follow. Given for one instant an intelligence which could comprehend all the forces by which nature is animated and the respective situation of the beings who compose it—an intelligence sufficiently vast to submit these data to analysis—it would embrace in the same formula the movements of the greatest bodies of the universe and those of the lightest atom; for it, nothing would be uncertain, and the future, as well as the past, would be present to its eyes." [78]. Under the conditions set out in this statement of the notion, perfect predictions and retrodictions regarding the behavior of individual physical systems are possible, in principle.

The determination of future (and past) events involved here is identified with *in-principle predictability* under the assumption that unlimited resources are available to the predictor, P: It involves in its application the existence of an intelligence with unlimited capabilities, both physical and computational. However, this is something beyond the scope of the practice of physics from within the universe: Finite beings—with finiteness assessed via the physical and/or cognitive resources at their disposal—are inherently limited in their ability to predict physical events, given that prediction requires computational resources; in any application of physical theory, such as prediction, any finite agent can exploit only *finite* physical resources for this purpose.

5.1. Predictability

Although the contraints on the resources of finite beings do not always present difficulties for precisely predicting a given future event among a given finite set of events in classical mechanics, even unlimited resources do not suffice for *quantum mechanics* to be considered a deterministic theory according to the above classic definition and that even when one among a *finite* set of alternatives need be distinguished. An alternative, more physically straightforward definition of determinism applicable to quantum mechanics would therefore be superior, in particular, one more suitable to physics and independent of radical assumptions about the availability of compuational and other resources. One such a definition that has been suggested is: A scientific theory is *deterministic* if and only if in that theory any two trajectories in the state space in models of systems overlap at one point do overlap at *every point*, and it is indeterministic if and only if it is not deterministic (cf. discussion in [79–81]).

Judging this within the (O, S, p) framework of general physics, one sees that quantum mechanics is *not* a theory supporting determinism of this latter kind *for individual, measured systems*, but at best provides precise predictions of the behavior of collections of identically prepared and subsequently measured systems: On the Copenhagen approach, quantum measurement of physical properties are understood to introduce a probabilistic change of physical state that precludes the state-trajectory overlap required by this definition. The "acausality"—as it was often called early in the history of the approach—of this quantum state evolution is responsible for the indeterminism of the states actually appearing from among those possible beforehand as the result of measurement. The states connected with measurement data are always those dictated by following the EE link rule, which offers an alternative to indeterminacy ("uncertainty") relations for capturing the objective indefiniteness of physical properties. As explained in Section 4.3 above, the Copenhagen approach provides an Aristotelian form of causation for this process even in the presence of chance.

5.2. Randomness

Despite the clarity gained by the move to a trajectory-based version of determinism, predictability remains relevant to the question of the randomness of the appearance of measurement outcomes. Note that a crucial distinction exists between indeterminism and unpredictability: Predictability hinges on the total context of prediction of *P*—be that a human or another sort of cognitive agent such as an artificially intelligent robot—including the all the conditions of the experimental context. Indeterminism, in the trajectory-based definition just discussed, involves only the theoretical character of the theory involved, here QM, specifically regarding the topology of state-space trajectories, independently of whether or how they could, if ever, come to be *known*. The randomness appearing in quantum mechanics should not be identified with essentially probabilistic state evolution, that is, indeterminism, as noted by Geoffrey Hellman [82], because indeterminism is not a necessary condition for the appearance of randomness in a theory.

Randomness can be defined as maximal unpredictability, and the Copenhagen approach provides a consistent understanding of the notions of indeterminism and unpredictability matching this conception well. A notion of predictability that accords with one of the distinguishing characteristics of the approach, namely, the involvement not only of the measurement apparatus but also the large world (environment) of the measurement apparatus in the very definition of proper measurement (cf. [5,10,14]), has been recently introduced by Anthony Eagle [81], namely, this physical process definition:

> "A prediction function $p(S,t)$ takes as input the current state S of a system described by a theory T as discerned by a predictor P and an elapsed time t, and yields a temporally indexed probability distribution $\Pr_t$ over the space of possible states of the system. A *prediction* is a specific use of some prediction function by some predictor on some initial state S_0 and time t_0 who adopts $\Pr_t$ as their posterior credence function conditionally on the evidence and the theory." [81]

This definition of predictability when applied using the probabilities of outcomes of quantum measurements allows their random character to be explicated.

The randomness in quantum mechanics in the Copenhagen approach can be understood specifically as follows. Let *P* be any experimenter performing a measurement, let *T* be the quantum mechanics, let $|a_0\rangle$ be the the quantum state attributed to the pertinent system via the preparation procedure used by *P* at time t_0, and let *t* be the time elapsed between its preparation and the completion of the measurement. Predictor *P* will use the appropriate choice of quantum probability as his/her $p(|a_0\rangle, t)$, according to the circumstances of the measurement it performs, that a given outcome *b* obtains, and use it in finding his posterior credence function given his entire background knowledge. In particular, one can apply the following definition of *unpredictability*: "An event *E* (at some temporal distance *t*) is unpredictable for a predictor *P* if and only if *P*'s posterior credence in *E* after conditioning on current evidence and the best prediction function available to *P* is not 1, that is, if the prediction function yields a posterior probability distribution that does not assign probability 1 to *E*" [81].

Given, in particular, that measurement results are unpredictable in this sense whenever incompatible observables are measured in succession, one then has a well defined sense of *randomness* that applies to quantum mechanics:

"An event E is random for a predictor P using theory T if and only if E is maximally unpredictable. An event E is maximally unpredictable for P and T if and only if the posterior probability of E yielded by the prediction functions that T makes available, conditional on current evidence, is equal to the prior probability of E. This also means that P's posterior credence in E, conditional on theory and current evidence (the current state of the system), must be equal to P's prior credence in E conditional only on theory."

where E is the appearance of the eigenvalue b as the measurement outcome. The quantum measurement process is then seen to be random when successive measurements of non-commuting sharp observables are made: Knowledge of the outcome obtained for state preparation via measurement of given quantum observable does not provide additional information about the outcomes of measurements of a sharp observable with which it does not commute, such as when the x-spin of a spin-1/2 particle is measured just after its z-spin is measured, for example, using a Stern-Gerlach apparatus.

The general notion of randomness is as an extrinsic property of events that is dependent on properties of agents such as P—and, more importantly, the scientific communities they form—and the theories they use pertaining to all elements involved in their scientific activity. However, it only requires that an account of how probability influences credence need not involve an interpretation of probability that itself depends on credence, and in no way requires a *subjective* interpretation of the probability attributed by quantum mechanics; it only recognizes that *predictability* expressed via $\Pr_t$ is *epistemic* in character, cf. [81]. This notion is one that the reinforces the Copehagen approach: The randomness in quantum theory is evidenced in relation to measurements when two non-commuting sharp observables are measured in succession. The result of the second measurement is maximally unpredictable using the law of motion to predict it on the basis of the result of the first measurement and any other information obtained from the world external to the system.

6. Conclusions

Results from the investigation of the behavior of quantum probability in a range of novel circumstances after Heisenberg's clarification of the elements of the Copenhagen interpretation in the mid-1950s were shown to clarify further the nature and identity of the forms of causation, indeterminacy, and randomness that the interpretation attributes to quantum mechanics. A number of novel trade-off relations quantifying complementarity in additional areas of mathematics and physics were reviewed and described, as was the introduction of a number of mathematical constructions making indeterminacy more precise and extending its application. Results showing that the chancy nature of the results of measurements on quantum systems, and hence of the appearance of the quantum state, can be explicated in a way that accords with the Aristotelian form of causation introduced by Heisenberg in his late explication of the Copenhagen interpretation were also reviewed. These various developments in theoretical physics, some of which also bear directly on experimental physics, that appeared subsequent to Heisenberg's clarification were in this way shown to demonstrate the continually increasing value to physics of the Copenhagen approach to quantum mechanics.

Conflicts of Interest: The author declares no conflict of interest.

References and Notes

1. Jammer, M. *The Philosophy of Quantum Mechanics*; John Wiley and Sons: New York, NY, USA, 1974.
2. Busch, P.; Grabowski, M.; Lahti, P.J. *Operational Quantum Physics*; Springer: Berlin, Germany, 1995.

3.   Beller, M. *Quantum Dialogue: The Making of a Revolution*; The University of Chicago Press: Chicago, IL, USA, 1999.
4.   Schlosshauer, M.; Kofler, J.; Zeilinger, A. A Snapshot of Foundational Attitudes Toward Quantum Mechanics. *Stud. Hist. Philos. Mod. Phys.* **2013**, *44*, 222. [CrossRef]
5.   Heisenberg, W. The Development of the Interpretation of the Quantum Theory. In *The Philosophy of Quantum Mechanics*; Pauli, W., Ed.; Pergammon: London, UK, 1955.
6.   Scully, M.O.; Englert, B.-G.; Walther, H. Quantum Optical Tests of Complementarity. *Nature* **1991**, *351*, 111–116. [CrossRef]
7.   Zeilinger, A. On the Interpretation and Philosophical Foundation of Quantum Mechanics. In *Vastakohtien Todelisuus*; Ketvel, U., et al., Eds; Helsinki University Press: Helsinki, Finland, 1996, pp. 167–178;
8.   Zeilinger, A. Experiment and the Foundations of Quantum Physics. *Rev. Mod. Phys.* **1999**, *71*, S288. [CrossRef]
9.   Faye, J. *Niels Bohr: His Heritage and Legacy*; Kluwer: Dordrecht, The Netherlands, 1991.
10. Plotnitsky, A. *Epistemology and Probability*; Springer: New York, NY, USA, 2010.
11. Plotnitsky, A. *The Principles of Quantum Theory, from Planck's Quanta to the Higgs Boson*; Springer International Publishing: Cham, Switzerland, 2016.
12. Jaeger, G. *Entanglement, Information, and the Interpretation of Quantum Mechanics*; Springer: Heidelberg, Germany, 2009; Chapter 3.
13. Howard, D. Who Invented the 'Copenhagen interpretation'? A study in mythology. *Philos. Sci.* **2004**, *71*, 669–682. [CrossRef]
14. Jaeger, G. Quantum Potentiality Revisited. *Proc. R. Soc. A* **2017**, *375*, 20160390. [CrossRef] [PubMed]
15. Jaeger, G. "Wave-Packet Reduction" and the Quantum Character of the Actualization of Potentiality. *Entropy* **2017**, *19*, 15. [CrossRef]
16. Bub, J. *Interpreting the Quantum World*; Cambridge University Press: Cambridge, UK, 1997.
17. Busch, P.; Lahti, P.; Mittelstaedt, P. *The Quantum Theory of Measurement*, 2nd ed.; Springer: Heidelberg, Germany, 1996.
18. Heisenberg, W. *Physics and Philosophy*; Harper and Row: New York, NY, USA, 1958.
19. Mackey, G.W. *Mathematical Foundations of Quantum Mechanics*; W.A. Benjamin, Inc.: New York, NY, USA, 1968.
20. Maczyński, M.J. A remark on Mackey's axiom system for quantum mechanics. *Bull. Acad. Pol. Sci. Ser. Sci. Math. Astronom. Phys.* **1967**, *15*, 583–587.
21. Gleason, A.M. Measures on the Closed Subspaces of a Hilbert Space. *J. Math. Mech.* **1957**, *6*, 885. [CrossRef]
22. Amann, A.; Primas, H. What is the Referent of a Nonpure Quantum State. In *Potentiality, Entanglement and Passion-at-a-Distance*; Cohen, R.S., Horne, M., Stachel, J., Eds; Kluwer: Dordrecht, The Netherlands, 1997; pp. 9–30.
23. Von Neumann, J. *Mathematical Foundations of Quantum Mechanics*; Princeton University Press: Princeton, NJ, USA, 1955.
24. Dirac, P.A.M. *The Principles of Quantum Mechanics*; Clarendon Press: Oxford, UK, 1930.
25. Born, M. Quantenmechanik der Stoßvorgänge. *Z. Phys.* **1926**, *38*, 803. [CrossRef]
26. Accardi, L. L'edificio Matematico della Meccanica Quantistica Non-relativistica: Situazione attuale. *Nuovo Cimento* **1995**, *110*, 685–721. [CrossRef]
27. Jaeger, G. What in the (Quantum) World is Macroscopic? *Am. J. Phys.* **2015**, *82*, 896. [CrossRef]
28. Jaeger, G. Grounding the Randomness of Quantum Measurements. *Proc. R. Soc. A* **2016**, *374*, 20150238. [CrossRef] [PubMed]
29. Heisenberg, W. *The Physical Principles of Quantum Theory*; University of Chicago Press: Chicago, IL, USA, 1930.
30. Born, M. Statistical Interpretation of Quantum Mechanics. *Science* **1955**, *122*, 675–679. [CrossRef] [PubMed]
31. Lahti, P. Uncertainty Principle and Complementarity in Axiomatic Quantum Mechanics. *Rep. Math. Phys.* **1980**, *17*, 287. [CrossRef]
32. Robertson, H.P. The Uncertainty Principle. *Phys. Rev.* **1929**, *34*, 163. [CrossRef]
33. Dummett, M. Is Time a Continuum of Instants? *Philosophy* **2000**, *75*, 497–515. [CrossRef]
34. Shimony, A. Physical and Philosophical Issues in the Bohr–Einstein debate. In *Niels Bohr: Physics and the World*; Feshbach, H., Matsui, T.; Olesonet, A., Eds.; Routledge: London, UK, 1988; pp. 285–304.
35. Jaeger, G. *Quantum Objects*; Springer: Heidelberg, Germany, 2014; Chapter 2.

36. Kastner, R.E.; Kauffman, S.; Epperson, M. Taking Heisenberg's Potentia Seriously. *Int. J. Quantum Found.* **2018**, *4*, 158.

37. Accardi, L. Quantum Probability: New perspectives for the laws of chance. *Milan J. Math.* **2010**, *2*, 481–502. [CrossRef]

38. Jaeger, G. The Double-slit Experiment. In *Compendium of Quantum Physics*; Greenberger, D., Hentschel, K., Weinert, F., Eds.; Springer: Heidelberg, Germany, 2009.

39. Khrennikov, A. *Probability and Randomness: Quantum versus Classical*; Imperial College Press: London, UK, 2016; Chapter 6.

40. Jaeger, G.; Shimony, A.; Vaidman, L. Two Interferometric Complementarities. *Phys. Rev. Lett.* **1995**, *51*, 54–67. [CrossRef]

41. Busch, P.; Shilladay, C. Complementarity and Uncertainty in the Mach–Zehnder interferomenter and Beyond. *Phys. Rep.* **2006**, *435*, 1–31. [CrossRef]

42. Schrödinger, E. Die gegenwärtige Situation in der Quantenmechanik. *Die Naturwissenschaften* **1935**, *23*, 807. (In German) [CrossRef]

43. Schrödinger, E. Discussion of Probability Relations Between Separated Systems. *Math. Proc. Camb. Philos. Soc.* **1935**, *31*, 555–563. [CrossRef]

44. Clauser, J.F.; Horne, M.; Shimony, A.; Holt, R.A. Proposed experiments to test local hidden-variable theories. *Phys. Rev. Lett.* **1973**, *23*, 880. [CrossRef]

45. Aspect, A.; Grangier, P.; Roger, G. Experimental Test of Realistic Theories via Bell's Inequality. *Phys. Rev. Lett.* **1981** *47*, 460. [CrossRef]

46. Horne, M.; Shimony, A.; Zeilinger, A. Two-particle interferometry. *Nature* **1990**, *347*, 429–430. [CrossRef]

47. Jaeger, G.; Horne, M.A.; Shimony, A. Complementarity of One-particle and Two-particle Interference. *Phys. Rev. Lett.* **1995**, *48*, 1023. [CrossRef]

48. Abouraddy, A.; Nasr, M.B.; Saleh, B.E.A.; Sergienko, A.V.; Teich, M.C. Demonstration of the Complementarity of One- and Two-photon interference. *Phys. Rev. Lett.* **2001**, *63*, 063803. [CrossRef]

49. Saleh, B.E.A., Abouraddy, A.F., Sergienko, A.V., Teich, M.C. Duality Between Partial Coherence and Partial Entanglement. *Phys. Rev. Lett.* **2000**, *62*, 043816.

50. Jaeger, G. *Entanglement, Information, and the Interpretation of Quantum Mechanics*; Springer: Heidelberg, Germany, 2009.

51. Crull, E.; Bacciagaluppi, G. Translation of: W. Heisenberg, 'Ist eine Deterministische Ergänzung der Quantenmechanik möglich?'. Available online: http://philsci-archive.pitt.edu/8590/ (accessed on 4 May 2011).

52. Heisenberg, W. Ist eine deterministische Ergänzung der Quantenmechanik möglich? In *Wissenschaftlicher Briefwechsel mit Bohr, Einstein, Heisenberg u.a., Band II: 1930–1939*; Meyenn, K.V., Hermann, A., Weisskopf, V.F., Eds.; Springer: Berlin, Germany, 1985; pp. 407–418. (In German)

53. Bacciagaluppi, G.; Crull, E. Heisenberg (and Schrödinger, and Pauli) on hidden variables. *Stud. Hist. Philos. Mod. Phys.* **2009** *40*, 374. [CrossRef]

54. Margenau, H. Quantum-mechanical description. *Phys. Rev.* **1936**, *49*, 240. [CrossRef]

55. Srinivas, M.D. Collapse Postulate for Observables with Continuous Spectra. *Commun. Math. Phys.* **1980**, *71*, 131–158. [CrossRef]

56. Kraus, K. *States, Effects and Operations*; Springer: Berlin, Germany, 1983.

57. Ludwig, G. Versuch einer axiomatischen Grundlegung der Quantenmechanik und allgemeinerer physikalischer Theorien. *Z. Phys.* **1964**, *181*, 233–260. [CrossRef]

58. Gilton, M.J.R. Whence the Eigenstate–eigenvalue Link? *Stud. Hist. Philos. Mod. Phys.* **2016**, *55*, 92–100. [CrossRef]

59. Dirac, P.A.M. *Quantum Mechanics*, 4th ed.; Oxford University Press: Oxford, UK, 1958.

60. Leibniz, G.W. Lettre à la princesse Elisabeth, de fin 1678, in Leibniz, 1940, p. 58. (In France).

61. Leibniz, G.W. Lettre à Jacquelot du 20 novembre 1702 (Raisons que M. Jacquelot m'a envoyeés pour justifier l'argument contesté de des-Cartes qui doit prouver l'existence de Dieu, avec mes réponses), GP III 444. (In France).

62. Heisenberg, W. Sprache und Wirklichkeit in der modernen Physik. *Wort Wirklichk.* **1960**, *1*, 32.

63. Birkhoff, G.; von Neumann, J. The Logic of Quantum Mechanics. *Ann. Math.* **1936** *37*, 823. [CrossRef]

64. Dalla Chiara, M.L. Quantum logic. In: *Handbook of Philosophical Logic, Synthese Library*; Gabbay, D., Guenther, F., Eds.; Springer: Dordrecht, The Netherlands, 1936; Volume 166.

65. Kochen, S.; Specker, E.P. Logical structures arising in quantum theory. In *The Theory of Models*; Addison, J., Henkin, L., Tarski, A., Eds.; North-Holland: Amsterdam, The Netherlands, 1965.

66. Stairs, A. Quantum logic, realism, and value-definiteness. *Philos. Sci.* **1983**, *50*, 578–602. [CrossRef]

67. Bell, J.S. *Speakable and Unspeakable in Quantum Mechanics*; Cambrige University Press: Cambridge, UK, 1987.

68. Isham, C. *Lectures on Quantum Theory: Mathematical and Structural Foundations*; Imperial College Press: London, UK, 1995.

69. Doring, A. Kochen-Specker theorem for von Neumann algebras. *Int. J. Theor. Phys.* **2005**, *44*, 139–160. [CrossRef]

70. Holik, F.; Fortin, S.; Bosyk, G.; Plastino, A. On the Interpretation of Probabilities in Generalized Probabilistic Models. In *Quantum Interaction QI 2016. Lecture Notes in Computer Science*; de Barros, J., Coecke, B., Pothos, E., Eds.; Springer: Cham, Switzerland, 2017; Volume 10106.

71. Pitowsky, I. George Boole's 'Conditions of Possible Experience' and the Quantum Puzzle. *Br. J. Philos. Sci.* **1994**, *45*, 95–125. [CrossRef]

72. Boole, G. On the Theory of probabilities. *Philos. Trans. R. Soc.* **1862**, *152*, 225. [CrossRef]

73. Pitowsky, I. From George Boole to John Bell: The Origins of Bell's Inequality. In *Bell's Theorem, Quantum Theory and the Conceptions of the Universe*; Kafatos, M., Ed.; Kluwer: Dordrecht, The Netherlands, 1989.

74. Pitowsky, I. Betting on the Outcomes of Measurements: A Bayesian theory of quantum probability. *Stud. Hist. Philos. Mod. Phys.* **2003**, *34*, 395–414. [CrossRef]

75. Barnum, H.; Caves, C.M.; Finkelstein, J.; Fuchs, C.A.; Schack, R. Quantum Probability from Decision Theory? *Proc. R. Soc. Lond. A* **2000**, *456*, 1175. [CrossRef]

76. Hemmo, M.; Pitowsky, I. Quantum Probability and Many Worlds. *Stud. Hist. Philos. Mod. Phys.* **2007**, *38*, 333–350. [CrossRef]

77. Jaeger, G. Uncertainty Relations and Possible Experience. *Mathematics* **2016**, *4*, 40. [CrossRef]

78. Laplace, P.-S. *A Philosophical Essay on Probabilities*; Truscott, F.W., Emory, F.L., Trans.; Dover Publications: New York, NY, USA ,1951; p. 4.

79. Earman, J. *A Primer on Determinism*; D. Reidel: Dordrecht, The Netherlands, 1986; pp. 137–138.

80. Montague, R. Deterministic Theories. In *Formal Philosophy*; Thomason, R.H., Ed.; Yale University Press: New Haven, RI, USA, 1974.

81. Eagle, A. Randomness is unpredictability. *Br. J. Philos. Sci.* **2005**, *56*, 749. [CrossRef]

82. Hellman, G. Randomness and Reality. In *PSA 1978*; Asquith. P.D., Hacking, I., Eds; University of Chicago Press: Chicago, IL, USA, 1978; Volume 2, pp. 79–97.

entropy

MDPI

Letter

Dimensional Lifting through the Generalized Gram–Schmidt Process

Hans Havlicek [1] and Karl Svozil [2,3,*] (ID)

1   Institute of Discrete Mathematics and Geometry, Vienna University of Technology, Wiedner Hauptstraße 8-10/104, A-1040 Vienna, Austria; havlicek@geometrie.tuwien.ac.at
2   Institute for Theoretical Physics, Vienna University of Technology, Wiedner Hauptstraße 8-10/136, A-1040 Vienna, Austria
3   Department of Computer Science, University of Auckland, Auckland 1142, New Zealand
*   Correspondence: svozil@tuwien.ac.at

Received: 26 March 2018; Accepted: 10 April 2018; Published: 14 April 2018

Abstract: A new way of orthogonalizing ensembles of vectors by "lifting" them to higher dimensions is introduced. This method can potentially be utilized for solving quantum decision and computing problems.

Keywords: orthogonality; quantum computation; Gram–Schmidt process

PACS: 03.65.Aa; 02.10.Ud; 02.30.Sa; 03.67.Ac

The celebrated Gram–Schmidt algorithm allows the construction of a system of *orthonormal* vectors from an (ordered) system of *linearly independent* vectors. Let us mention that there exist a wide variety of proposals to "generalize" the Gram–Schmidt process [1] serving many different purposes. In contrast to these generalizations, we construct a system of *orthogonal vectors* from an (ordered) system of *arbitrary* vectors, which may be linearly dependent. (Even repeated vectors are allowed.) This task is accomplished by what will be called "dimensional lifting".

Some quantum computation tasks require the orthogonalization of previously non-orthogonal vectors. This might be best understood in terms of mutually exclusive outcomes of generalized beam splitter experiments, where the entire array of output ports corresponds to an ensemble of mutually orthogonal subspaces, or, equivalently, mutually orthogonal perpendicular projection operators [2].

Of course, by definition (we may define a unitary transformation in a complex Hilbert space by the requirement that it preserves the scalar product [3] (§ 73)), any transformation or mapping of non-orthogonal vectors into mutually orthogonal ones will be non-unitary. However, we may resort to requiring that some sort of angles or distances (e.g., in the original Hilbert space) remain unchanged.

Suppose, for the sake of demonstration, two non-orthogonal vectors, and suppose further that somehow one could "orthogonalize" them while at the same time retaining structural elements, such as the angles between projections of the new, mutually orthogonal vectors onto the subspace spanned by the original vectors. For instance, the two non-orthogonal vectors could be transformed into vectors of some higher-dimensional Hilbert space satisfying the following properties with respect to the original vectors: (i) the new vectors are orthogonal, and (ii) the orthogonal projection along the new, extra dimension(s) of the two vectors render the original vectors. A straightforward three-dimensional construction with the desired outcome can be given as follows: suppose the original vectors are unit vectors denoted by $|\mathbf{e}_1\rangle$ and $|\mathbf{e}_2\rangle$; and $0 < |\langle \mathbf{e}_1 | \mathbf{e}_2 \rangle| < 1$. Suppose further a two-dimensional coordinate frame in which $|\mathbf{e}_1\rangle$ and $|\mathbf{e}_2\rangle$ are planar; thus, we can write in terms of some orthonormal basis $|\mathbf{e}_1\rangle = \left(x_{1,1}, x_{1,2} \right)$ as well as $|\mathbf{e}_2\rangle = \left(x_{2,1}, x_{2,2} \right)$. Suppose we "enlarge" the vector space to include an additional dimension, and suppose a Cartesian basis system in that greater space that includes the two vectors of the old basis (and an additional unit vector that is orthogonal with respect to the original plane spanned by the original basis vectors).

Ad hoc, it is rather intuitive how two (not necessarily unit) vectors can be found that project onto the original vectors, and which are orthogonal: "create" a three-dimensional vector space with one extra dimension, assign a non-zero extra coordinate (such as 1) associated with this dimension for the first vector, and use the extra coordinate of the second vector for compensating any nonzero value of the scalar product of the two original vectors— in particular, whose coordinates with respect to the new basis are

$$
\begin{aligned}
|f_1\rangle &= \left(x_{1,1}, x_{1,2}, 1\right), \\
|f_2\rangle &= \left(x_{2,1}, x_{2,2}, -\left(x_{1,1}x_{2,1} + x_{1,2}x_{2,2}\right)\right),
\end{aligned}
\tag{1}
$$

which are orthogonal by construction.

It is not too difficult to find explicit constructions for the more general case of k vectors $|e_1\rangle, \ldots, |e_k\rangle$ in $\mathbb{R}^n$ (cf. Ref. [2] for a rather inefficient method).

In the following, for the sake of construction, we shall embed $\mathbb{R}^n$ into $\mathbb{R}^{n+k}$, such that we fill all additional vector coordinates of $|e_1\rangle, \ldots |e_k\rangle$ with zeroes. For the new, mutually orthogonal, vectors we make the following *Ansatz* by defining

$$
\begin{aligned}
|f_1\rangle &= \left(e_1, 1, 0, \ldots, 0\right), \\
|f_2\rangle &= \left(e_2, x_{2,1}, 1, 0, \ldots, 0\right), \\
&\cdots \\
|f_k\rangle &= \left(e_k, x_{k,1}, x_{k,2}, \ldots, x_{k,k-1}, 1\right),
\end{aligned}
\tag{2}
$$

with yet to be determined coordinates $x_{i,j}$. (The symbols e_i stand for all the n coordinates of $|e_1\rangle$.)

The unit coordinates 1 ensure that the new vectors are linearly independent. By construction, the orthogonal projection of $|f_i\rangle$ onto $\mathbb{R}^n$ renders $|e_i\rangle$ for all $1 \le i \le k$.

What remains is the recursive determination of the unknown coordinates $x_{i,j}$. Note that all $|f_j\rangle$ must satisfy the following relations: for $j > 1$, orthogonality demands that $\langle f_1|f_j\rangle = 0$, and therefore $\langle e_1|e_j\rangle + 1 \cdot x_{j,1} = 0$, and therefore

$$
x_{j,1} = -\langle e_1|e_j\rangle.
\tag{3}
$$

In this way, all unknown coordinates $x_{2,1}, \ldots, x_{k,1}$ can be determined.

Similar constructions yield the remaining unknown coordinates in $|f_2\rangle, \ldots, |f_k\rangle$. For $j > 2$, $\langle f_2|f_j\rangle = 0$, and therefore $\langle e_2|e_j\rangle + x_{2,1}x_{j,1} + x_{j,2} = 0$, yielding

$$
x_{j,2} = -\langle e_2|e_j\rangle - x_{2,1}x_{j,1}.
\tag{4}
$$

In this way, all unknown coordinates $x_{3,2}, \ldots, x_{k,2}$ can be determined.

This procedure is repeated until one arrives at $j = k - 1$, and therefore at the orthogonality of $|f_{k-1}\rangle$ and $|f_k\rangle$, encoded by the condition $\langle f_{k-1}|f_k\rangle = 0$, and hence

$$
\begin{aligned}
x_{k,k-1} = -\left(\langle e_{k-1}|e_k\rangle + \right. \\
\left. + x_{k-1,1}x_{k,1} + \cdots + x_{k-1,k-2}x_{k,k-2}\right).
\end{aligned}
\tag{5}
$$

The approach has the advantage that, at each stage of the recursive construction, there is only a single unknown coordinate per equation. This situation is well known from Gaussian elimination. The *Ansatz* also works if one of the original vectors is the zero vector, and if some of the original vectors are equal.

The resulting system of orthogonal vectors is not the only solution of the initial problem—to find an orthogonal vector that projects onto the original ones—which can be explicitly demonstrated by multiplying all vectors $|\mathbf{f}_1\rangle, \ldots, |\mathbf{f}_k\rangle$ with the matrix

$$\text{diag}\left(\mathbb{I}_n, c\mathbf{T}\right),\tag{6}$$

whereby $\mathbb{I}_n$ stands for the n-dimensional unit matrix, c can be a real nonzero constant, and $\mathbf{T}$ is a k-dimensional orthogonal matrix. (For complex Hilbert space, the orthogonal matrix needs to be substituted by a unitary matrix, and by a complex constant $c \neq 0$.)

On the other hand, we may reinterpret our procedure as follows: Let $|\mathbf{e}_1\rangle, \ldots, |\mathbf{e}_k\rangle$ be a system of vectors in $\mathbb{R}^n$, not necessarily spanning $\mathbb{R}^n$, and not necessarily being linearly independent. (The ordering of the vectors in this system will be essential throughout.) We embed $\mathbb{R}^n$ in $\mathbb{R}^{n+k}$ as we did above and denote the orthogonal complement of $\mathbb{R}^n$ by $C \cong \mathbb{R}^k$. Therefore, the first n coordinates of all vectors in C vanish, and $\mathbb{R}^{n+k}$ can be represented by a direct sum $\mathbb{R}^{n+k} = \mathbb{R}^n \oplus \mathbb{R}^k$. Additionally, we choose some (ordered) orthonormal basis of C, say, $|\mathbf{g}_1\rangle, \ldots, |\mathbf{g}_k\rangle$.

Then, there is a *unique* system of orthogonal vectors $|\mathbf{f}_1\rangle, \ldots, |\mathbf{f}_k\rangle$ in $\mathbb{R}^{n+k}$ such that the following conditions are satisfied:

1. For all $1 \leq i \leq k$, the orthogonal projection of $\mathbb{R}^{n+k}$ onto $\mathbb{R}^n$ sends $|\mathbf{f}_i\rangle$ to $|\mathbf{e}_i\rangle$.
2. The orthogonal projection of $\mathbb{R}^{n+k}$ onto C sends $|\mathbf{f}_1\rangle, \ldots, |\mathbf{f}_k\rangle$ to some (ordered) basis of the subspace C. Applying the Gram–Schmidt process to this (ordered) basis gives the orthonormal basis $|\mathbf{g}_1\rangle, \ldots, |\mathbf{g}_k\rangle$.

Indeed, in our previous *Ansatz*, we tacitly assumed the orthonormal basis $|\mathbf{g}_1\rangle, \ldots, |\mathbf{g}_k\rangle$ of C to comprise the orthogonal projections of the last k vectors of the standard basis $|\mathbf{b}_1\rangle, \ldots, |\mathbf{b}_{n+k}\rangle$ of $\mathbb{R}^{n+k}$ onto C. Condition 2 enforces the presence of all the 1s and 0s in Formula (2), since the Gram–Schmidt process, applied to the vectors

$$|\mathbf{f}_1\rangle - |\mathbf{e}_1\rangle, \ldots, |\mathbf{f}_k\rangle - |\mathbf{e}_k\rangle,$$

has to result in $|\mathbf{b}_{n+1}\rangle, \ldots, |\mathbf{b}_{n+k}\rangle$. Notice that the usual Gram–Schmidt process gives merely an *orthogonal* basis, whose vectors can be normalized in a second step in order to obtain an *orthonormal* basis. In our setting, however, such a second step is not allowed. As we saw above, now Condition 1 guarantees that $|\mathbf{f}_1\rangle, \ldots, |\mathbf{f}_k\rangle$ are uniquely determined.

Besides uniqueness, this construction has the additional advantage that the dot product in $\mathbb{R}^{n+k}$ "decays" into the sum of dot products in $\mathbb{R}^n$ and in $\mathbb{R}^k$: any basis vector $\mathbf{f}_i \in \mathbb{R}^{n+k}$ can be uniquely written as $\mathbf{f}_i = \mathbf{e}_i + \mathbf{h}_i$, where $\mathbf{e}_i$ and $\mathbf{h}_i$ represent the projection of $\mathbf{f}_i$ along $\mathbf{h}_i$ onto the original subspace $\mathbb{R}^n$, and the projection of $\mathbf{f}_i$ along $\mathbf{e}_i$ onto C, respectively. Since $\mathbf{e}_i$ is orthogonal to $\mathbf{h}_i$, for $i \neq j$, $\mathbf{f}_i \cdot \mathbf{f}_j = \mathbf{e}_i \cdot \mathbf{e}_j + \mathbf{h}_i \cdot \mathbf{h}_j = 0$, and thus

$$\mathbf{e}_i \cdot \mathbf{e}_j = -\mathbf{h}_i \cdot \mathbf{h}_j.\tag{7}$$

Let us, for the sake of a physical example, study configurations associated with decision problems that can be efficiently (that is, with some speedup with respect to purely classical means [2]) encoded quantum mechanically. The *inverse problem* is the projection of orthogonal systems of vectors onto lower dimensions. This method renders a system of non-orthogonal rays, also called *eutactic stars* [4–8], which can be effectively levied to mutually exclusive outcomes in a generalized beam splitter configurations [9,10] reflecting the higher dimensional Hilbert space.

One instance of such a quantum computation involving the reduction to ensembles of orthogonal vectors (and their associated span or projection operators) is the Deutsch–Jozsa algorithm, as reviewed in Ref. [2]. Another, somewhat contrived, problem can be constructed in three dimensions from a eutactic star

$$\frac{1}{\sqrt{3}}\left\{\left(1,1\right), \left(\tfrac{1}{2}\left[\sqrt{3}i-1\right], \tfrac{1}{2}\left[-\sqrt{3}i-1\right]\right),\right.$$
$$\left.\left(\tfrac{1}{2}\left[-\sqrt{3}i-1\right], \tfrac{1}{2}\left[\sqrt{3}i-1\right]\right)\right\},\tag{8}$$

which is the projection onto the plane formed by the first two coordinates of a three-dimensional orthormal basis

$$
\begin{aligned}
\mathfrak{B}_3 = \frac{1}{\sqrt{3}} \Big\{ &\left(1,1,1\right), \\
&\left(\tfrac{1}{2}\left[\sqrt{3}i-1\right], \tfrac{1}{2}\left[-\sqrt{3}i-1\right], 1\right), \\
&\left(\tfrac{1}{2}\left[-\sqrt{3}i-1\right], \tfrac{1}{2}\left[\sqrt{3}i-1\right], 1\right) \Big\},
\end{aligned}
\tag{9}
$$

which, together with the Cartesian standard basis, forms a pair of unbiased bases [11].

Still another decision configuration is the eutactic star

$$
\begin{aligned}
\frac{1}{2} \Big\{ &\left(1,1,1\right), \left(1,1,-1\right), \\
&\left(1,-1,1\right), \left(1,-1,-1\right) \Big\},
\end{aligned}
\tag{10}
$$

which is the projection onto the subspace formed by the first three coordinates of a four-dimensional orthormal basis

$$
\begin{aligned}
\mathfrak{B}_4 = \frac{1}{2} \Big\{ &\left(1,1,1,1\right), \left(1,1,-1,-1\right), \\
&\left(1,-1,1,-1\right), \left(1,-1,-1,1\right) \Big\}.
\end{aligned}
\tag{11}
$$

More concretely, suppose some, admittedly construed, function f, and some quantum encoding $|xf(y)\rangle$, where x and y stand for (sequences of) auxiliary and input bits, respectively, would yield one of the basis systems $\mathfrak{B}_3$ or $\mathfrak{B}_4$. By reducing the auxiliary bits x, one might end up with the eutactic stars introduced above. Alas, so far, no candidate of this kind has been proposed.

In summary, a new method of orthogonalizing ensembles of vectors has been introduced. Thereby, the original vectors are "lifted" to or "completed" in higher dimensions. This method could be utilized for solving quantum decision and computing problems if the original problem does not allow an orthogonal encoding, and if extra bits can be introduced that render the equivalent of the extra dimensions in which the original state vectors can be lifted and orthogonalized.

Compared with methods that were introduced [12–14] previously to optimally differentiate between two non-orthogonal states, the scheme suggested here is similar in the sense that, in order to obtain a better resolution, the effective dimensionality of the problem is increased. However, our scheme is not limited to the differentiation between two states, as it uses arbitrary dimensionality. More importantly, whereas our scheme is capable of separating different states precisely, but in general is non-unitary—indeed, the original vectors are not mutually orthogonal, but the lifted vectors are, thereby changing the angles among vectors, resulting in transformations that cannot be unitary—the former methods are unitary and probabilistic.

Acknowledgments: This work was supported in part by the European Union, Research Executive Agency (REA), Marie Curie FP7-PEOPLE-2010-IRSES-269151-RANPHYS grant.

Author Contributions: Both authors have read and approved the final manuscript.

Conflicts of Interest: The authors declare no conflict of interest.

References

1. Leon, S.J.; Björck, Å.; Gander, W. Gram–Schmidt orthogonalization: 100 years and more. *Numer. Linear Algebra Appl.* **2013**, *20*, 492–532.
2. Svozil, K. Orthogonal Vector Computations. *Entropy* **2016**, *18*, 156.
3. Halmos, P. *Finite-Dimensional Vector Spaces*; Springer: Berlin/Heidelberg, Germany, 1974.
4. Schläfli, L. Theorie der vielfachen Kontinuität. In *Gesammelte Mathematische Abhandlungen*. Springer: Basel, Switzerland, 1950; pp. 167–387. (In German)

5. Hadwiger, H. Über ausgezeichnete Vektorsterne und reguläre Polytope. *Comm. Math. Helv.* **1940**, *13*, 90–107. (In German)
6. Coxeter, H.S.M. *Regular Polytopes*, 3rd ed.; Dover Publications: New York, NY, USA, 1973
7. Seidel, J.J. Eutactic stars. In *Colloquia Mathematica Societatis János Bolyai*; North-Holland Publishing Co.: Amsterdam, the Netherlands, 1976; Volume 18, pp. 983–999.
8. Haase, D.; Stachel, D. Almost-orthonormal vector systems. *Beitr. Algebra Geom.* **1996**, *37*, 367–382.
9. Reck, M.; Zeilinger, A.; Bernstein, H.J.; Bertani, P. Experimental realization of any discrete unitary operator. *Phys. Rev. Lett.* **1994**, *73*, 58–61.
10. Żukowski, M.; Zeilinger, A.; Horne, M.A. Realizable higher-dimensional two-particle entanglements via multiport beam splitters. *Phys. Rev. A* **1997**, *55*, 2564–2579.
11. Schwinger, J. Unitary operator bases. *Proc. Natl. Acad. Sci.* **1960**, *46*, 570–579.
12. Ivanovic, I.D. How to differentiate between non-orthogonal states. *Phys. Rev. A* **1987**, *123*, 257259.
13. Peres, A. How to differentiate between non-orthogonal states. *Phys. Rev. A* **1988**, *128*, 19.
14. Jaeger, G.; Shimony, A. Optimal distinction between two non-orthogonal quantum states. *Phys. Rev. A* **1995**, *197*, 83–87.

entropy

MDPI

Concept Paper

On Interpretational Questions for Quantum-Like Modeling of Social Lasing

Andrei Khrennikov [1,2,*], Alexander Alodjants [1], Anastasiia Trofimova [1] and Dmitry Tsarev [1]

[1] Mechanics and Optics (ITMO) Department, National Research University for Information Technology, 197101 St. Petersburg, Russia; alexander_ap@list.ru (A.A.); nastasiatm@gmail.com (A.T.); dmitriy_93@mail.ru (D.T.)

[2] International Center for Mathematical Modeling in Physics and Cognitive Sciences, Linnaeus University, SE-351 95 Växjö, Sweden

* Correspondence: Andrei.Khrennikov@lnu.se

Received: 27 October 2018 ; Accepted: 1 December 2018; Published: 2 December 2018

Abstract: The recent years were characterized by increasing interest to applications of the quantum formalism outside physics, e.g., in psychology, decision-making, socio-political studies. To distinguish such approach from quantum physics, it is called *quantum-like*. It is applied to modeling socio-political processes on the basis of the social laser model describing *stimulated amplification of social actions*. The main aim of this paper is establishing the socio-psychological interpretations of the quantum notions playing the basic role in lasing modeling. By using the Copenhagen interpretation and the operational approach to the quantum formalism, we analyze the notion of the social energy. Quantum formalizations of such notions as a social atom, s-atom, and an information field are presented. The operational approach based on the creation and annihilation operators is used. We also introduce the notion of the social color of information excitations representing characteristics linked to lasing coherence of the type of collimation. The Bose–Einstein statistics of excitations is coupled with the *bandwagon effect*, one of the basic effects of social psychology. By using the operational interpretation of the social energy, we present the thermodynamical derivation of this quantum statistics. The crucial role of information overload generated by the modern mass-media is emphasized. In physics laser's resonator, the optical cavity, plays the crucial role in amplification. We model the functioning of social laser's resonator by "distilling" the physical scheme from connection with optics. As the mathematical basis, we use the master equation for the density operator for the quantum information field.

Keywords: quantum-like models; operational approach; information interpretation of quantum theory; social laser; social energy; quantum information field; social atom; Bose–Einstein statistics; bandwagon effect; social thermodynamics; resonator of social laser; master equation for socio-information excitations

1. Introduction

From the very beginning, it has to be pointed out that we tried to make this paper readable for people working in psychology, decision-making, cognitive, social, and political science, and having minimal knowledge about the mathematical apparatus of quantum physics. Therefore, we try to minimize the number of mathematical expressions, except Sections 4.4 and 9. The introduction is very detailed and its aim is to describe the general state of the art in applications of quantum theory to humanities. One one hand, we want to convince experts in humanities that quantum theory can resolve the well known problems, in particular, in decision theory. On the other hand, we want to convince physicists and especially those who work in quantum foundations and quantum information and probability that applications of quantum theory to humanities are not an exoticities: many top level experts, e.g., psychologists, work actively on quantum-like modeling.

1.1. Quantum versus Quantum-Like Models

Since the first days of quantum mechanics, the analogy between quantum physical and psychical processes sporadically attracted attention of leading physicists, psychologists, and philosophers. There can be mentioned the Pauli–Jung correspondence [1], see also [2], and Whitehead's attempt to unify physical and mental processes within the quantum picture [3,4]. There should be also emphasized the numerous contributions to creation of quantum models of brain's functioning [5–15]. However, we point out that the topic of this paper has nothing to do with their attempts to model cognition and consciousness from the genuine quantum physical processes in the brain. In the present paper, we proceed with the quantum-like approach, the operational application of the mathematical formalism of quantum theory, especially quantum probability and information, outside physics.

1.2. Quantum-Like Modeling of the Process of Decision-Making

Nowadays, quantum-like modeling is widely used in mathematical modeling in cognitive science, psychology, decision-making, game theory, economics and finances, social and political sciences (see, for example, monographs [16–23] and a few representative papers [24–38]). These applications are typically based on the use of quantum probabilistic calculus, instead of the classical one. It was demonstrated that experimental statistical data collected in, cognitive psychology, decision-making, social science, game theory, molecular biology, and epigenetic demonstrates quantum probabilistic features [18,19,22,23]. For example, such data can violate the classical formula of total probability and this violation can be mathematically expressed in the form of quantum interference. Some cognitive experiments demonstrating violation of the Bell type inequalities have been performed [35–38]. The violation can be interpreted as *contextuality of mental observables* represented as questions or tasks.

In theoretical studies in cognitive psychology and decision-making, quantum probability provides the adequate mathematical models for the basic psychological effects such as conjunction, disjunction, and order effects. Its use also resolves the fundamental paradoxes of the decision theory such as Allais, Ellsberg, and Machina paradoxes [39–41]. We remark that this resolution of the paradoxes of the classical decision theory (including expected utility, subjective utility, and prospect theories) and creation of the paradox-free (at least up to now) decision theory played a very important role in justification of applications of quantum probability outside physics.

1.3. Operational Formalism: Creation and Annihilation Operators

We emphasize that the quantum formalism provides only a formal operational description of physical processes; in particular, the spontaneous emission and stimulated absorption and emission which play the fundamental role in lasing theory. One of the important mathematical representations of the operational formalism is given in terms of the *creation and annihilation operators* $\hat{a}^\star, \hat{a}$. In quantum optics, $\hat{a}^\star$ represents creation of a photon through its emission; $\hat{a}$ represents disappearance of a photon from the field resulting from absorption of this photon by an atom. These are linear operators acting in complex Hilbert space representing the states of the quantum field. The operator $\hat{a}^\star$ is adjoint to the operator $\hat{a}$. The operator $\hat{n} = \hat{a}^\star \hat{a}$ plays the important role in the quantum field theory. This is the operator of the photons number. Its eigenstates are states $|n\rangle$ of the quantum field with the fixed number n of photons. Similar operational description can be given for the processes of state transitions in atoms.

The operational formalism is so useful in quantum theory, since here one cannot construct a more detailed description in terms of classical-like variables, known as hidden variables. Therefore, one proceeds with the operational formalism, representing preparations of system and observations. In this paper, as well in papers cited in Section 1.2, the operational formalism, including creation and annihilation operators (see, especially, [20,34]), is used to model the process of social lasing, stimulated amplification of social actions.

One of the reasons for application of the quantum operational formalism to modeling social processes is that, similarly to quantum physics, it is (practically) impossible to present the detailed account of all socio-psychological factors involved in stimulated amplification of social actions. In this situation, the operational description in terms of formal absorption and emission of information excitations can be fruitful.

We can refer to stormy discussions in socio-political literature (see [42–46] and further references in [47–49]) on the origin of color revolutions. These discussions are characterized by the diversity of opinions and the impossibility to present the detailed account of interrelation of all social, psychological, political, and economical and financial factors leading to such protest waves. The operational quantum model of "social laser" is based on ignoring these factors (playing the role of hidden variables). We are not interested in difference in, e.g., the emotional states of people in different countries or difference in the political situations in these countries. The operational formalism describes formally the processes of absorption and emission of portions of social energy; see Sections 1.5 and 4.4 for further justification (from the information-theoretical and thermodynamical principles) of applying the quantum formalism to modeling of social lasing.

Nowadays, the big-data approach is widely applied for analysis of social processes. The deep learning is one of the cornerstones of the big-data project. We remark that the deep learning can be considered as a form of the operational description. The intermediate layers of networks performing deep learning are just the operational components having no real social or cognitive meaning. In contrast to the big-data approach, which is based on extensive consumption of computational resources, the quantum operational approach is endowed with the powerful mathematical formalism of analytic analysis. This formalism can be used for determination of the key-parameters playing the crucial role in stimulated amplification of social actions (see Section 9).

1.4. Social Laser as a Fruit of the Quantum Information Revolution

In this paper, we are interested in the information component of quantum theory. We recall that the recent quantum information revolution generated *the information interpretation of the quantum theory*. In fact, there were proposed a few often competing information interpretations [50–63]. However, all of them are characterized by the paradigm shift: quantum systems are treated as merely carriers of information. Thus, from this viewpoint, quantum theory is about information flows. Hence, it may be possible to apply the quantum information approach even outside physics by identifying the quantum-like features of processes under consideration.

In [47,48], the author proposed the quantum-like model of the social (or more generally information) laser, the social analog of the physical laser. The social laser theory was used for the mathematical description of *stimulated amplification of social actions* [49].

The model of the social laser is based on treatment of humans as carriers of *social energy*. In [47,48], the notion of the social energy was (rather schematically) discussed in the relation to the social lasing model. In this paper, we shall present the detailed analysis of this notion based on the operational approach to quantum theory and its information interpretation (Section 3).

In the simplest setting, we consider a two level cognitive system with the relaxation and excitation states. We call such a system a (two level) *social atom*, *s-atom*. In this paper, we restrict our considerations to two energy level systems. In the quantum-like modeling, the discrete structure of the social energy levels of *s*-atoms, energy quantization, has no straightforward relation to neurophysiology. In particular, the two levels structure is based on so to say "to be or not to be" scaling of the social energy. This is a kind of emotional quantity. Of course, emotions are coupled to physical and neurophysiological processes in human's body, but this coupling is very complicated. In addition, it is not important for us in the present paper.

A social gain medium is composed of *s*-atoms. They interact with the information field generated by the mass-media and the Internet. This field is modeled as a quantum information field carrying quantized portions of social energy, social excitations. This field was formally introduced in [47,48].

Now, we put essential efforts to propose the proper interpretation of excitations of this quantum field as carriers of the social energy (Section 3.2).

Additionally to the social energy, we introduce the notion of a social color of an information excitation (Section 4.2). This quantity represents additional characteristics of information excitations linked to lasing coherence of the type of polarization and collimation (Section 4.5). We couple the *bandwagon effect*, one of the fundamental effects of social psychology, with the Bose–Einstein statistics of indistinguishable social excitations (Section 5).

1.5. Powerful Information Flows as the Basic Condition of Social Laser Functioning

The basic feature of quantum information fields is indistinguishability of excitations carried by them. This is indistinguishability with respect to the social energy carried by excitations (see Section 4.3 for further consideration).

The Bose–Einstein statistics of information excitations playing the crucial role in social lasing can be derived (by appealing to thermodynamical considerations) from such energy based indistinguishability; see Section 4.4.

People overloaded by information do not put essential efforts to analyze contents of communications delivered the by mass-media. They function as absorbers of social energy delivered by communications emitted by TV, newspapers, and the Internet. In particular, to approach population inversion, the mass-media can pump into a human gain medium a flow of shock news about various catastrophes, violence, and political scandals. Agents absorb the social energy. Later, this energy can be liberated and directed through the laser-like process of stimulated emission. Therefore, *the power of information flows plays the crucial role in creation of quantum-like processes of absorption and emission of the social energy.* Nowadays, this flow is extremely powerful. Hence, it is easier to create social (information) lasers than say 150 years ago.

1.6. Resonators of Physical and Social Lasers

In physics, the laser's resonator in the form of the optical cavity plays the crucial role in the process of amplification of the output beam and making it coherent. In this paper, we model the functioning of social laser's resonator by "distilling" the physical scheme to exclude the straightforward connection with optcis. We proceed with the *quantum master equation for the density operator* describing excitations of the quantum information field inside the resonator of the social laser; see Section 9.3. The main aim is establishing the proper social interpretations of the basic quantities and parameters of this dynamical system.

The Internet based *Echo Chamber* is considered as the important example of social resonators. Its functioning is mathematically represented by the field of social excitations in the form of posts and comments interacting with the human gain medium.

One of the basic consequence of the quantum-like dynamics is the existence of the threshold value for the pump parameter (see Section 9.5): if the power of pumping is essentially higher than this threshold, then practically all social energy pumped into the human gain medium is transferred into the output beam of social actions.

This is the good place to emphasize once again the role of huge power of the information flows generated by modern mass-media.

2. Physical Laser: Schematic Presentation

Since we really hope that this paper would be interesting for researchers working in cognitive, social, and political sciences and having no educational background in quantum physics, we present schematically the basic scheme of the physical laser functioning. For the moment, we do not consider the optical cavity component of the physical laser.

2.1. Spontaneous and Stimulated Emission

We start with description of the processes of spontaneous and stimulated emission for excited two level atoms with energy levels $E = E_1, E_2$. Denote the resonance energy of an atom by $E_A = E_2 - E_1$. Typically, in the physical literature, one proceeds with frequencies and speaks about the resonance frequency $\omega_A = \frac{E_A}{2\pi\hbar}$, where h is the Planck constant. To proceed to the social modeling, we try to eliminate the space-time picture from the model. In particular, we want to operate solely with energies.

We consider the quantum electromagnetic field interacting with atoms and describe the physical processes generated by the interaction:

1. An atom in the excited state can spontaneously emit a photon. This process is irreducibly random, i.e., even for a single atom, it is impossible to predict neither the instance of time nor the direction of emitted photon.
2. Atoms in the ground state can absorb from this field only photons having the resonance energy $E_A = E_2 - E_1$.
3. An excited atom interacting with photons with energy E_A emits a photon of the same energy
4. This output flow of photons is coherent. All photons produced from the "seed-photon" have the same features: direction of flow, polarization, and energy.

Thus, one photon produces two, these two interact with two atoms and produce four photons, after n-step, this process generates $N = 2^n$ photons. We note that this description and its illustration by Figure 1, although typical for physics textbooks, is too straightforward. In fact, an atom interacts not with a single photon, but with all photons in the field (see Section 5 and Equation (10)).

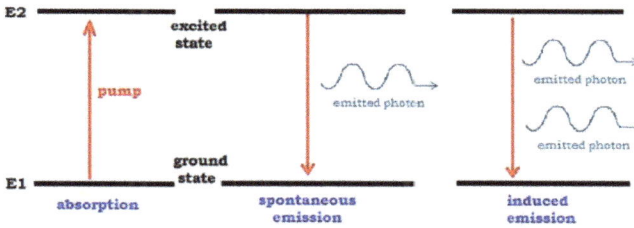

Figure 1. Emission and absorption of photons.

2.2. Population Inversion

The cascade process described above plays the crucial role in generating laser beams carrying huge energy. One of the problems is spontaneous emission. (Of course, the primary problem is that the thermodynamic heat-bath Boltzmann distribution leads to higher population in the lower levels than in the upper levels.)

Pumping photons into a gain medium (an ensemble of atoms in the ground state) transfers ground state atoms into excited atoms. However, spontaneously, they can fall back to the ground level. The basic step in generating lasing is approaching *population inversion*. This is transition from an ensemble of atoms in the ground state to an ensemble in which more than 50% of atoms are in the excited state.

We remark that, for a gain medium composed of two level atoms, population inversion cannot be approached (at least straightforwardly) because the transition probabilities are equal: $p(E_1 \rightarrow E_2) = p(E_2 \rightarrow E_1)$. These probabilities are known as the *Einstein coefficients*. Their equality for the electromagnetic field can be proven by using thermodynamics for indistinguishable systems following Bose–Einstein statistics. For the electromagnetic field, one needs a gain medium composed of atoms with at least three energy levels, $E_1 < E_2 < E_3$.

The above considerations are about the physical laser based on the quantum electromagnetic field. However, in general, information thermodynamics [47] does not imply the coincidence of the Einstein coefficients for a two-level system. In principle, the social laser can be based on the simplest *s*-atoms having two levels of social energy.

3. Social Energy

The notion of *social energy* plays the crucial role in our framework. We start considerations with a general remark that in quantum theory value a of observable A obtained in its measurement cannot be interpreted as the property of system S on which the measurement is performed. By the Copenhagen interpretation, this value is generated in the complex process of interaction of S and a device used for A-measurement. In particular, this interpretation has to be applied to energy observable E. Although one often says, e.g., "the energy of the electron", the correct meaning of this statement is about the output of the E-measurement. The Copenhagen interpretation is well accommodated to the notion of states' *superposition*. Quantum system S can be in state ψ of superposition of two different energy levels E_1, E_2

$$\psi = c_1|E_1\rangle + c_2|E_2\rangle, \tag{1}$$

where c_i are complex numbers, probability amplitudes, such that $|c_1|^2 + |c_2|^2 = 1$. If system S is in this state, then the probability to get the value E_i of the energy observable is equal to $p_i = |c_i|^2$. The states $|E_1\rangle, |E_2\rangle$ correspond to the definite energy levels.

Supported by the quantum interpretation of the notion of energy, we are ready to consider the very complex notion of a *social (mental) energy*. This notion has been actively used in cognition, psychology, social and political science (since the works of James [64] and Freud [65,66], later by Jung [67], see also [68]) and recently in economics and finances, multi-agent modeling, evolution theory and industrial dynamics [69–71], see also [47–49] for details. In addition, of course, these previous studies are supporting for our model. However, we emphasize once again that the application of the Copenhagen methodology simplifies and clarifies essentially the issue of the social energy.

3.1. Energy of s-Atoms

In this framework, the simplest quantification of the social (mental) energy can be done by the question (observable) $E =$ "Are you in the state of relaxation or excitement?" This observable takes two values, say $E_1 = 0, E_2 = 1$. The Copenhagen interpretation is strongly involved. Before being asked the E-question, *s*-atoms can be in superposition of these two states. Only by confronting the E-question *s*-atom determines his/her state. The social energy observable can be represented in different forms; for example, in the form of the question $E =$ "Shall you go to demonstration against Trump or Brexit?" Of course, we need not be restricted to the simplest "yes"–"no", to be or not to be, quantification.

Finer quantifications of the social energy can considered as well. Different types of *s*-atoms can emit and absorb social energy portions of different magnitudes. Here, a type of an *s*-atom is determined by her/his psyche and social environment. In the operational formalism, we can proceed with some grading of the possible social energy levels for *s*-atoms. Since we restrict consideration to two-level *s*-atoms, they are characterized by the social energy levels $E = E_1, E_2$, the ground and excited states, and the resonance energy $E_A = E_2 - E_1$. We remark that in principle the ground level energy for one type of *s*-atoms can be higher than the excitation level energy for another type. Social lasing is possible only for a gain medium with the homogeneous structure of energy levels.

This methodology demystifies the notion of social (mental) energy. Of course, the same Copenhagen methodology can be applied to any social (mental) observable. One should not be surprised that the methodology of quantum physics is applicable outside it. The Copenhagen interpretation presents the very general methodology which is applicable to any kind of measurement. We remark that the use of this measurement methodology does not imply that the whole apparatus of quantum theory can be applicable. One should be careful by checking constraints on the class of systems and observables leading to applicability of one or another part of quantum theory. The quantum-like approach does not mean to copy straightforwardly the complete quantum theory to say social science. For example, to derive the Bose–Einstein statistics, we have to assume indistinguishability of information quanta, excitations of the quantum information field (see Section 4.1).

3.2. Energy of the Quantum Information Field

In physics, energy can be assigned not only to atoms, material systems, but also to carriers of interactions, e.g., photons or neutrino, which are excitations of corresponding quantum fields. (Here "assigned" has the operational meaning: "can be measured"). In social lasing, the interaction processes are formally modeled with the aid of a quantum information field generated by a variety of information sources (see Section 4.1). Communications "emitted" by them carry portions of the social energy. These quanta of social energy are interpreted as excitations of the quantum information field.

Again, as in the case of s-atoms, we can proceed with "to be or not to be" quantification of the social energy carried by communications. If an s-atom in the ground state absorbs energy from communication $\mathcal{C}$ (and becomes excited), then $\mathcal{C}$ carries social energy $E_2 = 1$. If $\mathcal{C}$ cannot excite a social atom, then $\mathcal{C}$'s energy $E_1 = 0$.

This social energy quantification depends on the concrete ensemble of s-atoms, a social group. It is easy to give examples of social and political communications which would excite average Englishman or American, but not Russian or Chinese, and vice versa. Thus, the definition of information field's energy is purely operational. In some sense, it is even "more operational" than the definition of the energy for a quantum physical field. It is meaningless to speak about the social energy of the information field without to describe the class of "detectors"; in our case, these are s-atoms. As in the case of s-atoms, it is possible to proceed with models based on finer scales of the social energy assigned to excitations of the information field. Each communication $\mathcal{C}$ is characterized by social energy $E_{\mathcal{C}}$. It can be absorbed by an s-atom with the resonance energy $E_A = E_{\mathcal{C}}$.

A variety of communications can carry the same portion of the social energy. All communications with $E_{\mathcal{C}} = E$, where E is the fixed portion of the social energy are considered as equivalent from the viewpoint of energy delivering. They can be represented by the same field's state $|E\rangle$, the ket-vector in field's state space. We say that E determines *a mode of the quantum information field: E* is the analog of characteristic energy $E_\omega = h\omega$ of the electromagnetic mode with frequency ω.

How many elementary excitations of energy E are carried by the E-mode of the quantum information field? It depends on the power of information sources emitting communications belonging to the E-mode.

4. Social Laser

4.1. Quantum Field Representation of Information Flow Generated by Mass-Media

For the physical laser, the electromagnetic field is the basic energy source. The Bose–Einstein statistics of excitations (quanta) of this field plays the crucial role in laser's functioning.

For a social analog of the physical laser, the basic (social) energy source is the *information field* generated by the mass-media and the Internet. As was discussed in Section 3.2, this field can be considered as composed of information quanta field's excitations. These excitations are emitted by a variety of information channels and absorbed by humans, s-atoms. The information field is not a classical physical field defined on the physical space-time, as, e.g., the classical electromagnetic field given by the vector of the electric and magnetic fields $(E(x), B(x)), x \in \mathbf{R}^3$. We model the information field operationally with the aid of creation and annihilation operators, as a quantum field, see Section 1.3.

We remark that even in physics only the electromagnetic field has its classical counterpart with the space-time representation. If we consider, for example, the neutrino field, this field has only the quantum representation.

In the quantum field theory, a state of a field is mathematically described by a normalized vector belonging to the Fock space—the complex Hilbert space representing superpositions of possible excitations of the field. Consider the E-mode of the quantum information field. It represents excitations in the form of communications carrying the portion of social energy E: communications which are graded by the social group under consideration as having social energy E. In the Fock space representation, the quantum state of this field is represented as superposition

$$\Phi = \sum c_n |n, E\rangle = \sum_n \frac{c_n}{\sqrt{n!}} \hat{a}_E^{\star n} |0\rangle, \qquad (2)$$

where $\sum |c_n|^2 = 1$. Here, $\hat{a}_E^{\star}$ is the creation operator for E-excitations; $|0\rangle$ denotes the vacuum state of the field—no information excitations. For the real information sources, the sum is always finite. The use of infinite series is the price for mathematical consistency—the possibility to use a Hilbert space.

Generally, a social group is exposed to radiation of the quantum information field containing different modes of the social energy. It is natural to consider the discrete grading of the social energy. Here, we follow von Neumann who pointed out [72] that quantum observables with continuous spectrum are just mathematical idealizations of real measurement procedures. For such discrete grading, the Fock state can be written in the standard form

$$\Phi = \sum \frac{c_k}{\sqrt{k!}} \hat{a}_{E_1}^{\star k_1} \cdots \hat{a}_{E_n}^{\star k_n} \cdots |0\rangle, \qquad (3)$$

where $k = (k_j)$ and, for each multi-index k, only a finite number of k_j differs from zero; the squared sum of absolute values of the coefficients equals to one.

4.2. Coloring Information Excitations

For the quantum electromagnetic field, excitations are photons and photon's type is determined by index $\lambda = (E, \alpha)$. Here, E is photon's energy and α encodes additional characteristics linked to lasing coherence of the type of polarization and collimation. In the same way, for the quantum information field, the type of an excitation is determined by index $\lambda = (E, \alpha)$. Here, E is *the social energy* carried by an excitation of the information field and α is a *social color* of this excitation encoding its basic social characteristics. The color of an information excitation is linked to coherence of social actions generated by a social laser. Thus, creation and annihilation operators have to be labeled not by just the energy index E, but by index λ, including the state representation, cf. Equation (3): $|n, \lambda\rangle = \hat{a}_\lambda^{\star n} |0\rangle$.

We can mention a few examples of social colors of excitations of the quantum information field: war in Iraq, elections in USA, Brexit, tsunami in Japan, sex scandal, anti-globalism, climate change, racism, sexism, Trump, and so on. Determination of social colors is a socio-psychological phenomenon. It depends on a social group. Thus social colors are not internal characteristics of the information field. Their depend on a social group exposed to "information radiation" generated by mass media.

The social analog of lasing can be initiated only in a social group, a social gain medium, with sufficiently rough coloring structure—to approach high concentration of excitations of the same social color mode in the *output beam* of excitations. For example, social color α = "sex scandal in Conservative and Unionist Party (UK)" is appropriate for social lasing. This is the proper social color for the communication: "A BOMBSHELL dossier naming and shaming 36 Tories suspected of inappropriate sexual behaviour has emerged as Westminster remains engulfed by a sex abuse scandal". (Express, 30 October 2017). However, if somebody would spit this color and started to use 36 colors corresponding to concrete Tory-executives involved in the sex scandals, then such s-atom is not a proper subject for social lasing.

4.3. Indistinguishability from Information Overload and Complexity

For *s*-atoms, excitations of the information field carrying the same social energy and color, $\lambda = (E, \alpha)$, are *indistinguishable*. Indistinguishability is the basic condition leading to quantum statistics through thermodynamical analysis, both for physical and information systems (see Section 4.4). Indistinguishability of excitations generated by the mass-media is relative. It corresponds to coarse graining of coloring and depends on a social group, a collection of *s*-atoms. If *s*-atoms do not operate in the indistinguashability regime, i.e., they perform the detailed analysis of contents of communications, then the quantum statistical description is inapplicable for them. In this case, social lasing is impossible. We remark that the detailed analysis of communication's content preassumes the use of Boolean logic. Thus, the indistinguishability regime implies deviations from Boolean logic, cf. [73,74].

The *information overload* is one of the basic reasons for operation at the indistinguishability regime. People simply do not have time and information processing resources for the deep analysis of communications' contents. They proceed with coarse graining leading to relative indistinduishability. Here, "relative" has the meaning of relative with respect to a social group. In the modern human society, the information overload is combined with complexity of information delivered by a variety of information sources. The majority of population simply does not have mental capacity to perform the detailed analysis of complex socio-political, financial, and ecological problems.

Coarse graining, "rough coloring" and the indistinguishability regime are the basic features of human cognition. However, the information overload and complexity led to tremendous increase of their role.

4.4. From Statistical Mechanics to Thermodynamics of Indistinguishable Systems

Here, we follow [47], but with emphasis of the operational meaning of the notion of the social energy. In turn, the presentation in paper [47] on transition from statistical mechanics of indistinguishable systems to thermodynamics is based on Schrödinger's book [75].

Consider a system which is composed of *m* indistinguishable subsystems. Compound system will be denoted by $\mathcal{S}$ and its subsystems by S with indexes. It is assumed that an observer can assign to all systems some quantity called energy and satisfying the additivity requirement with respect to its distribution over subsystems of a system. We shall denote the energy of $\mathcal{S}$ by $\mathcal{E}$ and the energy of S by E. These quantities should not be interpreted as objective properties of systems. Energy $E = E(S)$ can be assigned to S as the output of observation performed by an observer on S. The same interpretation is used for energy $\mathcal{E} = \mathcal{E}(\mathcal{S})$ assigned to $\mathcal{S}$.

As always derivation of thermodynamical quantities from ensemble statistics is started with construction of *partition function Z*. The possible energy levels of a subsystem S are denoted by $E_1, ..., E_j, ...$ An energy level $\mathcal{E}_k$ of *m*-particle compound system $\mathcal{S}$ is characterized by a sequence of natural numbers $m_1, ..., m_j, ...$ of subsystems on the corresponding levels. The latter means "subsystems with measurement outputs" $E_1, ..., E_j, ...$ Here, it is crucial that subsystems with the same value of energy E_j are not distinguished one from another. This indistinguishability determines the form of Z and, hence, all thermodynamical quantities. We have

$$\mathcal{E}_k = \sum_s E_s m_s. \tag{4}$$

Thus, the partition function is given by the sum

$$Z = \sum_{(m_s)} e^{-\mu \sum_s E_s m_s}, \tag{5}$$

where symbol (m_s) denotes an admissible set of numbers m_s. For the moment, $\mu > 0$ is just a parameter of the model.

From ln Z, it is possible to deduce the basic thermodynamical quantities; in particular, the average value of m_s

$$\bar{m}_s = -\frac{1}{\mu}\frac{\partial \ln Z}{\partial E_s}.$$ (6)

Now, different statistics for systems composed of indistinguishable subsystems can be obtained by consideration of different possible ranges of values of natural numbers m_s :

1. $m_s = 0, 1, 2, ...$ (Bose–Einstein statistics),
2. $m_s = 0, ..., q$, where $q \geq 1$ is a natural number (parastatistics).

In physics, $q = 1$, i.e., $m_s = 0, 1$; this is the case of the Fermi–Dirac statistics. We remark that in quantum physics selection of the Fermi–Dirac statistics from a bunch of parastistics is just postulated. Of course, it is confirmed by the experimental situation. One cannot exclude that in social science parastistics with $q > 1$ can find applications.

By restricting considerations to the Bose–Einstein and Fermi–Dirac statistics and following Schrödinger [75], we find that the corresponding partition functions can be expressed as

$$Z = Z_{\text{BE}} = \prod_s \frac{1}{1 - e^{-\mu E_s}}, \ Z = Z_{\text{FD}} = \prod_s (1 + e^{-\mu E_s}).$$ (7)

This leads to the following basic expression for the average value of m_s

$$\bar{m}_s = \frac{1}{\frac{1}{\xi}e^{\mu E_s} \mp 1},$$ (8)

where $0 < \xi \leq 1$ is so called parameter of degeneration, $\xi = \xi(m)$. We remark that, for photons, $\xi = 1$, with the Bose–Einstein statistics.

Then, one gets the average energy as

$$U = \sum_s \frac{\alpha_s}{\frac{1}{\xi}e^{\mu E_s} \mp 1}.$$ (9)

In physics, the quantity T inverse to parameter μ is interpreted as temperature. In social modeling, we can speak about a kind of the *social temperature*. This is a complicated notion and we are not ready to discuss it in detail in this paper. On one hand, we can try to proceed as in classical thermodynamics. However, even by mimicking classical thermodynamics, it is important to remember that such a quantity is of the socio-emotional type. Thus, it cannot be considered as the objective feature of a social system. One can try to define social temperature through consideration of classes of equivalent thermometers, measurement procedures for the social temperature. However, the above thermodynamical considerations for indistinguishable systems represent the quantum situation. Hence, the classical definition of temperature does not match them. As in quantum physics, we can try to introduce a kind of social temperature as a parameter characterizing phase transitions. This is a complicated mathematical theory and we postpone such considerations to one of further publications.

4.5. Coloring Role: Pumping versus Emission

We also point to the following striking similarity in behavior of atoms and s-atoms. A portion of social energy absorbed by s-atom generally "lost its color".

Somebody, say Elena, living in Moscow absorbed a social excitation emitted by the Russian radio-station, *Echo of Moscow*. Typically, such excitations have the anti-corruption colors. However, this does not mean that her spontaneous relaxation would be directed against corruption. She can emit the portion of social energy absorbed from *Echo of Moscow* in a family scandal or another kind of private or social action.

Such behavior is similar to behavior of atoms interacting with photons. If an atom absorbs a photon carrying some concrete physical characteristics, "color", such as direction or polarization, then it immediately forgets about its pre-absorption value. Later, it can emit a photon with a different color, i.e., a photon flying in the direction different from the direction of the pre-absorption photon.

However, we remind that the quantum theory is about observational quantities. We do not know what happens inside the atom between absorption and emission of photons. We neither know what happens inside the head between hearing the news from *Echo of Moscow* and going to the kitchen to start scandaling or to the center of Moscow to protest.

This kind of memory lost can be very useful in approaching population inversion (see [47]). There is no need to pump in a gain medium the social energy of the same color as in laser's output beam. In principle, the social energy pumping need not be based solely on information about corruption and other dysfunctions of the government. Shocking news about catastrophes, tornado, killers are the important part of energy-pumping in human gain medias.

Now, we turn to physics. In contrast to the process of absorption-emission, the process of stimulated emission of photons is characterized by "color" conservation: the "colors" of the emitted photon and photons stimulating emission coincide. We also point out that in the process of stimulated emission, the crucial role is played by intensity of the stimulating electromagnetic field, so-called Bosonic effect: increase of probability of stimulated emission with increase of intensity of the flow bosonic excitations (see Section 5 and Equation (10)).

In Section 5, we compare this effect with the bandwagon effect in psychology. The operational identity of these two effects, physical and psychological, supports application of the quantum field formalism and methodology to social processes. Stimulated emission of social excitations by excited *s*-atoms can be considered as exhibition of the bandwagon effect: *s*-atom in the excited state exposed by radiation compounded of α-colored social excitations would emit a social excitations of the same color.

5. Comparing Stimulated Emission in Quantum Physics and Bandwagon Effect in Psychology and Social Science

We stress that the quantum field description of the stimulated emission is a *collective effect*, i.e., an atom interacts with a bunch of photons and not just with an individual photon. It interacts with all excitations of the quantum electromagnetic field having the resonance energy of this atom, $E_A = E_2 - E_1$.

The crucial role is played by the Bose–Einstein statistics of the photons. We consider the fixed energy (frequency) mode of the quantum electromagnetic field. For fixed color mode α, n-photon state $|n, \alpha\rangle$, can be represented in the form of the action of the photon creation operator $a_\alpha^\star$ corresponding to this mode on the vacuum state $|0\rangle$:

$$|n, \alpha\rangle = [(a_\alpha^\star)^n / \sqrt{n!}]|0\rangle. \tag{10}$$

This representation gives the possibility to find that the transition probability amplitude from the state $|n, \alpha\rangle$ to the state $|n + 1, \alpha\rangle$ equals to $\sqrt{(n + 1)}$. On the other hand, it is well known that the reverse process of absorption characterized by the transition probability amplitude from the state $|n, \alpha\rangle$ to the state $|(n - 1), \alpha\rangle$ equals to $\sqrt{n}$. .

Thus, in a quantum Bosonic field increasing photons' number leads to increasing the probability of generation of one more photon in the same state. This constitutes one of the basic quantum advantages of laser stimulated emission showing that the emission of a coherent photon is more probable than the absorption. This is the strong argument in favor of using the quantum modeling of lasing.

This behavior of photons or more generally excitations of any quantum Bosonic field matches *the cognitive bias known as the bandwagon effect* [76]. This effect is characterized by the probability of individual adoption increasing with respect to the proportion who have already done so. People are not interested in underlying rational justification based on Boolean logic. They "hop on the bandwagon" by taking into account only the number n of those who are already seating on it. It is important to stress that, for an agent interacting with bandwagon's population, personalities of people on bandwagon play no role: these people are indistinguishable.

This indistinguishability is only with respect to this concrete interaction: social, financial, racial, gender, or political action characterizing this "bandwagon." Of course, people seating on a bandwagon are individual agents who can differ crucially: biologically, mentally, culturally.

As we have already emphasized, indistinguishability is the crucial assumption leading to quantum statistics. In the case of bandwagon effect, this is the Bose–Einstein statistics. Of course, it need not be exactly the photon statistics.

Thus, the bandwagon effect can be considered as social exhibition of the Bose–Einstein statistics caused by indistinguishability.

To model social lasing, we consider the information version of the bandwagon effect: s-atom interacts mainly not with other s-atoms, but with excitations of the information field generated by the mass-media and the Internet as well as emitted by other agents. For the concrete social color, these excitations are indistinguishable. In addition, s-atom in the excited state surrounded by information excitations of social color α emits excitation of the same color with probability proportional to the number of excitations. The crucial role is played by the field coherence, with respect to the concrete color.

We can conclude that the formal mathematical model is the same for physical and social structures. This is the model of stimulated emission of a system, physical or human, interacting with some Bosonic field.

6. Social Lasing Schematically

Each class of information communications is characterized by the social energy. Coherence corresponds to social colour sharpness (ideally one single mode α) generating a coherent beam of social actions. People in the excited state interacting with α-colored excitations of the information field would also emit α-colored excitations.

For example, a gain medium consisting of agents in the excited state and stimulated by the anti-corruption coloured information field would "radiate" a wave of anti-corruption protests. The same gain medium stimulated by some information field carrying another social colour would generate the wave of actions corresponding this last colour.

The amount of the social energy carried by communications stimulating lasing should match the resonance energy of s-atoms in the human gain medium.

To approach the population inversion, the social energy is pumped into the gain medium. This energy pumping is generated by the mass-media and the Internet sources. The gain medium should be homogeneous with respect to the social energy spectrum. In the ideal case, all s-atoms in the gain medium should have the same spectrum, $E = E_1, E_2$. In reality, it is impossible to create such a human gain medium. As in physics, the spectral line broadening has to be taken into account.

Social colors of excitations in the energy pumping beam have no straightforward connection with the social color of excitations in the output beam.

7. Resonators of Physical Lasers

In laser physics one of the main problems in lasing initiation is approaching the population inversion. However, population inversion is not enough to generate lasing. Stimulated and spontaneous emissions are competing with each other. Thus, before becoming an amplifying device, a gain medium pumped by an external energy source is first radiated as a usual electric "lamp". Here, spontaneous emission is dominating. The light power is distributed over a variety of frequencies and directions of propagation, generally uniformly distributed. It is the optical cavity, laser's resonator, that creates the conditions necessary for stimulated emission to become predominant over spontaneous emission.

In further considerations, it is assumed that the gain medium has approached population inversion.

The cavity or resonator is composed of two mirrors that *bounce the beam back and forth through the gain medium.* The cascade process of increasing photons' number inside the cavity can be initiated either by spontaneous emission from an atom in the gain medium or by a bunch of photons injected in the optical cavity. (The photons carry the energy-quanta matching the energy levels of atoms in the gain medium.) In the latter case, these photons interact with atoms and generate stimulated emission. It is crucial that these photons have the same phase. One can imagine them as a cloud of exponentially increasing size moving between mirrors. *The concentration of the field inside the cavity increases the probability of stimulated emission rather than spontaneous emission occurring. This is the basic feature of bosons* (see Section 5). We repeat that *bosons' behavior is similar to human's behavior known as the bandwagon effect* (see again Section 5).

8. Resonators of Social Lasers

The same competition between spontaneous and stimulated emission plays the crucial role in social processes. People in the excited state may "radiate" social energy spontaneously, say in debates with relatives and friends about the political and social problems. Social colors of excitations in such spontaneous radiation are typically randomly distributed, often uniformly distributed. Such emission of social energy cannot lead to coherent social actions.

8.1. Structure and Functioning of Social Resonator

A social resonator consists of a gain medium composed of s-atoms which has already approached population inversion and say an Internet based communication system, e.g., some social network. We call such a system *Echo Chamber.* We restrict modeling to the Internet based Echo Chambers. Consider the following idealized model.

Each s-atom in the excited state can emit a quantum of social energy in the form of a post or a comment on some post. We call them, posts and comments, *excitations of the social resonator.* By posting or commenting, i.e., emitting an excitation, s-atom falls to the ground state. Resonator's excitations play the role of photons in the optical cavity, the resonator of the physical laser. Moreover, to simplify the model, we assume that the social resonator under consideration accepts only excitations of the concrete color α. This is the strong constraint that is in visible contradiction with functioning of typical Internet based social networks. We shall relax it in later modeling. The social color of an excitation plays the role of the direction of propagation of output beam of photons emitted by laser's resonator, the x-axis of the optical cavity.

Suppose that, at the fixed instance of time in the social resonator, there are n excitations. Each member of the gain medium interacts with all these excitations—with the information field. The boson behavior of excitations implies that the probability that the concrete agent would fall to the ground state and emit an excitation increases with n, Section 5. It is crucial that, if all excitations of the social resonator have fixed color α, the color of excitation emitted by this agent is also α. These dynamics lead to the exponential increasing the number n of excitations having the α-color inside this social resonator (see Section 9 for modeling of temporal dynamics). Excitations of colors different from α also can be spontaneously emitted by the gain medium. However, they cannot generate the cascade process, since in the present model they are simply blocked.

Output Beam from Echo Chamber

When n becomes sufficiently large, see Section 9, it is possible to open the output channel of the Echo Chamber and generate the stable flow of high intensity of excitations of the fixed α-color. In "outer space", this flow is realized in the form of meetings, demonstrations, and brutal protest actions.

8.2. Stimulated Initiation of Social Lasing

As was stressed, the straightforward blocking of excitations with colors different from one fixed color α is the strong assumption. Of course, moderators of social networks block some posts and comments, e.g., having extremist, racist, or sexist content. However, the proportion of filtered excitations seems not to be so high, in any event it is far from 100%. Therefore, we have to improve the above model. As was presented in Section 7, there are two possible scenarios for initiating lasing:

1. spontaneous emission and filtering photons with momentum vectors deviating from the *x*-axis by using the optical cavity;
2. stimulated emission generated by a coherently injected ensemble of photons with the *x*-momentum vector.

In social lasing, the second scenario is preferable because the social mechanism of *filtering* of "wrongly directed and spontaneously emitted social excitations" is not so straightforward as in optics. Actually, the components which are not coherent with the beam are not eliminated, but become insignificant and can be considered as "noise", as if they were in some way "ignored" by the mainstream social movement.

Thus, generation of the beam of social excitations having the same color *α* is started by injection a block of *α*-colored posts into the Internet Echo Chamber. They are injected in the same moment of time. (Of course, one has to take into account the temporal scale of Echo Chamber's functioning). This initializing block of excitations generates the cascade of stimulated emissions. After a few interactions, the propagating wave of excitations is so big that the probability of stimulated emission becomes very close to one. This is a good time to open the output channel of the Echo Chamber and to transform information excitations into physical social actions.

Of course, spontaneously posted excitations of colors different from *α* can also be generated in this Echo Chamber. However, they are generated in different moments and have a variety of colors. Even if such a post starts to generate its own cascade, its power is negligible comparing with the dominating cascade started with injection of *α*-posts.

9. Dynamics the Quantum Information Field in Social Laser's Resonator

We proceed with the standard formalism of theory of open quantum systems by using the quantum master equation for the state of a subsystem of a compound system. In our considerations, the latter consists of the quantum information field interacting with the *s*-atom gain medium. The basic dynamical equation is given by the quantum Markov approximation of the Schrödinger dynamics for the state of the compound system. This is the *Gorini–Kossakowski–Sudarshan–Lindblad* (GKSL) equation [77]. We point out that the using the Markov approximation is an important assumption. Its meaning and validity for social systems is a complex question (see [78] for the corresponding analysis of the general socio-political situation). Applicability of the quantum Markov approximation to modeling social lasing should be studied in more detail. We have no possibility to do this in the present paper but plan to turn to this problem in one of the future works.

As always in theory of open quantum systems, we should extend the notion of the state of a social system by considering mixed states represented by density operators.

We follow presentations in physical works, e.g., [79–83]. One of the essential differences in transition from quantum modeling of real physical processes to quantum-like modeling outside physics is that the Planck constant *h* cannot be considered as social action quantum. In quantum physics, the constant *h* couples photon's energy *E* and angular frequency *ω* as

$$E = \hbar\omega, \tag{11}$$

where $\hbar = h/2\pi$ is the reduced Planck constant. As was already mentioned, we were not able to find a natural social analog of angular frequency *ω*. We tried to proceed without it. Equation (11) can be written as

$$\tau = 1/\omega = \hbar/E. \tag{12}$$

Here, τ has the physical dimension of time and it can be interpreted as the time scale of the quantum dynamics for the E-mode of the quantum electromagnetic field.

By transition from genuine quantum to quantum-like modeling, we have to introduce an analog of the Planck constant, say γ. It is interpreted similarly to Equation (12), as a constant coupling the time scale and energy. There is no reason to assume that it is equal to the physical constant $\hbar$. Moreover, we cannot assume that γ is the same for all social processes. If it were the case, it would be really surprising! This is just the time scale of a social process modeled with the aid of the quantum formalism.

9.1. Creation-Annihilation Algebras for s-Atoms and Quantum Information Field

The quantum information field carrying the fixed amount of social energy E_F can be represented in the following operator form

$$\hat{E}(t) = u e^{-it\frac{E_F}{\gamma}} \hat{a} + \bar{u} e^{it\frac{E_F}{\gamma}} \hat{a}^\star,$$ (13)

where u is the complex field amplitude and $\hat{a}$ and $\hat{a}^\star$ are annihilation and creation operators for social excitations of $\hat{E}(t)$, i.e.,

$$\hat{a}|n\rangle = \sqrt{n}|n-1\rangle, \ \hat{a}^\star|n\rangle = \sqrt{n+1}|n+1\rangle.$$ (14)

Here, scaling constants $\sqrt{n}, \sqrt{n+1}$ are selected in such a way that the operator

$$\hat{n} = \hat{a}^\star \hat{a}$$ (15)

can be interpreted as the operator of the excitations' number: $\hat{n}|n\rangle = n|n\rangle$.

The creation–annihilation operators satisfy the canonical commutation relation:

$$[\hat{a}, \hat{a}^\star] = \hat{a}\hat{a}^\star - \hat{a}^\star \hat{a} = I,$$ (16)

where I is the unit operator.

The information field Hamiltonian is given by

$$\hat{H}_F = E_F \, \hat{n}.$$ (17)

The n-excitation state $|n\rangle$ is the eigenstate of the field Hamiltonian. In this state, the field energy equals nE_F. In the absence of interactions, it is preserved in the process of field's evolution.

As everywhere in this paper, we consider s-atoms with the two-level structure of the social energy, $E = E_1, E_2$ and transition energy $E_A = E_2 - E_1$. Energy lowering and rising can be formally described by creation and annihilation operators. In contrast to the information field, these are fermionic operators:

$$\hat{b}|E_1\rangle = 0, \ \hat{b}^\star|E_1\rangle = |E_2\rangle, \ \hat{b}|E_2\rangle = |E_1\rangle, \ \hat{b}^\star|E_2\rangle = 0.$$ (18)

Hence, they satisfy the fermionic canonical commutation relation:

$$\{\hat{b}, \hat{b}^\star\} = \hat{b}\hat{b}^\star + \hat{b}^\star\hat{b} = I.$$ (19)

These operators are also known as *level lowering and rising operators*. Here, the vectors $|E_1\rangle$ and $|E_2\rangle$ represent the ground and excited states, respectively. Similarly to the field's excitations number operator $\hat{n}$, we set $\hat{n}_A = \hat{b}^\star b$. We have $\hat{n}_A|E_1\rangle = 0$ and $\hat{n}_A|E_2\rangle = |E_2\rangle$.

We also note that the field and s-atom's operators commute:

$$[\hat{a}, \hat{b}] = [\hat{a}, \hat{b}^\star] = [\hat{a}^\star, \hat{b}] = [\hat{a}^\star, \hat{b}^\star] = 0.$$ (20)

The reader should not be disappointed that to model transitions between s-atom's states as well as information filed's states, we use the same operator algebra as in physics. The formalism of creation and annihilation operators can be used in all models describing transitions between states. For example, besides quantum physics, this formalism plays the important role in analysis of reaction–diffusion equations. This formalism is widely used to model human cognition, decision-making, in finances, and political studies [20,29,34].

Hamiltonian of an *s*-atom is given by

$$\hat{H}_A = E_1 + E_A \hat{n}_A. \tag{21}$$

Since $\hat{H}_A |E_\alpha\rangle = E_\alpha |E_\alpha\rangle, \alpha = 0, 1$, in this model, an *s*-atom who is isolated from the information field and being in the ground state cannot become excited by herself and being in the excited state cannot emit spontaneously a social excitation and relax.

In fact, the forms of the field and *s*-atom Hamiltonians are selected to preserve the states of the concrete social energy (in the absence of interactions). Thus, the forms of these Hamiltonians express the law of energy conservation, but only for the states of the concrete energy. The same energy conservation constraint is supposed in quantum physics. Therefore, the reader should not surprised that we proceed with Hamiltonians of the same form as in physics.

9.2. Dynamics of the Compound System s-Atom-Field

Interaction between the quantum information field and *s*-atoms is described by Hamiltonian

$$\hat{H}_I = -\gamma g (\hat{b}(t) + \hat{b}^\star(t)) \hat{E}(t), \tag{22}$$

where the parameter $g > 0$ expresses the strength of coupling between *s*-atoms and the information field.

The dynamics of *s*-atom's creation and annihilation operators is given by

$$\hat{b}(t) = e^{-it\frac{E_A}{\gamma}} \hat{b}, \quad \hat{b}^\star(t) = e^{it\frac{E_A}{\gamma}} b^\star. \tag{23}$$

This dynamics is just the Heisenberg picture of the state dynamics given by the Schrödinger equation.

Consider action of the *s*-atom component of the interaction Hamiltonian at $t = 0$,

$$(\hat{b} + \hat{b}^\star)|E_1\rangle = |E_2\rangle, \quad (\hat{b} + \hat{b}^\star)|E_2\rangle = |E_1\rangle. \tag{24}$$

This is the flipping-operator representing transitions between states of an *s*-atom. Such transitions should be compensated by modification of the information field state. Mathematically, this interaction process is represented as composition of two operators.

Formally, we should also use the time-dependent creation and annihilation operators in free Hamiltonians $\hat{H}_A, \hat{H}_F$. However, these Hamiltonians contain only compositions of operators of the forms $\hat{b}^\star(t)\hat{b}(t), \hat{a}^\star(t)\hat{a}(t)$ and complex exponents containing time dependence cancel each other.

Finally, the interaction Hamiltonian can be represented in the form:

$$\hat{H}_I = -\gamma g \left[u e^{-it\frac{E_F+E_A}{\gamma}} \hat{a}\hat{b} + \bar{u}e^{it\frac{E_F+E_A}{\gamma}} \hat{a}^\star \hat{b}^\star + \tag{25}\right.$$

$$\left. u e^{-it\frac{E_F-E_A}{\gamma}} \hat{a}\hat{b}^\star + \bar{u}e^{it\frac{E_F-E_A}{\gamma}} \hat{a}^\star \hat{b} \right].$$

The Hamiltonian of the compound system, the quantum information field carrying the social energy E_F and interacting with an *s*-atom having the social energy spectrum $E = E_1, E_2$, is given by

$$\hat{H} = \hat{H}_A + \hat{H}_F + \hat{H}_I.$$

Denote by ρ the state of this compound system. Its dynamics is described by the von Neumann equation:

$$i\gamma \frac{\partial \rho}{\partial t}(t) = [\hat{H}, \rho(t)],$$ (26)

with the initial condition

$$\rho(t_0) = \rho_A(t_0) \otimes \rho_F(t_0).$$ (27)

Here, it is supposed that at $t = t_0$ s-atom's and field's information states are uncorrelated. Mathematically, this absence of correlations is represented by factorization of the compound system state into the states of the s-atom and the field.

As typical in physical considerations, it is convenient to move to the interaction representation. In this representation, the two components of dynamics, one given by Hamiltonian $H_{\text{free}} = \hat{H}_A + \hat{H}_F$ and another by interaction Hamiltonian H_I, are spit. The first part is used for transformation of the density operator $\rho(t)$ of the form $\tilde{\rho}(t) = U^\star(t)\rho(t)U(t)$, where $U(t)$ is the unitary one parametric group describing the dynamics generated by H_{free}, i.e., $U(t) = e^{-iH_{\text{free}}/\gamma}$. This is the transformation to the interaction representation. In addition, in the latter state's dynamics is generated by H_I.

To simplify notations, in the interaction representation, we will use the symbol $\rho(t)$ to denote the state (instead of symbol $\tilde{\rho}(t)$). In this representation, the von Neumann equation has the form:

$$i\gamma \frac{\partial \rho}{\partial t}(t) = [\hat{H}_I, \rho(t)].$$ (28)

This equation can be solved approximately by iterated integration starting with

$$\rho^{(1)}(t) = \rho(t_0) + \frac{1}{i\gamma} \int_{t_0}^{t} [\hat{H}_I, \rho(s)]ds.$$ (29)

9.3. Gorini–Kossakowski–Sudarshan–Lindblad Equation for the State of the Quantum Information Field

By using the above integral iterations and under some assumptions (see Section 9.4), we can derive the approximate quantum master equation for the reduced density operator of the field,

$$\rho_F(t) \equiv \text{Tr}_A \text{æ}(t),$$

where the partial trace is with respect to basis $|E_1\rangle, |E_2\rangle$ in the state space of s-atom's social energy. This is the special case of the quantum Markov approximation for the dynamics of a subsystem—of a compound quantum system described by the GKSL-equation.

Our main task is to present the proper social interpretations for the parameters of the GKSL-equation. For s-atoms composing the gain medium, denote by $T_i, i = 1, 2$, the average times of being in the ground and excited states, respectively. We can call them the lifetimes of relaxation and excitation. Denote by r_2 the rate of excitation of s-atoms generated by social energy pumping into this gain medium and by r_1 the absorption rate.

The quantum master equation has the form:

$$\frac{\partial \rho_F}{\partial t}(t) = -A[\hat{a}\hat{a}^\star \rho_F(t) + \hat{a}^\star \rho_F(t)\hat{a} + \text{h.c}] + B[\hat{a}\hat{a}^\star \text{æ}_F(t) + 3\hat{a}\hat{a}^\star \text{æ}_F(t)\hat{a}\hat{a}^\star - 4\hat{a}^\star \hat{a}\hat{a}^\star \text{æ}_F(t)\hat{a} + \text{h.c}]$$ (30)

$$-C[\hat{a}^\star \hat{a}\rho_F(t) + \hat{a}\rho_F(t)\hat{a}^\star + \text{h.c}],$$

where h.c is the abbreviation for "Hermitian conjugate." Here,

$$A = r_2(gT_2)^2|u|^2$$ (31)

and

$$C = r_1(gT_1)^2|u|^2$$ (32)

are *the gain and loss coefficients*, respectively. These coefficients depend quadratically on the amplitude of the information field u, the interaction coefficient g, and excitation and relaxation lifetimes T_2 and T_1. They are linearly proportional to excitation and absorption rates, r_2, r_1. The term with the saturation coefficient

$$B = r_2 (g T_2)^4 |u|^4 \tag{33}$$

plays the crucial role in generation of exponential increase of the number of excitations in the field. We shall turn to this question later by considering the dynamics of probabilities.

We remark that, since in Equation (30) the operator coefficients are time-independent, the dynamical state update is not based on the long term memory. If we consider a discrete time approximation of this dynamics, $\rho_F(t_k), t_k = t_0 + k\delta t$, then the state at the moment t_{k+1} is completely determined by the state at $t = t_k$. Such property is known as the Markov property.

9.4. Social Interpretation of Assumptions for Derivation of Quantum Master Equation

There are two basic assumptions for derivation of master Equation (30) from von Neumann Equation (26) (see [83]). To formulate these two assumptions, it is convenient to introduce the time scale

$$\tau_F = \frac{\gamma}{E_F} \tag{34}$$

of the evolution of field's mode with the energy E_F and the time scale

$$\tau_A = \frac{\gamma}{E_A} \tag{35}$$

of transition between the states $|E_i\rangle, i = 1, 2$.

Assumption 1. *Analysis of dynamics (30) can be essentially simplified under the following condition:*

$$2\pi \frac{\gamma}{E_A} << T_2. \tag{36}$$

In this situation, the first two terms in the expression (25) of the interaction Hamiltonian H_I can be neglected in the process of integration with respect to time (see (29)).

Inequality (39) can be written as

$$\frac{T_2}{\tau_A} >> 2\pi. \tag{37}$$

Thus, the lifetime of the excited state of an s-atom should be long enough comparing with the transition time scale τ_A. We note that the latter is inversely proportional to the resonance energy $E_A = E_2 - E_1$.

For a gain medium with a large gap of the social energy, it is easier to satisfy the condition (37). For such a social group, the master Equation (30) gives a better approximation. Of course, the condition (37) expresses the complex interplay between the magnitudes of the lifetime for the excited state $|E_2\rangle$ and the size of the energy gap E_A.

A good gain medium is characterized by the long lifetime of the excited state and the big gap between the states of relaxation and excitation.

For such a gain medium, the master equation approximation gives the adequate picture of social processes in the gain medium (under additional Assumption 2).

Assumption 2. *The difference between the energies E_A and E_F is very small, so the social energy of quanta associated with communications compounding the field differs not so much from the resonance energy of s-atoms—the social energy for transition between s-atoms' states:*

$$|E_F - E_A| \frac{T_2}{\gamma} << 1. \tag{38}$$

This condition of matching social energies is natural for s-atoms interacting with the information field. The social energy E_F carried by excitations of the information field need not be exactly equal to the resonance energy of s-atoms in the gain medium. It can deviate from the latter, but not so much. The right formulation of this statement can be done in the probabilistic terms. The emission spectrum of the gain medium has the Gaussian distribution with the mean value E_F and sufficiently small standard deviation σ_A, where $\sigma_A \frac{T_2}{\gamma} \ll 1$.

Technically, the condition (38) justifies the following approximation:

$$e^{\pm it\frac{E_F-E_A}{\gamma}} \approx 1, \ t_0 \le t \le T_2, \tag{39}$$

in the last two terms of interaction Hamiltonian (25). By taking into account that the first two terms in expression (25) can be neglected due to Assumption 1 interaction Hamiltonian H_I, can be approximately treated as a time-independent operator.

The condition (38) can be rewritten as

$$|1 - \frac{E_F}{E_A}| \ll \frac{\tau_A}{T_2}. \tag{40}$$

Therefore, for the gain medium which is "good" from the viewpoint of Assumption 1, i.e., $\frac{\tau_A}{T_2} \ll 1$, the lifetime is long and the social energy gap is high, this condition is satisfied only for very sharp Gaussian distribution, with the mean value E_F, of the emission spectrum of the gain medium.

9.5. Probabilistic Consequences of the Quantum Markov Dynamics

Now, by using quantum master Equation (30), we can describe the dynamics of probability $p(t, n)$ to find n excitations in social laser's resonator, e.g., in the form of an Internet Echo Chamber. We have $p(t, n) = \langle n|\rho_F(t)|n\rangle$. By averaging Equation (30) with respect to the state $|n\rangle$, we obtain the probabilistic dynamics (see [84]):

$$\frac{\partial p}{\partial t}(t, n) = -A[(n+1)p(t, n) - np(t, n-1)] - C[np(t, n) - (n+1)p(t, n+1)] \tag{41}$$

$$+B[(n+1)^2 p(t, n) - n^2 p(t, n-1)].$$

As in physics, we are interested in the steady state of this dynamics, the state of equilibrium in the Echo Chamber. After approaching this state, the social resonator can be used to emit the powerful wave of social actions. To find a steady state, we set $\frac{\partial p}{\partial t}(t, n) = 0$ and obtain the recurrence equation:

$$p(n) = (A/C)(1 - nB/A)p(n-1). \tag{42}$$

In spite of simplicity of this recurrence equation, its solution cannot be represented in a compact analytic form. In physics, one considers approximations corresponding different ranges of values of the parameter A/C, *the pump parameter*, representing interrelation of the gain and loss. The equality $A/C = 1$ is *the threshold condition for the laser.*

1. If $A/C < 1$, then the solution of Equation (42) can be approximately represented in the form $p(n) \approx (1 - A/C)(A/C)^n$. Thus, for this region of variation of the parameter A/C, the field is characterized by a small number of excitations: the probability that, in the resonator of the social laser, there can be found n excitations decreases exponentially, $p(n) \sim e^{-(\ln C - \ln A)}$.
2. If $A/C \approx 1$, then the solution has no simple analytical representation. This range of variation of parameter A/C is characterized by large fluctuations of number n of field's excitations.
3. If $A/C \gg 1$, then $p(n)$ can be approximated by the Poission distribution:

$$p(n) \approx e^{-\bar{n}}\frac{\bar{n}^n}{n!}. \tag{43}$$

where $\bar{n} = A^2/CB$ is the average number of excitations. It is crucial that the standard deviation of the Poission distribution $\sigma = \sqrt{\bar{n}}$. This implies that the Gaussian distribution approximating the Poisson distribution with the mean value $\bar{n} = A^2/CB >> 1$ is concentrated around $\bar{n}$. Thus (with the high probability), the number of excitations n present in the resonator of the social laser is very large.

In some papers, one sets $A/C = e^b$, i.e., $b = \ln(A/C)$. Then, the threshold value equals to zero.

10. Conclusions

This paper is a new step towards clarification of the basic notions of the quantum-like social lasing model. We heavily refer to the information interpretations of quantum mechanics and the operational approach. The latter is based on the Copenhagen interpretation of quantum mechanics and Bohr's emphasis that quantum theory is about observations and not genuine physical processes in the micro-world. This approach is applied to formalization of the notion of the social (mental) energy. Although this notion has been discussed by psychologists and philosophers [64–66] for a few hundred years, its proper formalization was missed. It seems that this notion can be handled properly only in the quantum framework, cf., however, with other recent attempts [69–71].

The Bose–Einstein statistics of excitations of the quantum electromagnetic field, photons, plays the crucial role in generation of the cascade process of stimulated emission. Following Schrödinger [75] who used Gibbs ideal ensembles to derive thermodynamical quantities from statistical mechanics, in [47], one of the authors of this paper considered thermodynamics of excitations of the quantum information field. The Bose–Einstein statistics can be derived for them under the assumption of their indistinguishability with respect to the social energy. In this paper, we analyze the meaning of such indistinguishability and emphasize the crucial role of information overload in its creation. The Bose–Einstein statistics of information excitations matches well with one of the basic effect of social psychology, the bandwagon effect. We also present briefly thermodynamical derivation of the Bose–Einstein statistics for information excitations [47] by emphasizing the operational meaning of the notion of the social energy.

We discuss in a lot of detail the notion of the social color of an excitation of the quantum information field. This quantity represents characteristics of information excitations linked to social lasing coherence. The role of the social color in the process of energy pumping versus amplified coherent emission is clarified.

As in physical lasing, a resonator is the basic component of social lasing. The standard resonator of a physical laser is the optical cavity. Therefore, modeling of functioning of a physical laser resonator is typically presented in the framework of quantum optics. Our aim was to "distill" the physical scheme from connection with optics. We proceeded with a theory of open quantum systems and the quantum Markov approximation (given by the GKSL-equation) of the dynamics of the compound system, s-atoms interacting with the quantum information field. As is typical in quantum-like modeling, we borrowed the mathematical formalism of quantum physics, but assigned new (social) interpretations to the basic quantities and parameters. This interpretational analysis of the (standard) mathematical expressions highlights the following social lasing constraints on the human gain medium and the quantum information field:

- A gain medium is characterized by the long lifetime of the excited state and the big gap between the states of relaxation and excitation.
- The social energy carried by excitations of the information field has to match the resonance energy of s-atoms in the gain medium.
- The interrelation of the magnitudes of the excitation and absorption rates r_2, r_1 and the lifetimes of the corresponding levels T_2, T_1 has to imply inequality $A > C$, where A and C are gain and loss coefficients, respectively.
- The nonlinear character of interactions between excitations in a social laser resonator (encoded in the B-coefficient) plays the crucial role in initiation of stable social lasing.

- The quantum-like regime of lasing is characterized by the threshold value of the pump parameter.

The presented study on the foundational side of the quantum-like modeling of social lasing is important for its further development, since, as was mentioned in the Introduction, application of the quantum theory outside physics requests reanalyzing of quantum methodology.

This paper may even have some impact for quantum foundations, especially the information interpretations of quantum mechanics [51–63]. The application of the mathematical formalism of quantum mechanics outside physics supports (at least indirectly) the claim that this formalism basically reflects the special laws of information processing. (The latter is confirmed by its derivation from the natural informational principles [57–59].) In particular, the thermodynamical derivation of quantum statistics (Section 4.4) highlights the role of indistinguishability of information quanta.

Author Contributions: Conceptualization, A.K.; Methodology, A.K. and A.A.; Validation, A.T. and D.T.; Formal Analysis, A.K.; Investigation, A.K. and A.A.; Writing, Original Draft Preparation, A.K.; Writing, Review & Editing, A.A., A.T., and D.T.; Supervision, A.K.; Project Administration, A.A.; Funding Acquisition, A.A.

Funding: This research was funded by the Government of the Russian Federation, Grant 08-08 and by the Ministry of Education and Science of the Russian Federation within the Federal Program Research and development in priority areas for the development of the scientific and technological complex of Russia for 2014-2020, Activity 1.1, Agreement on Grant No. 14.572.21.0008 of 23 October, 2017, unique identifier: RFMEFI57217X0008.

Conflicts of Interest: The authors declare no conflict of interest.

References

1. Jung, C.G.; Pauli, W. *Atom and Archetype. The Pauli/Jung Letters 1932–1958*; Princeton University Press: Princeton, NJ, USA, 2014.
2. Atmanspacher, H.; Fuchs, C.A. (Eds.) *The Pauli–Jung Conjecture: And Its Impact Today*; Imprint Academic: Exeter, UK, 2014.
3. Whitehead, A.N. *Process and Reality*; Macmillan: New York, NY, USA, 1929.
4. Whitehead, A.N. *Adventures of Ideas*; Macmillan: New York, NY, USA, 1933.
5. Ricciardi, L.M.; Umezawa, H. Brain Physics and Many-Body Problems. *Kibernetik* **1967**, *4*, 44–48. [CrossRef]
6. Penrose, R. *The Emperor's New Mind*; Oxford University Press: Oxford, UK, 1989.
7. Umezawa, H. *Advanced Field Theory: Micro, Macro and Thermal Concepts*; AIP: New York, NY, USA, 1993.
8. Hameroff, S. Quantum coherence in microtubules. A neural basis for emergent consciousness? *J. Cons. Stud.* **1994**, *1*, 91–118.
9. Vitiello, G. Dissipation and memory capacity in the quantum brain model. *Int. J. Mod. Phys.* **1995**, *B9*, 973. [CrossRef]
10. Vitiello, G. *My Double Unveiled: The Dissipative Quantum Model of Brain*; Advances in Consciousness Research; John Benjamins Publishing Company: Amsterdam, The Netherlands, 2001.
11. Ezhov, A.A.; Berman, G.P. *Introduction to Quantum Neural Technologies*; Rinton Press: Paramus, NJ, USA, 2003.
12. Bernroider, G.; Summhammer, J. Can quantum entanglement between ion transition states effect action potential initiation? *Cogn. Comput.* **2012**, *4*, 29–37. [CrossRef]
13. Igamberdiev, A.U.; Shklovskiy-Kordi, N.E. The quantum basis of spatiotemporality in perception and consciousness. *Prog. Biophys. Mol. Biol.* **2017**, *130A*, 15–25. [CrossRef]
14. Bernroider, G. Neuroecology: Modeling neural systems and environments, from the quantum to the classical level and the question of consciousness. *J. Adv. Neurosc. Res.* **2017**, *4*, 1–9. [CrossRef]
15. Basti, G.; Capolupo, A.; Vitiello, G. Quantum field theory and coalgebraic logic in theoretical computer science. *Prog. Biophys. Mol. Biol.* **2017**, *130A*, 39–52. [CrossRef]
16. Khrennikov, A. *Information Dynamics in Cognitive, Psychological, Social, and Anomalous Phenomena*; Series Fundamental Theories of Physics; Kluwer: Dordreht, The Netherlands, 2004.
17. Pylkkänen, P. *Mind, Matter and the Implicate Order*; Springer Frontiers Collection: Berlin/Heidelberg, Germany, 2007.
18. Khrennikov, A. *Ubiquitous Quantum Structure: From Psychology to Finances*; Springer: Berlin/Heidelberg, Germany, 2010.

19. Busemeyer, J.R.; Bruza, P.D. *Quantum Models of Cognition and Decision*; Cambridge Press: Cambridge, UK, 2012.
20. Bagarello, F. *Quantum Dynamics for Classical Systems: With Applications of the Number Operator*; Wiley Ed.: New York, NY, USA, 2012.
21. Haven, E.; Khrennikov, A. *Quantum Social Science*; Cambridge University Press: Cambridge, UK 2013.
22. Asano, M.; Khrennikov, A.; Ohya, M.; Tanaka, Y.; Yamato, I. *Quantum Adaptivity in Biology: From Genetics to Cognition*; Springer: Dordrecht, The Netherlands, 2015.
23. Haven, E.; Khrennikov, A.; Robinson, T.R. *Quantum Methods in Social Science: A First Course*; WSP: Singapore, 2017.
24. Khrennikov, A. Classical and quantum mechanics on information spaces with applications to cognitive, psychological, social and anomalous phenomena. *Found. Phys.* **1999**, *29*, 1065–1098. [CrossRef]
25. Busemeyer, J.B.; Wang, Z.; Townsend, J.T. Quantum dynamics of human decision-making. *J. Math. Psych.* **2006**, *50*, 220–241. [CrossRef]
26. Haven, E. Private information and the 'information function': A survey of possible uses. *Theor. Decis.* **2008**, *64*, 193–228. [CrossRef]
27. Pothos, E.M.; Busemeyer, J.R. A quantum probability explanation for violation of rational decision theory. *Proc. Royal. Soc. B* **2009**, *276*, 2171–2178. [CrossRef] [PubMed]
28. Brandenburger, A. The Relationship between quantum and classical correlation in games. *Game Econ. Behav.* **2010**, *69*, 175–183. [CrossRef]
29. Bagarello, F.; Oliveri, F. A phenomenological operator description of interactions between populations with applications to migration. *Math. Models Methods Appl. Sci.* **2013**, *23*, 471–492. [CrossRef]
30. Plotnitsky, A. Are quantum-mechanical-like models possible, or necessary, outside quantum physics? *Phys. Scripta* **2014**, *T163*, 014011. [CrossRef]
31. Boyer-Kassem, T.; Duchene, S.; Guerci, E. Quantum-like models cannot account for the conjunction fallacy. *Theor. Decis.* **2015**, *10*, 1–32. [CrossRef]
32. Broekaert, J.; Basieva, I.; Blasiak, P.; Pothos, E.M. Quantum dynamics applied to cognition: A consideration of available options. *Phil. Trans. R. Soc. A* **2017**, *375*, 20160387. [CrossRef]
33. Khrennikov, A. Quantum Bayesianism as the basis of general theory of decision-making. *Phil. Trans. R. Soc. A* **2016**, *374*, 20150245. [CrossRef]
34. Khrennikova, P. Modeling behavior of decision makers with the aid of algebra of qubit creation-annihilation operators. *J. Math. Psych.* **2017**, *78*, 76–85. [CrossRef]
35. Conte, E.; Khrennikov, A.; Todarello, O.; Federici, A. A preliminary experimental verification on the possibility of Bell inequality violation in mental states. *Neuroquantology* **2008**, *6*, 214–221. [CrossRef]
36. Asano, M.; Khrennikov, A.; Ohya, M.; Tanaka, Y.; Yamato, I. Violation of contextual generalization of the Leggett-Garg inequality for recognition of ambiguous figures. *Phys. Scr.* **2014**, *T163*, 014006. [CrossRef]
37. Dzhafarov, E.N.; Kujala, J.V. On selective influences, marginal selectivity, and Bell/CHSH inequalities. *Top. Cogn. Sc.* **2014**, *6*, 121–128. [CrossRef] [PubMed]
38. Barros, J.; Toffano, Z.; Meguebli, Y.; Doan, B.-L. Contextual Query Using Bell Tests. In Proceedings of the Quantum Interaction, Leicester, UK, 25–27 July 2013; Atmanspacher, H., Haven, E., Kitto, K., Raine, D., Eds.; Springer: Berlin/Heidelberg, Germany; ISBN 978-3-642-54942-7.
39. Allais, M. Le comportement de l'homme rationnel devant le risque: Critique des postulats et axiomes de l' cole amricaine. *Econometrica* **1953**, *21*, 503–536. [CrossRef]
40. Ellsberg, D. Risk, ambiguity and the Savage axioms. *Q. J. Econ.* **1961**, *75*, 643–669. [CrossRef]
41. Machina, M. Risk, Ambiguity and the Dark-dependence Axiom. *Am. Econ. Rev.* **2009**, *99*, 385–392. [CrossRef]
42. Krastev, I. *In Mistrust We Trust: Can Democracy Survive When We Don't Trust Our Leaders?* TED Conferences Press: Monterey, CA, USA, 2013.
43. Mason, P. *Why It's Kicking off Everywhere: The New Global Revolutions*; Verso: London, UK, 2012.
44. Fukuyama, F. The middle-class revolution. *The Wall Street Journal*, 28 June 2013.
45. Schmidt, E.; Cohen, J. *The New Digital Age: Reshaping the Future of People, Nations and Business*; Knopf: New York, NY, USA, 2013.
46. Krastev, I. *Democracy Disrupted. the Global Politics of Protest*; Penn Press: Philadelphia, PA, USA, 2014.
47. Khrennikov, A. Towards information lasers. *Entropy* **2015**, *17*, 6969–6994. [CrossRef]

48. Khrennikov, A. 'Social laser': Action amplification by stimulated emission of social energy. *Phil. Trans. R. Soc.* **2016**, *374*, 20150094. [CrossRef] [PubMed]
49. Khrennikov, A. Social laser model: From color revolutions to Brexit and election of Donald Trump. *Kybernetes* **2018**, *47*, 273–278. [CrossRef]
50. Khrennikov, A. (Ed.) *Quantum Theory: Reconsideration of Foundations*; Serise Math. Modelling in Physics, Engineering, and Cognitive Science; Vaxjo University Press: Vaxjo, Sweden, 2002; Volume 2.
51. Zeilinger, A. A foundational principle for quantum mechanics. *Found. Phys.* **1999**, *29*, 631–643. [CrossRef]
52. Brukner, C.; Zeilinger, A. Information invariance and quantum probabilities. *Found. Phys.* **2009**, *39*, 677–689. [CrossRef]
53. Caves, C.M.; Fuchs, C.A.; Schack, R. Quantum probabilities as Bayesian probabilities. *Phys. Rev. A* **2002**, *65*, 022305. [CrossRef]
54. Fuchs, C.A. Quantum mechanics as quantum information (and only a little more). In *Proceedings of Quantum Theory: Reconsideration of Foundations*; Vaxjo University Press: Vaxjo, Sweden, 2002; pp. 463–543.
55. Fuchs, C.A.; Schack, R. QBism and the Greeks: Why a quantum state does not represent an element of physical reality. *Phys. Scr.* **2014**, *90*, 015104. [CrossRef]
56. Fuchs, C.A.; Mermin, N.D.; Schack, R. An introduction to QBism with and application to the locality of quantum mechanics. *Am. J. Phys.* **2014**, *82*, 749–754. [CrossRef]
57. Chiribella, G.; D'Ariano, G.M.; Perinotti, P. Probabilistic theories with purification. *Phys. Rev. A* **2010**, *81*, 062348. [CrossRef]
58. D'Ariano, G.M. Operational axioms for quantum mechanics. In Proceedings of the Foundations of Probability and Physics-4, Vaxjo, Sweden, 4–9 June 2006; Volume 889, pp 79–105.
59. Chiribella, G.; D'Ariano, G.M.; Perinotti, P. Informational axioms for quantum theory. In Proceedings of the Foundations of Probability and Physics-6, Vaxjo, Sweden, 14–16 June 2011; Volume 1424, pp. 270–279, ISBN 978-0-7354-1004-6.
60. Plotnitsky, A. *Reading Bohr: Physics and Philosophy*; Springer: Dordrecht, The Netherlands, 2006.
61. Plotnitsky, A. *Epistemology and Probability: Bohr, Heisenberg, Schrödinger, and the Nature of Quantum-Theoretical Thinking*; Springer: New York, NY, USA, 2009.
62. Plotnitsky, A. *Niels Bohr and Complementarity: An Introduction*; Springer: New York, NY, USA, 2012.
63. Khrennikov, A. External Observer Reflections on QBism, Its Possible Modifications, and Novel Applications. In *Quantum Foundations, STEAM-H: Science, Technology, Engineering, Agriculture, Mathematics & Health*; Khrennikov, A., Toni, B., Eds.; Springer: Cham, Switzerland, 2018; pp. 93–118.
64. James, W. *The Principles of Psychology*; Henry Holt and Co.: New York, NY, USA, 1890; Reprinted Harvard University Press: Boston, MA, USA, 1983.
65. Freud, S. *The Ego and the Id*; W.W. Norton and Company: New York, NY, USA, 1923.
66. Freud, S. *The Interpretation of Dreams*; In Standard edition 1954–1974; Hogarth Press: London, UK, 1900; Volumes 4, 5, pp. 1–627.
67. Jung, C.G. *On the Nature of the Psyche*; Routledge Classics: London, UK, 2001.
68. Khrennikov, A. Quantum-like modeling of cognition. *Front. Phys. Interdisciplin. Phys.* **2015**, *3*, 77. [CrossRef]
69. Grauwin, S.; Bertin, E.; Lemoy, R.; Jensen, P. Competition between collective and individual dynamics. *Proc. Natl. Acad. Sci. USA* **2009**, *106*, 20622–20626. [CrossRef] [PubMed]
70. Jaeger, C.; Horn, G.; Lux, T.; Fricke, D.; Frurst, S.; Lass, W.; Lin, L.; Mandel, A.; Meissner, F.; Schreiber, S.; et al. From the Financial Crisis to Sustainability? Potsdam, European Climate Forum. Available online: https: //globalclimateforum.org/?id=ecfreports (accessed on 1 December 2018).
71. Collins, R. *Interaction Ritual Chains*; Princeton University Press: Princeton, NJ, USA, 2004.
72. Von Neuman, J. *Mathematical Foundations of Quantum Mechanics*; Princeton University Press: Princeton, NJ, USA, 1955.
73. Khrennikov, A.; Basieva, I. Possibility to agree on disagree from quantum information and decision-making. *J. Math. Psychology* **2014**, *62*, 1–5. [CrossRef]
74. Dubois, F.; Toffano, Z. Eigenlogic: A Quantum View for Multiple-Valued and Fuzzy Systems. In *Lecture Notes in Computer Science, Proceedings of the Quantum Interaction, San Francisco, CA, USA, 20–22 July 2016*; de Barros J., Coecke, B., Pothos, E., Eds.; Springer: Cham, Switzerland, 2016; Volume 10106, ISBN 978-3-319-52288-3.
75. Schrödinger, E. *Statistical Thermodynamics*; Dover Publications: Mineola, NY, USA, 1989.
76. Colman, A. *Oxford Dictionary of Psychology*; Oxford University Press: Oxford, UK, 2003; p. 77.

77. Ingarden, R.S.; Kossakowski, A.; Ohya, M. *Information Dynamics and Open Systems: Classical and Quantum Approach*; Springer: Berlin/Heidelberg, Germany, 1997.

78. Khrennikova, P.; Haven, E.; Khrennikov, A. An application of the theory of open quantum systems to model the dynamics of party governance in the US political system. *Int. J. Theor. Phys.* **2014**, *53*, 1346–1360. [CrossRef]

79. Fleck, J.A., Jr. Quantum theory of laser radiation. Many-atom effects. *Phys. Rev.* **1966**, *149*, 309–319. [CrossRef]

80. Haken, H. *Light, Vol. II: Laser Light Dynamics*; North-Holland Phyics Publishing: Amsterdamn, The Netherlands, 1985.

81. Erneux, T.; Glorieux, P. *Laser Dynamics*; Cambridge University Press: Cambridge, UK, 2010.

82. Barlow, T.M.; Bennett, R.; Beige, A. A master equation for a two-sided optical cavity. *J. Mod. Opt.* **2015**, *62*, S11–S20. [CrossRef] [PubMed]

83. Priyashanka, K.M.; Wijewardena Gamalath, K.A.I.L. Coherent states in a laser cavity. *Int. Lett. Chem. Phys. Astron.* **2015**, *56*, 47–62. [CrossRef]

84. Greenberger, D.M.; Erez, N.; Scully, M.O.; Svidzinsky, A.A.; Zubairy, M.S. Planck, photon statistics and Bose–Einstein condensation. *Prog. Opt.* **2007**, *50*, 273–330.

MDPI

Article
Paths of Cultural Systems

Paul Ballonoff [ID]

Owner and Operator, Ballonoff Consulting, 9307 Kings Charter Drive, Richmond, VA 23116, USA; Paul@Ballonoff.net; Tel.: +1-703-780-1761

Received: 6 November 2017; Accepted: 17 December 2017; Published: 25 December 2017

Abstract: A theory of cultural structures predicts the objects observed by anthropologists. We here define those which use kinship relationships to define systems. A finite structure we call a partially defined quasigroup (or pdq, as stated by Definition 1 below) on a dictionary (called a natural language) allows prediction of certain anthropological descriptions, using homomorphisms of pdqs onto finite groups. A viable history (defined using pdqs) states how an individual in a population following such history may perform culturally allowed associations, which allows a viable history to continue to survive. The vector states on sets of viable histories identify demographic observables on descent sequences. Paths of vector states on sets of viable histories may determine which histories can exist empirically.

Keywords: quantum logic; groups; partially defined algebras; quasigroups; viable cultures

1. Ethnographic Foundation

The structures described here and their consequences imply much of what may be predicted about empirical cultures. Anthropologists very often draw illustrations of structures using methods discussed here, but based on intuition, thus have little notion of what their commonly used diagrams might predict. While [1,2] defined mathematical means to describe the current and future demographic organization of lineage organizations, with empirical examples, we here specify the demography of kinship-based systems [3–8] with some related definitions in our Appendix A; and empirical examples in [9–13]. We follow the inspiration of [14]. In an empirical culture many other relations may also occur; we note some of those in our Part 7, Discussion at the end.

Our notion of studying viable minimal structures—which are the smallest minimal cultural structures that can "reproduce" the ascribed social relations in one generation—follows from [15]. Our history describes how sustaining those relations allow the culture to reproduce the rules. Cultural rules describing histories may be stated in natural languages, which label the individuals in a descent sequence with a subset called a kinship terminology. Our term viable embodies what anthropologist Radcliff-Brown called "persistent cultural systems" ([10], p. 124). Radcliff-Brown and others often described histories using discrete generations, as do we. Empirical cultures are almost ubiquitously described by many anthropologists using viable histories, typically represented in an ethnography by its minimal structure. The nearly ubiquitous presence in ethnographies of viable histories implies they may be the only observed histories.

An example is Figure 1 (whose source is [9]) which is an actual viable minimal structure of a history (that is, a persistent cultural system). The triangles in the illustration are males, the circles are females, descent moves in the downward direction, the labels are names used in the kinship terminology. The sign "=" means "marriage" between the two individuals attached to it; the "=" on the far right in the illustration shows a marriage to the partner on the far left in each generation. The horizontal line which connects two individuals shows that those two are assigned descendants of the marriage above them. The fifth generation in this illustration is equal to the first (by its labels in vertical descent in the diagram), showing that minimal structure shown here reproduces the labelled culture here in four generations, but reproduces the minimal structure (the graphs without the kinship labels) in each generation. This minimal structure has 4 marriages in each generation hence has structural number $s = 4$. The minimal structure is not intended to illustrate the actual empirical relations of each empirical generation of individuals, instead it shows "how the rule operates"—it describes the minimal representation of the kinship and marriage rules (see Appendix A Definitions A3). The minimal structure describes the "principles" used in the rules of the culture.

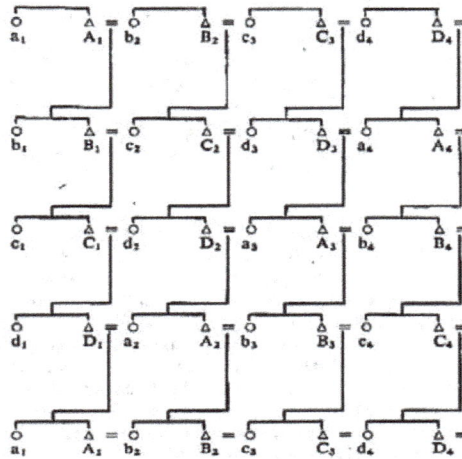

Figure 1. Example of a typical anthropological illustration of a persistent cultural system per Radcliff-Brown, which is also a group per Levi-Strauss and Weil.

Figure 1 is also obviously an example of a group. Claude Levi-Strauss [9] initially used groups to demonstrate histories in an appendix by A. Weil. Ref. [9] mainly discusses groups of orders 2, 4 and 8, though its illustration 1.17 shows a helical structure of order 4. Other studies include orders 3, 6 and others. Ref. [16] also shows a helical structure of order 7; helical minimal structures are also groups and have surjective descent sequences, so our modal demography discussed in [8] and Appendix A Definition A7 also apply to helices.

2. Basic Definitions

While lineage organizations [1,2] predicted examples of population measures including the "local" village size given the lineage structure, the kinship examples described here use values found in [8] to predict values associated to the structural number of the history, which apply no matter what the empirical size of the total population (so long as it is at or above the minimal size). Our definitions are stated in our Appendix A from previous articles (see also Appendix A) and those below.

Definition 1. *Let D be a finite non-empty set (called a dictionary) and let * be a partially defined binary operation on D, such that when $x, y \in D$:*

(1) *If there exists an $a \in D$ such that $a*x$ and $a*y$ are defined and $a*x = b$ and $a*y = b$ then $x = y$, we call such object $(D, *)$ a partially defined quasigroup, or pdq.*
(2) *If $(D, *)$ is a pdq and * is fully defined on D, then $(D, *)$ is a (complete) quasigroup.*
(3) *The pair $L = (D, *)$ is a natural language with dictionary D whenever $(D, *)$ is a pdq.*
(4) *If $L = (D, *)$ is a natural language, a kinship terminology is a quasigroup subset $k \subseteq L$.*

Definition 2. *Let X, Y and Z be non-empty finite sets and let (X, *), (Y, °) and (Z, •) be quasigroups with binary relations *, ° and • respectively. Then:*

(1) *A function f: X → Y is a homomorphism if f for all b, c ∈ X, f(b * c) = f(b) ° f(c).*
(2) *If f: X → Z and g: Y → Z are homomorphisms then f and g are isotopic.*

All empirical languages are natural languages [7]. Under Definition 2(2) if Y and Z are isotopic they are also istotopic dictionaries: two possibly different descriptions, thus typically of distinct natural languages, of the "same" objects or illustrations. Our definition of kinship terminologies based on quasigroups follows from [17], which refined the discussion of Weil in [9]. While empirical kinship systems are non-associative [18], a large class is associative and complete, form finite permutations, indeed groups [19–21] hence form kinship terminologies as defined here. Groups thus arise in anthropology because a set of all 1-1 mappings of a finite subset set k of a quasigroup (a kinship terminology k) onto itself forms a group (the symmetric group on k). If (X, *), (Y, °) and (Z, •) are complete pdqs then (isotopic) homomorphisms classify kinship terminologies by the form of the pdq onto which they are mapped. For example, the isotopic terminologies classified as Dravidian [12,22,23] and others are often discussed, in part because they have interesting group theoretical structures.

Definition 3. *Let H be a non-empty finite set of viable histories. Let G be a non-empty descent sequence using H, let $G_t \in G$ be a generation of G at t, and let $H_t \subseteq H$ be a subset of t. Then then for each $\alpha \in H_t$, the real numbers $0 \le v_\alpha(t) \le 1$ such that $\Sigma_\alpha v_\alpha(t) = 1$ is the vector state of G_t.*

Adopting a standard order for listing the histories, we write the vector state at t as $v(t) := (v_a(t), \dots, v_\chi(t))$, or when $|H_t| = h$, as $v(t) = (v_1(t), \dots, v_h(t))$. Let H be a finite non-empty set of viable histories, let $\alpha \in H$, let $G_t \in G$ be a generation of G, and let $v(t)$ be the vector state of G_t (see also Appendix A Definitions A2–A5). Then:

a. From [2,5,6,8] and Appendix A Definition A4 each structural number s has a set of values n_s and p_s where $n_s p_s = 2$, where n_s is the average family size of a pure system of structural number s and p_s is the proportion of reproducing adults of a pure system of structural number s. If history α has structural number s, then each α has modal demography $(n_\alpha, p_\alpha) = (n_s, p_s)$ (see Appendix A Definition A7) where $p_s = 2/n_s$; for $s \ge 3$ and $s_\alpha \ne s_\chi$ then $(n_\alpha, p_\alpha) \ne (n_\chi, p_\chi)$.

b. Determination of the (n_s, p_s) values are based on the Stirling Number of the Second Kind (SNSK) see [8,24]. We assume here the (n_s, p_s) pairs determined by [8].

c. Since H is finite, each non-empty set of viable histories H thus has a largest structural number s_{max} with modal demography (n_{max}, p_{max}), and a smallest structural number s_{min} with modal demography (n_{min}, p_{min}). Note that if n_{max} increases then p_{max} decreases (and as n_{min} decreases then p_{min} increases, since given s, $n_s p_s = 2$, with $0 < p \le 1$. Structural numbers $s = 2$ or 3 have identical modal demography $(n_s, p_s) = (2, 1)$; all others structural numbers have distinct modal demographies see [5,8].

d. The modal demography of history α with structural number s_α is $(n_\alpha, p_\alpha) = (n_s, p_s)$ is a set of values that represent the history α maintaining its modal demography with neither increase nor decrease in total empirical population size; it is prediction of n_α and p_α based on the determination that the structural number is s, and maintains the structural number s.

e. $n(t) = \Sigma_\alpha v_\alpha(t) n_\alpha$, $\alpha \in H$, is the predicted average family size of G_t at t, given the vector state at t see [8]. Note that this is the average family size of the population at time t, given the vector state of each $\alpha \in H_t$. This while the "size" of the minimal structure might be small, the size predicted by $n(t)$ is the predicted actual size of the total population at t, not of the minimal structure; the minimal structure illustration "size" is dependent on the rules, not on the empirical size of the population.

f. $p(t) = \Sigma_\alpha v_\alpha(t) p_\alpha$, $\alpha \in H$, is the predicted proportion of reproducing adults of G_t at t ascribed as married and reproducing, given the vector state of s_α at t [8].

g.   Thus, all of the "demographics" of cultural theory discussed here are predictions on the result of maintaining or changing the vector states of t, given the SNSK determined values for each modal demography (n_α, p_α) at time t. Thus, [8] defines

$$e^{r(t)} = 1/2n(t)p(t) \tag{1}$$

where $r(t) \in R$ predicts an average rate of change of total population size between two generations of G, based on the vector state of structural numbers of the histories $H_t \subseteq H$. [8] showed that $r(t)$ predicts changes in the probabilities $v(t)$ imply cultural change is adiabatic.

h.   Let H be a finite non-empty set of viable histories, and α, $\chi \in H$. Using $v_\alpha(t) = 1 - v_\chi(t)$, $n_\alpha = 2/p_\alpha$ and $n_\chi = 2/p_\chi$, then Equation (1) becomes

$$e^{r(t)} = 1 + (n_{\alpha\chi} - 2)v_\alpha(t) + (2 - n_{\alpha\chi})v_\alpha(t)^2 \tag{2}$$

where:

$$n_{\alpha\chi} := (n_\alpha^2 + n_\chi^2)/(n_\alpha n_\chi) \tag{3}$$

is a constant determined by the values of n_α and n_χ; note $n_{\alpha\chi} = n_{\chi\alpha}$.

3. Paths of Descent Sequences

Definition 4. *Let H be a finite set of non-empty viable histories. Let α, χ, etc $\in H_t \subseteq H$ and let the structural number of $\alpha \neq \chi$, etc. If for any such set, $|H_t| > 1$, $v_t(\alpha) = 1$ or 0, then H is not full; otherwise H is full.*

Definition 5. *If H is a finite non-empty set of viable histories, then $F \subseteq H$ is a face of H see [2,4,8]. Let $I = [1, 0]$. A path from point a to point b in a set X is a function $f: I \rightarrow X$ with $f(0) = a$ and $f(1) = b$, in which case a is called the initial point of the path and b is called the terminal point of the path. Given a path, in case $a = b$ then such path is a closed path. If $[x, y] \in I$ and $f(x) = a$ and $f(y) = b$ then $f[x, y]$ is called an interval and a sub-path of I. A reverse path from point a to point b in X is a function $f: I \rightarrow X$ with $f(1) = a$ and $f(0) = b$. Given a path (or reverse path) from t_0 to t_1, if $t_1 \geq t_k \geq t_0$ we say that t_k is in the path.*

Lemma 1. *Let H_t be a finite non-empty set of full viable histories and α, $\chi \in H_t$. Then:*

1.   *$r(t) \geq 0$, and $r(t) = 0$ if all structural numbers have the same modal demography or if all have structural numbers 2 or 3.*

2.   *If α and χ are distinct structural numbers and at least one has structural number >3, then $r(t) > 0$ and $r(t)$ has a maximum at $v(t) = (0.5, 0.5)$.*

3.   *Given any finite non-empty set H of two or more viable histories, there is a unique maximum $r(t)$, given by (ii).*

Proof 1. Assume first the modal demographies of histories α and χ are equal (which occurs if $s_\alpha = s_\chi$ or if s_α, $s_\chi < 4$). Then $n_\alpha = n_\chi$. Then $n_{\alpha\chi} = 2$, so the sum in Equation (2) is $e^{r(t)} = 1$, or $r(t) = 0$. Assume now $s_\alpha \neq s_\chi$ and at least one structural number is >3, so the modal demographies of α and χ are distinct, and we do not have $v_\alpha(t) = 0$ or 1. Then in Equation (2) $(n_{\alpha\chi} - 2) = -(2 - n_{\alpha\chi})$, but $v_\alpha(t) > 0$, so then $v_\alpha(t) > v_\alpha(t)^2$. Thus, $(n_{\alpha\chi} - 2)v_\alpha(t) > (2 - n_{\alpha\chi})v_\alpha(t)^2$, so Equation (2) states $r(t) > 0$. □

Proof 2. To show $r(t)$ has a maximum when $v(t) = (0.5, 0.5)$, we use the first two terms of a Taylor expansion from (1) to find $e^{r(t)} = 1 + r(t)$. Differentiating twice gives $a^2 \exp[r(t)]/\delta y^2 = 2 - 2n_{\alpha\chi}$ where $y = v(t)$. When $s_\alpha \neq s_\chi$ and at least one of s_α, s_χ is > 3, $n_{\alpha\chi} > 2$, then $\exp[r(t)]$ is concave down; so, $r(t)$ is also concave down. □

Proof 3. There is a finite number of two-history pairs in H. Since n_s increases as s increases, then Lemma 1 part 1 shows that the largest value of $r(t)$ will be set by that two-history subset α, χ of H having the largest difference between their structural numbers, hence the largest $n_{\alpha\chi}$ in Equation (3). Combining the n_s of the largest s with any other combination of n_s values will result in a smaller $n_{\alpha\chi}$ hence smaller $e^{r(t)}$, from Equation (3). □

Observation 1. Assume a finite non-empty set of viable histories H acting on a finite non-empty descent sequence G. Let α, β, $\chi \in H_t \subseteq H$ act on generation $Gt \in G$. See Figure 2:

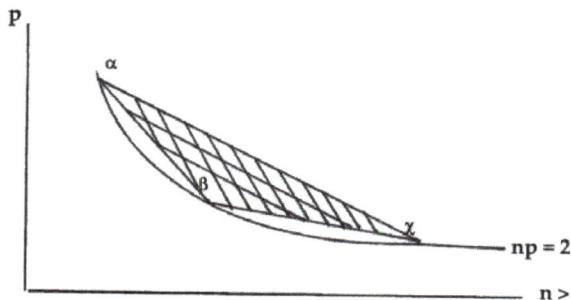

Figure 2. Illustration of the curve $np = 2$ showing three histories and connections among those three histories at or above that curve.

The curved bottom-line in Figure 2 is the locus wherever $np = 2$, so includes the modal demography of each history in H; that is, the modal demography for each of α, β and χ each appear on the line $np = 2$, since $n_s p_s = 2$. If for any α, $v_\alpha(t) = 1$, then $n(t) = n_\alpha$, $p(t) = p_\alpha$, so $e^{r(t)} = \frac{1}{2}n(t)p(t) = \frac{1}{2}n_\alpha p_\alpha = \frac{1}{2}2 = 1$ so $r(t) = 0$; this occurs if all histories in H_t have the same structural number (or all have structural numbers 2 and 3). When that does not occur we have a set of two or more histories in H_t each with $0 < v_\alpha(t) < 1$ and thus $r(t) > 0$; see also Lemma 1. When H is full, assume α, β and χ have distinct structural numbers, at least two of α, β, χ have structural numbers >3, and α is has the lowest structural number of those three histories. Then the modal demography of α, β, χ have $(n_\alpha, b_\alpha) \neq (n_\beta, b_\beta) \neq (n_\chi, b_\chi)$, and the computation of $r(t)$ appears in the values in the triangle area of Figure 2. However, if H_t is not full then events in the triangle area might not occur. Even if paths allow α with β, α with χ and β with χ (thus the boundaries of the triangle area), values of $r(t)$ within the triangle only occur if all three histories are allowed by H, which may be prohibited by not-full H. We call such area an un-accessed region. Thus, we study change using both full and non-full sets of histories.

4. Pictures of States on Descent Sequences

Definition 6. *Let H be a finite non-empty set of histories and let (n_α, p_α) be the modal demography for history $\alpha \in H$. Let G be a finite non-empty descent sequence using H, and let $G_t \in G$ be the generation at time t. Let H_t be a face of H specified at t. Let $S_t := \{ s_\alpha \mid \alpha \in H_i \}$ be the set of structural numbers $S_t \subseteq S$ of histories H_t available at t. Let $A_t := \{(n_\alpha, p_\alpha) \mid s_\alpha \in S_i, \alpha \in H_i, (n_\alpha, p_\alpha) = (n_s, p_s)\}$ be the set of modal demographies A_t of the histories in H_t; and let $v(t)$ be the vector state of G_t. List the histories in H in a defined order from α to χ. Then for all $\alpha \in H_t$ and all $(n_\alpha, p_\alpha) \in A_t$, let:*

1. *$(n_t| := (n_\alpha, \ldots, n_\chi)$ be a row vector;*
2. *$|p_t) := (p_\alpha, \ldots, \ldots, p_\chi)$ be a column vector;*
3. *for all α, $\chi \in H$, arranging the sum of the inner product $(n_t | p_t)$ as a square matrix then for all α, $\chi \in H$, $H(t) := [n_\alpha p_\chi]$ is a demographic picture (analogous to a Heisenberg picture in physics) at t;*

4. the square matrix we get by arranging the products of $v(t)v(t)^T$ as $V(t) := [v_{\alpha\chi}(t)]$ is a probability picture (analogous to a Schroedinger picture in physics) of the vector state of a descent sequence at t;

5. for $\varepsilon \geq 0$ let $V(\Delta(t)) := V(t+\varepsilon) - V(t) = [v_{\alpha\chi}(t+\varepsilon) - v_{\alpha\chi}(t)] := [\Delta_{\alpha\chi}(t)]$. (Notice that $-1 \leq \Delta_{\alpha\chi}(t) \leq 1$).

We note [25] for our analogy of terminology.

Then for all $\alpha, \chi \in H$ we can rewrite Equation (1) as:

$$e^{r(t)} = \frac{1}{2}\,(n \mid V(t) \mid p) \tag{4}$$

using a probability picture which focuses on the vector states; and

$$e^{r(t)} = \frac{1}{2}v(t)H(t)v(t)^T \tag{5}$$

using a demographic picture which focuses on demographic properties of the histories.

5. Comments on Demographic Pictures

Given a finite non-empty set of full viable histories H, observing a face $H_t \subseteq H$ at t produces a list of the available $H_t \subseteq H$ and thus creates a list of possible modal demographies $(n_\alpha, p_\alpha) \in A_t$ for all $\alpha \in H_t$. Let $|H_t| = h$. List the h histories in a fixed order from α to χ, with row vector $(n_t \mid = (n_\alpha, \ldots, n_\chi)$ and column vector $\mid p_t) = (p_\alpha, \ldots, p_\chi)^T$. So for histories $\alpha, \ldots, \chi \in H_t$, we can write the demographic picture for $|H_t| = h$ at t as $(n_t \mid p_t) = H(t)$ where:

$$H(t) = \begin{pmatrix} n_1 p_1 & \cdots & n_1 p_h \\ \vdots & \ddots & \vdots \\ n_h p_1 & \cdots & n_h p_h \end{pmatrix} = \begin{pmatrix} 2 & \cdots & n_1 p_h \\ \vdots & \ddots & \vdots \\ n_h p_1 & \cdots & 2 \end{pmatrix} \tag{6}$$

Each diagonal entry $= 2$ because a diagonal entry $n_\alpha p_\alpha$ is determined by the modal demography (n_s, p_s) for each history, and $n_s p_s = 2$. Thus, using $\frac{1}{2}H(t)$ we can restate Equation (5):

$$e^{r(t)} = \frac{1}{2}v(t)H(t)v(t)^T = \frac{1}{2}v(t)\begin{pmatrix} 2 & \cdots & n_1 p_h \\ \vdots & \ddots & \vdots \\ n_h p_1 & \cdots & 2 \end{pmatrix}v(t)^T$$

$$= v(t)\begin{pmatrix} 1 & \cdots & \frac{1}{2}n_1 p_h \\ \vdots & \ddots & \vdots \\ \frac{1}{2}n_h p_1 & \cdots & 1 \end{pmatrix}v(t)^T \tag{7}$$

Because the two-history case has some useful properties, we present much of our discussion on the two history version, which becomes:

$$e^{r(t)} = v(t)\begin{pmatrix} 1 & \frac{1}{2}n_1 p_2 \\ \frac{1}{2}n_2 p_1 & 1 \end{pmatrix}v(t)^T, \text{ where } H(t) = \begin{pmatrix} 1 & \frac{1}{2}n_1 p_2 \\ \frac{1}{2}n_2 p_1 & 1 \end{pmatrix}. \tag{8}$$

Lemma 2. *Let H be a finite non-empty set of full viable histories. Let G be a non-trivial descent sequence using H, let $H_t \subseteq H$ be the face of H observed at t, and let $G(t) \in G$ be the generation at t with vector state $v(t)$. Then $r(t) = 0$ only if $\frac{1}{2}H(t) = [1]$ at all entries.*

Proof of Lemma 2. Assume the premises. Equations (4)–(7) simply rearrange terms in $\frac{1}{2}\Sigma_i\Sigma_j v_i(t)v_j(t)n_i p_j$. From the definition of modal demography, $p_i = 2/n_i$ and $p_j = 2/n_j$. The values on the diagonal of $\frac{1}{2}H(t)$ are for each history α, $\frac{1}{2}n_\alpha p_\alpha = 1$. We thus examine the off-diagonal products $n_\alpha p_\chi$ and $n_\chi p_\alpha$. Then $n_i p_j = 2n_i/n_j$ and $n_j p_i = 2n_j/n_i$. Thus, $2n_i/n_j = 2n_j/n_i$ occurs only if $n_i = n_j$, in which case $n_i p_j = n_j p_i = 2$. This occurs only if all histories i and j have the same structural number or both have $s = 2$ or 3, and thus $\frac{1}{2}H(t) = [1]$ in all entries. Otherwise stated, in this case the value from Equation (3) is $n_{\alpha\chi} = 2$. $\square$

Implications of Lemma 2: knowing the modal demography of histories in H we can compute a proposed population growth rate $r(t)$.

1. The result $n_\alpha = n_\chi$ occurs if structural numbers s_α, s_χ are <4 or whenever $s_\alpha = s_\chi$; so $r(t) = 0$. Otherwise, then Lemma 2 implies Lemma 1, which says that $e^{r(t)} \neq 1$, and thus $r(t) > 0$. This occurs since $n_\alpha p_\chi$ does not equal $n_\chi p_\alpha$; thus from Lemma 1 and [8] the off-diagonal elements of $\frac{1}{2}H(t)$ implies adiabatic change in $r(t)$.

2. In discussions in physics, when $n_\alpha p_\chi \neq n_\chi p_\alpha$ some claim that the resulting $r(t)$ is "not commutative". In physics, the "non-commutative" result actually means switching which experiment is taken, then comparing their results; in physics when changing the order of the products it also means changing the experiment; but this comparison of the two results also creates an equation that looks like our Equations (1), (2), (7) or (8). However, in physics reversing the experiment causes different measurements, which causes the physical uncertainty between the two results. In contrast, the seemingly "non-commuting" values in culture theory exist because the equation for computing $r(t)$ requires computing both "directions" of the modal demography of histories in H (similar to comparing both directions of the physics model), and if any two (or more) of those have histories of distinct structural numbers (at least one >3), so that one or more $n_{\alpha\chi} > 2$ (see Equation (3)), then $n_\chi p_\alpha \neq n_\alpha p_\chi$. Culture theory thus predicts adiabatic demographic change, not uncertainty, from a mechanism similar to that which causes uncertainty in physics.

6. Comments on Probability Pictures

Lemma 3. *A probability picture V(t) is symmetric, $\Sigma_i\Sigma_j v_i(t)v_j(t) = 1$ and $\Sigma_i\Sigma_j(\Delta_{ij}(t)) = 0$.*

Proof of Lemma 3. In Equation (4) $V(t)$ is symmetric since each pair $v_i(t)v_j(t) = v_j(t)v_i(t)$. Since $\Sigma_\alpha v_\alpha(t) = 1$ then $v(t)v(t)^\mathrm{T} = \Sigma_i\Sigma_j v_i(t)v_j(t) = 1$. At $t + \varepsilon \geq t$ $(\varepsilon < t - (t - 1))$ then $\Sigma_\alpha v_\alpha(t + \varepsilon) = 1$: so $\Sigma_i\Sigma_j v_i(t + \varepsilon)v_j(t + \varepsilon) = 1$; so $\Sigma_i\Sigma_j(v_{ij}(t + \varepsilon) - v_{ij}(t)) = \Sigma_i\Sigma_j(\Delta_{ij}(t)) = 1 - 1 = 0$. $\square$

Since we discuss paths of histories, a frequency-domain representation of vector states is useful.

Definition 7. *Let r_1, r_2, r_3 be real numbers such that $r_1{}^2 + r_2{}^2 + r_3{}^2 = 1$. Let R be a set of 2 by 2 matrices with complex entries that forms a ring with respect to matrix addition and multiplication. Let $\mathbf{R} \subseteq R$ be a set of hermitian idempotent matrices of R; and let $R \in \mathbf{R}$ be such that $R = \frac{1}{2}[r_{ij}]$ where $r_{11} = 1 + r_3$, $r_{22} = 1 - r_3$, $r_{21} = r_1 + ir_2$, $r_{12} = r_1 - ir_2$. That is:*

$$R = \frac{1}{2}[r_{ij}] = \frac{1}{2}\begin{pmatrix} 1+r_3 & r_1 - ir_2 \\ r_1 + ir_2 & 1 - r_3 \end{pmatrix}$$

Following ([26], p. 30) we define matrices

$$1 = \begin{pmatrix} 1 & 0 \\ 0 & 1 \end{pmatrix}, \Sigma_1 = \begin{pmatrix} 0 & 1 \\ 1 & 0 \end{pmatrix}, \Sigma_2 = \begin{pmatrix} 0 & -i \\ +i & 0 \end{pmatrix}, \Sigma_3 = \begin{pmatrix} 1 & 0 \\ 0 & -1 \end{pmatrix}$$

and let z_0, z_1, z_2, z_3, be complex numbers such that $R = z_0 1 + z_1 \Sigma_1 + z_2 \Sigma_2 + z_3 \Sigma_3$ where:

$$z_0 = \frac{1}{2}(r_{11} + r_{22}), z_1 = \frac{1}{2}(r_{21} + r_{12}), z_2 = i\frac{1}{2}(r_{21} - r_{12}), z_3 = \frac{1}{2}(r_{11} - r_{22}).$$

The four matrices $1, \Sigma_1, \Sigma_2$, and Σ_3 are the standard Pauli spin matrices, where for R then $z_0 = 1$, $z_1 = r_1, z_2 = r_2$ and $z_3 = r_3$. Note that $-1 \le r_1, r_2, r_3 \le 1$. From ([27], p. 104) R is a set of non-trivial 2 by 2 version of R; a ring of such forms an orthomodular poset and indeed an atomic orthomodular lattice with the covering property, that is in 1-1 correspondence with the set of closed subspaces of a two-dimensional complex Hilbert space.

Definition 8. *Let H be a finite non-empty set of viable histories, let G be a non-trivial viable descent sequence using histories $H_t \in H$, let $G_t \in G$ be a generation of G using a face $H_t \in H$ at t, and let $v(t)$ be the vector state of G_t.*

1. Let $_2H_t = \{\alpha, \chi\} \subseteq H_t$ be a two-history subset of H_t. Let $R(t) = \frac{1}{2}[r_{ij}(t)]$ be a projection, let $r_1(t), r_2(t)$, and $r_3(t)$ be real numbers such that $r_1(t)^2 + r_2(t)^2 + r_3(t)^2 = 1$, such that $0 \le r_1(t) < 1, 0 \le r_2(t) < 1$, and such that $v_\alpha(t) = \frac{1}{2}r_1(t) = \frac{1}{2}(1 + r_3(t))$. Then $R(t)$ is the *status* of G_t.
2. A *unit circle C* is meant a set of points (x, y) in the plane R^2 which satisfy the equation $x^2 + y^2 = 1$.

Theorem 1. *Assume the premises of Definition 8. Let H be a finite non-empty set of viable histories having structural numbers $s < 152$ (see [8] for use of this limit). Let G be a descent sequence using H. Let $R(t)$ be the status of H_t, and let $v(t)$ be the vector state of H_t. Let $_2H_t = \{\alpha, \chi\} \subseteq H_t$ be a non-empty subset of H_t. Let $t_2 > t_1 > t_0$ define a path of $v_\alpha(t)$ from $t = t_0$ to $t = t_2$ such that $v_\alpha(t_0) = 1$ changes monotonically to $v_\alpha(t_1) = 0$ and then monotonically back to $v_\alpha(t_2) = 1$. That is, let r_3 move from $r_3(t_0) = 1$ to $r_3(t_1) = -1$ and then back to $r_3(t_2) = 1$. Let $O(t) = (n(t), p(t), r(t))$. Then:*

(1) $trR(t) = 1;$
(2) $v_\chi(t) = \frac{1}{2}(1 - r_3(t));$
(3)   the vector state $v(t)$ of $_2H_t$ is given by the main diagonal of $R(t)$;
(4) $r(t)$ is a maximum when $r_3 = 0$.

Theorem 2. *Let $r_1(t) = 0$. Then: (i) $R(t)$ has $\Sigma\Sigma_{ij}r_{ij}(t) = 1$; and (ii) the sum $\Sigma \int r(t)dv(t) = 0$ when summed over all paths (all variants of paths) of for pairs $_2H_t$.*

Theorem 3. *Let $r_2(t) = 0$. Then $\Sigma \int r(t)dv(t) = 0$ when summed over all paths (all variants of paths) of all pairs $_2H_t$.*

Proof of Theorem 1. Assume the premises of Theorem 1. In a two history system, $_2H_t = \{\alpha, \chi\} \subseteq H_t$ is the vector state $v(t) = (v_\alpha(t), v_\chi(t))$ where $v_\chi(t) = 1 - v_\alpha(t)$. $R(t)$ is a status and since in a status $v_\alpha(t) = \frac{1}{2}r_{11} = \frac{1}{2}(1 + r_3)$, and since $v_\alpha(t) + v_\chi(t) = 1$ in a two-history state, then $v_\chi(t) = \frac{1}{2}r_{22} = \frac{1}{2}(1 - r_3)$. In addition, also then $v_\alpha(t) + v_\chi(t) = \frac{1}{2}(1 + r_3) + \frac{1}{2}(1 - r_3) = 1 = trR(t)$, which establishes Theorems 1, 2, and 3. Establishing 4: We find $r(t)$ is a maximum when $r_3 = 0$, given Theorem 1(1) and 1(3), and Lemma 1(2), so when $r_3 = 0$ then $v(t) = (0.5, 0.5)$.

Let $r_1(t) = 0$ so

$$R(t) = \frac{1}{2}[r_{ii}] = \begin{pmatrix} 1 + r_3 & -ir_2 \\ ir_2 & 1 - r_3 \end{pmatrix}$$

and thus $\frac{1}{2}\Sigma_i\Sigma_j r_{ij}(t) = \frac{1}{2}2 = 1$ which establishes 1(1).

Let $H_t = \{\alpha, \chi\}$. At time t, H_t picks a set of modal demographies $A_t = \{(n_\alpha, p_\alpha), (n_\chi, p_\chi)\}$ and $v(t)$ acts as a linear operator on A_t; so we get

$$v(t)A_t = \Sigma_\alpha v(t)(n_\alpha, p_\alpha) = (n(t), p(t)) \text{ for all } \alpha \in H_t.$$

From Lemma 2, $O(t) = (n(t), p(t), e(t))$ are the predicted results at t; when $s_\alpha \neq s_\chi$ then $(n_\alpha, p_\alpha) \neq (n_\chi, p_\chi)$. $R(t)$ is an idempotent Hermitian matrix per Definition 7, and under the premises has $r_1 = 0$. Then $r_2^2 + r_3^2 = 1$. We have a two history system with vector state $v(t) = (v_\alpha(t), v_\chi(t))$ where $v_\chi(t) = 1 - v_\alpha(t)$, and where $v_\alpha(t) = \frac{1}{2}r_{11} = \frac{1}{2}(1 + r_3(t))$. We let $t_2 > t_1 > t_0$ define a path from $t = t_0$ to $t = t_2$ such that $v_\alpha(t_0) = 1$ changes monotonically to $v_\alpha(t_1) = 0$ and then monotonically again to $v_\alpha(t_2) = 1$, which occurs as $r_3(t)$ moves monotonically from $r_3(t) = 1$ to $r_3(t) = -1$ and then back to $r_3(t) = 1$. At each t, given $r_3(t)$, we compute $r_2(t)^2 = 1 - r_3(t)^2$. Then $r_2(t)^2 + r_3(t)^2 = 1$ traces a unit circle C. Theorems 2 and 3 then follow from Green's theorem. □

Observation 2. *Let G be a population, $G_t \in G$ with the sub-populations G_t using the set of histories using face $H_t \in H$, and let $v(t)$ be the vector state of G_t. Assume history $\alpha \in H_t$, $\alpha \in H_{t+1}$, and $v_\alpha(t) = v_\alpha(t+1) = 1$. Then from t to t + 1, v(t) forms a loop. That is, the minimal descent sequence of any viable pure system α also forms a loop, indicated also since the minimal structure of α is a group. So any pure system is a loop. Diagrams like Figure 1 could occur when v(t) is not simply a pure system. Describing probability pictures by complex Hilbert spaces (Definition 7) can assist predicting demographic pictures, using pure systems as the basis of computing n(t), p(t) hence r(t).*

7. Discussion

Following the examples of [1,2] we here study systems in which the cultural organization is based on kinship descriptions using natural languages. In both cases, our theory makes predictions on population measures on the observed outcome of the kinship systems at stated times. Our Observation 1 and Lemma 1 predict what is found empirically: either single history systems and specific (n_s, p_s) pairs by the structural number of the identified history; or systems undergoing change in their culture. In that case the (n_s, p_s) pairs are changing and we can predict that rate of change yielding both the $n(t)$ and $p(t)$ for the given t, and the value of the adiabatic growth rate $r(t)$ at t. An example of this prediction of rate of change in western Europe for about 1000 years from about AD 1000 to 1950 is given in [1]. The time period of that study was about 1000 years of human history in a defined area.

Thus, study of the homotopy groups resulting from Definitions 5, 7 and 8 may thus tell us a lot about the possible paths of the empirical demography of cultures. Definitions 7 and 8 assume no physical model, but we can use their math to study the changes in vector states on histories on the predicted $n(t)$, $p(t)$ and $r(t)$ of the society per generation. The methods of [28,29] and many other current works such as [30] in social sciences use complex Hilbert spaces to describe models of how "cognition" works, using much shorter time periods, and to otherwise interpret how societies of individuals can describe and change the world around them. Hilbert space probability models per [31], which is a foundation paper for [13], are quite close to the Pauli model used here to describe changes in cultural systems; they differ from ours in their application. In particular, [31], Postulate 4 does not apply here since the applications are distinct. However, the probabilities of [30,31] are averages of probabilities on a population, not predictions of individual probabilities. There may be thus be many ways to discuss evolution of cultural systems using complex Hilbert spaces that have simply not yet been tried.

In this paper, in [17], and in both [30,31] the mathematical foundation starts with representation of the basic objects as languages; ours are natural languages. Kinship systems are derived from non-associative algebras [20,32] which in natural languages may allow groups to occur. Cultural systems with different dictionaries but similar groups are studied as isotopic kinship terminologies for example [8,11], which is a separate topic mathematically and empirically from study of languages [9–13]. Ref. [33] says "...kinship organizations are based on terminologies, which have their own distinctive logical structure centered on a "self" or I position. Language does not have a structure of this kind ...". So while kinship terminologies occur as part of natural languages, kinship analysis is not the same as the study of the language.

Our study also helps identify what cannot be predicted by this method. For example, sociologists and anthropologists use relationship studies to describe how individuals are "related"; the minimal structure defined here based on assignments made based on the "principles" used to arrange or avoid marriages, given the natural language and the history; they do not define which specific individuals are in fact assigned to each relation. In contrast, in genetic inbreeding experiments Sewall Wright [34] at diagrams 7.1(a), 7.12, 7.16 and others used the minimal structure of kinship relations for illustrating inbreeding arrangements; but in those situations, the individuals are not "assigned"—they are the actual kin of the identified sources.

The ability to derive population measures from the language-based statement of rules is something new to science, and should be explored. Many other things also affect population change, and are not explored here.

Conflicts of Interest: The author declares no conflict of interest.

Appendix A. Mathematical Background from Previous Papers

For convenience of use here, this appendix adopts background previously presented mainly in [8].

Definition A1. *General mathematical usages. Let L be a finite non-empty set. A partial order $\leq$ is a binary relation on L such that for a, b, d $\in$ L, a $\leq$ a; a $\leq$ b and b $\leq$ a implies a = b; and a $\leq$ b and b $\leq$ d implies a $\leq$ d. Then (L, $\leq$) is a partially ordered set or poset. A lattice is a poset on which is defined two binary relations join $\cup$ and meet $\cap$ such that for a, b $\in$ L then a $\cup$ b $\in$ L and a $\cap$ b $\in$ L. A lattice (L, $\leq$) is bounded if L contains special elements 0 and 1 such that for b $\in$ L, 0 $\leq$ b $\leq$ 1, which we denote as (L, $\leq$, 0, 1). An involution is a unary relation ' on L such that for b $\in$ L, then b' $\in$ L, b = b'' and for b, d $\subseteq$ L if b $\leq$ d then d' $\leq$ b'. An object (L, $\leq$, ', 0, 1) is a bounded involution poset. If L is a bounded involution poset an orthogonality relation $\perp$ on L is a binary relation such that for b, d $\in$ L, b $\perp$ d if b $\leq$ d'.*

Definition A2. *Properties of populations and descent sequences:*

Definition A2.1. *Let G be a finite non-empty set called a population whose members d $\in$ G are called individuals. Let D (descent), B (sibling of) and M (marriage) be binary relations on G, satisfying these four axioms: (1) D is anti-symmetric and transitive; (2) M is symmetric; (3) if bDc and there exists no d $\in$ G, d $\neq$ b, c for which bDd and dDc, then we write cPb, and require bBc iff for b, c, d $\in$ G, dPb and dPc; (4) |bM| $\leq$ 2.*

Definition A2.2. *Let G(t) = {G_t | $G_t \subseteq$ G, t $\in$ T, $G_i \cap G_j = \emptyset$ for i $\neq$ j} is a family of subsets of G, indeed a partition of G, where t $\in$ T is a set of consecutive non-negative integers starting with 0; such G(t) is a descent sequence of G.*

Definition A2.3. *$G_t \subseteq$ G is called the tth generation of G, in case, for all $G_t \in$ G, each cell bB occurs in only one generation, each subset bM occurs in only one generation, and for t > 0 when $G_t \in$ G, b $\in G_t$, and cPb, then c $\in G_{t-1}$ (that is, G_t contains all of and only the immediate descendants of individuals in G_{t-1}). Let |G_t| = δ_t.*

Definition A2.4. *Let G_t be a descent sequence of G. Let B_t := {bB | b $\in G_t$, $G_t \in$ G, t > 0} be a partition of G_t. in which each bB is a sibship; and let M_{t-1} := {bM | b $\in G_t$, $G_t \in$ G, t $\geq$ 0} be a set of disjoint subsets of G^{t-1} in which each bM is a marriage. Let |B_t| := β_t, and |M_t| := μ_t.*

We allow that only at $t = 0$ may there be individuals in a generation that did not arise by descent from a previous generation of G. For b $\in$ G, any set bM is assumed to be reproducing. Other than $t = 0$, members of G_t arise from (assignment of offspring to) marriages in G_{t-1}, thus $\beta_t = \mu_{t-1}$.

Definition A3. *Properties of configurations:*

Definition A3.1. *Let G(t) be a descent sequence, $G_t \in G$ be a generation of G, and let a, b, c, ..., k $\in G_t$, $a \neq b \neq c \neq \ldots \neq k$ be individuals in G_t. Then a regular structure is a closed cycle aBb, bMc, cBd, ... , kMa of a finite number of alternating B and M relations, in which each $a \in G_t$ occurs exactly twice in such a list, being exactly once on the left of a B followed immediately by once on the right of an M, or once on the right of a B preceded immediately by once on the left of an M, and in such cycle each $|bB| = |bM| = 2$. If there are j instances of M in such a cycle, then the regular structure is of type Mj.*

Definition A3.2. *Given a finite positive integer k, we assume a set of unit basis vectors e_i, $1 \leq i \leq n$, such that in e_i the ith position = 1 and all others = 0; and write $c = (m_1, m_2, \ldots , m_k)$. If $G_t \in G$ is a generation and m_i is the number of regular structures of type Mi in G_t, then such a c is called a configuration.*

Definition A3.3. *Let $c = (m_1, m_2, \ldots , m_k)$ be a configuration. Then $\mu_c := \sum_i (m_i i)$ is the number of marriages of c.*

Definition A3.4. *Let $C := \{c \mid c$ is a configuration} be the set of configurations. Let $M \in R^+$. Let $C_M := \{c \mid \mu_c \leq M\}$ be a set of configurations of order M.*

For example, a configuration with a single M2 structure would be written $(0, 1, 0, 0, \ldots)$. Note that the null configuration $0 := (0, 0, \ldots , 0) \in C_M$. If $c \in C_M$, $c \neq 0$, then such a c is non-null, and if $B \subseteq C_M$ contains at least one non-null configuration, then such a B is non-null.

Configurations are often used in ethnography when describing kinship systems. A set bM identifies a reproducing marriage in a configuration which has offspring in the succeeding generation. In a configuration we ignore all non-reproducing individuals who may exist "empirically" in G_t. So while $|bB| \geq 1$ in general, $|bB| = 2$ is required in a configuration. If we let $|G_t| = \delta_t$, the number of individuals in a configuration on G_t is $\gamma_t = 2\mu_t$; so $\delta_t \geq \gamma_t$ and $\delta_t \geq 2\mu_t$; so $n_t = \delta_t / \beta_t = \delta_t / \mu_{t-1}$. Since individuals in G_t arise only from reproducing marriages among individuals of G_{t-1}, then $\beta_t = \mu_{t-1}$. Thus:

Definition A4. *If G_t is a generation of G, then $n_t := \delta_t / \beta_t$ is the average family size of G_t.*

Definition A5. *Properties of histories:*

Definition A5.1. *A history α is a binary relation on $P(C_M)$, that is, $(c,d) \in \alpha \subseteq P(C_M) \times P(C_M)$; such an α induces a function of the power set of $P(C_M)$ which we also call α, defined, for $B \in P(C_M)$, by $(c,d) \in \alpha(B) = \{d \in C_M \mid c \alpha d$ for some $c \in C_M\}$. We let $H_M := \{\alpha \mid \alpha$ is a history defined on $C_M\}$ be the set of all histories on C_M.*

Definition A5.2. *A configuration $c \in C_M$ is viable under α if there exists an integer $k > 0$ such that $c \in \alpha^k(c)$.*

Definition A5.3. *A history $\alpha \in H_M$ is viable if there is at least one $c \in C_M$, $c \neq 0$, such that c is viable under α. Let $V(\alpha) := \{c \mid c \in C_M$ and c is viable under $\alpha\}$ be the set of all viable configurations under α. If 0 is the only configuration viable under a history α, then such an α is called trivial; otherwise α is called non-trivial.*

Definition A5.4. *If α is a viable history then $s_\alpha := min(\{\mu_c \mid c \in V(\alpha)\backslash\{0\})$ is the structural number of $\alpha\}$, where $S := \{s_\alpha \mid \alpha \in H_M \}$ is a set of structural numbers of viable histories in H_M. If $c_\alpha \in V(\alpha)$ such that $\mu_c = s_\alpha$ then c_α is a minimal structure of α. If α is a viable history, let $Min_\alpha := \{c \in V(\alpha) \mid \mu_c = s_\alpha\}$ be the set of minimal structures.*

Definition A5.5. *Let $c, d \in C_M$, and let $\eta_c \in H_M$ be a viable history such that $\eta_c(c) = \{c\}$ and $\eta_c(d)$ for $c \neq d$ is not defined; then η_c is a pure system. Let $H_p := \{\eta_c \mid c \in C_M\}$ be the set of pure systems on C_M. If α is a history, G(t) is a descent sequence of α, then $c \in Min_\alpha$ and $c = c_t$ for all $G_t \in G$ then G is a pure system of α.*

With the usual set union ∪ and intersection ∩ then the power set $(P(C_M), \cup, \cap)$ is a Boolean algebra, and using ≤ as a partial order by set inclusion, $(P(C_M), \leq, 0, C_M)$ is a poset, indeed a bounded involution poset. A history is thus a natural language describing "how α works to create a history". Here, a history is a rule; specifically a marriage rule. The structural number s_α of a viable history α is simply the value of μ of a smallest configuration which is viable under α; so also $s_\alpha > 0$, indeed $|V(\alpha)| \geq 2$. Notice that $Min_\alpha \subseteq V(\alpha)$ and $|Min_\alpha| \geq 1$.

Definition A6. *Let $G_t \subseteq G$ be a non-empty generation of G at time t, let $|G_t| = n$, and let G_t be partitioned into $1 \leq k \leq n$ subsets. Then:*

(i)   *We call a pair (n, k) an assignment.*

(ii)   *A set of assignments is a selection denoted by A with subsets $A \subseteq \mathbf{A}$. To specify more detail of the membership of a set A we may also use subscripts or a square bracket notation [n, k] with subscripts as required.*

(iii)   *[n, k] := {(n, k) | given a positive integer n, (n, k) where $1 \leq k \leq n$}.*

(iv)   *$[n, k]_j$:= {(n, k) | given a finite positive integer j, (n, k) where $1 \leq n \leq j$ and for each n, $1 \leq k \leq n$}.*

(v)   *$[n, k]_{j,i}$:= {(n, k) | given finite integers i, j where $i \geq j$, (n, k) for $1 \leq n \leq j$ and $1 \leq k \leq i$}.*

(vi)   *P_j := $P([n, k]_j)$ denotes the set of subsets of $[n, k]_j$.*

(vii)   *If (n_1, k_1), (n_2, k_2) are assignments such that $n_1 \neq n_2$ or $k_1 \neq k_2$, then (n_1, k_1) and (n_2, k_2) are distinct assignments.*

(viii) *If A is a set of assignments, is a unary relation on $A \subseteq \mathbf{A}$ such that $A' := \mathbf{A} \backslash A$.*

If (n, k) is an assignment, then

(ix)   *$\bar{n}$:= n/k is the average family size of (n, k).*

(x)   *p := 2/n is the reproductive ratio of (n, k).*

Since np = 2, then $0 < p \leq 1$.

Definition A7. *Let (n, k) be an assignment.*

(i)   *S(n, k) is a Stirling Number of the Second Kind, where*

$$S(n, k) = \frac{1}{k!} \sum_{j=0}^{k} (-1)^{k-j} \binom{k}{j} j^n \qquad (9)$$

is the number of ways to partition a set of n distinct elements into k nonempty subsets. S(n, k) computes the number of ways to achieve an assignment (n, k) for n individuals in a generation partitioned into $k = \beta_t = \mu_{t-1} \geq s_\alpha$ non-empty subsets [J] which we call families. Then:

(ii)   *S[n, k] := {S(n, k) | for given n, S(n, k), k = 1, ... , n} is called a distribution.*

(iii)   *Given a distribution S[n, k], then [n, k] := {(n, k) | for given n, k = 1, ... , n} is the underlying selection of S[n, k].*

Since S(n, k) = 1 when n = k then (uniquely) for n = 2 the distribution S[2, k] is bimodal, with modes at k = 1, 2. Therefore for n > 2:

(iv)   *$n^\uparrow$:= {j | given n}.*

(v)   *$n^\uparrow_s$:= {$n^\uparrow$ | j = s, for a given s > 0}.*

(vi)   *$S[n, n^\uparrow_s]$:= {$S(n, n^\uparrow)$ | given s, $n^\uparrow \in n^\uparrow_s$}.*

(vii)   *N_s := n | $S(n, n^\uparrow) = max(S[n, n^\uparrow_s])$.*

(viii) *$A_{[s]}$:= ∪ [n, k] for n such that $n^\uparrow \in n^\uparrow_s$ and for each such n, $1 \leq k \leq n$, called the minimal collection of s.*

(ix)   *A_M := ∪ $A_{[s]}$, given a positive integer M, for structural numbers $1 < s \leq M$.*

(x) $m_s := (N_s, s)$ is the modal assignment for s.

(xi) $A_s := \{m_s \mid s \in S\}$ is the set of modal assignments for $s \in S$.

(xii) $n_s := N_s/s$ is the modal average family size for s.

(xiii) $p_s := 2/n_s$ is the modal reproductive ratio for s.

(xiv) (n_s, p_s) is the modal demography of s.

References

1. Ballonoff, P. Structural statistics: Models relating demography and social structure with applications to Apache and Hopi. *Soc. Biol.* **1973**, *10*, 421–426. [CrossRef]
2. Ballonoff, P. Theory of Lineage Organizations. *Am. Anthropol.* **1983**, *85*, 79–91. [CrossRef]
3. Ballonoff, P. Mathematical Demography of Social Systems. In *Progress in Cybernetics and Systems Research*; Trappl, R., Pichler, F.R., Eds.; Hemisphere Publishing: Washington, DC, USA, 1982; Volume 10, pp. 101–112.
4. Ballonoff, P. Mathematical Demography of Social Systems, II. In *Cybernetics and Systems Research*; Trappl, R., Ed.; North Holland Publishing Co: Amsterdam, The Netherlands, 1982; pp. 555–560.
5. Ballonoff, P. MV-Algebra for Cultural Rules. *Int. J. Theor. Phys.* **2008**, *47*, 223–235. [CrossRef]
6. Ballonoff, P. Some Properties of Transforms in Culture Theory. *Int. J. Theor. Phys.* **2010**, *49*, 2998–3004. [CrossRef]
7. Ballonoff, P. Cross-Comment on Terminologies and Natural Languages. In *Mathematical Anthropology and Cultural Theory*; 1and1.com: Philadelphia, PA, USA, 2012; Volume 3, pp. 1–15. .
8. Ballonoff, P. Manuals of Cultural Systems. *Int. J. Theor. Phys.* **2014**, *53*, 3613–3627. [CrossRef]
9. Levi-Strauss, C. *Elementary Structures of Kinship*; Beacon Press: Boston, MA, USA, 1969.
10. Radcliffe-Brown, A.R. *A Natural Science of Society*; Free Press: Glencoe, UK, 1948.
11. Special Issue on Australian Systems. In *Mathematical Anthropology and Cultural Theory*; 1and1.com: Philadelphia, PA, USA, 2013; Volume 5. Available online: http://mathematicalanthropology.org/toc.html (accessed on 18 December 2017).
12. Dravidian Kinship Analysis. In *Mathematical Anthropology and Cultural Theory*; 1and1.com: Philadelphia, PA, USA, 2010; Volume 3. Available online: http://mathematicalanthropology.org/toc.html (accessed on 18 December 2017).
13. White, H.C.; Coleman, J. *An Anatomy of Kinship*; Prentice Hall: Englewood Cliffs, NJ, USA, 1963.
14. Chiara, D.M.; Guintini, R.; Greechie, R. *Reasoning in Quantum Theory, Sharp and Unsharp Quantum Logics*; Kluwer Academic Publishers: Dordrecht, The Netherlands, 2004.
15. Hirshleifer, J. Economics from a biological viewpoint. *J. Law Econ.* **1977**, *20*, 1–52. [CrossRef]
16. Denham, W. Kinship, Marriage and Age in Aboriginal Australia. In *Mathematical Anthropology and Cultural Theory*; 1and1.com: Philadelphia, PA, USA, 2012; Volume 4, pp. 1–79.
17. Cargal, J.M. An Analysis of the Marriage Structure of the Murngin Tribe of Australia. *Behav. Sci.* **1983**, *23*, 157–168. [CrossRef]
18. Greechie, R.; Ottenheimer, M. An Introduction to a Mathematical Approach to the Study of Kinship. In *Genealogical Mathematics*; Ballonoff, P., Ed.; Mouton: Paris, France, 1974; pp. 63–84.
19. Gottscheiner, A. On some classes of kinship systems, I: Abelian systems. In *Mathematical Anthropology and Cultural Theory*; 1and1.com: Philadelphia, PA, USA, 2008.
20. Gottscheiner, A. On some classes of kinship systems, II: Non-Abelian systems. In *Mathematical Anthropology and Cultural Theory*; 1and1.com: Philadelphia, PA, USA, 2008.
21. Kronenfeld, D. (Ed.) *A New System for the Formal Analysis of Kinship*; University Press of America: Lanham, MD, USA, 2000.
22. Barbosa de Almeida, M.W. On the Structure of Dravidian Relationship Systems. In *Mathematical Anthropology and Cultural Theory*; 1and1.com: Philadelphia, PA, USA, 2010; Volume 3, pp. 1–43.
23. Barbosa de Almeida, M.W. Comment on Vaz' "Relatives, Molecules and Particles". In *Mathematical Anthropology and Cultural Theory*; 1and1.com: Philadelphia, PA, USA, 2014; Volume 6, pp. 1–9.
24. Hildon, P.; Peterson, J.; Stiger, J. On Partitions, Surjections and Stirling Numbers. In *Bulletin of the Belgian Mathematical Society*; Belgian Mathematical Society: Brussels, Belgium, 1994; Volume 1, pp. 713–735.
25. Foulis, D.J. Effects, Observables, States and Symmetries in Physics. *Found. Phys.* **2007**, *37*, 1421–1446. [CrossRef]

26. Jordan, T.F.; Jagannathan, K. *Quantum Mechanics in Simple Matrix Form*; Dover Publications: Mineola, NY, USA, 1986.

27. Beltrametti, E.; Cassinelli, G. *The Logic of Quantum Mechanics*; Addisson-Wessley: London, UK, 1981.

28. Beim Graben, P. Comment on Gil's "What are the best hierarchical organisations for the success of a common endeavor". In *Mathematical Anthropology and Cultural Theory*; 1and1.com: Philadelphia, PA, USA, 2016; Volume 9, pp. 1–4.

29. Gil, L. What are the best hierarchical organizations for the success of a common endeavor? In *Mathematical Anthropology and Cultural Theory*; 1and1.com: Philadelphia, 2016; Volume 9, pp. 1–24.

30. Blutner, R.; Beim Graben, P. Dynamic Semantics and the Geometry of Meaning. Available online: https://pdfs.semanticscholar.org/a4ce/22edc62ff93305dde18d70edaad96fc220d3.pdf (accessed on 21 December 2017).

31. Wang, Z.; Busemeyer, J.R. A Quantum Question Order Model Supported by Empirical Texts of an A Priori and Precise Prediction. *Top. Cogn. Sci.* **2013**, *5*, 689–710. [PubMed]

32. Cargal, J.M. Reflections on the Algebraic Representations of Kinship Structure. In *Mathematical Anthropology and Cultural Theory*; 1and1.com: Philadelphia, PA, USA, 2017; Volume 11, pp. 1–5.

33. Leaf, M. Personal communication, 2015.

34. Wright, S. Evolution and the Genetics of Populations. In *The Theory of Gene Frequencies*; University of Chicago Press: Chicago, IL, USA, 1969.

MDPI

St. Alban-Anlage 66

4052 Basel

Switzerland

Tel. +41 61 683 77 34

Fax +41 61 302 89 18

www.mdpi.com

Entropy Editorial Office

E-mail: entropy@mdpi.com

www.mdpi.com/journal/entropy

www.ingramcontent.com/pod-product-compliance
Lightning Source LLC
Chambersburg PA
CBHW051722210326
41597CB00032B/5574